Introduction to Data Governance for Machine Learning Systems

Fundamental Principles, Critical Practices, and Future Trends

Aditya Nandan Prasad

Apress®

Introduction to Data Governance for Machine Learning Systems: Fundamental Principles, Critical Practices, and Future Trends

Aditya Nandan Prasad
Bangalore, Karnataka, India

ISBN-13 (pbk): 979-8-8688-1022-0   ISBN-13 (electronic): 979-8-8688-1023-7
https://doi.org/10.1007/979-8-8688-1023-7

Managing Director, Apress Media LLC: Welmoed Spahr
Acquisitions Editor: Shaul Elson
Editorial Assistant: Gryffin Winkler

Cover designed by eStudioCalamar

Cover image designed by Freepik (www.freepik.com)

Distributed to the book trade worldwide by Springer Science+Business Media New York, 1 New York Plaza, Suite 4600, New York, NY 10004-1562, USA. Phone 1-800-SPRINGER, fax (201) 348-4505, e-mail orders-ny@ springer-sbm.com, or visit www.springeronline.com. Apress Media, LLC is a California LLC and the sole member (owner) is Springer Science + Business Media Finance Inc (SSBM Finance Inc). SSBM Finance Inc is a **Delaware** corporation.

For information on translations, please e-mail booktranslations@springernature.com; for reprint, paperback, or audio rights, please e-mail bookpermissions@springernature.com.

Apress titles may be purchased in bulk for academic, corporate, or promotional use. eBook versions and licenses are also available for most titles. For more information, reference our Print and eBook Bulk Sales web page at http://www.apress.com/bulk-sales.

Any source code or other supplementary material referenced by the author in this book is available to readers on GitHub. For more detailed information, please visit https://www.apress.com/gp/services/source-code.

If disposing of this product, please recycle the paper

Table of Contents

About the Author

Aditya Nandan Prasad is an experienced analytics leader with a strong track record in driving business intelligence and recommendations for operational and strategic decision-making. He excels at leading and developing high-performing teams and collaborating to identify growth strategies. With a passion for complex data analysis and a tool-agnostic approach, he brings a data-driven perspective to solving business problems. Aditya has successfully led data migration projects and implemented innovative analytics solutions to support strategic business initiatives, and his experience in leading and collaborating with cross-functional teams has helped him become an expert on implementing data governance practices within organizations.

About the Technical Reviewer

 **Vinoth Nageshwaran** is a distinguished data engineering leader based in Frisco, Texas, with extensive experience in designing and implementing scalable data architectures using modern technologies like Snowflake, AWS, Google Cloud, DBT, and Apache Airflow. He excels in managing containerized applications with Kubernetes, utilizing Looker for comprehensive data analytics, and integrating Python for advanced data manipulation. Vinoth's expertise includes building dynamic ETL workflows and optimizing data pipelines using Apache Airflow, along with establishing robust data governance frameworks. Certified in AWS, Python, Snowflake, and Generative AI, he effectively leads cross-functional teams, driving innovation and ensuring that his data solutions remain cutting edge and future proof.

Vinoth has also been featured on several podcasts, sharing his knowledge on topics such as Large Language Models (LLMs), Retrieval-Augmented Generation (RAG) Transformers, data engineering trends, and leveraging data for strategic business decisions. Additionally, he has reviewed various IEEE conference manuscripts, contributing his expertise to the academic and professional community.

Introduction

This book is an essential resource for professionals and organizations looking to implement or enhance their machine learning (ML) data governance practices. As machine learning becomes increasingly central to business operations, ensuring that data is managed effectively throughout its lifecycle is crucial. This book provides a comprehensive guide to understanding and applying data governance principles specifically tailored for ML environments.

What the Book Is About

The book delves into the complexities of data governance within the context of machine learning, highlighting the unique challenges and considerations that arise when managing data for ML systems. It emphasizes the importance of data quality, security, privacy, and ethical considerations, all of which are critical for developing reliable and trustworthy ML models. The book also covers the regulatory landscape, offering insights into how organizations can ensure compliance with data privacy laws and industry standards.

Who the Book Is For

This book is designed for a broad audience, including data scientists, ML engineers, data governance professionals, IT managers, and business leaders. It is particularly valuable for those responsible for overseeing data governance initiatives in organizations that leverage machine learning. Whether you are new to data governance or looking to deepen your understanding of ML-specific governance challenges, this book provides practical guidance and best practices that can be applied across various industries.

Structure of the Book

The book is structured to guide readers through the fundamental concepts of data governance, progressing to more advanced topics and practical applications. It is divided into ten chapters, each focusing on a different aspect of ML data governance:

1. **Introduction to Machine Learning Data Governance**: Establishes foundational concepts, emphasizing the critical role of data governance in successful ML initiatives.

2. **Establishing a Data Governance Framework**: Explores the core principles and components necessary to build a robust data governance framework within an organization.

3. **Data Quality and Preprocessing**: Focuses on ensuring data quality and preprocessing techniques that are vital for training accurate and reliable ML models.

4. **Data Privacy and Security Considerations**: Discusses the importance of protecting sensitive data and complying with privacy regulations in the context of ML.

5. **Ethical Implications and Bias Mitigation**: Examines the ethical challenges of ML, particularly around bias, and provides strategies for mitigating these risks.

6. **Model Transparency and Interpretability**: Addresses the need for transparency in ML models, exploring methods to make models more interpretable and trustworthy.

7. **Monitoring and Maintaining Machine Learning Systems**: Emphasizes the importance of continuous monitoring to maintain the accuracy and fairness of ML models over time.

8. **Regulatory Compliance and Risk Management**: Covers the regulatory requirements and risk management strategies essential for organizations using ML.

9. **Organizational Culture and Change Management**: Discusses the cultural shifts and change management strategies needed to successfully implement ML data governance.

10. **Future Trends and Emerging Challenges**: Looks ahead to the evolving landscape of ML data governance, identifying emerging trends and challenges that organizations need to prepare for.

Key Takeaways for Readers

Before diving into the book, readers should understand that the field of ML data governance is dynamic and complex, requiring continuous adaptation and learning. The book serves as both an introduction and a practical guide, offering readers the tools and knowledge needed to navigate the challenges of data governance in ML environments. It emphasizes the importance of collaboration across departments, the need for robust policies and frameworks, and the critical role of ethical considerations in the development of AI systems.

By the end of the book, readers will have a comprehensive understanding of how to implement effective data governance practices in their ML projects, ensuring that their models are not only accurate and reliable but also ethical and compliant with regulatory standards.

CHAPTER 1

Introduction to Machine Learning Data Governance

The ever-growing field of machine learning (ML) holds immense promise for revolutionizing various aspects of our lives. From automating tasks and personalizing experiences to uncovering hidden patterns and driving better decision-making, ML applications are transforming industries and shaping the future. However, this transformative potential hinges on a critical aspect: machine learning data governance.

This chapter dives into machine learning data governance, outlining what machine learning data governance is, why it's critical for machine learning projects, and the core principles for effective data management in this context. Understanding both the complexities and potential of machine learning data governance empowers organizations to build trustworthy and high-performing AI applications.

Definition and Importance of Data Governance

Data governance refers to a set of principles, practices, and processes that ensure the effective management of data throughout its lifecycle within an organization. It encompasses data quality, accessibility, security, privacy, and compliance with relevant regulations. Effective data governance fosters a data-driven culture where data is treated as a valuable asset, managed responsibly, and leveraged to achieve organizational goals.

Machine learning is especially dependent on data governance. These algorithms rely heavily on the quality, reliability, and representativeness of the data they're trained on. "Garbage in, garbage out" is a well-known adage that perfectly applies to machine

1

© Aditya Nandan Prasad 2024
A. Nandan Prasad, *Introduction to Data Governance for Machine Learning Systems*,
https://doi.org/10.1007/979-8-8688-1023-7_1

learning. Poor quality data can lead to inaccurate models, biased predictions and, ultimately, unreliable and potentially harmful outcomes.

Data governance has become an essential function for organizations of all sizes in today's data-driven landscape. As we increasingly rely on data to inform critical decisions, ensure operational efficiency, and gain a competitive edge, the effective management of this valuable asset becomes paramount.

Data governance establishes a framework for handling data throughout its lifecycle, from creation and storage to analysis and utilization. The cornerstone of machine learning data governance is ensuring **high-quality and reliable data**. In the realm of ML, this translates to building models on a foundation of data accuracy and consistency. Inaccurate or inconsistent data leads to unreliable models with misleading predictions, potentially derailing ML projects and incurring significant costs. Data governance for ML ensures data quality, providing a trustworthy base for robust and reliable ML models.

Data accessibility is another crucial element of data governance in ML projects. Data scientists, engineers, and business stakeholders require access to the right data at the right time. Data governance in such projects establishes clear access controls while ensuring data availability for authorized individuals. This fosters collaboration across teams. Data scientists can leverage data from various sources, while domain experts can provide valuable insights for model development. Accessibility empowers stakeholders to understand and interpret model outputs, facilitating informed decision-making based on ML insights.

Data security is paramount in any ML project. Data breaches can expose sensitive information and compromise the integrity of ML models. Data governance plays a critical role in safeguarding data assets by implementing robust security measures. This includes encryption of sensitive data, access controls to prevent unauthorized access, and clear protocols for data handling and disposal. By prioritizing data security, organizations can maintain trust with stakeholders and ensure compliance with relevant data privacy regulations.

Data governance in the context of ML extends beyond data security and accessibility. **Ethical considerations** are vital. Biases present in training data can lead to biased ML models, resulting in discriminatory outcomes. Machine learning data governance frameworks can help mitigate bias by promoting responsible AI practices. This includes building diverse data pipelines, implementing fairness checks during model development, and establishing clear guidelines for ethical data use in ML projects.

However, the importance of data governance in ML field extends beyond just managing internal processes. Let's delve into the critical challenges and risks associated with **poor data governance**, particularly its impact on the increasingly prevalent field of machine learning.

Challenges and Risks of Poor Data Governance

Machine learning (ML) has become a transformative force across industries, promising groundbreaking advancements in everything from healthcare diagnostics to financial forecasting. However, the power of ML hinges on a fundamental requirement: high-quality data. Poor data governance in ML projects creates a critical vulnerability in ML systems, leading to a cascade of challenges and risks.

Garbage In, Garbage Out: The Issue of Data Quality

Data quality is the cornerstone of effective machine learning (ML). The entire process of building reliable, accurate, and ethical ML models hinges on the quality of the data used to train these models. When data is inaccurate, incomplete, or inconsistent, it directly impacts the performance of the ML algorithms, resulting in flawed models that produce unreliable outputs. This is because ML models learn patterns from the data they are trained on; if the data is compromised, the model's understanding of the underlying patterns will be skewed, leading to erroneous predictions and decisions.

Consider an example where a financial institution is using an ML system to predict loan defaults. The goal is to assess whether a potential borrower is likely to default on a loan based on various factors such as credit score, income, employment history, and outstanding debts. If the data fed into the model is inaccurate—say, it contains incorrect credit scores, outdated income information, or incomplete employment history—the model will learn from this flawed data. As a result, it might incorrectly classify creditworthy individuals as high-risk borrowers.

For instance, if a customer's credit score is erroneously recorded as significantly lower than it actually is, the ML model might flag this individual as likely to default. Consequently, the bank might reject the loan application, denying the customer access to credit they rightfully deserve. This not only harms the individual but also results in lost business opportunities for the bank. The issue is compounded if the data inaccuracies are widespread, leading to systemic biases in lending decisions that disproportionately affect certain groups of people.

In the healthcare sector, the quality of data used to train ML models can have life-or-death consequences. Consider an ML system developed to assist in diagnosing diseases based on patient data, including medical history, lab results, and imaging scans. If the data used for training this system is incomplete or biased—such as lacking sufficient representation of certain demographic groups—the model's diagnostic accuracy can be severely compromised.

For instance, if the training data primarily consists of medical records from younger patients, the model might perform well in diagnosing conditions in younger individuals but fail to accurately diagnose older patients. This could lead to underdiagnosis or misdiagnosis of diseases in older adults, who may present symptoms differently. Incomplete data might also result in the model overlooking critical factors, such as comorbidities or genetic predispositions, which are essential for making accurate diagnoses. The consequences of such errors can be dire, leading to inappropriate treatments, delayed interventions, and even preventable deaths.

The Silo Effect: Data Fragmentation and Accessibility

Data fragmentation, where relevant information is scattered across disparate systems and departments, poses a significant challenge for effective data governance in machine learning (ML) projects. This siloed data environment makes it difficult to create a unified, holistic view of the data, which is essential for developing robust and accurate ML models. Fragmented data not only leads to data duplication and inconsistency but also complicates data management and increases the risk of errors. Additionally, limited access controls can exacerbate these issues by restricting the availability of data to the teams that need it most for model development and training. This can stifle innovation and prevent organizations from fully leveraging the data they possess, ultimately hindering the success of their ML initiatives.

In the financial sector, credit risk modeling is a critical application of machine learning. Banks and financial institutions rely on ML models to assess the creditworthiness of borrowers and make informed lending decisions. However, data fragmentation within these organizations can significantly hinder the development of accurate credit risk models.

For instance, customer data might be dispersed across various departments such as retail banking, mortgage lending, and credit card services. Each department may maintain its own separate database, with little to no integration between them. As a result, critical information about a customer's financial history, such as their payment behavior on different types of loans or their credit utilization across multiple products, remains isolated within these silos.

When data scientists attempt to build a credit risk model, they might only have access to partial data from one department, leading to an incomplete picture of the borrower's financial profile. This incomplete data can result in models that are less accurate and fail to capture the true risk of default. Moreover, data duplication across different systems can lead to inconsistencies, where the same customer may be represented differently in each database, further complicating the model-building process.

For example, a customer might have a mortgage with one department and a credit card with another. If these records are not integrated, the ML model might assess the customer's risk based solely on their credit card history, overlooking their mortgage payment behavior, which could provide valuable insights into their overall creditworthiness. This fragmented approach not only undermines the accuracy of the model but also exposes the institution to greater financial risk by potentially misclassifying borrowers.

Limited access controls can prevent data scientists from accessing the full range of data necessary for comprehensive model development. If strict access policies restrict data availability to specific teams or departments, the ML models will be built on an incomplete dataset, leading to suboptimal outcomes. This lack of access can stifle innovation, as data scientists are unable to explore and experiment with the full breadth of data available within the organization. In turn, this limits the institution's ability to develop advanced predictive models that could enhance decision-making and improve customer service.

In the healthcare sector, ML models are increasingly used to optimize patient care, from predicting disease outbreaks to personalizing treatment plans. However, data fragmentation poses a significant challenge to these efforts, as patient information is often spread across different systems, such as electronic health records (EHRs), lab results, imaging systems, and insurance databases. Each of these systems may be managed by different departments or even different organizations, such as hospitals, labs, and insurance companies, with varying degrees of integration.

For instance, a patient might have their primary care records stored in one EHR system, while their specialist visits and lab results are stored in another system managed by a separate hospital network. Additionally, imaging results might be stored in a third system, with insurance claims data housed in yet another. This fragmented data environment makes it challenging to develop ML models that provide a comprehensive view of the patient's health.

When healthcare providers attempt to use ML models to predict patient outcomes, such as the likelihood of readmission or the effectiveness of a treatment plan, the fragmented data can lead to incomplete or biased predictions. For example, if an ML model only has access to EHR data from primary care visits but lacks information about specialist consultations or lab results, it may underestimate the severity of a patient's condition or fail to identify critical risk factors. This can lead to suboptimal treatment recommendations, potentially compromising patient care.

Data duplication across different systems can create inconsistencies in patient records. For instance, a patient's medication list might be updated in their primary care EHR but not in the specialist's records, leading to conflicting information. If an ML model is trained on these inconsistent records, it may produce unreliable predictions, such as recommending a medication that the patient is no longer taking or overlooking a potential drug interaction.

Limited access controls can further exacerbate the issue by restricting the availability of data to the healthcare professionals and data scientists who need it most for model development. For example, if only certain departments have access to imaging data or lab results, the ML models developed will be based on incomplete datasets, leading to less accurate and less useful predictions. This restricted access can hinder innovation in patient care, as healthcare providers are unable to leverage the full range of available data to develop more effective and personalized treatment plans.

In both the financial and healthcare sectors, data fragmentation and limited access controls present significant obstacles to effective data governance in ML projects. Addressing these challenges requires organizations to invest in data integration, ensure consistent data management practices, and establish access controls that balance security with the need for data availability. By overcoming these hurdles, organizations can unlock the full potential of their data, enabling more accurate, reliable, and innovative ML models that drive better decision-making and outcomes.

Lack of Transparency: The Black Box Problem

The complex inner workings of some ML models, particularly deep learning algorithms, can be opaque, making it difficult to understand how they arrive at their predictions. This lack of transparency becomes a major challenge when dealing with biased or erroneous outputs. Without clear explanations for a model's decisions, it's challenging to identify and rectify issues within the data or the algorithm itself. This can erode trust in ML systems and create difficulties in regulatory compliance.

In the criminal justice system, machine learning models are increasingly used to assess the likelihood that an individual will reoffend, also known as recidivism risk. These assessments play a crucial role in decisions related to parole, sentencing, and bail. However, when deep learning algorithms are used for this purpose, their opacity can become a significant issue, particularly when the model's predictions are biased or difficult to interpret.

Imagine a deep learning model designed to predict recidivism risk based on factors such as an individual's criminal history, demographic information, and behavioral patterns. If the model is consistently flagging certain racial or ethnic groups as higher risk, this could be due to underlying biases in the training data, such as historical disparities in arrest rates or sentencing. However, because deep learning models operate as black boxes, it's challenging for criminal justice professionals to understand why the model is making these predictions.

Without transparency, it's difficult to identify whether the bias is a result of flawed data inputs or a consequence of how the model weighs different features. This lack of interpretability hinders the ability to address and correct these biases, potentially leading to unfair treatment of certain groups. For example, if the model disproportionately assigns higher risk scores to individuals from minority communities, they might face harsher sentencing, less favorable parole outcomes, or higher bail amounts compared to others with similar backgrounds but from different racial or ethnic groups.

The inability to explain the model's decisions can erode trust in the criminal justice system, leading to public outcry, legal challenges, and scrutiny from civil rights organizations. Additionally, the lack of transparency can complicate compliance with legal standards that require fairness and accountability in judicial decision-making. Courts and law enforcement agencies may find it difficult to justify the use of such models if they cannot provide clear, understandable reasons for the decisions being made, ultimately undermining the potential benefits of using machine learning in the criminal justice system.

In the field of human resources (HR), machine learning models are increasingly used to streamline recruitment and hiring processes. Companies often employ deep learning algorithms to screen resumes, assess candidate suitability, and predict employee performance. These models analyze a wide range of data, including work experience, education, skills, and even behavioral traits extracted from video interviews or social media profiles. However, the opacity of deep learning models can create significant challenges, particularly when the outputs are biased or difficult to interpret.

Consider a scenario where a deep learning model is used to screen job applicants. Suppose this model consistently ranks male candidates higher than equally qualified female candidates or favors candidates from certain educational backgrounds over others, regardless of actual job performance potential. If the HR team cannot understand how the model arrives at its decisions, it becomes nearly impossible to identify and correct the biases that are driving these outcomes.

The model might be picking up on subtle cues in the training data, such as historical hiring patterns where men were disproportionately favored, or it might be overemphasizing certain keywords commonly found in resumes from candidates who attended prestigious universities. Without transparency, HR professionals cannot determine whether these biases are due to the data used to train the model, the algorithm's design, or other factors.

The lack of explainability in this context can have serious consequences. Qualified candidates may be unfairly rejected, perpetuating existing biases and limiting diversity within the organization. This can harm the company's reputation and result in missed opportunities to hire top talent. Additionally, regulatory bodies and industry standards increasingly require organizations to demonstrate that their hiring practices are fair and non-discriminatory. If the company cannot explain how the ML model makes its decisions, it could face legal challenges, including accusations of discrimination, fines, or even litigation.

To mitigate these risks, companies may need to invest in developing more transparent models or implementing explainability tools that provide insights into how decisions are made. This could involve using interpretable machine learning techniques, conducting bias audits, or creating hybrid models that balance the predictive power of deep learning with the interpretability of more traditional approaches. By doing so, HR departments can ensure that their recruitment processes are both fair and effective, building trust among applicants and complying with legal requirements.

Biased Decisions and Algorithmic Discrimination

Poor data governance practices in ML projects can exacerbate societal biases present in the real world and amplify them within ML models. Biased datasets can lead to discriminatory outcomes, potentially perpetuating inequalities in areas like loan approvals, hiring decisions, or criminal justice predictions. These biases can have significant social and economic consequences, unfairly disadvantaging certain demographics.

In the education sector, ML models are increasingly being used to assess student performance, predict academic success, and even assist in university admission decisions. However, poor data governance practices can result in the use of biased datasets, leading to discriminatory outcomes that disproportionately affect certain groups of students.

Consider an ML model used by a university to predict the likelihood of a student's success in a particular academic program based on historical data, including past student grades, standardized test scores, extracurricular activities, and demographic information. If the training data used by the model reflects historical biases—such as lower grades or standardized test scores for students from underrepresented minority groups due to systemic inequalities in education—the model may inadvertently predict lower success rates for these students.

As a result, students from these groups may be unfairly denied admission or placed in lower-level courses, limiting their academic opportunities and reinforcing existing educational disparities. This bias can have long-term social and economic consequences, as students who are unfairly disadvantaged may have fewer opportunities for higher education and, consequently, reduced access to high-paying jobs and career advancement. Poor data governance that fails to identify and mitigate these biases can thus perpetuate inequalities in education, further entrenching societal divisions.

In the public services, ML models are increasingly used to allocate resources and administer welfare programs. These models often rely on data such as income levels, employment history, family size, and geographic location to determine eligibility for benefits or to prioritize services. However, poor data governance practices can result in the use of biased data, leading to inequitable distribution of resources and services.

For instance, consider an ML model used by a government agency to determine eligibility for housing assistance programs. If the training data reflects historical biases—such as lower access to housing in certain minority neighborhoods due to discriminatory lending practices or zoning laws—the model may unfairly deprive individuals from these communities of much-needed assistance. Similarly, if the data used to allocate resources to schools or healthcare facilities is biased, certain regions or demographic groups may receive fewer resources, exacerbating existing inequalities in access to public services.

The consequences of such biases can be severe, leading to further marginalization of already disadvantaged groups. For example, families in underprivileged areas may struggle to access adequate housing, healthcare, or education, perpetuating cycles of poverty and social exclusion. Poor data governance that fails to address these biases can therefore have profound social and economic implications, reinforcing structural inequalities and hindering efforts to achieve greater social equity.

In each of these examples—education and public services—poor data governance practices can lead to the use of biased data in ML models, resulting in discriminatory outcomes that exacerbate societal biases and inequalities. To prevent such outcomes, it is crucial to implement robust data governance frameworks that identify, mitigate, and rectify biases in data, ensuring that ML models contribute to fairer and more equitable decision-making processes across various sectors.

Security Breaches and Data Privacy Concerns

ML systems often rely on vast amounts of sensitive data, making them prime targets for cyberattacks. Inadequate data security practices can lead to data breaches, exposing private information and potentially compromising the integrity of the data itself. Furthermore, poorly defined machine learning data governance frameworks can raise concerns about data privacy. Without clear guidelines for data collection, storage, and usage, organizations risk violating user privacy and eroding trust with customers and stakeholders.

In the retail sector, ML models often rely on vast amounts of customer data, including purchase history, browsing behavior, payment information, and personal details such as names, addresses, and contact information. This data is used to develop predictive models that can forecast demand, recommend products, and create personalized shopping experiences.

However, if the data security practices governing this information are inadequate, the organization becomes vulnerable to cyberattacks. For example, imagine a scenario where a major retail chain uses an ML model to personalize customer interactions but fails to implement strong encryption protocols and access controls for the data it collects. A hacker could exploit these vulnerabilities to gain unauthorized access to the retailer's database, resulting in a significant data breach.

Such a breach could expose sensitive customer information, including payment details and personal identifiers, to malicious actors. The consequences of this breach would be far-reaching: customers whose data was compromised could become victims of identity theft, and the retailer would likely face legal repercussions, including fines for violating data protection regulations such as the General Data Protection Regulation (GDPR) or the California Consumer Privacy Act (CCPA).

Beyond the immediate financial impact, the data breach would also severely damage the retailer's reputation. Customers might lose trust in the company's ability to protect their personal information, leading to a loss of business and customer loyalty. The erosion of trust would affect not only the retailer's relationship with its customers but also its standing with partners and stakeholders, who might question the company's commitment to data security and privacy.

If the retailer lacks a well-defined ML data governance framework, the risks extend beyond data breaches. Poorly governed data collection, storage, and usage practices could lead to inadvertent privacy violations. For instance, if the retailer's ML models use customer data without proper anonymization or fail to obtain explicit consent for data usage, they could be in breach of privacy laws. These violations could result in additional fines, legal challenges, and further damage to the company's reputation.

Erroneous Decisions and Flawed Outcomes

Inaccurate or incomplete data can lead to unreliable outputs from ML models. This poses a significant risk in applications where ML decisions have real-world consequences. Imagine an ML system used for medical diagnosis trained on faulty medical records. Inaccurate diagnoses could lead to delayed or improper treatment, potentially impacting patient health. Similarly, flawed ML models in financial markets could result in erroneous investment decisions, causing economic losses.

In the manufacturing sector, machine learning models are widely used for predictive maintenance—predicting when equipment or machinery is likely to fail so that maintenance can be performed before a breakdown occurs. This approach can significantly reduce downtime, extend the lifespan of equipment, and lower maintenance costs. However, the effectiveness of predictive maintenance models hinges on the accuracy and completeness of the data they are trained on.

Imagine a scenario where an ML model is used to predict the failure of critical machinery in a large manufacturing plant. If the historical data on which the model is trained is incomplete—perhaps missing records of past maintenance activities, equipment usage patterns, or environmental conditions—the model's predictions will be unreliable. For instance, the model might fail to recognize patterns leading to certain types of failures, or it might overestimate the reliability of the machinery, leading to delayed maintenance.

In a real-world setting, such a flawed model could result in unexpected equipment failures, leading to costly production halts and even safety hazards. For example, if the model inaccurately predicts that a machine is in good working condition when it is, in fact, on the verge of breaking down, the resulting failure could cause an entire production line to shut down. This not only incurs financial losses due to halted operations but could also pose risks to worker safety, especially if the machinery involved is hazardous.

Relying on an inaccurate predictive maintenance model can erode trust in the technology, leading to resistance from plant operators and managers who may revert to more traditional, reactive maintenance approaches. This resistance could stifle innovation and prevent the adoption of more advanced, data-driven maintenance practices that have the potential to transform the manufacturing industry.

Poor machine learning data governance poses a significant threat to the responsible and trustworthy development of machine learning. By understanding the challenges of data quality, accessibility, and transparency, organizations can take proactive steps to mitigate the risks of biased decisions, security breaches, and flawed outcomes. As ML continues to shape our world, robust data governance practices in ML projects will be paramount in ensuring its ethical and responsible application.

Overview of Machine Learning and Data Requirements

Machine learning is a subfield of artificial intelligence (AI) that focuses on algorithms that can learn from data without explicit programming. These algorithms are trained on large amounts of data, allowing them to identify patterns and make predictions. The data used for training ML models comes from a diverse range of sources, each offering unique advantages and considerations.

Sensor Data

Sensor data, collected from devices like smartphones, wearables, and Internet of Things (IoT) devices, provides a window into the physical world. This data can include everything from location coordinates and accelerometer readings to temperature and pressure measurements. For example, sensor data collected from fitness trackers can be used to train ML models that predict health risks or recommend personalized exercise plans. However, the sheer volume and real-time nature of sensor data can pose challenges in terms of storage, processing, and ensuring data privacy.

Transactional Data

Transactional data is generated from the lifeblood of any business—its transactions. This can include customer purchases, financial records, and clickstream data that captures user interactions with websites or applications. Analyzing transactional data allows ML models to identify purchasing patterns, predict customer churn, and optimize marketing campaigns. For example, an e-commerce platform might use transactional data to train a model that recommends products to customers based on their past purchases. However, ensuring data security and compliance with privacy regulations are crucial considerations when dealing with sensitive transactional data.

Social Media Data

Social media platforms are a treasure trove of data, offering insights into public opinion, brand sentiment, and emerging trends. Text, images, and other information publicly shared on these platforms can be harnessed to train ML models. For example, social media data can be used to train models that analyze customer sentiment toward a new product launch or track the effectiveness of a social media campaign. However, ethical considerations around data privacy and the potential for bias present in social media data require careful consideration when utilizing this source for ML training.

Public Datasets

Publicly available datasets from government agencies, research institutions, and other sources offer a valuable starting point for many ML projects. These datasets cover a wide range of topics, from weather patterns and demographics to scientific research findings. They can be particularly beneficial for exploring new applications of ML or for projects with limited access to proprietary data. However, the accuracy and completeness of public datasets can vary, requiring careful evaluation before using them for model training.

The specific data requirements for an ML project are not a one-size-fits-all proposition. Several factors come into play, including the chosen algorithm, the complexity of the task at hand, and the desired level of accuracy. Despite this variability, several general principles hold true when selecting data for ML project.

The cornerstone of successful ML training is **high-quality data**. This means ensuring the data used is accurate, consistent, and free from errors or inconsistencies. Imagine training a model to predict loan defaults based on customer financial data riddled with errors. The resulting model would likely be unreliable, potentially leading to inaccurate predictions and costly consequences. Data quality practices like data cleaning and validation are crucial to ensure the training data provides a solid foundation for robust and reliable ML models.

The **reliability** of the data sources is paramount. Training data should originate from trustworthy sources to ensure the integrity of the entire ML project. For instance, using social media data with questionable authenticity to build a model that predicts political sentiment could lead to unreliable and potentially misleading results. Organizations must carefully evaluate the provenance of their training data and prioritize sourcing information from reliable and verifiable sources.

The training data should be **representative of the real-world** scenario for which the model will be used. Imagine building a model to predict customer churn based on data that only reflects the demographics and behavior of young adults while neglecting older demographics. This model would likely fail to accurately predict churn for the entire customer base. Data representation is crucial—the training data needs to closely resemble the actual population or situation the model will be applied to for generating accurate and generalizable predictions.

Machine learning algorithms often have an insatiable appetite—they require large amounts of data for effective training. This is particularly true for complex tasks or deep learning models. For instance, training a model for image recognition might require millions of labeled images, while a model predicting customer churn in a large retail chain might necessitate vast amounts of customer transaction data. Organizations need to consider the **data volume requirements** of their chosen algorithm and ensure they have access to sufficient quantities of high-quality training data to fuel their ML projects.

Machine learning data governance practices ensure that the data used for ML projects meets these quality standards and is managed effectively throughout the machine learning lifecycle.

How Machine Learning Data Governance Differs from Traditional Data Governance

Machine learning (ML) data governance is a specialized subset of traditional data governance, tailored to the unique requirements and challenges of managing data in machine learning environments. While both traditional data governance and ML data governance share common goals—such as ensuring data quality, compliance, and security—there are several key differences that arise due to the nature of machine learning processes, the complexity of ML models, and the specific risks associated with deploying these models in real-world applications.

Data Lifecycle Management

In traditional data governance, data lifecycle management involves the standard processes of data collection, storage, processing, and archiving or deletion. These processes are generally linear and well-defined, with clear stages and transitions. The primary focus is on ensuring that data is accurately captured, securely stored, and appropriately used according to business rules and regulatory requirements.

In contrast, ML data governance involves a more complex and iterative data lifecycle. The data used in machine learning projects often goes through multiple rounds of processing, transformation, and feature engineering before it is ready for model training. This iterative nature of the ML data lifecycle introduces additional challenges in tracking and managing data across its different versions and states. For example, as new data becomes available, it may need to be integrated with existing datasets, requiring careful management of data versioning and lineage to ensure that the model remains consistent and reliable.

Moreover, ML data governance must account for the continuous feedback loop inherent in machine learning systems. Unlike traditional systems where data is typically static after initial processing, ML systems may continuously ingest new data to retrain models, adapt to changing conditions, and improve accuracy. This ongoing cycle requires governance mechanisms that can dynamically manage and monitor data throughout its evolving lifecycle, ensuring that it remains relevant, accurate, and compliant with regulations.

Data Quality Considerations

Data quality is a cornerstone of both traditional and ML data governance, but the criteria for assessing data quality can differ significantly between the two. In traditional data governance, data quality is often evaluated based on attributes such as accuracy, completeness, consistency, timeliness, and relevance. These criteria are applied uniformly across all data types and use cases, with the goal of ensuring that data meets business requirements and supports decision-making processes.

In ML data governance, the concept of data quality extends beyond these traditional attributes to include considerations specific to the performance and fairness of machine learning models. For instance, data balance and representativeness are critical factors in ML data quality. An imbalanced dataset—where certain classes or features are overrepresented—can lead to biased models that perform poorly in real-world scenarios. Therefore, ML data governance must include rigorous checks for data balance and mechanisms for addressing imbalances, such as re-sampling, re-weighting, or synthetic data generation.

Another key difference is the importance of feature engineering in ML data governance. Feature engineering involves transforming raw data into a format that is more suitable for model training, often by creating new features or modifying existing ones. This process can significantly impact the quality and performance of the ML model, making it a critical aspect of data governance in machine learning. Traditional data governance does not typically account for the complexities of feature engineering, as it is not a common practice outside of ML contexts.

The iterative nature of ML model development means that data quality checks must be continuously applied throughout the model training process. Unlike traditional data governance, where quality checks are often performed at discrete stages, ML data governance requires ongoing monitoring and validation to ensure that the data remains suitable for training and does not introduce biases or errors into the model.

Model Governance

One of the most significant differences between traditional data governance and ML data governance is the concept of model governance. Traditional data governance primarily focuses on managing and protecting the data itself, with little consideration given to the algorithms or systems that use the data. In contrast, ML data governance must extend beyond data management to include the governance of the machine learning models that are trained on the data.

Model governance encompasses several key areas, including model versioning, performance monitoring, explainability, and compliance. In ML data governance, it is essential to track and manage different versions of a model, ensuring that the correct version is deployed in production and that previous versions can be easily accessed for auditing or rollback purposes. This requires robust version control systems and processes that are specifically tailored to the needs of machine learning projects.

Performance monitoring is another critical aspect of model governance. Unlike traditional systems where performance metrics are relatively static, ML models require continuous monitoring to ensure that they maintain their accuracy and reliability over time. This is particularly important in dynamic environments where data patterns may change, potentially leading to model drift or degradation. ML data governance must include mechanisms for detecting and addressing these issues, such as automated retraining or model recalibration.

Explainability is a unique challenge in ML data governance that is not typically addressed in traditional data governance. Machine learning models, especially those based on complex algorithms like deep learning, can be difficult to interpret, making it challenging to understand how decisions are being made. This lack of transparency can be a significant barrier to trust and acceptance, particularly in regulated industries where explainability is a legal requirement. ML data governance must include strategies for improving model interpretability, such as using simpler models, implementing post hoc explanation techniques, or developing Explainable AI (XAI) tools.

Compliance is another area where ML data governance differs from traditional data governance. In addition to ensuring that data is managed in accordance with regulatory requirements, ML data governance must also ensure that the models themselves comply with relevant laws and ethical standards. This includes verifying that models do not introduce bias or discrimination, that they respect privacy rights, and that they are used in a manner that is consistent with the organization's ethical guidelines.

Ethical Implications and Bias Mitigation

Ethical considerations play a more prominent role in ML data governance than in traditional data governance. While both approaches must address issues such as data privacy and security, ML data governance must also contend with the ethical implications of how machine learning models are developed and deployed.

One of the most significant ethical challenges in ML data governance is the potential for bias in machine learning models. Bias can arise at various stages of the ML lifecycle, from the selection of training data to the design of the model architecture and the interpretation of results. If left unchecked, these biases can lead to unfair or discriminatory outcomes, particularly in high-stakes applications such as hiring, lending, or law enforcement.

ML data governance must include robust mechanisms for detecting and mitigating bias at every stage of the ML lifecycle. This can involve implementing bias audits, using fairness-enhancing algorithms, and engaging with diverse stakeholders to ensure that different perspectives are considered in the model development process. Traditional data governance does not typically address these issues, as they are specific to the use of machine learning and other advanced analytics techniques.

ML data governance must also address broader ethical concerns related to the use of AI and machine learning. This includes ensuring that ML models are used in a manner that is consistent with the organization's values and ethical principles, and that they do not cause harm to individuals or society. This may involve developing ethical AI frameworks, conducting impact assessments, and establishing governance bodies to oversee the ethical use of AI.

Continuous Monitoring and Adaptation

Another key difference between ML data governance and traditional data governance is the need for continuous monitoring and adaptation. In traditional data governance, once data is collected, processed, and stored, it is relatively stable, with periodic reviews and updates to ensure ongoing compliance and relevance. In contrast, ML data governance requires continuous monitoring of both data and models to ensure that they remain accurate, relevant, and free from bias.

Continuous monitoring is essential in ML data governance because machine learning models are often deployed in dynamic environments where data patterns can change over time. This phenomenon, known as concept drift, can lead to a decline in model performance if the model is not regularly retrained or updated to reflect new data. ML data governance must include processes for detecting concept drift and other performance issues, as well as strategies for addressing them, such as automated retraining or model recalibration.

Adaptation is another critical aspect of ML data governance. As new data becomes available or as the regulatory landscape evolves, organizations may need to adapt their data governance practices to ensure ongoing compliance and model accuracy. This may involve updating data quality standards, revising model governance procedures, or implementing new tools and technologies to support more effective data management.

Traditional data governance does not typically require this level of continuous monitoring and adaptation, as the data is often static and the governance processes are well-established. In contrast, ML data governance must be agile and responsive to changes in the data environment, ensuring that both data and models remain fit for purpose.

Integration with AI and ML Workflows

Traditional data governance is often integrated with broader IT governance and business processes, focusing on data management activities such as data warehousing, database management, and business intelligence. The integration of data governance into these workflows is relatively straightforward, with well-defined processes and tools.

In contrast, ML data governance must be tightly integrated with AI and ML workflows, which are inherently more complex and dynamic. This integration involves ensuring that data governance processes align with the iterative nature of ML model development, which includes data preprocessing, model training, validation, deployment, and monitoring. ML data governance must also account for the use of advanced analytics techniques, such as feature engineering and hyperparameter optimization, which are not typically addressed in traditional data governance.

Integration with AI and ML workflows also requires a greater emphasis on collaboration between data scientists, ML engineers, and data governance professionals. This collaboration is essential for ensuring that data governance practices are effectively implemented throughout the ML lifecycle, from data collection to model deployment and beyond. Traditional data governance does not typically require this level of cross-functional collaboration, as the focus is primarily on managing static data assets.

While ML data governance shares some commonalities with traditional data governance, it is distinguished by the unique challenges and requirements associated with managing data in machine learning environments. These differences are evident in areas such as data lifecycle management, data quality considerations, model governance, ethical implications, and the need for continuous monitoring and adaptation. As organizations continue to adopt machine learning and AI technologies, it is essential that they develop and implement robust ML data governance frameworks that address these challenges, ensuring that their models are not only accurate and reliable but also ethical, transparent, and compliant with regulatory standards.

Goals and Objectives of Machine Learning Data Governance

Effective data governance for machine learning (ML) entails establishing a robust framework of principles and practices designed to manage data responsibly and ethically throughout the ML lifecycle. This framework is crucial not only for harnessing the full potential of ML technologies but also for mitigating the risks associated with data misuse, biases, and breaches of privacy. Understanding the goals and objectives of such a framework is fundamental for organizations aiming to leverage ML technologies effectively while maintaining trust and compliance.

Fostering Data Quality and Integrity

The cornerstone of successful ML projects is high-quality data. Machine learning data governance strives to ensure data accuracy, completeness, and consistency. This involves establishing data quality standards, implementing data cleansing and validation processes, and monitoring data quality metrics throughout the ML pipeline. By ensuring data integrity, machine learning data governance minimizes the risk of "garbage in, garbage out" scenarios, where flawed data leads to unreliable and potentially harmful ML outputs.

Data Standardization and Definition

Data governance in ML projects is critical for establishing clear definitions and standardized formats for the data elements used in machine learning models. By defining these elements and enforcing standardization, organizations can ensure that data from different sources is consistent, easily integrated, and aligned with the specific needs of the ML models. This process simplifies data integration across various datasets, enabling the ML models to "speak the same language" regardless of the data's origin. Standardization not only improves the quality and reliability of the data but also facilitates smoother collaboration between different teams, as everyone adheres to the same definitions and formats, reducing the likelihood of errors and misinterpretations.

Poor data standardization and a lack of clear definitions can lead to significant issues in ML models. For example, if different datasets use varying formats or naming conventions for the same data element—such as representing dates in different formats (MM/DD/YYYY vs. DD/MM/YYYY) or using inconsistent labels for categorical data— the ML model may misinterpret this information, leading to incorrect or biased outputs. Inconsistent data can cause the model to make faulty assumptions or fail to recognize patterns that are critical for accurate predictions. This can be particularly problematic in complex ML applications where the model needs to integrate and analyze data from multiple sources, such as in supply chain management or fraud detection. In these scenarios, even small discrepancies in data definitions or formats can result in significant errors, undermining the effectiveness of the ML model and leading to potentially costly or dangerous decisions.

For instance, in a global logistics company, poor data standardization across different regions could lead to a scenario where the ML model misinterprets delivery times, stock levels, or customer demands because the underlying data was not consistently formatted or defined. This could cause delays, overstocking, or stockouts, ultimately impacting customer satisfaction and operational efficiency. Additionally, a lack of standardized data definitions can make it difficult to audit the ML model's decisions, as the inconsistencies in the data might obscure the reasoning behind specific outputs. This not only complicates troubleshooting and optimization efforts but also poses challenges in maintaining compliance with regulatory requirements, particularly in industries that demand high levels of accuracy and transparency, such as finance, healthcare, and transportation.

Data Lineage and Traceability

Tracking the origin and journey of data throughout the ML lifecycle is crucial for maintaining the integrity and reliability of machine learning models. Machine learning data governance plays a vital role in facilitating data lineage by meticulously documenting the source of the data, the transformations it undergoes, and how it is used within the model. This transparency enables organizations to trace any data-related issues back to their root cause, making it easier to identify potential errors and facilitating efficient debugging when problems arise. For instance, if a model begins producing unexpected outputs, data lineage allows data scientists to quickly pinpoint whether the issue stems from a specific data source, a flawed transformation process, or a misconfiguration in how the data is fed into the model. By having a clear record of each step in the data journey, teams can confidently address issues without having to re-examine the entire data pipeline, thus saving time and reducing the risk of compounding errors.

Without a clear understanding of where data comes from, how it has been transformed, and how it is used in the model, organizations may struggle to diagnose and correct problems when they occur. For example, if an ML model starts to produce biased or inaccurate results, and there is no comprehensive data lineage in place, it can be incredibly challenging to determine whether the issue lies in the original data source, an intermediate transformation step, or the way the data was integrated into the model. This lack of traceability can lead to prolonged troubleshooting processes, during which the model may continue to produce flawed outputs, potentially leading to erroneous business decisions. In industries where decisions based on ML models have critical consequences—such as autonomous driving, where safety is paramount, or legal tech, where decisions impact people's lives—the inability to trace and rectify data issues swiftly can have severe repercussions.

Poor data lineage can hinder an organization's ability to comply with regulatory requirements and industry standards. Many regulations, such as the General Data Protection Regulation (GDPR), require organizations to maintain detailed records of how data is collected, processed, and used, particularly when it involves personal or sensitive information. If an organization cannot demonstrate a clear lineage of the data used in its ML models, it may face legal challenges, fines, or reputational damage. The lack of traceability makes it difficult to audit the model's decision-making process, reducing transparency and potentially eroding stakeholder trust. In the worst-case scenario, poor data lineage could result in the deployment of a flawed model that produces unreliable outputs, leading to decisions that have negative financial, operational, or ethical consequences.

Ensuring Data Security and Privacy

Robust data governance is imperative for safeguarding data security and privacy, particularly in ML projects where sensitive data is often utilized. Establishing rigorous access controls based on user roles and responsibilities helps protect sensitive information from unauthorized access and prevents potential data breaches. This role-based access control is crucial not only for protecting data integrity but also for ensuring that data is utilized ethically and in compliance with relevant data protection regulations.

Poor management of data access and permissions can lead to severe breaches and loss of data integrity. If unauthorized personnel gain access to sensitive data, the repercussions can include legal liabilities, loss of customer trust, and significant reputational damage. Moreover, improper handling of data access can lead to data corruption or unintentional biases being introduced into the ML models, further compromising their effectiveness and fairness.

Effective data governance in ML ensures that data is not only used to power innovative solutions but is also managed responsibly to uphold the values of trust, transparency, and compliance essential in today's data-driven world.

Data Access Controls and Permissions

Data governance for ML is essential in establishing a clear hierarchy of data access, ensuring that individuals within an organization have appropriate access to data based on their roles and responsibilities. By implementing role-based access controls, organizations can restrict access to sensitive data, such as personally identifiable information (PII) or proprietary business data, to only those who need it for model development and training. This approach not only protects sensitive data from unauthorized access but also helps to maintain data integrity by limiting the number of people who can modify or manipulate critical datasets. Effective data access controls ensure that data is used responsibly and ethically, reducing the risk of data breaches and ensuring compliance with data protection regulations.

When access to sensitive data is not properly managed, it increases the risk of data breaches, where unauthorized individuals might gain access to confidential information. This can have severe consequences, such as the exposure of private customer data or trade secrets, leading to legal repercussions and damage to the organization's reputation. If too many individuals have broad access to data, it becomes challenging to maintain data quality and consistency. Multiple users may inadvertently introduce errors, inconsistencies, or biases into the data, which can then propagate through the ML model, resulting in flawed outputs. For example, if a model is trained on a dataset that has been unknowingly altered by an unauthorized user, the predictions made by the model may be skewed, leading to poor decision-making and potentially costly outcomes.

Inadequate access controls can also lead to compliance violations, particularly in industries governed by strict data privacy regulations. For instance, regulations like GDPR and the California Consumer Privacy Act (CCPA) require organizations to implement stringent data access controls to protect personal data. If an organization fails to enforce proper access permissions, it may inadvertently expose sensitive data to unauthorized individuals, resulting in non-compliance with these regulations. This can lead to hefty fines, legal challenges, and a loss of trust among customers and stakeholders. Furthermore, poor access controls can complicate audit processes, making it difficult for organizations to demonstrate compliance with data governance standards. Without clear records of who accessed data and when, organizations may struggle to provide the necessary transparency required during audits, further exposing them to regulatory risks.

Compliance with Data Privacy Regulations

Data governance for machine learning (ML) practices is crucial to ensuring that ML deployments comply with relevant data privacy regulations, such as the General Data Protection Regulation (GDPR) and the California Consumer Privacy Act (CCPA). These regulations require organizations to implement specific measures to protect individuals' privacy, including obtaining informed consent for data collection, allowing individuals to access and control their personal data, and ensuring that data is deleted when no longer needed or when requested by the individual. Compliance with these regulations not only protects individuals' privacy rights but also helps build trust with customers and stakeholders by demonstrating a commitment to responsible data handling practices.

Poor compliance with data privacy regulations can lead to significant issues in ML models, potentially causing legal, financial, and reputational damage. For example, if an organization fails to obtain proper consent for data collection and uses this data to train ML models, it risks violating privacy laws, which can result in hefty fines and legal actions. These fines can be substantial, particularly under GDPR, where penalties can reach up to 4% of annual global turnover or €20 million, whichever is higher. Moreover, using data without consent can lead to biased or skewed models if the dataset does not accurately represent the broader population or if it includes biases that were not addressed because the data was not gathered transparently.

Non-compliance can erode trust between the organization and its customers. If customers discover that their data is being used without their consent or not being handled according to the regulations, they may lose confidence in the organization's ability to protect their privacy. This loss of trust can lead to a decrease in customer engagement, a reluctance to share data, and damage to the organization's brand reputation. In the age of social media and instant communication, such reputational damage can spread quickly, potentially leading to a broader public relations crisis. Moreover, failure to comply with data privacy regulations can also hinder the organization's ability to operate in certain markets or partner with other entities that require compliance as a prerequisite, thereby limiting business opportunities and growth potential. Therefore, maintaining robust compliance with data privacy laws is not only a legal necessity but also a critical component of sustainable business practices in any data-driven industry.

Promoting Transparency and Explainability

Transparency and explainability are pivotal in ML data governance as they enable stakeholders to understand how ML models make decisions. This is particularly important in sectors where decisions have significant impacts on individuals' lives, such as healthcare, employment, and law enforcement. By ensuring that models are explainable, organizations can provide justifications for decisions made by AI, facilitating ethical reviews, and regulatory compliance.

The absence of model explainability can lead to mistrust and skepticism among end users and regulators. If stakeholders cannot understand or verify the basis of algorithmic decisions, they may question the fairness and accuracy of these systems, potentially leading to resistance against their deployment. Additionally, in regulated environments, the inability to explain model decisions can result in non-compliance with laws mandating transparency, further complicating legal standings and public relations.

Model Explainability and Interpretability

Machine learning data governance encourages the use of techniques that enhance the interpretability of ML models. This improved interpretability allows stakeholders—

including developers, end users, and regulatory bodies—to better understand the factors influencing model predictions. By making the decision-making process of ML models more transparent, stakeholders can identify and address potential biases within the data or the algorithm itself, ensuring that the models perform as intended without unfair or discriminatory outcomes. This level of transparency is crucial for maintaining trust in automated systems, particularly in applications that affect individual lives or business-critical decisions, such as credit scoring, hiring, or healthcare diagnostics.

Poor model explainability and interpretability can lead to significant issues, particularly when models make decisions that have serious consequences. Without a clear understanding of how decisions are made, it becomes nearly impossible to verify the accuracy of the model's outputs or to justify those decisions in contexts where accountability is crucial. For example, if an ML model used in hiring processes inadvertently discriminates against certain groups of applicants, but the factors driving its decisions are not interpretable, the organization may continue unknowingly endorsing unfair practices. This not only poses ethical and legal risks but also damages the organization's reputation and trustworthiness. Furthermore, in regulated industries, such as finance or healthcare, where decisions need to be explained to comply with legal standards, lack of interpretability can result in non-compliance with regulatory requirements, leading to fines, sanctions, or restrictions on the use of ML models. Thus, ensuring model interpretability is not just about enhancing understanding and trust; it's also about fulfilling ethical obligations and complying with legal standards that govern the use of artificial intelligence and machine learning technologies.

Auditing and Monitoring of ML Systems

Regularly auditing and monitoring ML models is essential to maintain their accuracy, fairness, and effectiveness over time. Effective machine learning data governance frameworks provide structured processes to continually track model performance, identify potential biases, and ensure that models are operating in alignment with their intended outcomes. These processes help organizations detect any degradation in model performance or shifts in data patterns—known as concept drift—that might affect model accuracy. Monitoring also enables the ongoing assessment of how models impact different demographic groups, helping to prevent biases from influencing the models' decisions. This continuous oversight is crucial for maintaining the reliability and trustworthiness of ML systems, especially in dynamic environments where data and conditions can change rapidly.

Poor auditing and monitoring of ML systems can lead to significant issues that may compromise the integrity and utility of machine learning models. Without robust monitoring practices, an organization may not detect when a model starts to perform poorly or when biases begin to emerge due to changes in underlying data or market conditions. For example, a credit scoring model that was initially unbiased might begin to show discriminatory tendencies as economic conditions change, potentially leading to unfair treatment of certain groups of applicants. Similarly, models used in predictive maintenance could start to miss critical faults in machinery if not regularly recalibrated to reflect new operational data. This lack of timely detection and correction can result in flawed decisions, which can have severe consequences, from financial losses to reputational damage and legal challenges. Moreover, inadequate monitoring and auditing undermine an organization's ability to comply with regulatory standards that increasingly require transparency and fairness in automated decision-making systems. Such regulatory non-compliance can lead to fines, sanctions, and a loss of public trust, further exacerbating the challenges faced by organizations in competitive and highly regulated sectors.

Enabling Collaboration and Knowledge Sharing

Effective data governance fosters a culture of collaboration and knowledge sharing within organizations. This is especially crucial in environments where interdisciplinary teams work together on ML projects. Clear governance frameworks help delineate roles and responsibilities, ensuring that all stakeholders have access to the necessary data and insights to contribute effectively. This collaborative environment encourages the sharing of expertise and promotes innovative solutions, which are vital for the successful implementation of complex ML systems.

Without a unified framework that encourages open communication and information exchange, silos can develop within an organization. These silos prevent the flow of crucial data and insights necessary for the comprehensive training and refining of ML models. As a result, models may be built on incomplete or skewed datasets, leading to inaccurate predictions and inefficiencies that could have been avoided with better interdepartmental cooperation.

Data Ownership and Stewardship

Machine learning data governance assigns clear ownership of data assets within the organization, which is crucial for maintaining the integrity and utility of these assets throughout their lifecycle. Data owners, who are typically key stakeholders within the organization, are charged with ensuring that their respective data domains are managed properly in terms of quality, security, and compliance. This ownership involves not only overseeing how data is collected, stored, and used but also ensuring that it meets the necessary standards for feeding into ML models. Effective data stewardship helps prevent issues such as data decay, unauthorized access, and non-compliance with regulatory standards, all of which can undermine the performance and trustworthiness of ML systems.

Poorly defined data ownership and stewardship can lead to significant issues in ML models. Without clear responsibilities and accountability, data assets may suffer from neglect or mismanagement, resulting in poor data quality—such as outdated, incomplete, or incorrect data being used to train and operate ML models. For instance, if it's unclear who is responsible for updating a customer data database, critical updates might be overlooked, leading to ML models that make predictions based on outdated information. Additionally, a lack of proper stewardship can lead to inconsistent data security practices, increasing the risk of data breaches that not only compromise sensitive information but also damage the credibility of ML applications. Furthermore, unclear data ownership complicates compliance with data protection regulations, as there is no clear point of contact to ensure that data handling practices adhere to legal standards. This organizational ambiguity can result in fines, legal challenges, and a loss of stakeholder confidence. Therefore, establishing well-defined roles and responsibilities for data ownership and stewardship is essential to safeguarding the quality and security of data used in ML models, thereby enhancing their effectiveness and compliance.

Data Catalogs and Knowledge Management

Data governance promotes the creation and maintenance of data catalogs that are essential for organizing and managing data across an organization. These catalogs serve as comprehensive directories documenting data definitions, locations, and access controls, and they play a crucial role in facilitating knowledge sharing among various teams and streamlining data discovery for ML projects. By providing a clear and accessible overview of available data assets, data catalogs help ensure that data scientists and ML engineers can quickly find and utilize the correct datasets for their models, significantly speeding up development times and reducing the likelihood of errors caused by using inappropriate or outdated data.

Poor management of data catalogs and overall knowledge management practices can lead to significant issues in the development and performance of ML models. When data catalogs are not properly maintained or updated, they may contain inaccurate or outdated information about the data assets. This misinformation can mislead data scientists into using incorrect datasets or old versions of data, which can compromise the integrity of ML models. For example, an ML model trained on outdated customer demographic information may fail to reflect recent market trends or shifts in customer behavior, leading to ineffective or misguided business strategies. Additionally, without proper knowledge management, crucial information about data preprocessing, feature engineering, and model parameters may not be adequately documented or shared within the organization. This lack of documentation makes it difficult to replicate or build upon previous work, hindering collaboration and innovation. It also complicates efforts to audit and review ML models, as there is no clear record of how data was handled or decisions were made. This scenario can expose the organization to risks of non-compliance with industry regulations that require transparent and accountable data handling practices, such as GDPR in Europe or HIPAA in the United States. Therefore, robust management of data catalogs and effective knowledge sharing are critical to the success and reliability of ML initiatives.

Enabling Ethical and Responsible Use of Data

Data governance frameworks are pivotal in ensuring that ML systems are used ethically and responsibly. They enforce policies that require fairness, accountability, and transparency in all machine learning processes. This includes implementing checks for bias in datasets, making sure that data is used in a manner that respects privacy and ensures security, and providing clear documentation of the decision-making processes to satisfy regulatory and ethical standards.

The absence of strong ethical guidelines and responsible data use policies can lead to the development of ML systems that unintentionally perpetuate biases, invade privacy, or otherwise harm individuals. This not only leads to ethical breaches but also exposes the organization to regulatory penalties and severe damage to its reputation.

Furthermore, if stakeholders perceive that ML systems are developed without a strong ethical foundation, it could undermine trust in the technology and deter adoption, both internally and by customers.

Bias Detection and Mitigation

Machine learning data governance practices emphasize the importance of detecting and mitigating bias in data and algorithms to ensure fairness and accuracy in model outputs. Bias can originate from a variety of sources, including skewed data collection processes, historical inequalities reflected in the data, or subjective decisions made during data labeling. To counteract these biases, governance frameworks typically include procedures for employing diverse datasets that accurately reflect the target population and using fairness metrics to assess and correct biases during model training. Additionally, continuous evaluation of models is crucial to identify and address biases that may become apparent as the model is exposed to new data or as societal norms and values evolve. These practices help maintain the ethical integrity of ML systems and ensure they perform fairly across different groups.

Problems with bias detection and mitigation can lead to significant issues in ML models, which can manifest as discriminatory outcomes or flawed decisions that disproportionately affect certain groups. For instance, a recruitment model that has not been adequately checked for gender bias might consistently undervalue resumes from female applicants due to biased training data or biased feature selection. This not only perpetuates existing societal inequalities but also exposes the organization to legal and reputational risks. Similarly, a credit scoring model biased against certain racial or ethnic groups could deny individuals fair access to financial services, leading to regulatory scrutiny and potential sanctions.

Failure to effectively address biases in ML models can erode trust in these technologies. Public and consumer trust is crucial for the widespread adoption of AI and ML systems, and visible failures that suggest bias or unfairness can lead to public backlash and loss of consumer confidence. This is particularly critical in sectors like healthcare, law enforcement, and public services, where decisions made by ML models can have profound impacts on individuals' lives. For businesses, this loss of trust can translate into reduced customer engagement, boycotts, or loss of market share, especially if competitors are perceived to be more committed to fairness and ethical considerations. Therefore, robust bias detection and mitigation are not only about compliance and fairness but are also integral to maintaining public trust and the long-term viability of ML deployments.

Accountability and Explainability for Algorithmic Decisions

Data governance frameworks are crucial for establishing clear lines of accountability for the development, deployment, and outcomes of ML models. By defining who is responsible for each stage of the machine learning lifecycle, these frameworks ensure that there is transparency in how data is used and that stakeholders can track decisions back to the individuals or teams who made them. This level of transparency is essential not only for maintaining trust in ML systems but also for ensuring that these systems are used responsibly and ethically. Moreover, clear accountability helps facilitate the process of troubleshooting and optimizing ML models, as it becomes easier to identify who should address any issues that arise and who should make decisions about changes or improvements.

Without clear accountability, it becomes difficult to determine who is responsible for any errors or unintended outcomes of an ML model. This can result in a lack of ownership, with no specific party taking charge of rectifying issues, leading to prolonged periods of poor performance and potentially severe consequences depending on the application area. For instance, in critical applications like autonomous driving or medical diagnostics, unaddressed errors due to unclear accountability could lead to life-threatening situations, not only putting individuals at risk but also exposing the organization to legal liabilities and damaging its reputation.

The lack of explainability compounds these challenges. If stakeholders, including users, regulators, and decision-makers, cannot understand the rationale behind algorithmic decisions, they are less likely to trust and adopt the technology. This is particularly true in sectors where decisions significantly affect people's lives, such as in judicial sentencing, credit scoring, or hiring processes. If affected individuals cannot obtain clear explanations for decisions made by algorithms, this can lead to perceptions of unfairness and opacity, eroding trust and potentially leading to public backlash. Additionally, in many jurisdictions, there are increasing legal requirements for algorithmic transparency, such as the right to explanation under GDPR. Failure to provide this level of explainability can result in non-compliance with regulations, leading to fines, restrictions, or mandatory changes to the ML system. Thus, ensuring robust accountability and explainability is not only a matter of ethical responsibility and regulatory compliance but also a critical component of building and maintaining public confidence in machine learning technologies.

Key Stakeholders and Roles

Data governance for projects like ML is a collaborative effort that requires the involvement of various stakeholders within an organization. Here are some key players and their roles:

Executive Sponsors

Executives provide high-level leadership, championing the importance of data governance for ML. They allocate resources, secure funding, and establish a clear vision for ethical and responsible data use within the organization. Executive buy-in sets the tone for machine learning data governance and ensures it aligns with broader organizational goals. Some of their responsibilities include providing overall machine learning data governance strategy for ML, secure buy-in and commitment from stakeholders across departments, and advocate for the allocation of resources for machine learning data governance initiatives.

Data Governance Council

This cross-functional body oversees the development and implementation of machine learning data governance policies and standards. It provides strategic direction and ensures alignment between data governance practices and broader organizational goals. The council acts as a central decision-making body for machine learning data governance frameworks within the organization. Members of this council may include representatives from IT, legal, compliance, data science, business units, and security teams. The council is responsible to develop and maintain machine learning data governance policies for ML, oversee the implementation and ongoing effectiveness of data governance practices, and address challenges and emerging issues related data governance.

Data Owners

Data owners are accountable for specific data assets used in ML models. They ensure the data is accurate, complete, secure, and used according to established policies. They may be subject matter experts from business units or data management professionals. Data ownership defines clear lines of responsibility for data quality within the organization by defining boundaries and access controls. They also oversee the quality and integrity of their assigned data assets and collaborate with data stewards and data scientists to ensure data meets ML project requirements.

Data Stewards

Data stewards act as the day-to-day champions for machine learning data governance within their area of responsibility. They translate data governance policies related to machine learning into actionable practices, promoting data quality and user awareness. Data stewards serve as the bridge between data owners, data users, and technical teams. They implement data governance policies for ML projects for their assigned datasets, monitor data quality and identify potential issues, and promote data literacy and awareness among data users.

Data Scientists and ML Engineers

These technical experts are responsible for building and deploying ML models. They collaborate with data owners and stewards to ensure data quality and ethical considerations are factored into the development process. Data scientists and ML engineers are the architects who transform raw data into actionable insights through machine learning models. They understand and adhere to machine learning data governance policies and guidelines and actively participate in data quality assessments and bias detection procedures. One of the key responsibilities they carry is to document the data lineage and usage within their ML models and communicate model limitations and potential biases to stakeholders. It is their responsibility to ensure models are interpretable and explainable for ethical considerations.

Business Users

Business users represent the stakeholders who leverage ML models to inform decision-making across the organization. They provide valuable insights into business needs and how ML models can be used to address them. Business users are the ultimate beneficiaries of successful ML initiatives. They collaborate with data scientists to define project requirements and success metrics.

Legal and Compliance Teams

These teams ensure machine learning data governance practices comply with relevant data privacy regulations and legal requirements. They guide the organization on data collection, storage, usage, and potential risks associated with ML models. The legal and compliance teams act as safeguards to ensure ethical and responsible data use within the organization. Some of the key responsibilities for them include provide legal and compliance guidance on machine learning data governance practices, help assess the potential risks and biases in ML models, and ensure compliance with data privacy regulations such as GDPR and CCPA.

Security Teams

Security teams play a vital role in protecting sensitive data used in ML systems. They implement robust security measures to prevent unauthorized access, data breaches, and potential misuse. They develop and maintain data encryption protocols for sensitive data. Security teams are the gatekeepers, ensuring data is protected throughout its lifecycle. They are expected to conduct security audits and penetration tests to identify vulnerabilities.

Effective data governance is not a solo performance but rather a collaborative symphony. Each stakeholder plays a crucial role, and their combined efforts ensure the success of ML initiatives. Regular communication, shared goals, and a commitment to ethical data use are essential for a harmonious data governance ecosystem.

Data governance in ML is not just about setting rules but also about fostering a data-driven culture within the organization. This involves promoting data literacy, encouraging collaboration between technical and business teams, and empowering individuals to leverage data for informed decision-making. Stakeholders across the organization must share a common understanding of the importance of data quality, security, and ethical use.

As the field of ML continues to evolve, the roles and responsibilities of data governance stakeholders may adapt. New technologies like Explainable AI (XAI) will require collaboration between data scientists and legal teams to ensure model transparency and fairness. Emerging data privacy regulations will necessitate ongoing efforts from compliance teams to keep data governance practices up-to-date.

Machine Learning Data Lifecycle

The data used in machine learning projects goes through a well-defined lifecycle, and data governance for such projects plays a vital role at each stage. Here's a breakdown of the key phases:

Data Acquisition

This stage involves identifying and collecting data from various sources, considering factors like data relevance, quality, and legal compliance. Data governance for ML project practices ensures data is obtained through authorized channels and adheres to ethical data collection principles.

Data Preprocessing

Raw data often needs cleaning, transformation, and feature engineering before it can be used for training machine learning models. Data governance promotes the use of well-defined data preprocessing steps and documentation to ensure consistency and reproducibility.

Model Training and Development

The preprocessed data is used to train the machine learning model. Data governance ensures the training data is representative of the target population and reflects the intended use case of the model.

Model Deployment

Once trained, the model is deployed into production and used to make predictions or generate insights. Data governance practices focus on ensuring model security, monitoring model performance, and establishing processes for model versioning and rollback if necessary.

Model Monitoring and Evaluation

Continuously monitoring the deployed model's performance is crucial to ensure it remains accurate and unbiased over time. Data governance promotes the development of monitoring frameworks to track model performance metrics and detect potential issues like concept drift (changes in the underlying data distribution).

Data Retention and Archiving

Data governance establishes guidelines for how long data used in machine learning projects needs to be retained and how it should be archived after its useful life has expired. This ensures compliance with data privacy regulations and minimizes storage costs.

By implementing data governance practices at each stage of the machine learning data lifecycle, organizations can ensure the responsible, efficient, and trustworthy use of data for their AI initiatives.

Data Governance Challenges Specific to Machine Learning

Machine learning (ML) holds immense promise for revolutionizing various sectors. However, unlocking this potential hinges on a crucial element: effective data governance. Unlike traditional data management, ML presents unique challenges that require specialized data governance practices. Let's delve into these challenges and explore the complexities they introduce.

Large and Complex Datasets

The cornerstone of successful ML models is vast amounts of data, which often come in large and complex datasets. Managing these datasets presents significant governance hurdles, as traditional data governance practices designed for smaller, structured datasets may struggle to handle the scale and variability inherent in ML data. These large datasets not only include structured data but also unstructured data such as images, videos, and text, which can exponentially increase their size and complexity.

For example, a company specializing in facial recognition technology may gather millions of images and videos from diverse sources around the globe. Each of these files can vary greatly in size, quality, and format, contributing to a complex dataset that can quickly reach the scale of petabytes. Similarly, a company analyzing consumer behavior might collect vast amounts of data from social media, including text posts, user interactions, and multimedia content. These datasets are not only large but also highly variable, with a range of attributes and formats that can differ drastically from one piece to another.

Merging data from such diverse sources can be challenging and often leads to inconsistencies and biases in the dataset. For instance, if data collected from different regions uses varying formats for dates and times, or if image data is collected under different lighting conditions, it can lead to discrepancies that must be resolved before effective analysis can occur. Data governance needs to ensure rigorous quality checks, standardization, and data cleansing procedures to address these issues. For example, aligning date and time formats across datasets from different regions ensures that time-based data analysis is accurate and reliable.

Storing and managing these massive datasets requires robust infrastructure and efficient data storage solutions. As datasets grow in size, the need for scalable storage solutions becomes critical. Traditional data storage solutions may become cost-prohibitive or inefficient, prompting the need for more sophisticated technologies such as distributed databases or cloud-based storage solutions that can dynamically scale according to the needs of the organization.

Data governance frameworks must address these scalability concerns and ensure efficient data access for authorized users. For example, a cloud-based storage solution might employ data lakes where raw data is stored in its native format, and data lakes utilize big data processing tools like Apache Hadoop or Spark to manage and process data efficiently. These tools allow for handling large volumes of data at high velocities, which is essential for ML applications where timely data processing can be critical to performance.

Bias and Fairness

ML models are susceptible to perpetuating biases present within the data they're trained on. Unaddressed bias can lead to discriminatory outcomes, eroding trust and potentially causing harm. Data governance needs to implement methods for detecting and mitigating bias in datasets and algorithms. This may involve employing diverse training data, using fairness metrics, and continuously evaluating models for potential biases. Understanding how ML models arrive at their decisions is crucial for addressing bias. Data governance practices should encourage the use of Explainable AI (XAI) techniques to provide insights into model reasoning and identify potential bias sources.

Model Interpretability

The complex nature of some ML models, particularly deep learning algorithms, creates a "black box" effect. This lack of interpretability makes it difficult to understand how models reach their conclusions, hindering trust and raising ethical concerns.

Without interpretability, pinpointing errors and troubleshooting issues within ML models becomes challenging. Data governance frameworks should promote practices that improve model interpretability, facilitating debugging and ensuring accurate outcomes. Data governance needs to establish clear processes for human oversight of ML models, particularly in high-stakes applications. This ensures human comprehension of model decisions and allows for intervention when necessary.

Data Privacy and Security

ML systems often process sensitive data, making them prime targets for cyberattacks and privacy breaches. Balancing innovation with data privacy necessitates robust data governance practices.

Data governance frameworks should promote data minimization principles, collecting only the data necessary for specific ML projects. Furthermore, establishing clear access controls restricts access to sensitive data to authorized personnel.

Data governance can encourage data anonymization techniques where appropriate to protect individual privacy. Additionally, it should emphasize robust data encryption practices to safeguard sensitive information from unauthorized access.

Evolving Regulatory Landscape

The legal landscape surrounding data privacy and machine learning is in a state of flux, with regulations continually evolving to address the challenges posed by new technological advancements. This dynamic regulatory environment poses significant challenges for companies attempting to navigate the complexities of data governance in machine learning contexts. To remain compliant, data governance frameworks must not only adhere to existing regulations like the General Data Protection Regulation (GDPR) in the European Union and the California Consumer Privacy Act (CCPA) in the United States but also be flexible enough to adapt to new laws as they emerge.

For example, GDPR requires companies to obtain informed consent for data collection, which can be particularly challenging for ML applications where data usage can be extensive and not always anticipated at the point of collection. Additionally, GDPR and CCPA grant individuals substantial control over their personal data, including rights to access, rectify, and delete their information, which necessitates robust mechanisms within ML systems to handle these requests efficiently. Companies must also ensure that data used in ML models can be erased in compliance with the right to be forgotten, which requires advanced data tracking and management systems to trace data lineage accurately.

However, as new regulations continue to emerge globally, companies face ongoing challenges. For instance, Brazil's General Data Protection Law (LGPD) and China's Personal Information Protection Law (PIPL) introduce additional compliance nuances that can differ significantly from GDPR and CCPA. Navigating these differences requires a data governance framework that is not only robust but also exceptionally adaptable to change.

An example of the practical challenges faced by companies is seen in the financial sector, where AI models used for credit scoring or fraud detection must comply with both privacy regulations and sector-specific regulations like those enforced by the Consumer Financial Protection Bureau (CFPB) in the United States. These models must be designed not only to protect consumer data but also to ensure that their decision-making processes do not result in unfair or biased outcomes, which are subject to regulatory scrutiny.

Data Provenance and Lineage

Tracking the origin, transformations, and usage of data throughout the ML lifecycle is crucial for ensuring data quality, identifying potential errors, and understanding model behavior.

Data governance needs to establish robust data lineage management practices. This involves documenting the source, transformations, and usage of data within ML models, facilitating traceability and accountability.

Data governance frameworks should promote data version control practices to track changes made to datasets and models. This enables backtracking in case of issues and facilitates auditability for regulatory compliance.

Model Drift and Continuous Monitoring

Over time, the real-world environment and data distribution can change, causing ML models to drift from their original performance. Left unchecked, this drift can lead to inaccurate and unreliable model outcomes.

Data governance frameworks should outline processes for regularly monitoring ML models and tracking their performance over time. This allows for early detection of model drift and enables timely retraining or adjustments as necessary.

Data governance practices should ensure that data used for training and testing ML models is regularly refreshed to reflect real-world changes. This may involve scheduling periodic retraining cycles to maintain model accuracy and effectiveness.

Cost and Resource Constraints

Implementing and maintaining effective data governance for ML requires dedicated resources and budget allocation. Organizations may face challenges in balancing these needs with other priorities. Data governance frameworks must prioritize data governance activities based on risk and potential impact. Optimizing resource allocation and leveraging cost-effective data governance tools can help manage expenses. Developing in-house data governance expertise or partnering with external specialists can be crucial. This ensures the organization possesses the necessary skills and knowledge to manage data governance effectively.

Data governance challenges in ML are complex and multifaceted. However, by recognizing these hurdles and implementing robust data governance frameworks, organizations can navigate the thorny path toward responsible and ethical ML practices. Effective data governance fosters trust in ML systems, empowers stakeholders to leverage their potential for positive change, and ultimately paves the way for a future powered by responsible and ethical artificial intelligence.

Benefits and Value Proposition of ML Data Governance

While traditional data governance practices exist, ML introduces unique challenges that necessitate a specialized approach. This specialized approach, referred to as ML data governance, offers a compelling value proposition, unlocking a multitude of benefits for organizations embracing ML.

Traditional data governance focuses on data security, privacy, and accessibility within an enterprise. While these priorities remain crucial for ML data governance, the focus expands to encompass additional dimensions critical for successful ML initiatives.

Data Quality and Standards

Traditional data governance focuses on ensuring data security, privacy, and accessibility within an enterprise. These core principles remain crucial for ML data governance as well. However, the very nature of ML algorithms introduces additional complexities. Unlike traditional data analysis, ML models rely on complex algorithms to identify patterns and make predictions. These algorithms are highly sensitive to the quality of the data they are trained on. Inaccurate, inconsistent, or biased data can lead to unreliable and potentially discriminatory ML model outputs. For instance, an ML model trained on customer purchase data with inaccurate income levels might make biased recommendations for financial products.

Therefore, ML data governance expands its focus beyond just basic data security and accessibility. It emphasizes meticulous data quality and validation procedures to ensure the data used for training is free from errors, inconsistencies, and biases. This heightened focus on data quality is crucial for building robust and reliable ML models that deliver accurate and trustworthy results.

Focus on Lineage and Provenance

Traditional data governance practices often overlook the intricate details of data lineage. While ensuring data security, privacy, and accessibility remains essential, ML data governance demands a more meticulous approach. It emphasizes tracking the origin, transformations, and usage of data throughout the entire ML lifecycle. This focus on data lineage unlocks several valuable benefits.

Firstly, it facilitates debugging. Imagine an ML model producing unexpected results. By meticulously tracing the data lineage, data scientists can pinpoint the source of the issue, whether it's an error during data collection or an unintended consequence of a data transformation step. This allows for quicker troubleshooting and ensures the model is trained on clean, reliable data.

Secondly, data lineage empowers model explainability. Understanding the data journey—from its origin to its final use in training the model—allows organizations to explain how the model arrives at its predictions. This transparency fosters trust and acceptance of ML models by stakeholders, crucial for successful implementation across the organization.

Finally, meticulous data lineage tracking helps ensure compliance with data privacy regulations. In today's data-driven world, organizations are held accountable for how they handle user data. By having a clear understanding of data lineage, organizations can demonstrate responsible data practices and ensure they are adhering to relevant regulations, mitigating potential legal and reputational risks.

Model Monitoring and Performance

Traditional data governance practices often treat data as a static entity, focusing on its security and accessibility at a single point in time. However, machine learning (ML) data governance recognizes data as a dynamic element within the larger ML lifecycle. A crucial aspect of ML data governance is the regular monitoring of ML models to detect performance drift.

Performance drift occurs when the real-world data distribution that the model was trained on begins to shift over time. This can happen due to various factors, such as changes in customer behavior, market trends, or even seasonal variations. Imagine training a model to predict loan defaults based on historical customer data. If economic conditions worsen, the model's performance might degrade as it fails to account for the new financial landscape.

Regular monitoring of ML models through ML data governance helps identify performance drift early on. This allows data scientists to take corrective actions, such as retraining the model with fresh data that reflects the updated real-world distribution. By proactively addressing performance drift, organizations can ensure their ML models continue to deliver accurate and reliable results, maximizing the value they provide.

Focus on Fairness and Bias Mitigation

Traditional data governance practices often operate under the assumption that data is neutral. However, machine learning (ML) data governance recognizes the inherent potential for bias within datasets. This bias can stem from various sources, such as historical prejudices reflected in the data itself or biases present within the algorithms themselves. ML data governance requires proactive measures to identify and mitigate these biases, ensuring fair and unbiased model outcomes.

Imagine an ML model used for loan approvals that is trained on historical data biased against a particular demographic group. This could lead to unfair rejections for qualified individuals, perpetuating existing inequalities. ML data governance tackles this challenge by incorporating bias detection techniques. Data scientists can analyze datasets for patterns that might lead to biased outcomes and implement techniques like data balancing to mitigate these biases. Additionally, they can choose algorithms less susceptible to bias or even develop custom algorithms designed for fairness.

By prioritizing bias mitigation, ML data governance ensures that ML models function ethically and responsibly, delivering accurate and fair results that benefit everyone. This fosters trust in ML technology and paves the way for its responsible use across various domains.

Collaboration and Knowledge Sharing

Traditional data governance practices often operate within silos, with data ownership and access tightly controlled. This approach proves inadequate for the collaborative nature of machine learning projects. ML data governance, on the other hand, champions the concept of data democratization. It encourages breaking down data barriers and fostering collaboration across teams. Data scientists, engineers, and domain experts can readily access and share relevant datasets, facilitating knowledge exchange and accelerating the development of effective ML models.

This collaborative environment fostered by ML data governance unlocks the full potential of data. By bringing together diverse perspectives and expertise, teams can leverage data more effectively throughout the ML lifecycle—from choosing the right data for training to interpreting and deploying models successfully. This collaborative approach leads to the development of more robust, reliable, and ultimately impactful ML solutions.

By addressing these distinct requirements for ML, data governance unlocks a unique value proposition:

Enhanced Data Quality and Trustworthy Results

Data governance ensures that high-quality data feeds into ML models, which is crucial for the accuracy and effectiveness of these models. By implementing rigorous data quality assurance processes, organizations can minimize errors and inconsistencies in the data, leading to improved model performance and more reliable outputs. This, in turn, fosters trust among users and stakeholders in the ML systems, thereby empowering data-driven decision-making.

For example, a financial institution implementing ML models for credit scoring can significantly benefit from enhanced data quality. By using a robust data governance framework to clean, standardize, and verify the integrity of data from various sources (such as credit bureaus, transaction histories, and customer interactions), the institution ensures that the credit scoring models are trained on accurate and comprehensive data. This leads to more precise credit risk assessments, reducing the risk of default and enhancing the institution's ability to make informed lending decisions.

Robust Security and Privacy Safeguards

Data governance plays a critical role in protecting sensitive information used in ML models by enforcing strong access controls and adhering to strict data privacy regulations. This is vital in mitigating the risk of data breaches, which can have severe reputational and financial consequences for organizations. Robust data governance ensures that sensitive data, such as personal identifiers or proprietary information, is accessed only by authorized personnel and that all access is logged and auditable, further enhancing security.

For instance, a healthcare provider using ML models to predict patient outcomes must handle sensitive health data with utmost care. Through effective data governance, the provider implements encrypted data storage, role-based access controls, and regular audits, ensuring compliance with healthcare regulations like HIPAA in the United States. This not only protects patient data but also builds trust in the healthcare provider's data handling practices, crucial for patient confidence and regulatory compliance.

Increased Transparency and Explainability

Data governance promotes the use of Explainable AI (XAI) techniques, which are essential for improving the transparency of ML models. XAI helps stakeholders understand how decisions are made by ML models, which is particularly important in sectors where these decisions have significant impacts, such as healthcare, finance, and criminal justice. Implementing XAI techniques facilitates debugging, enhances trust, and ensures that there is human oversight in critical decision-making processes.

A practical example of this can be seen in an insurance company that uses ML models to assess claims. By incorporating XAI, the company can provide clear explanations to customers about the reasons for claim approvals or denials based on specific data points and model reasoning. This transparency helps in resolving disputes, complying with industry regulations, and maintaining customer trust.

Fosters Collaboration and Knowledge Sharing

Effective data governance breaks down data silos within organizations and promotes collaboration across different teams. By establishing comprehensive data catalogs and enforcing knowledge management practices, data governance frameworks make it easier for teams to discover, access, and utilize data efficiently. This is crucial for fostering innovation and ensuring that ML models are built on the most complete and relevant data available.

Consider a multinational corporation with multiple business units. Through robust data governance, the corporation maintains a centralized data repository that allows teams from different geographical locations and functional areas to access and share data seamlessly. This collaborative environment enables data scientists and business analysts across the corporation to develop more accurate and holistic ML models by leveraging a broader range of insights and expertise.

Mitigates Operational Risks and Ensures Efficiency

Data governance addresses data drift and optimizes ML models, improving ROI and preventing inaccurate model outputs. Effective data lineage tracking facilitates easier auditing and regulatory compliance.

While ensuring high-quality data and building trust in ML systems are fundamental benefits of data governance, it offers even more strategic advantages for organizations. One key aspect is its ability to future-proof ML initiatives.

The field of machine learning is constantly evolving, and so are the ethical considerations surrounding it. Biases present in training data can lead to discriminatory or unfair model outputs. Data governance tackles this challenge by prioritizing fairness and ethical considerations throughout the ML lifecycle. This includes employing techniques to identify and mitigate biases within datasets, as well as choosing algorithms less susceptible to such biases. By proactively addressing these issues, organizations can ensure their ML practices are responsible and compliant with emerging regulations in the data privacy landscape. This forward-thinking approach positions them to leverage the power of ML responsibly and avoid potential roadblocks in the future.

Furthermore, data governance fosters a data-driven culture within organizations. By establishing clear guidelines and practices for data management, it empowers employees to understand the value of data and utilize it responsibly. This creates an environment where data is not just collected but actively analyzed and leveraged to drive innovation and improve decision-making across all levels. Imagine a marketing team empowered with insights from customer data models to develop targeted campaigns with greater effectiveness. Data governance fosters this kind of data-driven thinking, allowing organizations to continuously unlock the potential of ML and achieve better outcomes.

While traditional data governance serves a critical role, ML data governance takes it a step further. By recognizing the unique challenges of ML and implementing a data governance framework specifically tailored for those challenges, organizations unlock the full potential of their data assets. ML data governance empowers them to build and deploy successful ML models ethically, responsibly, and efficiently, ultimately driving innovation and competitive advantage. In a world driven by data, effective ML data governance acts as a catalyst for unlocking the true power of machine learning.

Case Studies and Real-World Examples

As this introduction to machine learning data governance winds down, it's worth digging into a few brief examples that illustrate how data governance frameworks work in practice.

Case Study 1: Financial Fraud Detection with Machine Learning

A large financial institution utilizes machine learning models to identify fraudulent transactions in real time. Effective data governance practices ensure the training data for these models is accurate, representative of real-world fraudulent activities, and free from historical biases. The data governance framework also includes procedures for model monitoring to detect potential biases or performance degradation over time. This ensures the machine learning models can effectively detect fraudulent transactions while minimizing false positives that could inconvenience legitimate customers.

Case Study 2: Personalized Recommendations with Machine Learning

An e-commerce platform leverages machine learning to recommend products to customers based on their past purchase history and browsing behavior. Data governance plays a critical role in ensuring the quality and security of customer data used for these recommendations. Additionally, data governance practices address potential biases in the data to prevent discriminatory recommendations. This not only enhances the customer experience but also mitigates the risk of legal issues related to biased recommendation systems.

These are just a few examples of how effective machine learning data governance can benefit organizations in various industries. By implementing robust data governance practices, organizations can unlock the full potential of machine learning while ensuring responsible and trustworthy AI development and deployment.

Summary

In this chapter, we explored the foundational principles and importance of data governance within the context of machine learning. We highlighted how effective data governance is critical to ensuring the quality, security, and ethical use of data throughout the ML lifecycle. Central to this discussion was the idea that high-quality data is the cornerstone of successful ML models, and without rigorous data governance practices, organizations risk producing unreliable and potentially harmful outcomes.

We also examined the key elements of data governance, including the importance of data accessibility, security, and compliance with evolving privacy regulations. The challenges associated with poor data governance were discussed in depth, particularly the risks of fragmented data, lack of transparency, and the perpetuation of biases in ML models. To mitigate these risks, the chapter emphasized the need for practices that enhance data quality, promote transparency through Explainable AI (XAI) techniques, and ensure the ethical use of data.

Additionally, we considered the evolving regulatory landscape and the necessity for adaptable data governance frameworks that can keep pace with new legal requirements. The chapter underscored the importance of continuous monitoring and evaluation of ML models to maintain compliance and uphold trust in data-driven decision-making.

In summary, this chapter set the stage for understanding the critical role of data governance in machine learning. It provided a comprehensive overview of the challenges and opportunities that effective data governance presents, laying the groundwork for a deeper exploration of these topics in the subsequent chapters.

CHAPTER 2

Establishing a Data Governance Framework

In today's data-driven landscape, where machine learning (ML) initiatives are becoming increasingly prevalent, organizations must implement robust data governance frameworks to ensure the responsible and effective use of their data assets. A comprehensive data governance framework provides the necessary structure, policies, processes, and tools to manage data throughout its lifecycle, enabling organizations to leverage the full potential of their ML endeavors while mitigating risks and fostering trust.

This chapter delves into the key components of a data governance framework tailored for machine learning initiatives. It explores the organizational structures and roles involved, outlines essential policies and standards, defines critical processes and workflows, and discusses the tools and technologies that support data governance activities. Additionally, this chapter introduces the concept of a data governance maturity model and emphasizes the importance of stakeholder engagement and change management for successful implementation.

Components of a Data Governance Framework

Machine learning (ML) holds immense potential, but its success hinges on a critical foundation—data governance. A data governance framework is a systematic approach to managing an organization's data assets, encompassing policies, processes, and technologies to ensure responsible data use throughout its lifecycle. In the realm of ML, this framework plays a vital role in addressing unique challenges, like handling massive complex datasets, mitigating bias, and ensuring model transparency.

© Aditya Nandan Prasad 2024
A. Nandan Prasad, *Introduction to Data Governance for Machine Learning Systems*,
https://doi.org/10.1007/979-8-8688-1023-7_2

The core components of a data governance framework for ML initiatives can be categorized into several key pillars. Each of these pillars is discussed in detail later in the chapter.

Foundational Policies and Standards

Clearly defined policies and standards establish the guiding principles for data management. These address data quality, security, privacy, and compliance with relevant regulations. Imagine a healthcare organization leveraging ML to predict patient readmission risks. Robust data governance ensures patient privacy is protected, while adhering to regulations like HIPAA. A detailed discussion on policies and standards required for ML data governance is provided under section "Data Governance Policies and Standards."

Streamlined Processes and Workflows

Well-defined processes govern activities like data acquisition, validation, lineage tracking, access management, and issue resolution. These workflows streamline data governance practices, ensuring consistency and efficiency across the organization. For instance, data lineage tracking allows data scientists to understand the origin and transformations of data used in ML models, facilitating debugging and ensuring data integrity. Further discussion on processes and workflows used in ML data governance is provided under section "Data Governance Processes and Workflows."

Defined Roles and Responsibilities

Effective data governance assigns clear roles and responsibilities to various stakeholders. These include data owners (responsible for data accuracy), data stewards (overseeing data access and usage), data consumers (those who use data for analysis), and data governance committees (providing oversight and guidance). This clear delineation of roles ensures accountability and fosters a collaborative data governance environment. A detailed discussion on this topic is provided under "Data Governance Organizational Structure."

Specialized Tools and Technologies

Data governance leverages specialized tools and technologies to streamline its activities. These include data catalogs to locate relevant datasets, data lineage tools to track data origin and transformations, data quality tools to identify and rectify inconsistencies, and metadata management systems to organize data descriptions. A detailed discussion on this topic is provided under "Data Governance Organizational Structure."

Metrics and KPIs

Data governance effectiveness is measured using relevant metrics and key performance indicators (KPIs). These metrics track aspects like data quality improvement, access control efficiency, and user satisfaction with data governance procedures. KPIs allow organizations to continuously assess and improve their data governance program for optimal effectiveness. A detailed discussion on metrics and KPIs applicable for ML data governance is provided under section "Data Governance Metrics and Key Performance Indications (KPIs)."

Maturity Model and Roadmap

A data governance maturity model provides a structured roadmap for organizations to advance their data governance practices from initial, ad hoc stages to fully optimized frameworks. This model allows organizations to assess their current capabilities, identify gaps, and prioritize investments to improve data governance systematically. For example, a large retail corporation began at the "Initial" maturity level, where data governance was fragmented and inconsistent across departments. Recognizing the risks of this approach, the company established a centralized data governance team and standardized data practices, moving into the "Developing" stage. By implementing a data catalog and conducting training sessions, they began fostering a data-driven culture and improving data quality.

As the company progressed to the "Defined" stage, they introduced automated data quality checks and integrated data governance with enterprise processes like risk management and compliance. This integration reduced data errors and enhanced the reliability of business insights. Eventually, the company reached the "Optimized" stage, where data governance was fully embedded into the organization's culture. Advanced analytics and machine learning were employed to monitor data quality proactively, leading to improved operational efficiency, better customer insights, and a stronger competitive position. This example demonstrates how a structured approach using a data governance maturity model can significantly enhance an organization's data governance capabilities and business outcomes.

A detailed discussion on ML data governance maturity model is provided under section "Data Governance Maturity Model."

Stakeholder Engagement and Change Management

Implementing a data governance framework is a complex process that hinges on successful stakeholder engagement. This engagement involves more than just informing stakeholders about new practices; it requires actively involving them in the process, fostering a data-driven culture, and ensuring that all employees—from leadership to frontline staff—understand the value of responsible data use. For instance, a multinational financial institution effectively engaged its stakeholders by conducting workshops and training sessions that highlighted the strategic benefits of data governance. They maintained continuous dialogue with stakeholders, addressing concerns and incorporating feedback, which helped build trust and reduce resistance to the new practices. The institution's phased rollout strategy allowed them to refine their approach and demonstrate the tangible benefits of improved data quality and decision-making capabilities, ultimately securing broad support for the initiative.

Another example can be seen in a global healthcare provider that focused on engaging healthcare professionals, who were traditionally more concerned with patient care than data management. By creating cross-functional teams that included doctors, nurses, and IT staff, the provider was able to identify and address data governance challenges specific to clinical settings. The introduction of a rewards system to recognize excellence in data management further motivated employees to embrace the new practices, leading to significant improvements in data accuracy, compliance, and patient outcomes. These examples highlight that by fostering a data-driven culture and employing effective change management strategies, organizations can ensure that their data governance frameworks are not only adopted but also fully integrated into their operations, leading to sustained success.

A detailed discussion on this topic is provided under section "Stakeholder Engagement and Change Management."

By building a comprehensive data governance framework tailored for ML initiatives, organizations can unlock the full potential of ML. This framework not only ensures the effective and responsible use of data assets but also fosters trust, transparency, and compliance, creating a solid foundation for successful and impactful machine learning projects.

Data Governance Organizational Structure

Effective data governance requires the involvement and collaboration of various stakeholders within an organization. A well-defined organizational structure, with clearly defined roles and responsibilities, is essential for ensuring the successful implementation and maintenance of a data governance framework.

Centralized Model

The centralized model for machine learning (ML) data governance places the primary responsibility for overseeing all aspects of data governance on a dedicated team. This team acts as the central hub, collaborating with various departments and stakeholders across the organization. The structure of this model typically consists of two key components:

Data Governance Council, a high-level committee, serves as the guiding force for data governance within ML. Comprised of executives, legal representatives, and IT specialists, the council provides strategic direction and oversees the overall data governance framework. Imagine a large financial institution implementing ML models to assess loan risks. The council, with its diverse expertise, would ensure the data governance framework aligns with the institution's strategic goals, legal requirements, and IT infrastructure.

Data Governance Office (DGO) acts as the central unit responsible for translating the strategies set by the council into actionable practices. This office houses data governance specialists who develop and implement policies, standards, and procedures specifically tailored for ML initiatives. Their core focus lies on ensuring data quality, security, and compliance with relevant regulations throughout the ML lifecycle. For instance, the DGO might develop guidelines for data anonymization to protect user privacy while training ML models for customer behavior analysis.

The centralized model offers several key strengths. The centralized structure establishes a clear chain of command for data governance practices. This ensures accountability and facilitates efficient decision-making regarding ML data management.

A dedicated team overseeing data governance ensures consistent application of policies and procedures across all ML projects within the organization. This consistency minimizes the risk of variations in data handling practices that could compromise the validity of ML models.

The DGO acts as a central hub for data governance expertise. This pool of specialists can be leveraged by various departments within the organization, ensuring they have access to the necessary knowledge and guidance when implementing ML initiatives.

However, the centralized model also comes with considerations. As the number and complexity of ML projects within an organization increase, the centralized model might struggle to scale effectively. The dedicated team may become overwhelmed, leading to potential delays or inconsistencies in data governance practices for new initiatives.

The centralized approach might create a disconnect between the DGO and departments actively working on ML projects. Departments might feel a lack of ownership over data governance practices, potentially hindering their full collaboration and engagement with the DGO's efforts.

Decentralized Model

The decentralized model for machine learning (ML) data governance distributes responsibilities across different departments and business units. Each department or unit has a designated data governance lead, often referred to as a data steward, who champions data governance practices within their specific domain. This approach fosters a sense of ownership and empowers departments to actively participate in ensuring responsible data management for their ML initiatives.

The structure of this model mirrors the centralized model in one key aspect. **Data Governance Council**, a high-level committee, similar to the one in the centralized model, provides strategic oversight and direction for data governance within ML. This council, comprised of executives and representatives from legal and IT departments, establishes overall data governance policies and ensures alignment with the organization's broader goals.

However, the decentralized model deviates in how these policies are implemented. Each department or business unit designates **a data steward who acts as the local champion for data governance within their area**. These stewards collaborate with the central Data Governance Office (DGO) and data owners to translate general data governance policies into specific practices relevant to their departmental ML projects. For instance, the data steward for the marketing department might develop departmental guidelines for customer data anonymization practices specific to their marketing campaign models.

The decentralized model offers several advantages: Distributing data governance responsibilities fosters a sense of ownership among departments. This promotes buy-in and active participation in data governance practices, leading to a more collaborative and engaged approach to data management. Departments feel invested in the success of data governance and are more likely to actively identify and address data quality or compliance issues within their areas.

Data stewards within each department possess a deep understanding of the specific needs and challenges associated with their departmental data. This contextual knowledge empowers them to tailor data governance practices to address these unique requirements. They can develop departmental data standards and procedures that effectively manage data specific to their domain, ensuring its quality and suitability for building robust ML models.

However, there are also considerations to address with this approach. Decentralization can lead to inconsistencies in how data governance practices are applied across different departments. Without a central governing body ensuring consistent implementation, there's a risk of variations in data handling practices, potentially impacting the overall effectiveness of data governance across the organization.

Individual departments may lack the dedicated resources required to manage all aspects of data governance effectively. This could lead to situations where departments struggle to keep up with the evolving requirements of data governance, potentially hindering their ability to implement best practices for their ML projects.

Hybrid Model

The centralized model offers a structured approach with clear lines of responsibility, while the decentralized model fosters departmental ownership and context-specific practices. The hybrid model emerges as a solution that bridges the gap between these two approaches, combining elements of both for a balanced and effective approach to ML data governance.

The hybrid model's structure relies on three key components. Similar to the previous models, a high-level council, **Data Governance Council**, provides strategic direction and oversight for data governance within the organization's ML initiatives. This council, comprised of executives and representatives from legal and IT departments, establishes overall data governance policies and ensures alignment with the organization's broader goals and compliance requirements.

Data Governance Office (DGO) acts as the central hub for data governance within the hybrid model. It provides essential oversight and resources to departmental data stewards. The DGO's core responsibilities include developing and maintaining central data governance policies, standards, and procedures, offering training and support to departmental data stewards, and collaborating with them on addressing data governance challenges specific to their ML projects. Imagine a large retail organization leveraging ML for product recommendation systems. The DGO might develop a central data governance framework outlining data quality standards for customer purchase data, while also collaborating with the marketing department's data steward to address specific data privacy considerations related to their recommendation models.

Each department or business unit designates a **data steward who acts as the local champion for data governance within their area**. These data stewards collaborate closely with the DGO to translate central data governance policies into specific practices relevant to their departmental ML projects. They leverage their departmental knowledge to identify and address data governance challenges specific to their domain, ensuring that data quality, security, and compliance considerations are effectively addressed within their ML initiatives. For instance, the data steward for the finance department might establish departmental data lineage tracking procedures to ensure transparency and traceability of data used in their credit risk assessment models.

The hybrid model offers several key advantages. The hybrid model strikes a balance between centralized oversight and departmental autonomy. The DGO ensures consistent implementation of core data governance practices across the organization, while departmental data stewards have the flexibility to tailor these practices to address their specific departmental needs and challenges. This balance fosters a more comprehensive and effective approach to data governance for ML initiatives.

The hybrid model capitalizes on the strengths of both centralized and decentralized approaches. The DGO's centralized expertise in data governance best practices complements the departmental knowledge of data stewards. This combined expertise fosters a collaborative environment where the DGO provides guidance and support, while data stewards leverage their departmental understanding to implement effective data governance practices specific to their ML projects.

However, the hybrid model also presents some considerations. The success of the hybrid model hinges on clear communication and collaboration between the DGO and departmental data stewards. Establishing well-defined communication channels and fostering a collaborative environment are crucial to ensure that departmental data stewards understand and effectively implement central data governance policies within their respective ML projects.

Ultimately, the optimal organizational structure for ML data governance depends on several factors specific to each organization. Larger organizations with numerous and complex ML projects may benefit from a centralized or hybrid approach to ensure consistent data governance practices across all initiatives. Smaller organizations with fewer ML projects might find a decentralized model to be more efficient.

The sensitivity of data involved in ML projects also plays a role. If projects involve highly sensitive data, such as personal health information, a centralized model with robust security controls may be preferred to ensure the highest level of data protection.

Organizations with existing data governance programs may be able to leverage a decentralized or hybrid model more readily, as they already possess a foundation of data governance expertise and established communication channels.

Building a well-defined organizational structure for ML data governance lays the foundation for successful and responsible ML initiatives. By carefully considering the factors outlined above and fostering collaboration among key stakeholders, organizations can choose the structure that best suits their needs and empowers them to leverage the untapped potential of ML. With a well-orchestrated data governance framework in place, organizations can ensure the ethical and effective use of data, unlocking valuable insights and driving innovation through the power of machine learning.

Data Governance Policies and Standards

At the core of an effective data governance framework are well-defined policies and standards that govern data management practices within an organization. These policies and standards ensure consistency, quality, security, privacy, and compliance with relevant regulations, while also addressing the unique challenges posed by machine learning initiatives.

The following are some essential data governance policies and standards that organizations should consider implementing:

Data Quality Policy

The Data Quality Policy establishes the standards and requirements to ensure this quality. Imagine training an ML model to predict customer churn, but the data used is riddled with errors. The resulting model would likely be unreliable, leading to inaccurate predictions and potentially causing the business to lose valuable customers.

This policy outlines various aspects of data management for ML initiatives. It might include guidelines for data validation, a process to identify and correct errors. Data cleansing procedures ensure inconsistencies are removed, and enrichment techniques might involve adding valuable information to improve the data's usefulness for ML models. The policy should also address how to handle missing or incomplete data, a common challenge in many datasets.

Furthermore, the Data Quality Policy defines metrics and thresholds for measuring data quality. These metrics allow organizations to assess the overall health of their data and identify areas for improvement. By setting clear thresholds for acceptable data quality, organizations can ensure their ML models are built upon a solid foundation, leading to more accurate and reliable results.

Data Security and Privacy Policy

The Data Security and Privacy Policy serves as a critical document outlining the measures and controls organizations must implement to protect sensitive data, ensure user privacy, and comply with relevant regulations. This policy acts as a shield, safeguarding data throughout its lifecycle within ML projects.

The data used in ML models likely contains highly sensitive patient information. The Data Security and Privacy Policy would ensure this data is protected through robust measures like encryption, access controls, and data handling procedures. Encryption scrambles data into an unreadable format, rendering it useless in the event of a breach. Access controls dictate who can access the data and what they can do with it, minimizing the risk of unauthorized access or misuse. Furthermore, the policy would outline specific data handling procedures, ensuring data is only used for authorized purposes and disposed of securely when no longer needed.

Compliance with data privacy regulations like the General Data Protection Regulation (GDPR) or the California Consumer Privacy Act (CCPA) is another key aspect of this policy. These regulations grant individuals certain rights regarding their personal data, such as the right to access, rectify, or erase their data. The Data Security and Privacy Policy ensures organizations understand and adhere to these regulations, protecting user privacy and building trust with their customers.

The policy goes beyond simply outlining these measures; it establishes a framework for ongoing monitoring and improvement. Regular security audits and penetration testing identify potential vulnerabilities in data security practices. Additionally, the policy should mandate employee training on data security and privacy best practices, ensuring everyone within the organization understands their role in protecting sensitive data.

By implementing a comprehensive Data Security and Privacy Policy, organizations can build trust with their users and stakeholders. This policy demonstrates a commitment to responsible data handling practices, fostering transparency and ensuring compliance with evolving data privacy regulations. Ultimately, a strong Data Security and Privacy Policy is not just a regulatory requirement, but a cornerstone of ethical and trustworthy machine learning initiatives.

Data Classification and Handling Policy

The Data Classification and Handling Policy acts as a roadmap, defining clear criteria for classifying data based on its sensitivity, criticality, and regulatory requirements. This classification system serves a critical purpose—it dictates how data is treated and protected throughout its lifecycle within ML projects.

Highly sensitive data, such as social security numbers, would likely be classified at the highest level, requiring the most stringent security measures. The policy would outline appropriate handling procedures for this data, including robust encryption and access controls. In contrast, less sensitive data, like loan amounts or repayment history, might be classified at a lower level, with correspondingly less stringent handling requirements.

This policy ensures data is treated with the appropriate level of protection based on its inherent risks. By following these guidelines, organizations can minimize the risk of data breaches and ensure compliance with relevant regulations. Furthermore, the policy fosters transparency within the organization. Data users understand the sensitivity of the data they are working with and can make informed decisions about its handling within ML projects. This classification system empowers responsible data use and builds trust with stakeholders concerned about data privacy and security.

Data Access and Usage Policy

The Data Access and Usage Policy serves as a gatekeeper, governing who has access to what data and for what purposes within ML initiatives. This policy establishes clear roles, responsibilities, and procedures for granting, reviewing, and revoking data access privileges. Imagine a large retail organization leveraging ML models to optimize product recommendations. The Data Access and Usage Policy would ensure that only authorized personnel, such as data scientists working on the recommendation models, have access to customer purchase data.

The policy goes beyond simply granting access. It defines the acceptable use cases for data within the context of ML projects. This ensures data is used for its intended purpose and not misused for unauthorized activities. The policy might also outline procedures for data anonymization, a technique that removes personally identifiable information from data, allowing for analysis while protecting user privacy.

Furthermore, the Data Access and Usage Policy establishes a clear chain of responsibility. It defines roles within the organization, such as data owners (responsible for data accuracy) and data stewards (overseeing data access), ensuring everyone understands their role in safeguarding data access. Regular reviews of access privileges are mandated by the policy, ensuring that only those who still require access maintain it.

Data Lifecycle Management Policy

The success of machine learning (ML) initiatives hinges on not just the quality of data used, but also its responsible management throughout its entire lifecycle. The Data Lifecycle Management Policy serves as a roadmap, outlining the processes and procedures for effectively managing data from the moment it's acquired to its final archiving or disposal. Imagine a social media platform leveraging ML models to personalize user feeds. This policy ensures the vast amount of user data collected is handled securely and responsibly throughout its lifecycle.

The policy addresses key aspects of data management, ensuring data is readily available for use while minimizing risks associated with its storage and disposal. Data retention guidelines define how long different types of data need to be stored, considering legal requirements and business needs. For instance, financial data might require longer retention periods compared to anonymized clickstream data used for website optimization models. Version control ensures that different versions of the data are tracked and accessible, allowing for easy rollback in case of errors during the model development process.

Regular backups are crucial to safeguard against data loss due to hardware failures or other unforeseen events. The policy outlines secure backup procedures, ensuring data can be readily recovered in case of emergencies. Finally, the policy dictates procedures for secure deletion or archiving of data that has reached the end of its lifecycle. Secure deletion ensures data is permanently erased and cannot be retrieved, while archiving allows for long-term storage of data that might be needed for future use, but with stricter access controls in place.

By implementing a comprehensive Data Lifecycle Management Policy, organizations can ensure their data is managed responsibly and securely throughout its lifecycle. This not only minimizes risks associated with data breaches and unauthorized access, but also fosters trust with stakeholders concerned about data privacy and responsible data use within ML projects.

Metadata Management Policy

The Metadata Management Policy acts as a decoder ring, establishing standards and guidelines for documenting and managing metadata—the information that describes the data itself. Imagine a team building ML models to predict customer churn. Without proper metadata, it might be unclear what specific fields within customer data represent factors like purchase history or recent interactions. The Metadata Management Policy ensures everyone understands the data by establishing clear guidelines for documenting its meaning and origin.

This policy outlines essential metadata requirements. It might specify that data attributes like "customer ID" or "last purchase date" be clearly defined, including details about their format, units of measurement, and any transformations applied. Naming conventions are another crucial aspect. The policy might mandate consistent and descriptive names for data fields, such as "customer_age_in_years" instead of simply "age." This consistency ensures everyone working with the data can easily understand its meaning.

The policy establishes processes for maintaining metadata repositories or catalogs. These centralized locations store all the metadata associated with different datasets, making it readily accessible to anyone working on ML projects. Regular reviews and updates of metadata are also mandated by the policy, ensuring it remains accurate and reflects any changes made to the data itself.

Clear and well-documented metadata fosters transparency within ML projects. Data scientists and other stakeholders can understand the data they are working with, leading to more accurate models and better decision-making. This empowers responsible data use and builds trust with stakeholders concerned about data privacy and interpretability within ML initiatives.

Data Governance and Stewardship Policy

The effective use of data in machine learning (ML) initiatives necessitates a clear understanding of roles and responsibilities. The Data Governance and Stewardship Policy serves as a roadmap, defining who does what and how decisions are made regarding data management within the organization. Imagine a large e-commerce platform leveraging ML models for fraud detection. This policy ensures everyone involved, from data scientists to executives, understands their role in safeguarding data and ensuring its responsible use.

The policy clearly outlines the roles of various stakeholders within the data governance framework. Data owners, typically department heads or data experts, are responsible for the accuracy and integrity of data within their domain. Data stewards, often appointed within departments, act as local champions for data governance, ensuring adherence to data handling policies within their specific ML projects. The policy also defines the decision-making authority for various data governance issues. This clarity ensures timely and efficient decision-making when it comes to data access, usage, or potential conflicts arising from competing departmental needs.

Escalation procedures are also outlined in the policy. These procedures define the pathway for addressing concerns or disagreements related to data governance practices. By establishing a clear escalation chain, the policy ensures any issues are promptly addressed by the appropriate authorities within the organization.

Accountability measures are a crucial aspect of this policy. It defines how individuals and departments will be held accountable for their data governance actions. This fosters a culture of data responsibility, where everyone understands the importance of following established best practices for data handling within ML projects.

Ethical AI Policy

As machine learning (ML) continues to evolve, the ethical implications of artificial intelligence (AI) systems come to the forefront. The AI Ethics Policy serves as an ethical compass, guiding the responsible development and deployment of AI systems within ML initiatives. This policy ensures that AI is not just powerful, but also fair, transparent, and used for good.

Imagine a financial institution developing an AI system to assess loan applications. The AI Ethics Policy would address issues like bias mitigation. Historical lending data might contain biases that could unfairly disadvantage certain demographics. The policy would outline practices to identify and mitigate such biases, ensuring the AI system makes fair and objective loan decisions.

Transparency is another key principle addressed by the policy. It encourages organizations to explain how AI systems arrive at their decisions. This is particularly important for building trust with users and stakeholders. For instance, the policy might mandate that the loan approval AI system provides explanations for its decisions, allowing human loan officers to understand the rationale behind the AI's recommendations.

Accountability is paramount when dealing with AI. The policy defines who is accountable for the actions and outcomes of AI systems. This ensures clear ownership and responsibility for potential issues or unintended consequences. Furthermore, the policy addresses the ethical use of data and algorithms. It might restrict the use of certain types of data or algorithms that could lead to discriminatory or harmful outcomes.

By implementing a comprehensive AI Ethics Policy, organizations can ensure their ML initiatives are developed and deployed in a responsible and ethical manner. This fosters trust with stakeholders, promotes fairness in AI decision-making, and paves the way for a future where AI benefits everyone. The policy goes beyond simply outlining principles; it encourages ongoing evaluation and improvement of AI development practices, ensuring they remain aligned with evolving ethical considerations in the field of AI and machine learning.

When developing and implementing data governance policies and standards, it is crucial to involve relevant stakeholders, such as data owners, data stewards, legal and compliance teams, and subject matter experts. This collaborative approach ensures that the policies and standards are comprehensive, aligned with business objectives, and tailored to the specific needs and challenges of the organization's machine learning initiatives.

Best practices for developing and implementing data governance policies and standards include:

Conducting Policy Gap Analysis

The initial assessment involves a thorough evaluation of organization's existing policies. Imagine a company embarking on its first ML project to predict customer churn. A data governance audit would analyze relevant policies like data security, access control, or data retention.

This audit is crucial for identifying any gaps or areas where existing policies fall short in supporting responsible ML practices. For instance, the data security policy might not address the specific requirements for protecting sensitive customer data used in the churn prediction model. Identifying these gaps allows to prioritize the development of new policies or refine existing ones to ensure comprehensive coverage for all aspects of data used within ML initiatives.

By conducting a comprehensive assessment, one gains a clear picture of organization's data governance maturity and can chart a course for implementing the policies and procedures necessary to ensure responsible and ethical ML development.

Aligning with Industry Standards and Regulations

The ever-evolving world of data governance necessitates alignment with a complex web of regulations and best practices. This goes beyond simply creating internal policies; it ensures ML initiatives operate within the legal and ethical frameworks that govern the industry or domain.

Imagine a social media platform developing an ML model to personalize user feeds. Compliance with data privacy regulations like GDPR (General Data Protection Regulation) in Europe or CCPA (California Consumer Privacy Act) in the United States would be crucial. Here, the data governance framework would ensure policies and standards for data access, usage, and anonymization are not only robust but also adhere to the specific requirements of these regulations to protect user privacy.

Following relevant industry standards is another crucial aspect. For instance, a financial technology company might need to comply with data security standards established by organizations like FFIEC (Federal Financial Institutions Examination Council) in the United States. These industry standards often set the bar for data security practices, and aligning data governance policies with them demonstrates a commitment to best-in-class data handling within sector.

Establishing Policy Review and Update Processes

Business needs shift, regulations are updated, and technological advancements occur rapidly. A core principle of data governance involves regularly reviewing and updating policies and standards to address these evolving dynamics.

Consider a retail company's data governance policies for their product recommendation models. Initially, these policies might focus on protecting customer purchase history data. However, if the company incorporates social media sentiment or location data into their models, the existing policies would need revisions to address the security and privacy considerations associated with these new data types.

Regulatory requirements necessitate regular policy reviews as well. New data privacy laws might emerge, or existing regulations might be revised. The data governance framework should ensure policies are updated to reflect these changes, maintaining compliance. For instance, a new regulation requiring user consent for specific data uses within ML models would necessitate revising the data access and usage policy to incorporate this requirement.

Technological advancements also play a part. As new data storage and processing technologies are adopted, data security and lifecycle management policies might need to be updated to address the specific risks and opportunities associated with these new technologies.

By implementing a process for regular review and updates, an organization's ML DG framework remains agile and adaptable. This ensures that policies and standards continue to effectively support responsible and ethical ML development in the face of constant change within the broader business and technological landscape.

Providing Policy Training and Awareness

Developing training programs and communication strategies ensures all stakeholders understand and adhere to the established data governance policies and standards.

This training goes beyond simply informing employees about the existence of policies. Training programs should delve into the specifics of data governance, educating stakeholders on their roles and responsibilities in handling data responsibly. For instance, data scientists might receive training on data quality best practices to ensure the models they build are based on reliable and trustworthy data. Similarly, marketing teams using ML for targeted advertising campaigns would benefit from training on data privacy regulations and user consent requirements.

Effective communication strategies are crucial for reinforcing these training efforts. Regular communication can take various forms, from internal newsletters highlighting key aspects of data governance to Q&A sessions where employees can address any concerns or questions they might have. Furthermore, communication should be tailored to specific stakeholder groups. Executives might need a high-level overview of the importance of data governance, while data engineers would benefit from in-depth discussions on specific data handling procedures.

By prioritizing training and communication, organizations foster a culture of data responsibility within their workforce. Stakeholders gain the knowledge and understanding necessary to adhere to data governance policies, ultimately leading to more robust and ethical ML initiatives.

Monitoring and Enforcing Policy Compliance

A critical aspect of data governance involves implementing mechanisms for monitoring and enforcing policy compliance. These mechanisms ensure everyone within the organization adheres to the established data governance framework.

One key tool is regular auditing. Audits assess how effectively data governance policies are being implemented across different departments and ML projects. This might involve reviewing data access logs to identify potential misuse or analyzing data storage practices to ensure compliance with security protocols. Audits help identify any gaps in policy adherence and allow for timely corrective measures.

Compliance reporting is another crucial mechanism. Regular reports track key metrics related to data governance practices. These reports might include details on data access requests, security incidents, or employee training completion rates. By monitoring these metrics, organizations can identify areas for improvement and demonstrate their commitment to data governance stakeholders like regulators or auditors.

Accountability measures play a vital role in fostering a culture of data compliance. These measures might involve defining clear consequences for non-compliance with data governance policies. This could range from additional training for employees who make minor mistakes to disciplinary actions for intentional violations. By establishing clear accountability, organizations ensure everyone takes data governance seriously and understands the potential consequences of non-compliance.

Through a combination of monitoring, reporting, and accountability, organizations can ensure their ML data governance framework remains effective. This fosters an environment of responsible data handling and ethical ML development, ultimately leading to trustworthy and successful ML initiatives.

By establishing and adhering to comprehensive data governance policies and standards, organizations can foster a culture of data stewardship, ensure consistent and responsible data management practices, and mitigate risks associated with machine learning initiatives.

Data Governance Processes and Workflows

Effective data governance requires well-defined processes and workflows that govern the various activities involved in managing data throughout its lifecycle. These processes and workflows ensure consistency, efficiency, and accountability in data management practices, enabling organizations to leverage the full potential of their machine learning initiatives while mitigating risks and fostering trust.

The following are some essential data governance processes and workflows that should be established:

Data Acquisition Process

Data acquisition goes beyond simply obtaining data. The process starts with evaluating potential data sources, such as internal databases, external APIs, Internet of Things (IoT) devices, or third-party providers. This evaluation considers factors like data quality, relevance to the intended ML project, and potential legal or ethical issues associated with the data source. For instance, data from third-party providers might require careful scrutiny to ensure it complies with user privacy regulations.

Once a data source is chosen, data ingestion methods come into play. These methods define how the data will be extracted and transferred into the organization's ML environment. Data governance dictates best practices for secure data transfer protocols and data transformation processes to ensure the integrity of the data during ingestion.

Documentation is paramount for responsible data acquisition. The data governance framework should mandate clear data provenance and lineage documentation. Data provenance tracks the origin of the data, while data lineage documents the transformations it undergoes throughout the ML pipeline. This meticulous documentation allows for better data quality control, facilitates troubleshooting potential issues, and fosters transparency within the ML development process.

Data Validation Process

Machine learning data governance prioritizes data quality validation, a crucial process that ensures the data used in ML models is accurate, complete, and reliable. The data governance framework defines steps and methods for comprehensive data quality assessment. This might involve utilizing automated data validation rules to identify inconsistencies or errors within the dataset. For instance, an automated rule could flag customer records with missing zip codes or negative values for purchase amounts.

Data profiling techniques are another valuable tool. These techniques analyze the data to understand its statistical properties, like the range of values for numerical attributes or the distribution of categories within nominal attributes. Data profiling helps identify potential anomalies or biases within the data that could negatively impact the performance of ML models.

The data governance framework also establishes procedures for handling data quality issues. This might involve data cleaning techniques like correcting errors, filling in missing values, or removing outliers. In some cases, depending on the severity of the data quality issues and the specific ML project, the data source itself might need to be revisited or even replaced.

By prioritizing data quality validation, organizations ensure their ML projects are not built on a shaky foundation. The data governance framework empowers them to identify and address data quality issues, ultimately leading to the development of more accurate, reliable, and trustworthy machine learning models.

Data Lineage and Provenance Tracking

Data lineage tracks the path data takes. Imagine a company building an ML model to predict customer churn. Data lineage would document the source of the customer data (e.g., customer relationship management system), any transformations applied (e.g., anonymization), and how it's used within the model (e.g., feature engineering). This detailed record allows data scientists to understand the origin and evolution of the data used in the model.

Data provenance, on the other hand, focuses on the origin of the data itself. Continuing with the customer churn example, data provenance would pinpoint the specific customer records used to train the model. This information is crucial for ensuring the data was obtained ethically and complies with relevant regulations.

By meticulously documenting both data lineage and provenance, organizations achieve several benefits. Firstly, it enables traceability. If issues arise with the ML model's performance, data lineage allows data scientists to trace the data back to its source and identify potential problems. Secondly, it fosters reproducibility. Clear documentation allows others to recreate the model using the same data and transformations, facilitating validation and further development. Finally, robust data lineage and provenance are essential for auditability. Regulatory bodies or stakeholders can easily understand the data used in the model and ensure compliance with data governance practices.

Data Access Management Process

This process governs how data access privileges are granted, reviewed, and revoked within the organization. It includes procedures for role-based access control, data masking or anonymization, and monitoring and auditing data access activities.

The foundation of this process is role-based access control (RBAC). RBAC defines access permissions based on an individual's role within the organization. For instance, data scientists working on building the ML model might be granted access to the complete customer dataset, while marketing teams using the model for targeted campaigns might only require access to anonymized customer segments. RBAC ensures access is granted on a need-to-know basis, minimizing the risk of unauthorized data exposure.

The data governance also addresses the use of data masking or anonymization techniques. Data masking involves replacing sensitive data elements with fictitious values, while anonymization removes personally identifiable information (PII) altogether. These techniques allow data scientists to work with real-world data while protecting user privacy, particularly when dealing with sensitive data like customer information or financial records.

The data governance framework mandates procedures for monitoring and auditing data access activities. This might involve tracking who accessed what data, when they accessed it, and for what purpose. Regular audits of access logs help identify potential misuse of data and ensure adherence to established access control policies. Furthermore, ongoing monitoring allows for timely detection of suspicious activity, enabling a swift response to potential data security breaches.

Data Issue Resolution Workflow

This workflow outlines the steps and escalation procedures for identifying, reporting, and resolving data-related issues, such as data quality problems, security incidents, or policy violations. It defines the roles and responsibilities of stakeholders involved in the resolution process.

The workflow starts with issue identification. Data scientists working with the data might identify quality problems like missing entries or inconsistencies. Security personnel might detect a potential data breach attempt. Alternatively, an employee might report a suspected violation of data governance policies. The ML DG framework outlines clear channels for reporting these issues, empowering everyone within the organization to contribute to data security and quality.

Escalation procedures are crucial when dealing with complex issues. The workflow defines different levels of escalation depending on the severity of the problem. For instance, a minor data quality issue might be addressed by a data engineer within the ML team. However, a major security incident would require immediate escalation to the security team and potentially even regulatory authorities. Clear escalation procedures ensure the right people are notified promptly, allowing for a swift and effective response.

The workflow also defines the roles and responsibilities of stakeholders involved in the resolution process. Data scientists might be responsible for investigating data quality issues and implementing corrective actions. The security team would take the lead in handling security incidents, following established protocols for data breach response. Senior management plays a vital role in providing resources and ensuring adherence to data governance policies throughout the resolution process.

Data Quality Assessment Workflow

The data governance places significant emphasis on a Data Quality Assessment Workflow. This ongoing process ensures the data used in ML models remains fit for purpose, fostering accurate and reliable results.

The workflow outlines a series of activities for regular data quality assessment and monitoring. Data profiling techniques play a crucial role. These techniques analyze the data to understand its statistical properties, identifying potential issues like missing values, inconsistencies, or unexpected data distributions. For instance, data profiling might reveal a high number of customer records with missing income data, potentially impacting the performance of an ML model predicting customer spending habits.

Quality rule validations are another vital aspect of the workflow. The data governance framework establishes specific data quality rules that the data must adhere to. These rules might define acceptable ranges for numerical attributes or specify the format for specific data fields. Automated validation tools are employed to continuously check the data against these established rules, promptly identifying any deviations that could negatively affect the performance of ML models.

The workflow doesn't stop at simply identifying issues. It also mandates reporting mechanisms to ensure data quality remains within acceptable thresholds. Regular reports track key data quality metrics, allowing data scientists and stakeholders to monitor trends and identify areas for improvement. Proactive notifications can be set up to alert relevant personnel when data quality metrics fall below predefined thresholds, enabling timely intervention and corrective actions.

Data Remediation Workflow

This workflow outlines the procedures for addressing and remediating identified data quality issues, such as data cleansing, transformation, or enrichment. It may also include guidelines for triggering model retraining or updates in response to significant data quality issues.

The workflow outlines specific procedures for data remediation. Techniques like data cleansing take center stage. This might involve correcting errors like typos or inconsistencies, filling in missing values using appropriate methods, or removing outliers that could skew the data. Imagine an ML model predicting customer churn; data cleansing might involve correcting customer addresses with invalid zip codes or removing duplicate entries within a dataset.

Data transformation plays a crucial role as well. The workflow might dictate specific transformations needed to ensure the data aligns with the requirements of the ML model. This could involve data normalization (scaling numerical attributes to a common range) or encoding categorical variables for better machine learning processing. For instance, an ML model predicting loan defaults might require income data to be converted into a consistent format (e.g., annual salary) before feeding it into the model.

Data enrichment can also be employed within the workflow. This involves integrating additional data sources to enhance the quality and completeness of existing data. For instance, an ML model predicting customer churn might benefit from enriching customer data with social media sentiment analysis to gain a more holistic view of customer satisfaction.

The workflow also establishes guidelines for triggering model retraining or updates in response to significant data quality issues. If data remediation efforts significantly alter the underlying data used to train an ML model, the workflow might mandate retraining the model to ensure its continued accuracy and effectiveness. This ensures that the models adapt to changes in the data and continue to deliver reliable results.

Policy Enforcement Workflow

This workflow defines the processes and mechanisms for enforcing adherence to data governance policies and standards. It may include procedures for conducting audits, monitoring policy compliance, and implementing corrective actions or disciplinary measures in cases of non-compliance.

The workflow outlines various mechanisms for enforcing policy compliance. Regular audits play a critical role. These audits assess how effectively data governance policies are being implemented across different departments and ML projects. This might involve reviewing data access logs to identify potential misuse or analyzing data storage practices to ensure compliance with security protocols. Audits act as a red flag, highlighting any gaps in policy adherence and allowing for timely corrective actions.

Continuous monitoring is another key aspect of the workflow. This might involve utilizing automated tools to track key data governance metrics in real time. Metrics like data access requests, security incidents, or employee training completion rates can be monitored to identify potential issues and ensure everyone is adhering to the established policies.

The workflow defines clear procedures for implementing corrective actions or even disciplinary measures depending on the severity of the violation. Corrective actions might involve additional training for employees who make minor mistakes, while intentional violations could necessitate disciplinary measures. Establishing clear consequences for non-compliance ensures everyone takes data governance seriously and understands the potential ramifications of disregarding the policies.

Change Management Process

The Change Management Process within ML data governance ensures a smooth and controlled approach to modifying policies, processes, or technologies. This minimizes disruption to ongoing ML projects and mitigates potential risks associated with change.

The process starts with proposing a change. Stakeholders across the organization, from data scientists to legal teams, can propose updates to existing data governance policies or suggest the adoption of new technologies for data handling. The proposal clearly outlines the rationale behind the change, its potential benefits, and any anticipated impact on existing ML initiatives.

Evaluation is the next crucial step. The data governance framework establishes a review committee responsible for assessing the proposed change. This committee analyzes the proposal, considering factors like its alignment with overall data governance goals, potential impact on data security and privacy, and compatibility with existing infrastructure.

Following a thorough evaluation, the committee decides to approve, reject, or recommend modifications to the proposed change. Approved changes move forward to the implementation phase. This involves a well-defined plan outlining the steps for rolling out the change. Training programs might be developed to educate stakeholders on new policies or technologies, and communication plans ensure everyone is informed about the upcoming changes.

The Change Management Process doesn't end with implementation. Monitoring the impact of the change is crucial. Metrics are tracked to assess the effectiveness of the implemented change and identify any unforeseen challenges. This allows for course correction if necessary, ensuring the change ultimately benefits the organization's ML data governance practices.

When defining data governance processes and workflows, it is essential to involve relevant stakeholders, such as data owners, data stewards, data scientists, and subject matter experts. This collaborative approach ensures that the processes and workflows are tailored to the specific needs and challenges of the organization's machine learning initiatives while also promoting cross-functional alignment and accountability.

By establishing well-defined data governance processes and workflows, organizations can ensure consistent and efficient data management practices, promote accountability, and enable the successful implementation and maintenance of their data governance framework for machine learning initiatives.

Data Governance Tools and Technologies

Effective data governance for machine learning initiatives requires the support of specialized tools and technologies that enable organizations to manage, monitor, and govern their data assets effectively. These tools and technologies streamline data governance activities, automate processes, and provide visibility and control over data throughout its lifecycle.

The following are some common data governance tools and technologies that organizations should consider implementing:

Data Catalogs and Metadata Management Systems

The data governance framework leverages Data Catalogs and Metadata Management Systems to shed light on this data, fostering better understanding, discovery, and, ultimately, more effective governance.

These systems act as centralized hubs for storing, managing, and accessing metadata—the critical information that describes the data itself. Popular data catalog solutions include Alation and Informatica, while open-source options like Apache Atlas are also gaining traction. Metadata management tools like Azure Purview complement data catalogs by providing a more granular view of data quality, lineage, and ownership.

The benefits of these systems are manifold. Data catalogs offer a comprehensive view of an organization's data sources, acting as a searchable directory that simplifies data discovery for ML projects. This empowers data scientists to find relevant datasets efficiently, saving time and minimizing the risk of using outdated or inaccurate data.

Furthermore, data catalogs store information about the lineage of the data, essentially tracking its journey from origin to use within an ML model. Tools like Apache Atlas excel at capturing this lineage, providing a clear audit trail that facilitates troubleshooting and regulatory compliance. Additionally, data catalogs can house business glossaries, which define the meaning and context of specific data elements. This shared understanding ensures everyone within the organization interprets the data consistently, leading to more reliable and trustworthy ML models.

Data Catalogs and Metadata Management Systems are the cornerstones of effective data discovery and understanding within ML data governance. By providing a centralized platform for managing data's metadata, these tools empower organizations to leverage their data assets effectively, ultimately leading to more successful and trustworthy ML initiatives.

Data Lineage and Provenance Tracking Tools

These tools facilitate the tracking and visualization of data lineage, allowing organizations to understand the flow and transformations of data from its source to its final state. They provide visibility into data dependencies, enabling impact analysis and supporting compliance and auditing requirements.

Popular solutions like Amundsen provide clear visualizations of data lineage. Imagine an ML model predicting customer churn. Lineage tracking tools would map the data's journey, revealing its origin (e.g., customer relationship management system) and any transformations it undergoes (e.g., anonymization, feature engineering). This transparency empowers data scientists to understand the data's background and potential biases, leading to more robust model development.

Beyond visualization, these tools offer valuable functionalities for impact analysis. If changes are made to the source data or transformations, lineage tracking tools can pinpoint which ML models might be affected. This allows data scientists to assess the potential impact and take proactive measures, such as retraining models or monitoring their performance closely.

Compliance and auditing are also bolstered by lineage and provenance tracking. Regulatory bodies often require organizations to demonstrate the origin and handling of data used in ML models. Tools like Azure Purview, alongside lineage tracking solutions, provide a comprehensive audit trail. This allows organizations to easily demonstrate adherence to data privacy regulations and internal data governance policies.

Data Quality Tools

Data quality tools automate the process of profiling, validating, and monitoring the quality of data assets. They enable organizations to define data quality rules, identify and remediate data quality issues, and ensure that data meets the necessary standards for machine learning initiatives.

Solutions like Open Refine or Alteryx Designer Cloud empower organizations to define data quality rules. These rules act as checkpoints, specifying the expected format, range of values, and other characteristics for each data element. Imagine an ML model predicting loan defaults; a data quality rule might dictate that income data must be a positive numerical value. These tools then automate the process of validating the data against these established rules, promptly identifying inconsistencies, missing values, or any other anomalies that could negatively impact the performance of the ML model.

Data profiling is another area where these tools excel. Solutions like Talend Open Studio or Apache Spark Profiler can analyze vast datasets, automatically generating reports on statistical properties like data distribution, presence of null values, and identification of outliers. This comprehensive data profile empowers data scientists to understand the data's characteristics and potential biases, allowing for informed decisions during model development.

Data remediation, the process of fixing data quality issues, also benefits from automation. Tools like OpenDP or Actian Vector can automate tasks like data cleaning (e.g., correcting typos, filling missing values) and data transformation (e.g., converting formats, normalizing values). This not only saves data scientists valuable time but also ensures consistency and reduces the risk of human error during data preparation.

By automating the process of profiling, validating, and monitoring data quality, data quality tools are a cornerstone of effective data governance. These tools empower organizations to proactively identify and address data quality issues, ultimately leading to the development of more robust, reliable, and, ultimately, trustworthy machine learning models.

Data Masking and Anonymization Tools

These tools help organizations protect sensitive data by masking or anonymizing personally identifiable information (PII) or other confidential data elements. They ensure compliance with data privacy regulations while enabling secure data access for authorized users and ML initiatives.

Solutions like IBM InfoSphere Guardium or Cloudera data masking offer data masking functionalities. Masking replaces sensitive data elements with fictitious values, such as replacing customer names with random identifiers. This allows data scientists to work with realistic datasets while protecting personally identifiable information (PII) like social security numbers or email addresses. This is particularly crucial when dealing with data subject to regulations like GDPR or CCPA.

For scenarios demanding complete anonymization, tools like Anonymizer or Apache Griffin come into play. Anonymization goes beyond masking; it removes PII altogether. Imagine anonymizing customer purchase data for an ML model predicting buying trends. Anonymization tools would remove names and addresses, while preserving anonymized purchase details like product categories and transaction amounts. This allows data scientists to leverage valuable insights from the data without compromising individual privacy.

The benefits of these tools extend beyond just regulatory compliance. Secure data access for authorized users is another key advantage. When sensitive data is masked or anonymized, organizations can grant broader access for data analysis or ML initiatives without exposing users to confidential information. This fosters collaboration and innovation within the organization while maintaining robust data security practices.

Data Access Management and Security Tools

These tools facilitate the implementation and enforcement of data access controls, role-based permissions, and data governance policies. They provide audit trails, monitoring capabilities, and automated policy enforcement to ensure data security and compliance.

Tools like Apache Ranger or Azure Active Directory offer robust data access controls. These tools enable organizations to define role-based permissions (RBAC). Imagine a scenario where data scientists working on building the ML model require full access to customer data, while marketing teams might only need anonymized customer segments for targeted campaigns. RBAC allows for granular control, granting access based on a user's role within the organization and minimizing the risk of unauthorized data exposure.

Audit trails are another crucial aspect of data access management. Tools like Splunk or ELK Stack provide detailed logs of who accessed what data, when they accessed it, and for what purpose. This transparency allows for easy identification of potential misuse and ensures adherence to established data access control policies. Additionally, these tools offer real-time monitoring capabilities. Security Information and Event Management (SIEM) solutions like Palo Alto Cortex XDR or MacAfee Endpoint Security can continuously monitor data access activity, detecting suspicious behavior and enabling a swift response to potential security breaches.

Data Access Management and Security Tools go beyond simply logging access. They can also enforce data governance policies automatically. For instance, these tools can integrate with Data Loss Prevention (DLP) solutions like Forcepoint DLP or McAfee Data Loss Prevention to prevent unauthorized data transfers or downloads. This automated enforcement ensures consistent adherence to data governance policies and minimizes the risk of human error.

Data Visualization and Reporting Tools

These tools provide visual representations and dashboards for data governance metrics, KPIs, and other relevant information. They enable stakeholders to monitor the effectiveness of data governance practices, identify areas for improvement, and make informed decisions.

Tableau or Power BI allows for the creation of interactive dashboards that display key data governance metrics in real time. Imagine a dashboard tracking data quality issues identified within the organization. Data scientists and governance teams can monitor trends, identify data sources with recurring problems, and take proactive steps to improve data quality. These visualizations not only provide a clear picture of the current state of data governance but also make it easier to track progress over time.

Beyond dashboards, reporting tools enable the creation of comprehensive reports for stakeholders across the organization. Tools like Qlik Sense or Zoho Analytics can generate reports on data access activity, compliance audits, or the effectiveness of data lineage tracking. These reports provide valuable insights into the overall health of data governance practices and potential areas for improvement. For instance, a report might reveal a department struggling with consistent data lineage tracking, prompting targeted training or policy adjustments.

The benefits of these tools extend beyond internal communication. Regulatory bodies often require organizations to demonstrate adherence to data privacy regulations. Data visualization and reporting tools can generate reports that clearly showcase an organization's data governance practices, fostering trust and transparency with regulators and the public.

Workflow Automation and Orchestration Tools

These tools automate and orchestrate data governance processes and workflows, streamlining activities such as data validation, issue resolution, and policy enforcement. They improve efficiency and consistency and reduce the risk of manual errors.

Prefect or Kubeflow Pipelines excel at automating data governance workflows. Imagine a scenario where data validation is a crucial step before data can be used for model training. Workflow automation tools can orchestrate this process, automatically triggering data validation checks upon data ingestion. In case of inconsistencies, the workflow can be designed to notify relevant personnel and initiate remediation processes. This automation eliminates the need for manual intervention, saving valuable time for data scientists and governance teams.

Beyond automation, these tools excel at orchestration. Complex data governance processes often involve multiple steps and stakeholders. Workflow orchestration tools ensure everything runs smoothly, coordinating tasks and data flow between different systems. For instance, an issue resolution workflow might involve data engineers fixing data quality problems and security teams addressing potential breaches. Orchestration tools ensure seamless handoffs between these teams, facilitating a swift and efficient resolution process.

The benefits of automation and orchestration extend beyond just efficiency. Consistency is another key advantage. By automating tasks, these tools ensure data governance processes are followed uniformly across the organization, minimizing the risk of human error and potential inconsistencies. This fosters a culture of responsible data use and builds trust in the data used for ML models.

Workflow Automation and Orchestration Tools are the backbone of efficient and scalable machine learning data governance. By automating repetitive tasks and orchestrating complex workflows, these tools empower organizations to streamline data governance processes, improve data quality, and, ultimately, unlock the full potential of their ML initiatives.

Challenges of a Unified Machine Learning Data Governance Engine

The allure of a single, unified platform to manage all aspects of machine learning (ML) data governance is undeniable. For organizations that rely heavily on data-driven decision-making, such a platform promises streamlined operations, simplified workflows, and a consistent approach to data management across the board. However, the reality of implementing this vision is fraught with technical complexities and practical limitations that make it an incredibly challenging—and perhaps impractical—pursuit. To understand these challenges more deeply, it is instructive to examine real-world examples where companies have attempted to implement such unified systems, only to face significant hurdles that highlight the inherent difficulties of this approach.

One major challenge in developing a unified ML data governance platform is the vast and varied data ecosystem that companies must navigate. In the real world, ML projects often pull data from a multitude of sources—ranging from traditional databases and data lakes to APIs and external third-party datasets. A single platform would need to integrate seamlessly with these disparate elements, a task that is easier said than done. For instance, a large telecommunications company attempted to create a unified platform to manage customer data across its various divisions, including mobile, broadband, and television services. The goal was to streamline data governance across all departments, ensuring that data quality, security, and compliance were maintained consistently. However, the platform struggled to integrate the diverse data sources effectively, particularly when dealing with legacy systems that used outdated formats or proprietary data structures. This led to data inconsistencies and integration challenges, ultimately forcing the company to revert to a more flexible, multi-tool approach.

Another significant hurdle is the complexity of data preparation workflows in ML projects. These workflows often involve intricate processes such as data ingestion, transformation, cleansing, and feature engineering, each of which requires a high degree of customization depending on the specific use case. Consider the example of a financial services firm that aimed to deploy a unified platform to govern data across its various ML models used for risk assessment, fraud detection, and customer analytics. The platform needed to support a wide range of data transformations and feature engineering tasks, but it quickly became apparent that a one-size-fits-all solution could not adequately handle the nuances of each ML project. For instance, the requirements for cleansing and transforming transaction data differed significantly from those needed for customer sentiment analysis. The platform's workflow engine struggled to accommodate these differences, leading to delays in model development and suboptimal performance. This experience highlighted the limitations of a unified platform in managing the diverse and complex workflows inherent in ML projects.

Scalability is another critical concern when dealing with ML, particularly as the volume of data continues to grow exponentially. For example, an e-commerce giant sought to implement a unified data governance platform to manage its customer behavior data across multiple regions and product lines. The platform was expected to scale seamlessly as the company expanded its operations and collected increasingly larger datasets. However, the sheer volume of data—spanning billions of transactions and customer interactions—overwhelmed the platform's processing capabilities. As a result, the company faced significant delays in data processing, which in turn impacted the accuracy and timeliness of its ML models. This challenge underscores the importance of scalability in any data governance solution, particularly in industries where data growth is rapid and relentless.

Data security and privacy are paramount in any ML project, and a unified platform must incorporate robust features to protect sensitive information from unauthorized access or breaches. A global healthcare provider, for instance, attempted to implement a unified platform to manage patient data across its various hospitals and clinics. The platform was designed to enforce strict access controls, encryption, and audit trails to comply with healthcare regulations such as HIPAA in the United States. However, as data privacy regulations evolved and became more stringent, the platform struggled to keep up with the new compliance requirements. In particular, the platform's rigid structure made it difficult to integrate new compliance tools and adapt to changing regulations, leading to gaps in compliance and increased risk of data breaches. This example illustrates the difficulty of maintaining compliance in a dynamic regulatory environment, particularly when relying on a single, inflexible platform.

Even if the technical hurdles of building a unified platform could be overcome, practical challenges remain. For instance, many organizations already have established data management and governance systems in place, which may not integrate seamlessly with a new unified platform. A global retail chain, for example, had been using a combination of data cataloging and quality management tools across its various business units. When the company attempted to transition to a unified platform, it encountered significant integration issues, particularly with its existing data catalog solution. The platform could not easily import and manage the data from the existing catalog, leading to data silos and inefficiencies. The company ultimately decided to continue using its established tools rather than fully adopting the unified platform, illustrating the difficulties of integrating a new system into an existing data governance infrastructure.

A single platform may struggle to cater to the diverse skillsets of data scientists and governance professionals within an organization. A multinational corporation with a large and varied ML team attempted to implement a unified platform that would be accessible to both technical and non-technical users. However, the platform's user interface was either too simplistic for data scientists or too complex for business users, leading to frustration on both sides. Data scientists found the platform limiting in terms of customization and flexibility, while business users struggled with its complexity and technical jargon. This experience highlighted the challenge of designing a single platform that can meet the needs of a diverse user base, further questioning the practicality of a unified approach.

Given these challenges, many organizations find that a multi-tool approach offers a more robust and adaptable solution for ML data governance. By leveraging specialized tools that excel in specific functionalities, companies can tailor their data governance strategy to their unique needs. For example, a media company dealing with large volumes of video and audio data might use a dedicated data cataloging tool like Alation for metadata management, while simultaneously employing workflow automation tools like Apache Airflow to orchestrate its complex data processing pipelines. This approach allows the company to harness the best features of each tool rather than compromising on a single, all-encompassing platform.

A multi-tool approach mitigates the risk of vendor lock-in and single points of failure. Organizations can combine open-source solutions with commercial tools, ensuring flexibility and reducing dependence on any one vendor. For example, a tech startup might use an open-source tool like Prefect for workflow management while integrating it with a commercial data governance solution for enhanced data security. This strategy not only reduces the risk of disruption if one tool encounters issues but also provides the organization with the flexibility to adapt to new technologies and regulatory requirements as they arise.

While the idea of a unified ML data governance platform is appealing, the practical challenges associated with its implementation—ranging from technical integration and scalability issues to user adoption and compliance challenges—suggest that a more flexible, multi-tool approach may be a better solution. By carefully selecting and integrating specialized tools, organizations can build a robust and adaptable data governance framework that meets their evolving needs and maximizes the value of their ML initiatives.

Collaboration and Integration

The pursuit of robust ML data governance necessitates a collaborative approach, with various tools and platforms working together seamlessly. Integration with cloud providers like Microsoft Azure, Amazon Web Services (AWS), and Google Cloud Platform (GCP) allows organizations to leverage the data governance functionalities offered by these platforms alongside best-of-breed open-source solutions like Apache Airflow or Apache Atlas. This creates a comprehensive data governance ecosystem that caters to the specific needs of the organization.

Open Standards and Interoperability

Open standards and interoperability are crucial for fostering collaboration between different tools and platforms. Standardized data formats and APIs ensure a smooth exchange of information between different components of the data governance ecosystem. For instance, the adoption of open standards like Apache Parquet for data storage or Open Data Protocol (ODP) for data discovery facilitates seamless integration between various data management tools.

The quest for a single platform for data governance stems from a desire for simplicity and streamlined orchestration. However, considering the technical complexities and the dynamic nature of the ML landscape, a multi-tool approach offers greater advantages. By embracing a well-integrated ecosystem of specialized tools, organizations can build a robust and flexible data governance foundation that empowers them to navigate the ever-evolving world of ML with confidence. This approach ensures responsible data use, fosters trust and transparency, and ultimately unlocks the full potential of ML initiatives for driving organizational success.

Evaluating and Selecting Data Governance Tools

Evaluating and selecting data governance tools for machine learning (ML) initiatives requires careful consideration of several key criteria. Here are some critical factors to keep in mind:

Scalability and Performance

ML projects often generate massive amounts of data. The chosen data governance tools must be able to handle these ever-growing volumes efficiently. Look for solutions that offer scalable storage and processing capabilities to ensure smooth performance as ML initiatives evolve. Popular cloud platforms like Microsoft Azure, Amazon Web Services (AWS), and Google Cloud Platform (GCP) provide scalable data governance functionalities. Additionally, open-source tools like Apache Spark can be integrated to leverage distributed processing for large-scale data analytics within data governance framework.

Integration Capabilities

A siloed approach to data governance hinders efficiency. The chosen tools and technologies should seamlessly integrate with existing data ecosystem, including databases, data lakes, and other analytics tools. Look for solutions that offer open APIs and support for common data formats to facilitate smooth data exchange. For instance, integrating a data cataloging tool like Alation with organization's cloud storage platform ensures efficient data discovery and lineage tracking across all data assets.

Security and Compliance

Data security and adherence to data privacy regulations are paramount. Organization's data governance tools and technologies should offer robust security features, including access controls, data encryption, and detailed audit trails. Furthermore, they should be adaptable to evolving regulations like GDPR and CCPA to ensure continuous compliance. Consider solutions that integrate with established security frameworks and offer functionalities for data masking or anonymization to protect sensitive information used in ML models.

Workflow Automation and Orchestration

ML data governance involves complex workflows encompassing data quality checks, lineage tracking, and access control enforcement. Implementing automation capabilities within data governance framework can streamline these processes and minimize manual intervention. Look for tools like Apache Airflow or Prefect that excel at workflow orchestration, allowing to automate data pipelines and ensure consistent governance practices throughout ML lifecycle.

Usability and User Management

Data governance needs to cater to a diverse group of users, from data scientists with advanced technical knowledge to governance professionals with broader data management responsibilities. The chosen tools should offer a user-friendly interface with role-based access control to ensure ease of use for all stakeholders. Consider solutions with intuitive dashboards and visualization tools that provide clear insights into data quality and governance metrics for informed decision-making.

Implementing the appropriate data governance tools and technologies is crucial for enabling effective data management, ensuring compliance, and unlocking the full potential of machine learning initiatives. However, it's important to note that tools alone are not sufficient; they must be complemented by well-defined data governance policies, processes, and organizational structures to achieve a comprehensive and effective data governance framework.

Data Governance Metrics and Key Performance Indicators (KPIs)

To effectively measure the success and impact of a data governance program, organizations must establish relevant metrics and key performance indicators (KPIs). These metrics and KPIs provide quantifiable measures of the effectiveness of data governance practices, enabling organizations to monitor progress, identify areas for improvement, and demonstrate the value of their data governance initiatives.

Data Quality Metrics

Data quality is the cornerstone of successful machine learning (ML) initiatives. Inaccurate, incomplete, or inconsistent data can lead to biased models with poor performance. Data quality metrics play a crucial role in ML data governance, providing valuable insights into the health and reliability of the data used in the models.

Here's how these key metrics can be used for data governance:

Completeness

This metric measures the percentage of data records with no missing or null values for critical fields. In an ML dataset containing customer information, the "customer name" field is likely considered critical. A low completeness score for this field could indicate missing data that might hinder the model's ability to accurately identify customer segments or predict purchase behavior. Data governance practices can be implemented to address completeness issues, such as data cleansing procedures to fill missing values or defining data validation rules to enforce data entry requirements.

Accuracy

Accuracy reflects the percentage of data records that accurately represent real-world entities or events. Imagine an ML model predicting equipment failure based on sensor data. Inaccurate sensor readings would lead to a low accuracy score for the data. Data governance processes can incorporate data validation techniques to compare data points with historical trends or external sources, ensuring the accuracy of the information used for model training.

Consistency

Consistency measures the degree to which data records conform to defined data standards and formats. For instance, an ML model might require customer addresses to be formatted in a specific way (e.g., "Street Address, City, State Zip Code"). Inconsistent formatting can lead to errors during data processing and ultimately impact model performance. Data governance practices can enforce data formatting standards through data validation rules and data cleansing procedures within processing pipelines. This ensures consistency across the dataset and avoids potential errors.

Timeliness

Timeliness measures the percentage of data records that meet defined freshness or currency requirements. In a financial trading model, real-time data on stock prices is crucial. Stale data would lead to a low timeliness score. Data governance practices can be implemented to ensure data pipelines are configured for timely data ingestion and updates, ensuring the model is trained and operates on the most current information available.

By monitoring and analyzing these data quality metrics, organizations can identify areas where their ML data might be susceptible to issues and take corrective actions within their data governance framework. This proactive approach ensures data used in ML initiatives is reliable and trustworthy, leading to more robust and accurate models that drive successful outcomes.

Data Accessibility and Usage Metrics

Effective ML data governance goes beyond ensuring data quality; it also focuses on accessibility and usage. Data accessibility and usage metrics provide valuable insights into how effectively data is being discovered, utilized, and shared within an organization.

Here's how these key metrics can be used for ML data governance:

Data Discovery and Reuse

This metric tracks the number of data assets successfully discovered and reused across different teams and projects. A high data discovery and reuse rate indicates effective data governance practices that enable users to locate relevant datasets for their ML projects. Data catalogs like Alation can be leveraged to improve data discovery by creating a central repository for all data assets with detailed descriptions and lineage information. Additionally, promoting data reuse across projects helps reduce redundancy and ensures efficient use of valuable data resources.

Data Access Requests

This metric measures the number of data access requests received and processed by the data governance team. A high volume of requests might indicate a need for streamlined data access procedures. Data governance can address this by implementing self-service access controls through user roles and permissions. This empowers authorized users to access required data without needing approval for every request, fostering efficiency and reducing the workload on the data governance team.

Data Consumption

This metric captures the volume or frequency of data consumption by various teams or applications. Low data consumption might indicate data assets are not being utilized to their full potential. Data governance can address this through user education initiatives, highlighting the value propositions of specific data assets and showcasing successful use cases within the organization. Furthermore, data governance can leverage data usage metrics alongside data quality metrics to identify potentially problematic datasets. For instance, a high-quality dataset with low consumption might indicate challenges related to data accessibility or user awareness.

By monitoring and analyzing data accessibility and usage metrics, organizations can gain valuable insights into how effectively data is being leveraged within their ML initiatives. This data-driven approach allows them to optimize data governance practices, fostering a culture of data sharing and promoting responsible data use that maximizes the value extracted from machine learning initiatives.

Data Governance Process Metrics

Effective ML data governance involves not just data quality and accessibility, but also efficient processes to manage and maintain data integrity. Data governance process metrics provide valuable insights into the effectiveness of data governance practices, allowing to identify areas for improvement and streamline operations.

Here's a breakdown of how these key metrics can be used for ML data governance:

Issue Resolution Time

This metric measures the average time taken to resolve data-related issues or incidents. A low resolution time indicates a well-oiled data governance process that can efficiently address data quality problems and minimize disruptions to ML initiatives. Data governance frameworks can be enhanced by implementing clear escalation procedures and establishing a dedicated team to handle data-related issues promptly. This ensures data quality issues don't hinder the progress of ML projects.

Policy Compliance

This metric reflects the percentage of data assets or processes compliant with defined data governance policies. A high compliance rate indicates that data is managed and used responsibly according to established guidelines. Data governance can be strengthened by clearly outlining data access controls, data security protocols, and data retention policies. Additionally, implementing automated data validation rules can help enforce these policies throughout the data lifecycle, reducing the risk of human error and ensuring consistent compliance.

Audit Findings

This metric tracks the number and severity of audit findings related to data governance. A low number of findings, particularly those classified as high-severity, suggest a robust data governance framework that effectively mitigates data security risks and ensures adherence to regulations. Regular data governance audits are crucial for identifying potential gaps and vulnerabilities. These audits can be leveraged to refine data governance practices, close security loopholes, and continuously improve the overall data governance posture of the organization.

By monitoring and analyzing data governance process metrics, organizations can gain valuable insights into the efficiency and effectiveness of their data governance practices. This data-driven approach empowers them to identify bottlenecks, optimize workflows, and ensure their data governance framework can effectively support the evolving needs of their ML initiatives.

Data Security and Privacy Metrics

Data security and privacy are paramount concerns in any organization leveraging ML. Data security and privacy metrics provide crucial insights into the effectiveness of data governance framework in mitigating risks and ensuring responsible data utilization.

Here's how these key metrics can be used for ML data governance:

Data Breaches

This metric tracks the number of data breaches or unauthorized access incidents involving ML data assets. A low number of breaches indicate robust data security measures in place. Data governance practices can be strengthened by implementing robust access controls through user authentication and authorization procedures. Additionally, encryption practices can be used to safeguard sensitive data at rest and in transit. Monitoring data access logs and user activity can also help identify potential anomalies and suspicious access attempts.

Sensitive Data Exposure

This metric reflects the percentage of sensitive data assets with appropriate access controls and protection measures in place. Sensitive data, such as customer information or financial data, requires heightened security protocols. Data governance practices can address this by implementing data classification policies to identify sensitive datasets. Once classified, these data assets can be subject to stricter access controls, encryption, and anonymization techniques where appropriate. This minimizes the risk of unauthorized access or accidental disclosure of sensitive information used in ML initiatives.

Privacy Compliance

This metric measures the percentage of data assets compliant with relevant data privacy regulations such as GDPR (General Data Protection Regulation) or CCPA (California Consumer Privacy Act). Data governance plays a crucial role in ensuring organizations adhere to these regulations. Practices can include implementing data subject access request (DSAR) procedures, allowing individuals to request access to their data and have it corrected or deleted. Additionally, data governance can foster a culture of privacy awareness within the organization, ensuring responsible data collection, use, and storage practices throughout the ML lifecycle.

By monitoring and analyzing data security and privacy metrics, organizations can gauge the effectiveness of their data governance approach in safeguarding sensitive information. This proactive data-driven approach allows them to identify and address potential security vulnerabilities, minimize data breaches, and ensure compliance with evolving data privacy regulations. This promotes trust with data subjects and strengthens the overall foundation for responsible and ethical use of data within ML initiatives.

Data Lineage and Traceability Metrics

Data lineage and traceability are crucial aspects of ML data governance. These metrics provide insights into the origin, transformation, and flow of data used in ML models. Understanding this data journey is essential for ensuring transparency, accountability, and responsible use of data within ML initiatives.

Here's how these key metrics can be used for ML data governance:

Data Lineage Coverage

This metric measures the percentage of data assets with documented lineage and provenance. A high coverage rate indicates a comprehensive understanding of where data originates, how it is transformed through various processing stages, and ultimately how it is used in ML models. Data lineage tools, like Amundsen, can be used to capture and visualize data provenance across the entire ML data lifecycle. This documented lineage allows to identify potential biases or data quality issues that might have been introduced during data processing stages.

Impact Analysis

This metric assesses organization's ability to trace and understand the impact of data changes on downstream systems or processes, particularly the ML models. Effective data governance practices ensure the ability to track how changes in source data (e.g., corrections, updates) propagate through the data pipeline and potentially impact ML models. This allows organization to retrain or recalibrate models efficiently when necessary to maintain their accuracy and performance in the face of evolving data. Additionally, impact analysis helps identify potential downstream consequences of data changes, allowing to proactively mitigate risks associated with erroneous data updates.

By monitoring and analyzing data lineage and traceability metrics, organizations can gain a deeper understanding of their data flows within ML initiatives. This data-driven approach empowers them to identify and address potential data quality issues at their source, ensure transparency and auditability of data usage within ML models, proactively manage the impact of data changes on model performance, and foster trust and accountability within the organization by demonstrating responsible data governance practices.

A strong focus on data lineage and traceability fosters a culture of data awareness within the organization, leading to more reliable ML models and ultimately driving successful outcomes from machine learning initiatives.

Business Value and ROI Metrics

Effective ML data governance isn't just about technical processes; it's about driving tangible business value. Business value and ROI metrics provide insights into the financial impact of data governance practices on ML initiatives.

Here's how these key metrics can be used for machine learning data governance:

Cost Savings

This metric tracks the cost savings achieved through improved data governance practices. Data governance can help reduce costs associated with data redundancy by eliminating duplicate copies of datasets. Additionally, efficient data storage and processing practices can be implemented to optimize resource utilization and minimize infrastructure expenses. Data governance can also streamline data access management, reducing the need for manual access control procedures and associated administrative overhead.

Revenue Generation

This metric measures the revenue generated or potential opportunities unlocked by leveraging high-quality, governed data assets. Effective data governance ensures data quality and accessibility, empowering data scientists to develop more accurate and reliable ML models. These models can then be used to identify new customer segments, predict market trends, or personalize user experiences, ultimately driving revenue growth for the organization. Additionally, improved data governance can foster collaboration between data science and business teams, leading to the development of innovative data-driven products and services that generate new revenue streams.

Risk Mitigation

This metric reflects the potential financial losses avoided through effective data governance and compliance practices. Data breaches, non-compliance with data privacy regulations, or reputational damage caused by biased or inaccurate ML models can all lead to significant financial losses. Data governance helps mitigate these risks by prioritizing data security, ensuring compliance with regulations, and promoting responsible data utilization within ML initiatives. This proactive approach minimizes the risk of financial penalties or litigation costs associated with data misuse.

By showcasing the tangible benefits of data governance, organizations can foster a culture of data responsibility and secure ongoing support for robust data governance practices throughout the ML lifecycle. This ultimately leads to more successful and impactful ML initiatives that drive business growth and achieve strategic objectives.

Stakeholder Engagement and Adoption Metrics

Effective ML data governance relies heavily on active participation and adoption from stakeholders across the organization. Stakeholder engagement and adoption metrics provide valuable insights into the effectiveness of data governance program in fostering a culture of data awareness and responsibility.

Here's how these key metrics can be used for ML data governance:

Training Attendance

This metric tracks the number of employees who have participated in data governance training programs. High attendance indicates a commitment to educating stakeholders on data governance policies, best practices, and user roles within the data ecosystem. Training programs should be tailored to different user groups, ranging from data scientists requiring in-depth governance knowledge to business users needing a foundational understanding of responsible data usage. Monitoring training attendance allows to identify areas where additional training initiatives might be needed to ensure comprehensive awareness across the organization.

User Satisfaction

This metric measures feedback and satisfaction levels from data consumers and stakeholders regarding the data governance program. User satisfaction surveys and feedback mechanisms can be implemented to gather insights into the accessibility, usability, and overall effectiveness of the data governance tools and processes. Understanding user pain points can guide improvements to data governance practices, ensuring they are user-friendly and cater to the needs of diverse stakeholders. A high level of user satisfaction fosters a culture of collaboration and trust within the data ecosystem.

Data Literacy

This metric reflects the improvement in data literacy and data-driven decision-making across the organization. Data literacy programs can be implemented to teach employees how to access, understand, and utilize data effectively. The impact of these programs can be measured through surveys or assessments that gauge employee comfort levels with data analysis and their ability to integrate data insights into decision-making processes. Improved data literacy empowers stakeholders to actively participate in the data governance ecosystem and leverage data responsibly to drive business outcomes.

By fostering stakeholder engagement and adoption, organizations can ensure their data governance program is not just a set of policies, but a living and evolving ecosystem that drives responsible data use within their ML initiatives. This ultimately leads to a more informed and data-driven organization, enabling them to extract greater value from their machine learning efforts.

Establishing Data Governance metrics

Data governance metrics and KPIs should be tailored to the specific needs and objectives of the organization's machine learning initiatives, taking into account the data governance maturity level, industry-specific requirements, and regulatory landscape.

Here, we delve into the essential and critical factors to consider when establishing data governance metrics and KPIs:

Alignment with Business Objectives

Data governance exists to serve the broader organizational goals. Metrics and KPIs should be directly tied to these overarching business objectives. The key is to avoid getting bogged down in tracking every metric imaginable and, instead, focus on those that demonstrate how data governance contributes to achieving strategic objectives.

For example, if a company aims to improve customer segmentation through data-driven initiatives, metrics around data quality and accessibility specific to customer data would be crucial. Tracking the percentage of missing values in customer attributes or the time it takes for business users to access relevant customer datasets would provide valuable insights into data governance's effectiveness in supporting this specific business objective.

Stakeholder Needs

Data governance impacts various stakeholders within an organization. Business leaders might prioritize metrics showcasing the financial benefits of improved data quality, such as increased revenue generated through data-driven marketing campaigns.

Data analysts, on the other hand, might be more interested in metrics related to data completeness and accuracy for their data analysis tasks.

When selecting metrics and KPIs, it's crucial to consider the different stakeholders involved and choose those that are relevant to each group. This ensures that the metrics provide insights valuable to their role and ultimately contribute to a more comprehensive understanding of the data governance landscape.

Data Availability and Measurability

The chosen metrics should be measurable and supported by readily available data. Don't fall into the trap of selecting metrics that are difficult or impossible to track. Ensure the data required to calculate these metrics is accessible through data governance tools or existing data sources.

For instance, tracking data lineage, which maps the origin and flow of data throughout processing stages, might be challenging without a dedicated data lineage tool in place. Similarly, measuring user satisfaction with data governance processes might necessitate implementing user feedback mechanisms to gather reliable data.

Focus on Actionable Insights

Effective metrics provide more than just numbers; they offer actionable insights that can be used to identify areas for improvement or gauge the effectiveness of current data governance practices. The goal is to use the data to inform decisions about enhancing data governance framework.

For example, a low data reuse rate, reflecting the number of times a specific dataset is utilized for various purposes, might indicate a need to improve data discovery functionalities. The data governance team could then implement user education initiatives or enhance data cataloging tools to make relevant datasets more easily discoverable by different teams.

Achieving Balance

A well-rounded set of metrics encompasses different aspects of data governance. Don't get fixated on a single metric category like data quality, even though it's a crucial element. Include metrics from various categories such as data accessibility, process efficiency, security, and business value to have a holistic view of the data governance landscape.

For instance, consider including metrics that track the average time taken to resolve data-related issues (process efficiency), the number of data breaches (security), and the cost savings achieved through optimized data storage (business value). This comprehensive approach ensures that data governance strategy addresses all critical aspects of data management.

Start Simple, Scale Up

Building a robust data governance framework takes time and continuous improvement. It's best to begin with a core set of essential metrics that are easy to track and understand. These essential metrics might focus on high-level aspects like data quality, data accessibility, and user adoption of data governance practices.

As organization's data governance program matures and organization accumulates data, one can gradually add more complex or specialized metrics to gain deeper insights. For example, one might introduce metrics related to specific data governance tools or user behavior patterns within the data ecosystem.

Regular Monitoring and Evaluation

Metrics and KPIs are only valuable if they are reviewed and analyzed regularly. Schedule periodic reviews to assess the effectiveness of chosen metrics and whether they are still aligned with business objectives. This ongoing evaluation process allows to identify potential gaps in data governance strategy and adapt as the organization's needs evolve.

Don't be afraid to modify metrics and KPIs over time as data governance program evolves and business needs change. The data landscape itself is constantly evolving, and new technologies or regulations might necessitate adjustments to data governance strategy. Regularly reviewing and updating metrics ensures they remain relevant and continue to provide valuable insights for continuous improvement.

Selecting the right metrics and KPIs is critical for the success of any data governance program. They act as the compass and the map, guiding the organization toward achieving its data-driven goals and ensuring responsible data utilization. Actionable insights gleaned from data governance metrics empower stakeholders to make informed decisions about data management practices. These decisions can range from prioritizing investments in data quality initiatives to streamlining data access procedures for better user experience.

Quantifiable metrics showcase the positive impact of data governance on different aspects of the organization. Metrics like cost savings achieved through reduced data redundancy or increased revenue generated through data-driven decision-making can be used to demonstrate the business value of data governance efforts. This helps secure ongoing support and resources for maintaining a robust data governance framework. Metrics provide transparency into the effectiveness of data governance practices. This fosters accountability among stakeholders and ensures everyone involved is working toward achieving established data governance objectives.

By regularly monitoring and evaluating metrics, organizations can identify areas where their data governance program needs strengthening. This data-driven approach fosters a culture of continuous improvement, allowing to refine processes, implement new technologies, and adapt to evolving data management challenges. Data governance metrics can be used to demonstrate compliance with relevant data privacy regulations. For example, metrics tracking data access control procedures or data lineage can provide evidence that data is managed and utilized responsibly according to regulatory requirements.

Data Governance Maturity Model

A data governance maturity model provides a framework for assessing an organization's current state of data governance practices and capabilities while also serving as a roadmap for continuous improvement and evolution. This model recognizes that data governance is a journey, and organizations may progress through different levels of maturity as they enhance their data management practices and align them with their machine learning initiatives.

The data governance maturity model typically consists of several levels, each characterized by specific attributes, capabilities, and best practices. While the specifics may vary across different frameworks, a common approach is to define five levels of maturity:

Initial or Ad Hoc Level

At this level, data governance practices are informal, inconsistent, and largely reactive. Data management activities are siloed, and there is a lack of standardized processes, policies, and organizational structures. Data quality, security, and compliance issues are addressed on a case-by-case basis.

The Initial level represents a nascent stage where data governance practices are in their early development. While not entirely absent, these practices are likely to be informal, inconsistent, and largely reactive. While some semblance of structure exists, it's akin to navigating murky waters where visibility is limited. Data management activities might have some established procedures within individual departments or teams. However, these efforts lack coordination across the organization, leading to siloed approaches. Standardized processes, comprehensive policies, and a clear, centralized data governance structure are often underdeveloped or entirely missing. Data quality, security, and compliance are addressed as issues arise, lacking a proactive and preventative approach.

Imagine data scientists encountering scattered data sources, each with potentially unique access protocols and quality standards. While some rudimentary processes for data management might exist within specific teams, there's likely no centralized data catalog or established data lineage tracking. This lack of consistent governance creates several challenges. Data inconsistencies and varying quality across sources can lead to unreliable results in ML models, compromising their effectiveness. Data scientists may spend significant effort navigating different data access procedures and cleaning up inconsistencies, hindering project timelines and overall efficiency. Without standardized data governance practices, collaboration on ML projects can be hampered due to varying data access and utilization methods across departments.

The Initial level highlights the need to mature data governance practices. By implementing standardized processes, establishing a central data governance structure, and proactively addressing data quality, security, and compliance, organizations can transform this murky landscape into a clear and navigable environment. This shift empowers data scientists with readily accessible, consistent data, allowing them to focus on building robust ML models and drive innovation.

Developing or Repeatable Level

Organizations at this level begin to recognize the importance of data governance and establish basic processes, policies, and roles. Data management activities become more repeatable, and some standards and guidelines are defined. However, adoption and enforcement of these practices may still be inconsistent across the organization.

The Developing level signifies a crucial step forward. Organizations at this stage recognize the importance of data governance and are actively establishing the building blocks for a more controlled environment. Imagine emerging from the murky waters of the Initial level and starting to see the outlines of a structured system.

Basic processes and policies are being developed, outlining data access procedures, data quality standards, and rudimentary data security measures. Think of these as stepping stones being laid across the data landscape. Roles and responsibilities for data governance might be assigned, creating a rudimentary organizational structure. Data management activities become more repeatable, with some departments or teams following established procedures.

However, challenges remain. Adoption and enforcement of these practices might be inconsistent across the organization. Imagine some teams diligently following the newly laid data governance path, while others continue to operate in their own silos. This lack of uniformity that can lead to inconsistent application of data governance practices can lead to inefficiencies as data scientists navigate varying procedures across departments. Without consistent data management practices, it can be difficult to gain a comprehensive view of data quality, lineage, and overall data health within the organization. Inconsistencies in data security measures and access controls could leave the organization vulnerable to potential data breaches or non-compliance with regulations.

The Developing level represents a significant step toward mature data governance. By focusing on broader adoption and enforcement of established practices, organizations can solidify the foundation and move toward a more standardized and optimized data governance environment, ultimately empowering their ML initiatives.

Defined or Managed Level

At this level, data governance practices are well-defined, documented, and consistently applied across the organization. There is a formal data governance framework with established policies, roles, and responsibilities. Data quality, security, and compliance are actively managed, and data governance processes are integrated into broader organizational processes.

Data governance practices are no longer informal or reactive. They are well-defined, documented, and consistently applied across the organization. Organization has established a comprehensive data governance policy manual that serves as a blueprint for data management activities. This manual outlines established roles and responsibilities, clearly defining who owns, manages, and utilizes data for ML initiatives.

Data quality, security, and compliance are no longer afterthoughts; they are actively managed. Think of dedicated teams monitoring data quality metrics and implementing proactive data cleansing procedures. Robust security protocols safeguard sensitive data, and adherence to data privacy regulations becomes a core tenet of the data governance framework. This emphasis on data integrity fosters trust and empowers data scientists to utilize data with confidence.

Furthermore, data governance processes are integrated seamlessly into broader organizational processes. Imagine data scientists collaborating with data governance teams to ensure proper data access and utilization throughout the ML project lifecycle. This integration removes inefficiencies and fosters a collaborative environment where data governance is not seen as a hurdle, but rather as an enabler for successful ML initiatives.

The Defined level represents a significant milestone in the data governance journey. However, there's still room for further optimization. By leveraging data governance tools and technologies, and continuously monitoring and improving established practices, organizations can evolve toward an even more mature and optimized data governance environment, propelling their ML initiatives to even greater heights.

Quantitatively Managed or Optimized Level

Organizations at this level have implemented robust data governance processes and tools, enabling continuous monitoring, measurement, and optimization of data management practices. Data governance metrics and KPIs are established, and data-driven decision-making is embedded into the organizational culture. There is a strong focus on continuous improvement and alignment with evolving business needs and industry best practices.

The Quantitatively Managed level signifies a state of continuous refinement. Organizations at this stage have transformed their data governance framework from a sturdy bridge into a sophisticated network of data pipelines, constantly monitored and optimized for peak performance.

Robust data governance processes are well-established and supported by powerful tools and technologies. Imagine a central data catalog readily accessible to all stakeholders, allowing seamless data discovery and lineage tracking. Automated data quality checks and cleansing processes ensure the integrity of data used in ML models. Data governance metrics and key performance indicators (KPIs) are clearly defined, providing valuable insights into the effectiveness of data management practices.

These metrics and KPIs are not mere numbers; they form the foundation for data-driven decision-making. Imagine data governance teams and data scientists collaborating to analyze data quality trends and identify areas for improvement. This data-centric approach allows for continuous optimization of data management practices, ensuring they remain aligned with evolving business needs and industry best practices.

The emphasis is on continuous improvement. Think of dedicated resources allocated to staying abreast of the latest data governance tools and technologies, ensuring the framework remains cutting-edge. Furthermore, the culture fosters ongoing feedback loops, where insights from data scientists and other stakeholders are incorporated to refine data governance practices and adapt to changing data usage patterns.

Advanced or Optimizing Level

At the highest level of maturity, data governance is fully integrated into the organization's strategic decision-making processes. Data governance practices are proactively aligned with emerging technologies, regulatory changes, and industry trends. The organization demonstrates leadership in data governance and serves as a benchmark for others in the industry. At this stage, data governance is not merely a support system; it's woven into the very fabric of the organization's strategic decision-making processes.

Data governance practices are no longer reactive; they are proactive and forward-thinking. Think of dedicated teams actively researching emerging technologies, regulatory changes, and industry best practices in data governance. This foresight allows the organization to proactively align its data governance framework to capitalize on new opportunities and mitigate potential risks before they arise.

Furthermore, the organization sets the bar for others. Imagine being recognized as a leader in data governance, sharing best practices, and influencing the broader industry. This leadership role can take many forms—contributing to industry standards, publishing thought leadership content, or collaborating with other organizations on data governance initiatives.

The focus remains on continuous optimization, but with an even broader scope. It's no longer just about data management practices; it's about optimizing the entire data ecosystem to maximize its potential. Think of exploring advanced data governance tools like AI-powered data lineage tracking or leveraging automation to further streamline data management processes. This relentless pursuit of optimization ensures the data governance framework remains agile and adaptable, readily responding to the ever-evolving landscape of data and machine learning.

It's important to note that achieving the highest level of maturity is not necessarily the goal for all organizations. The desired level of maturity should be aligned with the organization's specific needs, resources, and the complexity of its machine learning initiatives.

Data governance maturity reflects the level of control and structure an organization exerts over its data. It encompasses various aspects, including data quality, access control, data lineage tracking, security protocols, and compliance with relevant regulations.

Organizations typically progress through different maturity levels within a data governance framework. These levels represent the increasing sophistication and effectiveness of data management practices. Frameworks like the data governance maturity model offer a valuable tool for self-assessment, enabling organizations to identify their current stage and establish a roadmap for improvement.

Conducting a Self-assessment

The first step toward improving data governance posture is to assess the current maturity level. Conduct a comprehensive review of existing data management practices. Analyze data governance policies, assess the level of data quality control, and evaluate data security protocols currently in place. Identify gaps between established policies and actual practice.

Interview key stakeholders involved in data initiatives, including analysts, business users, and data governance team members. These interviews can provide valuable insights into user experiences with data access, data quality, and overall data governance practices.

Analyze industry best practices and compare data governance practices to those of organizations considered leaders in the field. Benchmarking exercises can help identify potential gaps and areas for improvement.

Strategies for Improvement

Once organization has pinpointed the areas requiring improvement within maturity level, it can implement targeted strategies to enhance data governance practices. Here are some potential areas to focus on:

Develop and Implement Data Governance Policies: Establish clear and comprehensive data governance policies that address data quality, access control, data security, and data lineage. These policies should be readily available to all stakeholders involved in data initiatives and enforced consistently across the organization.

Invest in Data Governance Tools: Leverage technology to streamline data management processes. Consider implementing data quality tools to automate data validation and cleansing tasks. Additionally, data lineage platforms can provide valuable insights into the origin and flow of data, enhancing traceability and facilitating impact analysis.

Prioritize Data Quality: Data quality is the cornerstone of reliable insights. Implement robust data quality checks and data cleansing processes to ensure consistent and accurate data usage. Regularly monitor data quality metrics to identify and address potential issues.

Foster a Culture of Data Responsibility: Educate stakeholders on data governance practices and their role in ensuring data integrity. Training programs can equip users with the knowledge and skills necessary to access, utilize, and share data responsibly within the organization.

Establish a Data Governance Team: Consider establishing a dedicated data governance team to oversee policy implementation, data quality monitoring, and user education initiatives. This team can also collaborate with data science teams to ensure data governance practices cater specifically to the needs of data-driven projects.

Continuously Monitor and Review: Data governance is an ongoing process. Regularly monitor data governance metrics to assess the effectiveness of implemented strategies and identify areas for further improvement. Review data governance policies and procedures periodically to ensure they remain aligned with evolving business needs and regulatory requirements.

Embracing Continuous Improvement

The process of improving data governance maturity is a journey, not a destination. Organizations should embrace a culture of continuous improvement, constantly seeking ways to refine and optimize their data governance framework. Here are some additional considerations to maximize efforts:

Align Data Governance with Business Strategy: Ensure data governance strategy aligns with overall business goals and objectives. Data governance practices should not be implemented in isolation; they should serve as an enabler for success within broader data initiatives.

Adapt to Evolving Technologies: The data landscape is constantly evolving. Stay up-to-date on emerging data governance tools and technologies and explore opportunities to integrate them into the framework for improved efficiency and effectiveness.

Promote Collaboration and Communication: Foster open communication channels between data scientists, data governance teams, and business stakeholders. Encourage open dialogue regarding data needs, challenges, and opportunities for improvement. This collaborative approach fosters a shared understanding of data governance and its importance in achieving successful outcomes.

Measure the Impact of Data Governance: Track and measure the impact of data governance efforts on the success of data initiatives. Quantify improvements in data quality, reduction in errors, and overall efficiency of data workflows. This data-driven approach allows to demonstrate the value proposition of data governance and justify further investments in tools and resources.

Embrace a Data-Driven Culture: Building a data-driven culture is essential for the long-term success of data governance efforts. Embed data-centric principles into organizational DNA, empowering all stakeholders to leverage data insights for informed decision-making. This cultural shift fosters a sense of ownership and responsibility for data quality and utilization.

By adopting a structured approach to assessing and improving data governance maturity, organizations can unlock the full potential of their data-driven initiatives. Remember, the journey toward a mature data governance environment requires commitment, collaboration, and a continuous pursuit of improvement. By implementing the strategies outlined in this article, and fostering a data-driven culture, one can ensure organization leverages data responsibly and ethically. This ultimately results in more reliable, impactful, and successful data-driven projects, empowering organization to make informed decisions with confidence.

Stakeholder Engagement and Change Management

Implementing an effective data governance framework for machine learning initiatives requires more than just technical solutions and processes. It also necessitates a strong focus on stakeholder engagement and change management to ensure successful adoption and sustained organizational commitment.

Building Trust and Transparency

Clear articulation of the rationale behind data governance initiatives helps to establish the groundwork. Explain how improved data quality and management practices directly translate into more reliable ML models and, ultimately, better business outcomes. This empowers stakeholders to understand the value proposition of data governance and fosters buy-in.

Ensure open and transparent communication with stakeholders at all levels. Regularly share updates on data governance policies, progress made, and any potential challenges encountered. This transparency builds trust and demonstrates the organization's commitment to responsible data utilization.

Don't wait for questions or concerns to arise. Proactively engage with stakeholders to understand their needs, challenges, and expectations regarding data governance. This two-way communication fosters a sense of ownership and collaboration.

Addressing Concerns and Fostering Collaboration

Data governance initiatives can sometimes raise concerns among stakeholders, particularly regarding data access restrictions or perceived burdens on their workflows. Recognize that different stakeholders have varying levels of data expertise and needs. Tailor communication strategies accordingly, using clear and concise language that resonates with each audience. For example, technical explanations for data scientists may differ from those for business users.

Be prepared to address concerns regarding restrictions on data access. Explain the rationale behind data access controls and emphasize how they ultimately promote data security, enhance data quality, and protect user privacy. Explore alternative access options, such as data sandboxes or anonymized datasets, where appropriate.

Involve stakeholders in the development and implementation of data governance policies. This collaborative approach allows stakeholders to feel heard and helps ensure the policies are practical and address their specific needs.

Continuously emphasize the benefits of data governance for all stakeholders. Highlight how improved data quality leads to more accurate and reliable ML models, ultimately impacting business outcomes positively. Showcase success stories and concrete examples of how data governance has benefited other departments or projects.

Managing Change and Building Capacity

Implementing data governance often necessitates changes in existing workflows and user behaviors. Develop a comprehensive change management strategy that outlines communication plans, training programs, and support measures to assist stakeholders in adapting to the new data governance practices.

Provide targeted training programs on data governance policies, data quality practices, and access control protocols. Offer different training formats to cater to diverse learning styles and expertise levels. This empowers stakeholders with the knowledge and skills necessary to navigate the new data governance environment effectively.

Invest in initiatives to build data literacy across the organization. This can include workshops, online resources, and knowledge-sharing platforms. The goal is to foster a deeper understanding of data management best practices and empower all stakeholders to become responsible data stewards.

Identify and leverage internal data governance champions. These individuals can act as peer-to-peer mentors and advocates, providing support and guidance to colleagues as they navigate the changes associated with data governance.

Establish mechanisms for gathering continuous feedback from stakeholders regarding data governance practices and their impact. Actively listen to concerns, address them promptly, and demonstrate a willingness to adapt and improve the data governance framework based on user needs.

Successful stakeholder engagement and change management are not one-time events but rather ongoing processes. By implementing the strategies outlined above, organizations can navigate the human element of implementing ML data governance and create an environment where data is treated responsibly and ethically, ultimately leading to more reliable, impactful, and successful ML initiatives.

Building trust, fostering transparency, addressing stakeholder concerns, and managing change effectively contribute to a sense of ownership and collaboration regarding data governance within the organization. This collaborative culture not only ensures successful implementation of data governance practices but also empowers all stakeholders to leverage data as a valuable asset for driving informed decision-making and achieving strategic objectives.

Summary

An in-depth exploration of the critical components required for establishing a robust data governance framework tailored specifically to machine learning (ML) initiatives is provided. The discussion emphasizes the necessity of a comprehensive approach that integrates organizational structures, clearly defined roles, policies, processes, and technologies to effectively manage data throughout its lifecycle.

The foundational policies and standards essential to effective data governance are delineated, focusing on key areas such as ensuring data quality, security, privacy, and adherence to relevant regulatory requirements. The importance of well-defined processes and workflows is highlighted as a means of ensuring consistency and efficiency in data management practices. Additionally, there is an emphasis on assigning clear roles and responsibilities to various stakeholders, such as data owners, stewards, and consumers, to foster accountability and facilitate collaboration within the organization.

Specialized tools and technologies that are instrumental in supporting data governance activities are introduced, including data catalogs, lineage tools, and data quality management systems, all of which contribute to the effective governance of data in ML contexts. The role of metrics and key performance indicators (KPIs) in assessing the efficacy of the data governance framework is also addressed.

The concept of a data governance maturity model is presented as a structured roadmap for organizations to systematically enhance their data governance practices. The importance of stakeholder engagement and change management is emphasized, highlighting the need for active participation and collaboration across the organization to ensure the successful implementation and sustainability of the data governance framework.

A comprehensive discussion on the essential components and best practices for establishing a data governance framework specifically designed for ML initiatives is provided. The discussion underscores the significance of adopting a structured and systematic approach to data management, which is crucial for optimizing the potential of ML technologies while mitigating associated risks and fostering organizational trust.

CHAPTER 3

Data Quality and Preprocessing

In the fast-paced world of machine learning, data truly is king. While complex algorithms can uncover patterns, make predictions, and even automate processes, their success hinges on one crucial element: the quality of the data they're fed. This chapter kicks off by diving into the essential role that data quality and preprocessing play in building reliable and accurate machine learning models. It makes a clear point that the effectiveness of these models is directly tied to the quality of the data they're trained on. If the data is flawed, the results can be too—leading to biased outcomes, reduced accuracy, and even ethical issues. From there, the chapter transitions into why it's so important to have strong data governance frameworks in place. These frameworks help ensure that data remains consistently accurate, complete, and tailored to the specific needs of the ML tasks at hand.

As we move further into the chapter, the focus shifts to the nuts and bolts of data preprocessing, including critical steps like data cleaning, transformation, and feature engineering. The importance of keeping version control and meticulously documenting every step of the process is also emphasized. These practices are not just about improving data quality—they also play a key role in making sure the process is transparent, reproducible, and compliant with regulations. The chapter wraps up by highlighting the ongoing need for monitoring and refining data quality practices. As new data is brought in, it's essential to maintain the integrity and reliability of ML models throughout their lifecycle, ensuring they continue to deliver trustworthy results.

© Aditya Nandan Prasad 2024
A. Nandan Prasad, *Introduction to Data Governance for Machine Learning Systems*,
https://doi.org/10.1007/979-8-8688-1023-7_3

Importance of Data Quality for Machine Learning

Data quality refers to the overall health of the information used to train an ML model. It encompasses factors like accuracy, completeness, consistency, and relevance. High-quality data ensures that the model learns from reliable patterns and relationships, leading to accurate predictions and effective decision-making. Conversely, poor data quality can have a disastrous impact, introducing biases, skewing results, and ultimately rendering the model useless or even harmful.

Understanding the detrimental effects of bad data is essential. Imagine an ML model designed to predict loan defaults. If the training data contains a significant number of inaccurate income figures, the model might misclassify creditworthy borrowers as high risk. This can lead to unfair loan denials and have severe consequences for individuals. Similarly, a facial recognition system trained on biased datasets might struggle to identify people of color, raising ethical concerns and potentially leading to security breaches.

There are several ways in which bad data can infiltrate an ML project. Missing values, where crucial data points are simply absent, create gaps in the information the model needs to learn from. Duplicates, on the other hand, inflate the influence of certain data points, skewing the model's understanding of the underlying patterns. Inconsistencies, such as variations in data formatting or units of measurement, further muddy the waters and make it difficult for the model to identify meaningful relationships.

Beyond these technical shortcomings, data quality can also be compromised by biases. If the training data primarily reflects a specific demographic or set of circumstances, the model might inherit those biases and perpetuate unfair or discriminatory outcomes. For instance, an algorithm used for recruiting purposes could favor candidates from certain educational backgrounds if the training data was sourced from a limited pool of universities.

Machine learning algorithms are sensitive to the quality of the data they process. Dirty data, containing outliers, inconsistencies, and irrelevant features, can hinder the model's ability to learn effectively. This can manifest in low accuracy, poor generalization, and ultimately, a model that fails to deliver on its promises.

Fortunately, there are steps that data scientists and ML engineers can take to ensure data quality. Data cleaning, a crucial preprocessing step, involves identifying and rectifying errors and inconsistencies. This can involve techniques like filling in missing values, removing duplicates, and standardizing formatting. Data validation, another essential practice, involves checking the data against predefined rules and expectations to ensure its accuracy and compliance.

Techniques like data profiling can be used to understand the characteristics of the data and identify potential issues. Data visualization can also be a powerful tool, allowing data scientists to spot anomalies and patterns that might be missed by traditional methods.

However, data quality is not a one-time fix. It's an ongoing process that requires constant monitoring and vigilance. As new data is collected and integrated, it's essential to maintain the high standards established during the initial stages of the project. Additionally, the definition of "good data" can evolve as the goals of the ML project shift. The data that was perfectly adequate for one purpose might not be sufficient for another, requiring further cleaning and validation efforts.

Data quality is the cornerstone of successful machine learning. By prioritizing data quality, we ensure that ML models are built on a solid foundation, capable of delivering accurate, reliable, and unbiased results. This not only safeguards against potential harm but also unlocks the true potential of ML to create a future driven by intelligent and responsible technology. As we continue to invest in and develop ML solutions, a relentless focus on data quality will be paramount in ensuring that these powerful tools are used for good.

Data Profiling and Exploratory Data Analysis (EDA)

Data quality encompasses the overall health of information used for training, encompassing factors like accuracy, completeness, consistency, and relevance. To ensure reliable patterns and relationships are learned, data profiling and exploratory data analysis (EDA) become crucial first steps. Data profiling paints a comprehensive picture of the data's characteristics, highlighting potential issues that may hinder the learning process of an ML model. It involves summarizing key statistics, identifying data types, and uncovering patterns within the data. This preliminary analysis allows data scientists to assess the suitability of the data for the intended ML task and identify areas requiring further cleaning and transformation. There are several techniques employed in data profiling to gain a deep understanding of the data.

Central Tendencies and Dispersion

Understanding central tendencies and dispersion is essential for effective data governance in machine learning (ML) initiatives, particularly during the exploratory data analysis (EDA) phase. These concepts help summarize and describe the main features of a dataset, providing insights that are vital for developing robust ML models. Central tendencies—such as the mean, median, and mode—represent the center or typical value of a dataset, offering different perspectives on the data's overall behavior. The mean, as the arithmetic average, is widely used but is highly sensitive to outliers. For instance, in a dataset of incomes where a few individuals earn significantly more than the rest, the mean may not accurately represent the typical income level, leading to skewed interpretations. The median, on the other hand, provides a better indication of central tendency in such skewed distributions by focusing on the middle value, unaffected by extreme outliers. Meanwhile, the mode, which identifies the most frequently occurring value, is particularly useful for categorical data, offering insights into common occurrences within the dataset. From a data governance perspective, understanding and applying these measures appropriately is crucial to ensuring accurate data analysis and model development.

Dispersion, which describes the spread of data points around the central tendency, is equally critical in ML data governance. Measures of dispersion, such as range, variance, and standard deviation, offer insights into the variability and consistency of a dataset, directly impacting the performance and reliability of ML models. For example, while the range provides a simple measure of dispersion by indicating the difference between the maximum and minimum values, it can be misleading if the dataset contains significant outliers. Variance offers a more comprehensive view by measuring the average squared deviation from the mean, highlighting how spread out the data points are. High variance suggests a wide distribution, which can indicate instability in the dataset, while low variance implies that data points are closely clustered around the mean. Standard deviation, the square root of variance, is easier to interpret and is commonly used to assess the risk or uncertainty associated with ML model predictions. In practice, managing variance and standard deviation is vital for ensuring that ML models are not unduly influenced by noise or outliers, a key aspect of data governance.

In addition to these fundamental measures, more advanced concepts such as skewness and kurtosis provide deeper insights into the shape of a data distribution, which is crucial for ML data governance. Skewness measures the asymmetry of a distribution, with positive skewness indicating a long tail on the right and negative skewness a long tail on the left. This measure is particularly important for ML algorithms that assume normally distributed data, as significant skewness can affect model performance. For example, income distributions are often positively skewed, and applying transformations like logarithmic scaling may be necessary to normalize the data. Kurtosis, which measures the "tailedness" of a distribution, indicates whether data points are more or less concentrated in the tails compared to a normal distribution. High kurtosis suggests a higher probability of extreme events, which can be critical for accurately modeling risks in financial data. Understanding these distribution shapes is vital for data governance, as it helps in identifying potential outliers and assessing the reliability of ML models.

Multivariate analysis further underscores the importance of understanding central tendencies and dispersion across multiple variables simultaneously. Techniques such as covariance and correlation are essential for exploring relationships between variables in ML models. Covariance measures how two variables change together, indicating whether they increase or decrease in tandem, while correlation standardizes this measure, making it easier to interpret. In an ML context, these measures are critical for feature selection and dimensionality reduction, ensuring that models are built on a solid understanding of variable interactions. For instance, in a retail dataset, a strong positive correlation between marketing spend and sales revenue might guide feature selection, emphasizing the importance of these variables in predictive models. Data governance practices should ensure that covariance and correlation are analyzed systematically to avoid issues like multicollinearity, which can destabilize model coefficients.

Outlier detection is another critical aspect of ML data governance, closely linked to the understanding of central tendencies and dispersion. Outliers can significantly distort model outcomes if not properly managed, making their detection and treatment a key governance concern. Z-scores, which measure how many standard deviations a data point is from the mean, are commonly used to identify outliers. Points with z-scores beyond a certain threshold, such as ±3, are typically flagged for further investigation. Similarly, boxplots and the interquartile range (IQR) are effective tools for visualizing data distribution and highlighting outliers. These methods allow data governance frameworks to establish protocols for addressing outliers, whether through removal, adjustment, or closer examination, ensuring that ML models are not skewed by extreme values.

Time-series analysis, which often involves central tendencies and dispersion, plays a crucial role in understanding trends, seasonality, and volatility over time—factors that are critical for building accurate ML models. Moving averages, both simple and exponential, are used to smooth out short-term fluctuations, revealing underlying trends that might not be immediately apparent. Volatility, often measured as the standard deviation of returns, is particularly important in financial data, where understanding the degree of fluctuation can significantly impact risk management and model accuracy. From a data governance perspective, ensuring that moving averages and volatility measures are consistently applied and interpreted correctly is vital for maintaining model reliability.

Central tendencies and dispersion also have applications in handling categorical data, where they help in understanding the distribution and variability of non-numeric variables. For instance, the mode is particularly relevant for categorical variables, indicating the most common category in a dataset, which can inform feature encoding decisions in ML models. Additionally, metrics like entropy and the Gini index, commonly used in decision tree algorithms, assess the purity of splits in categorical data, guiding model development and ensuring that data diversity is appropriately managed. Data governance frameworks should ensure that these measures are correctly applied to categorical data, supporting the creation of well-structured and unbiased ML models.

Multimodal distributions, where data exhibits multiple peaks, present unique challenges in EDA and ML data governance. Identifying and analyzing multimodal distributions is crucial for understanding the presence of distinct subgroups within a dataset, which may require different modeling approaches. For example, a bimodal distribution in employee salaries might indicate two distinct groups, such as junior and senior staff, necessitating tailored ML models for each subgroup. Techniques like Gaussian Mixture Models (GMM) can be employed to model such distributions, clustering data into distinct components that reflect the underlying structure. Data governance practices should include protocols for identifying multimodal distributions and applying appropriate modeling techniques to ensure that all subgroups are adequately represented.

Feature engineering, another critical aspect of ML model development, relies heavily on an understanding of central tendencies and dispersion. Normalization and standardization are common preprocessing steps that ensure features contribute equally to the model, particularly when dealing with variables on different scales. Interaction terms, which capture the combined effect of two or more variables, are also engineered by understanding the relationships between central tendencies and dispersion. For instance, an interaction term between ad spend and customer age in a marketing dataset might reveal how ad effectiveness varies across age groups, providing deeper insights into consumer behavior. Data governance frameworks should establish best practices for feature engineering, ensuring that these techniques are applied consistently and that the resulting features enhance model performance without introducing new biases or complexities.

Dimensionality reduction, through techniques like Principal Component Analysis (PCA), often depends on understanding central tendencies and dispersion to reduce the number of features while retaining the most critical information. PCA transforms the original features into a new set of components, ordered by the variance they capture, allowing for the reduction of dimensionality while preserving the data's essential characteristics. This process is particularly useful in datasets with a large number of features, where reducing complexity can improve model interpretability and performance. Data governance practices should ensure that dimensionality reduction techniques are used appropriately, with clear documentation of the variance explained by each component and the implications for model development.

The evaluation of ML models also relies on a thorough understanding of central tendencies and dispersion, particularly in error analysis and model validation. Residual analysis, which examines the differences between predicted and actual values, helps assess model accuracy by highlighting any discrepancies or patterns in the residuals. Ideally, residuals should be centered around zero with minimal dispersion, indicating that the model is accurately capturing the underlying data patterns. Cross-validation, which involves splitting the data into training and validation sets multiple times, provides insights into model stability by assessing the variance in performance across different folds. Data governance frameworks should ensure that these evaluation techniques are consistently applied, supporting the development of reliable and accurate ML models.

Data Distribution

Data distribution analysis provides a snapshot of the overall structure of the data, allowing data scientists to assess whether the dataset is suitable for machine learning tasks. Different types of distributions—such as normal, uniform, skewed, and multimodal—can dramatically influence the choice of algorithms, preprocessing steps, and model evaluation strategies. For instance, the normal distribution, often referred to as the Gaussian distribution, is a bell-shaped curve where most data points cluster around the mean. This distribution is particularly relevant for many statistical techniques and ML algorithms, which assume normality in the data. In practice, however, real-world data often deviates from this idealized form. Consider a dataset of housing prices in a large metropolitan area; it is unlikely that prices would follow a perfect normal distribution. Instead, prices might exhibit positive skewness, where a small number of extremely high-priced homes pull the distribution to the right, creating a long tail. This skewed distribution has implications for ML models that rely on assumptions of normality, necessitating transformations such as log or square root scaling to stabilize the variance and make the data more amenable to linear models.

From a data governance perspective, understanding the distribution of data is crucial for identifying potential biases that could impact the fairness and accuracy of ML models. For example, in a hiring algorithm designed to screen job applicants, if the training data is heavily skewed toward a particular demographic group, the resulting model may inadvertently favor candidates from that group, perpetuating existing inequalities. A deep analysis of the data distribution might reveal that the dataset underrepresents certain age groups or ethnicities, signaling a need for rebalancing or augmentation. This process could involve generating synthetic data to fill in the gaps or collecting additional data to ensure a more representative sample. Addressing these issues at the data distribution stage is a critical aspect of data governance, as it helps prevent biased outcomes and aligns the ML model with ethical and legal standards.

Another key aspect of data distribution in the context of ML data governance is the identification of outliers—data points that deviate significantly from the rest of the dataset. Outliers can arise due to errors in data collection, entry, or processing, or they might represent rare but significant events. For example, in a financial dataset tracking daily stock returns, an extreme outlier might correspond to a market crash or a sudden surge in stock prices due to an unexpected event. While outliers can provide valuable information, they can also distort model training if not handled appropriately. In regression models, for example, outliers can disproportionately influence the model's parameters, leading to biased predictions. From a governance perspective, it is important to establish guidelines for outlier detection and treatment. Techniques such as Z-scores, IQR (interquartile range), and robust statistics like the Median Absolute Deviation (MAD) can be employed to systematically identify and manage outliers. Depending on the context, outliers might be removed, capped, or transformed to minimize their impact on the model, ensuring that the ML outcomes remain reliable and valid.

The concept of data distribution also plays a pivotal role in feature selection and engineering, two critical steps in the ML process. Feature selection involves identifying the most relevant variables that contribute to the predictive power of the model, while feature engineering entails creating new variables or transforming existing ones to better capture the underlying patterns in the data. Analyzing the distribution of features helps in understanding their importance and how they might interact with other variables. For example, in a dataset predicting customer churn for a subscription service, the distribution of the "tenure" variable (the length of time a customer has been with the service) might show a bimodal pattern, with one peak representing long-term customers and another peak representing recent sign-ups. This bimodal distribution could indicate that churn behavior differs between these two groups, suggesting the need for interaction terms or separate models to accurately capture the different dynamics. From a data governance standpoint, documenting and validating these feature transformations is essential to ensure that the models are interpretable, reproducible, and aligned with the overall data strategy.

Data distribution analysis also includes the use of Kernel Density Estimation (KDE) and Quantile-Quantile (Q-Q) plots, which offer more nuanced insights into the shape and spread of data distributions. Kernel density estimation is a non-parametric technique that smooths the data to create a continuous probability density function, providing a more detailed view of the data's distribution compared to traditional histograms. This method is particularly useful when dealing with complex, multimodal distributions that are not easily captured by simple parametric models. For instance, in a dataset of online purchase amounts, KDE might reveal multiple peaks corresponding to different spending behaviors, such as low-cost impulse buys versus high-cost planned purchases. Understanding these distinct distributions can inform more targeted marketing strategies and improve the accuracy of predictive models. Quantile-quantile plots, on the other hand, compare the quantiles of the dataset against the quantiles of a theoretical distribution, such as the normal distribution. Q-Q plots are a powerful tool for assessing whether the data deviates from expected distributions and identifying specific areas where the data might need transformation. For example, in a dataset of employee performance scores, a Q-Q plot might reveal that the data is heavily skewed or has heavier tails than a normal distribution, indicating the need for log transformation or Winsorization to mitigate the impact of extreme values.

Data distribution analysis is also critical in the context of time-series data, where the temporal ordering of data points adds an additional layer of complexity. Time-series data often exhibits patterns such as trends, seasonality, and autocorrelation, which can significantly impact the performance of ML models if not properly accounted for. For instance, in a time-series dataset of electricity consumption, the distribution of data might show a clear seasonal pattern with peaks during winter and summer months due to heating and cooling needs. Understanding these seasonal distributions is essential for developing models that can accurately predict future consumption and manage energy resources effectively. From a data governance perspective, ensuring that time-series data is correctly analyzed and preprocessed is crucial for maintaining the accuracy and reliability of the models. This might involve de-seasonalizing the data, applying differencing to remove trends, or using Fourier transforms to capture periodic patterns.

An important consideration in data distribution analysis is the concept of data imbalances, particularly in classification tasks where the number of instances in different classes is not equal. Class imbalance is a common issue in domains such as fraud detection, where fraudulent transactions represent a small fraction of the total dataset, or medical diagnostics, where certain diseases are rare compared to the overall population. Imbalanced data distributions can lead to models that are biased toward the majority class, failing to accurately predict the minority class. To address this, data governance frameworks should incorporate strategies for dealing with imbalanced distributions, such as resampling techniques (e.g., oversampling the minority class or undersampling the majority class), cost-sensitive learning, or the use of synthetic data generation methods like SMOTE (Synthetic Minority Over-sampling TEchnique). These techniques help create a more balanced training set, improving the model's ability to detect rare but critical events.

Understanding data distribution is essential for ensuring compliance with data privacy regulations and ethical considerations. In the era of big data and AI, organizations must navigate a complex landscape of privacy laws, such as the General Data Protection Regulation (GDPR) in Europe and the California Consumer Privacy Act (CCPA) in the United States. These regulations impose strict requirements on how personal data is collected, processed, and shared, with significant penalties for non-compliance. Data distribution analysis can help organizations identify and mitigate privacy risks by revealing patterns that could inadvertently expose sensitive information. For example, a dataset containing location data might show a distribution that allows individuals to be re-identified based on their unique movement patterns, even if the data is anonymized. By analyzing the distribution of such data, organizations can implement techniques like differential privacy, which adds noise to the data to protect individual identities while still preserving the overall distribution for analysis. This balance between data utility and privacy is a key aspect of data governance, ensuring that ML models are both effective and compliant with legal and ethical standards.

Data distribution analysis plays a critical role in detecting and mitigating biases that can arise in ML models. Bias in ML can manifest in various forms, including selection bias, where certain groups are underrepresented in the training data, or algorithmic bias, where the model systematically favors one group over another. By analyzing the distribution of different demographic variables, such as age, gender, or race, data scientists can identify potential biases in the dataset and take corrective action. For instance, in a facial recognition dataset, if the distribution of skin tones is skewed toward lighter tones, the model may perform poorly on individuals with darker skin. To address this, data governance frameworks should include guidelines for conducting bias audits during the EDA phase, ensuring that the data used for training is representative of the target population and that the resulting models are fair and unbiased.

The concept of data distribution extends to the evaluation and validation of ML models, where it is crucial to ensure that the training and test datasets are representative of the real-world scenarios in which the model will be deployed. Distributional shifts, where the distribution of data changes between the training and deployment phases, can lead to a significant drop in model performance. For example, a model trained on historical sales data might perform well in a stable economic environment but fail when deployed during a recession, where consumer behavior changes dramatically. To mitigate this risk, data governance frameworks should include protocols for monitoring and detecting distributional shifts over time, such as through the use of concept drift detection algorithms. These tools help ensure that models remain accurate and reliable even as the underlying data distributions evolve.

Data Types and Formatting

At the most basic level, data can be categorized into numerical and categorical types, each with its own set of characteristics and challenges. Numerical data, which includes both integers and floating-point numbers, is quantitative and can be subjected to arithmetic operations. For instance, data on the ages or incomes of individuals are typically represented as numerical data. In ML, numerical data is often used directly as input features for models such as linear regression or neural networks. However, even within numerical data, nuances exist that can affect model performance. For example, integers are discrete, meaning they can only take on whole values, while floating-point numbers are continuous, capable of representing a broad range of values with fractional precision. This distinction is important when considering how data will be processed or when determining the appropriate algorithm. From a data governance perspective, ensuring that numerical data is correctly classified and formatted is crucial for maintaining the integrity of subsequent analyses and model development.

Categorical data, on the other hand, is qualitative and represents characteristics or labels rather than quantities. Examples include variables such as gender, occupation, or customer preferences. Categorical data can be further divided into nominal data, where categories have no inherent order, and ordinal data, where categories have a defined order. For instance, "low," "medium," and "high" are ordinal categories that might represent the level of satisfaction in a survey. Handling categorical data in ML requires special consideration, as algorithms typically expect numerical input. As such, categorical data often needs to be encoded into numerical form, which can be done through various techniques such as one-hot encoding, label encoding, or binary encoding. From a governance standpoint, the choice of encoding method can significantly impact model interpretability and performance. For example, while one-hot encoding is straightforward and widely used, it can lead to a high-dimensional feature space when dealing with variables that have many categories, potentially leading to overfitting. Data governance frameworks should therefore include guidelines on when and how to apply different encoding techniques, ensuring that the transformation of categorical data is appropriate for the specific ML task.

A critical aspect of data governance is ensuring that data types are consistent across different datasets and systems. Inconsistent data types can lead to errors in data processing and analysis, particularly when integrating data from multiple sources. For example, consider a scenario where a date field is represented as a string in one dataset but as a date object in another. When these datasets are merged, discrepancies in data types can cause parsing errors or incorrect interpretations of the data. To avoid such issues, data governance policies should enforce standardized data formats across the organization. This might involve setting rules for how dates, times, and other commonly used variables should be stored and processed, ensuring that all systems and teams adhere to the same standards. Such standardization not only prevents errors but also facilitates easier data sharing and collaboration across different departments or projects.

Data formatting is another crucial element in EDA and ML data governance. Proper formatting ensures that data is stored in a way that is both accessible and interpretable by ML algorithms. For instance, numerical data should be consistently formatted with the correct number of decimal places to avoid rounding errors or inconsistencies in analysis. Similarly, categorical data should be consistently labeled, with categories clearly defined and free from typographical errors. Consider a scenario in which customer gender is recorded as "Male" and "Female" in one dataset but as "M" and "F" in another. This inconsistency can cause issues when combining the datasets or when using the data for analysis, as the ML algorithm might treat these as separate categories rather than the same entity. Data governance frameworks should therefore include rules for data labeling and formatting, ensuring that all data is stored in a consistent and standardized manner. This helps maintain data integrity and prevents errors that could compromise the accuracy of ML models.

Mixed data types can arise due to errors in data entry, inconsistencies in data collection methods, or differences in how data is stored across systems. For example, a variable that is supposed to represent customer ages might contain both numerical values (e.g., "25," "30") and categorical descriptions (e.g., "Twenty-five," "Thirty"). Such inconsistencies can cause issues during EDA and model training, as ML algorithms may not be able to process the data correctly. To address this, data governance frameworks should include procedures for detecting and resolving mixed data types, such as converting all values to a consistent format or flagging and correcting anomalies. In some cases, automated tools can be used to identify and correct mixed data types, but manual review may also be necessary to ensure that the data is accurate and consistent.

Data governance also plays a critical role in ensuring that data types and formatting are aligned with regulatory requirements and industry standards. For example, financial institutions must adhere to strict regulations regarding how financial data is recorded, processed, and reported. Similarly, healthcare organizations must comply with standards such as the Health Level Seven (HL7) International guidelines for the exchange, integration, sharing, and retrieval of electronic health information. Failure to comply with these standards can result in legal penalties, data breaches, and loss of trust among stakeholders. To ensure compliance, data governance frameworks should include specific guidelines on how data should be formatted and stored, including the use of standardized data types and formats that align with regulatory requirements. This might involve adopting industry-standard formats for certain types of data, such as using the ISO 8601 standard for date and time representations, or ensuring that personal data is encrypted and stored securely.

Data types and formatting also have significant implications for data privacy and security, which are critical components of ML data governance. Properly managing data types and formatting helps ensure that sensitive information is protected and that data is handled in a way that complies with privacy regulations such as the General Data Protection Regulation (GDPR) or the California Consumer Privacy Act (CCPA). For example, when handling personal data such as Social Security numbers or credit card information, it is important to ensure that this data is stored in a secure format and that appropriate access controls are in place to prevent unauthorized access. Data governance frameworks should include protocols for encrypting sensitive data, as well as guidelines for how data should be formatted to minimize the risk of exposure. This might involve masking certain parts of the data, such as displaying only the last four digits of a credit card number, or using secure hashing algorithms to protect passwords and other sensitive information. By implementing these measures, organizations can ensure that their data handling practices align with legal and ethical standards, reducing the risk of data breaches and other security incidents.

The use of advanced data types, such as spatial data, temporal data, and textual data, presents additional challenges and considerations for ML data governance. Spatial data, which represents geographical locations and features, requires specialized formats and handling techniques. For example, latitude and longitude coordinates must be accurately recorded and stored in a consistent format to ensure that spatial analyses and ML models, such as those used for location-based services, are accurate. Temporal data, which involves time-related information, also requires careful handling, particularly when dealing with different time zones, daylight saving time adjustments, and time-series analyses. Textual data, often unstructured and complex, poses unique challenges in terms of preprocessing and formatting. Techniques such as tokenization, stemming, and vectorization are often used to convert textual data into a format suitable for ML models. Data governance frameworks should provide specific guidelines for handling these advanced data types, ensuring that they are processed and stored in a way that preserves their integrity and usability for ML applications.

The concept of data interoperability is closely linked to data types and formatting, particularly in the context of data governance. Data interoperability refers to the ability of different systems and organizations to exchange and use data seamlessly. Ensuring interoperability requires that data is formatted and structured in a way that is compatible across different platforms and systems. For example, in a healthcare setting, ensuring that patient data can be shared between different hospitals, clinics, and laboratories requires adherence to standardized data formats and protocols. This might involve using formats such as XML or JSON for data exchange, or adopting industry-specific standards such as the Fast Healthcare Interoperability Resources (FHIR) for healthcare data. From a data governance perspective, ensuring data interoperability is essential for enabling collaboration, reducing data silos, and maximizing the value of data assets across the organization.

Missing Values

Missing data refers to the absence of certain values in a dataset where they were expected to be present. This can occur for a variety of reasons, including data entry errors, incomplete data collection processes, or even system glitches. For instance, in a healthcare dataset, missing values might arise if a patient's medical history was not fully recorded due to oversight or if certain diagnostic tests were not performed. In financial datasets, missing values might occur when transactions are not recorded in a timely manner or when there are gaps in historical financial data due to mergers or system upgrades. Regardless of the cause, the presence of missing values poses significant challenges for data scientists and ML practitioners, as most machine learning algorithms require a complete dataset to function optimally.

From a data governance perspective, the first step in addressing missing values is to understand the nature and extent of the missing data. Missing values can be categorized into three types: Missing Completely At Random (MCAR), Missing At Random (MAR), and Missing Not At Random (MNAR). MCAR occurs when the likelihood of a value being missing is independent of both observed and unobserved data, meaning that the missing data is a completely random event. MAR occurs when the likelihood of missing data is related to the observed data but not the missing data itself. For example, in a survey, older respondents might be more likely to skip questions about technology use, leading to missing data that is related to age but not to the technology use question itself. MNAR occurs when the likelihood of missing data is related to the missing data itself, such as when patients with more severe symptoms are less likely to have follow-up visits, leading to missing data that is directly related to the severity of their condition. Understanding the nature of the missing data is crucial for choosing the appropriate method to address it, as different types of missing data require different handling strategies.

One of the most common approaches to dealing with missing values is imputation, where missing data is filled in using estimates based on the observed data. Simple imputation methods include mean or median imputation, where the missing values are replaced with the mean or median of the observed data for that variable. For example, if a dataset of student grades has some missing values, these could be replaced with the mean grade for the entire class. While this approach is straightforward, it can introduce bias, particularly if the missing data is not MCAR. More sophisticated imputation methods include multiple imputation, where multiple versions of the dataset are created with different imputations for the missing data, and the results are averaged to produce a final estimate. Another advanced technique is model-based imputation, where a predictive model is built using the observed data to estimate the missing values. For instance, in a customer dataset where income data is missing for some individuals, a regression model could be used to predict the missing income values based on other variables such as education, occupation, and location.

However, imputation is not always the best solution, particularly if the missing data is MNAR. In such cases, imputing the missing values might obscure important patterns in the data or lead to incorrect conclusions. For example, if higher-income individuals are less likely to disclose their income in a survey, imputing these values based on the data from lower-income respondents could lead to a significant underestimation of the true income distribution. In such cases, a more appropriate approach might be to use techniques such as pattern analysis to understand the structure of the missing data and to make decisions about how to handle it based on this analysis. For instance, rather than imputing missing values, it might be more appropriate to create a separate category for missing data or to use methods such as inverse probability weighting to account for the missing data in the analysis.

Missing data can introduce biases into ML models, particularly if the missing data is not evenly distributed across different subgroups. For example, if a healthcare dataset has more missing data for certain demographic groups, such as older patients or patients from minority backgrounds, this could lead to models that are biased against these groups. This is particularly concerning in applications such as predictive policing or credit scoring, where biased models can have significant real-world consequences. To mitigate these risks, data governance frameworks should include guidelines for conducting bias audits and ensuring that missing data is handled in a way that does not disproportionately impact certain groups. This might involve techniques such as stratified imputation, where the imputation model is trained separately for different subgroups, or the use of fairness constraints in model training to ensure that the model's predictions are not biased by the missing data.

Data governance also plays a crucial role in ensuring compliance with regulatory requirements related to missing data. In many industries, regulations such as the General Data Protection Regulation (GDPR) in Europe or the Health Insurance Portability and Accountability Act (HIPAA) in the United States impose strict requirements on how data, including missing data, is handled. For example, under

GDPR, organizations are required to maintain accurate and complete records of personal data, and missing values in these records could be considered a violation of this requirement. Similarly, in the healthcare industry, HIPAA mandates that healthcare providers maintain complete and accurate patient records, and missing values in these records could lead to legal and financial penalties. To ensure compliance, data governance frameworks should include protocols for auditing data completeness, documenting the reasons for missing data, and implementing appropriate strategies for handling missing values. This might involve the use of automated tools for monitoring data quality, as well as the establishment of clear guidelines for data collection and entry to minimize the occurrence of missing data.

The treatment of missing values should be guided by the overall goals and objectives of the ML project. In some cases, it might be acceptable to exclude observations with missing data, particularly if the proportion of missing data is small and does not significantly impact the analysis. However, in other cases, excluding missing data might lead to biased results or loss of valuable information. For example, in a clinical trial, excluding patients with missing follow-up data could lead to an overestimation of the treatment effect if the missing data is related to the severity of the patient's condition. In such cases, it might be more appropriate to use techniques such as multiple imputation or pattern analysis to account for the missing data and to ensure that the results are robust and reliable.

Missing values are a pervasive challenge in data analysis and have significant implications for machine learning and data governance. Properly managing missing values during the EDA phase is crucial for ensuring data quality, model reliability, and compliance with regulatory requirements. By understanding the nature of the missing data, selecting appropriate imputation methods, and considering the impact of missing values on model interpretability and fairness, organizations can ensure that their ML models are built on a solid foundation of high-quality data. Advanced techniques such as multiple imputation, model-based imputation, and pattern analysis offer powerful tools for handling missing data, but these techniques must be used judiciously and in accordance with the overall goals and objectives of the ML project. By prioritizing the careful management of missing values, data governance frameworks can help organizations mitigate risks, enhance model performance, and ensure that their ML initiatives are both ethical and compliant with legal and regulatory standards.

Cardinality

Cardinality, a fundamental concept in data analysis, refers to the uniqueness of data values in a column or feature within a dataset. Understanding and managing cardinality during the Data Profiling phase is critical for machine learning (ML) and data governance. High or low cardinality in data can significantly influence model performance, data quality, and the interpretability of the results. From a data governance perspective, effectively addressing cardinality is essential to ensuring robust, fair, and compliant ML models.

In the context of ML, cardinality can be categorized into low, medium, and high cardinality. Low cardinality refers to columns or features with a limited number of unique values relative to the total number of records. An example of low cardinality data might be a "gender" column with only two possible values: "Male" and "Female." In contrast, high cardinality refers to columns with a large number of unique values, such as a "user ID" or "transaction ID" in a financial dataset, where each value might be unique to a single record. High cardinality features pose particular challenges in ML, as they can lead to overfitting, increased computational complexity, and difficulties in feature engineering. From a data governance perspective, managing high cardinality is crucial to maintaining model performance and ensuring that the data is used effectively and responsibly.

One of the primary concerns with high cardinality features is their potential to cause overfitting in ML models. Overfitting occurs when a model becomes too complex, capturing noise rather than the underlying data patterns. High cardinality features can exacerbate this issue, as the model may learn to rely on these unique identifiers rather than generalizable trends in the data. For example, in a dataset containing customer IDs, if the model overfits to these high cardinality features, it might perform well on the training data but poorly on unseen data, as it has learned specific customer ID patterns rather than broader customer behaviors. Data governance practices must include guidelines for identifying and mitigating the risks of overfitting due to high cardinality. This might involve techniques such as dimensionality reduction, feature hashing, or even the removal of high cardinality features if they do not contribute meaningfully to the model's predictive power.

Dimensionality reduction techniques, such as Principal Component Analysis (PCA) or Singular Value Decomposition (SVD), are often employed to address the challenges posed by high cardinality features. These methods reduce the number of features in a dataset by transforming the original features into a smaller set of new features that retain most of the original variance. For example, in a high cardinality dataset with thousands of unique product IDs, PCA could be used to create a few principal components that capture the most significant variations in product characteristics, thereby reducing the complexity of the model. However, from a data governance perspective, it is important to ensure that the application of dimensionality reduction techniques is well-documented and transparent. Stakeholders must understand how these techniques impact the interpretability of the model and whether the reduced features still align with the original business objectives.

Another approach to managing high cardinality is feature hashing, also known as the "hashing trick." Feature hashing converts high cardinality categorical variables into a fixed number of columns, reducing the feature space and computational requirements. This technique is particularly useful when dealing with text data, such as in Natural Language Processing (NLP) tasks, where each word or phrase can be considered a high cardinality feature. For instance, in a sentiment analysis model analyzing customer reviews, feature hashing can reduce the dimensionality of the text data by mapping each word to a fixed-size vector, making the model more scalable and efficient. However, feature hashing introduces the possibility of hash collisions, where different values map to the same hash, potentially leading to loss of information or interpretability. Data governance frameworks should therefore include considerations for when and how feature hashing is applied, ensuring that the trade-offs between efficiency and accuracy are carefully evaluated.

Low cardinality features, while generally easier to manage, also present specific challenges in ML and data governance. For instance, in a binary classification task, low cardinality features might lead to models that are too simplistic or fail to capture the nuances in the data. Consider a healthcare dataset where the "smoker" variable is a low cardinality feature with only two values: "Yes" or "No." If the ML model relies heavily on this feature, it might overlook other important variables, such as age or comorbidities, leading to biased or incomplete predictions. From a data governance perspective, it is crucial to ensure that low cardinality features are used appropriately and that their impact on the model is understood in the context of the overall data landscape. This might involve feature interaction analysis, where the relationships between low cardinality features and other variables are explored to identify potential interactions or dependencies that could improve model accuracy.

The concept of cardinality also plays a significant role in feature engineering, particularly when creating new features or transforming existing ones. In some cases, high cardinality features can be binned or grouped into categories to reduce cardinality and improve model performance. For example, in a retail dataset, individual product IDs might be grouped into broader product categories, reducing the cardinality of the feature while preserving important information about product types. However, binning must be done carefully to avoid introducing bias or losing valuable information. Data governance frameworks should provide guidance on best practices for binning and grouping high cardinality features, ensuring that these transformations are applied consistently and aligned with the business goals.

An advanced topic related to cardinality is the use of embedding techniques, particularly in deep learning models. Embeddings are dense vector representations of high cardinality categorical features that capture the relationships between different categories in a lower-dimensional space. For example, in a recommendation system for an e-commerce platform, product IDs and user IDs can be converted into embeddings, allowing the model to learn complex relationships between users and products based on past interactions. Embeddings are particularly powerful because they enable the model to handle high cardinality features more effectively, improving both scalability and accuracy. However, from a data governance perspective, the use of embeddings introduces additional complexity, as the relationships learned by the model may not be easily interpretable or transparent. Governance practices should include protocols for monitoring and validating embeddings, ensuring that the learned relationships are meaningful and do not introduce unintended biases.

Cardinality also has implications for data privacy and security, particularly in the context of high cardinality features that can be used to uniquely identify individuals. For instance, in a dataset containing social security numbers, each value is unique to an individual, making it a high cardinality feature with significant privacy risks. If such data is inadvertently exposed or misused, it could lead to severe privacy breaches and legal consequences. Data governance frameworks must therefore include strict guidelines for handling high cardinality features that contain personally identifiable information (PII). This might involve techniques such as anonymization, where high cardinality features are replaced with pseudonyms or encrypted to protect individual identities, or differential privacy, where noise is added to the data to prevent re-identification while preserving the overall distribution of values.

The handling of cardinality is closely tied to regulatory compliance, particularly in industries such as finance and healthcare, where strict regulations govern how data is managed and used. For example, the General Data Protection Regulation (GDPR) in Europe imposes stringent requirements on how personal data, including high cardinality features, is processed and stored. Data governance frameworks should ensure that all aspects of cardinality management, from feature engineering to data storage, comply with relevant legal and regulatory requirements. This might involve regular audits, documentation of data processing activities, and the implementation of data minimization principles, where only the necessary data is collected and processed to achieve the intended purpose.

EDA for Deeper Insights

Exploratory data analysis (EDA) provides a framework for understanding the data landscape, enabling the identification of data quality issues that could undermine the reliability of ML models. Data governance emphasizes the importance of data quality, and EDA plays a vital role in this by highlighting inconsistencies, missing values, and outliers that need to be addressed. For instance, consider a healthcare dataset used to predict patient outcomes. During EDA, a data scientist might discover that certain patient demographic fields, such as ethnicity or age, contain a high percentage of missing values. If not managed correctly, these gaps could introduce bias into the model, particularly if the missing data is not randomly distributed. Addressing these issues during the EDA phase ensures that the dataset is complete and representative, which is crucial for building fair and accurate models. Data governance frameworks should mandate comprehensive EDA to guarantee that datasets meet the necessary standards for quality and completeness before proceeding to the modeling phase.

One of the key aspects of EDA is data visualization, which transforms raw data into interpretable charts, graphs, and plots that reveal underlying trends and patterns. Visualization tools, such as histograms, scatter plots, and box plots, are essential for identifying relationships between variables and detecting outliers. For example, in an e-commerce dataset analyzing customer purchase behavior, scatter plots might reveal a strong correlation between the frequency of purchases and customer age, indicating that older customers tend to shop more frequently. From a data governance perspective, these visualizations not only aid in understanding the data but also provide a transparent means of communicating findings to stakeholders. This transparency is critical for ensuring that all parties involved in the ML process, including data scientists, governance professionals, and business leaders, have a clear understanding of the data's characteristics and potential issues.

Advanced EDA techniques, such as multivariate analysis and dimensionality reduction, offer deeper insights into complex datasets. Multivariate analysis examines the relationships between three or more variables simultaneously, providing a more comprehensive view of the data. For instance, in a financial dataset used to predict loan defaults, multivariate analysis could uncover interactions between variables such as income, credit score, and employment status that might not be apparent in a univariate or bivariate analysis. These interactions can be critical for identifying high-risk borrowers and improving the accuracy of the predictive model. From a data governance standpoint, it is essential to document these relationships and ensure that the model's assumptions are based on a thorough understanding of the data's multidimensional nature.

Dimensionality reduction techniques, such as Principal Component Analysis (PCA), are used to reduce the number of variables in a dataset while preserving as much of the original information as possible. This is particularly useful when dealing with high-dimensional data, where the sheer number of features can lead to overfitting and reduced model performance. PCA, for example, transforms the original variables into a smaller set of uncorrelated components that capture the maximum variance in the data. In a marketing dataset with hundreds of variables related to customer behavior, PCA might reduce the dataset to a handful of principal components that still explain most of the variation in customer preferences. From a data governance perspective, the use of dimensionality reduction techniques must be carefully monitored and documented, as these transformations can impact the interpretability and transparency of the model. Governance frameworks should include guidelines for selecting and validating dimensionality reduction methods, ensuring that the reduced dataset remains relevant and meaningful for the ML task.

EDA also plays a crucial role in feature engineering, the process of creating new features or modifying existing ones to improve model performance. Effective feature engineering requires a deep understanding of the data, which EDA provides by highlighting the relationships between variables and identifying key predictors. For instance, in a retail dataset, EDA might reveal that the day of the week has a significant impact on sales volume, leading to the creation of a new feature that captures this temporal trend. From a data governance perspective, feature engineering must be conducted in a way that preserves the integrity and relevance of the data. This involves not only creating new features but also evaluating their impact on the model and ensuring that they do not introduce bias or overfitting. Governance frameworks should provide best practices for feature engineering, including guidelines for testing and validating new features to ensure that they contribute positively to the model's performance.

Handling categorical data during EDA is another area where data governance plays a critical role. Categorical variables, such as gender, region, or product type, require special treatment to ensure that they are correctly represented in the model. EDA techniques, such as frequency analysis and cross-tabulation, are used to examine the distribution of categorical variables and their relationships with other variables. For example, in a survey dataset, cross-tabulation might reveal that respondents from different regions have significantly different preferences for certain products, which could inform marketing strategies. From a governance perspective, it is important to ensure that categorical data is properly encoded and that the chosen encoding method aligns with the model's objectives. This might involve one-hot encoding, label encoding, or more advanced techniques like target encoding, depending on the nature of the data and the requirements of the ML model. Governance frameworks should include protocols for handling categorical data, ensuring that these variables are accurately and fairly represented in the model.

Outlier detection is another critical component of EDA that has significant implications for data governance. Outliers are data points that deviate significantly from the rest of the dataset and can skew the results of an ML model if not properly managed. EDA techniques, such as box plots, Z-scores, and clustering algorithms, are used to identify and analyze outliers. For example, in a sales dataset, a few transactions with unusually high values might be identified as outliers. These outliers could be the result of data entry errors, fraudulent activity, or legitimate but rare events. From a data governance perspective, it is important to establish protocols for handling outliers, including criteria for determining whether to remove, adjust, or retain them in the dataset. Governance frameworks should ensure that outlier detection and treatment are conducted in a transparent and consistent manner, with clear documentation of the decisions made and their rationale.

EDA also provides the foundation for addressing data privacy and security concerns, which are critical components of data governance. During EDA, data scientists may encounter sensitive information, such as personally identifiable information (PII) or financial data, that requires special handling to comply with privacy regulations like the General Data Protection Regulation (GDPR) or the California Consumer Privacy Act (CCPA). For example, in an HR dataset used to predict employee turnover, EDA might reveal that certain features, such as social security numbers or medical records, contain sensitive information that must be protected. Data governance frameworks should include guidelines for anonymizing or encrypting sensitive data during EDA, ensuring that privacy is maintained while still allowing for meaningful analysis. Additionally, governance practices should address the need for secure storage and access controls, preventing unauthorized access to sensitive data during the EDA process.

Another advanced topic related to EDA in the context of ML data governance is the concept of data drift and concept drift. Data drift occurs when the statistical properties of the data change over time, while concept drift refers to changes in the underlying relationships between the input features and the target variable. EDA can help detect these drifts by monitoring the distribution of variables and the performance of the model over time. For instance, in a credit scoring model, EDA might reveal that the average income of loan applicants has increased significantly over the past year, indicating a potential data drift. Similarly, a decrease in the model's accuracy might suggest concept drift, where the relationship between income and creditworthiness has changed. From a data governance perspective, it is crucial to establish protocols for monitoring and addressing data and concept drift, ensuring that the model remains accurate and relevant in a changing environment. This might involve retraining the model with updated data, adjusting the model's parameters, or implementing automated drift detection mechanisms.

EDA also contributes to model interpretability, which is a key concern in ML data governance. Understanding the relationships between variables and their impact on the model's predictions is essential for building transparent and explainable models. Techniques such as partial dependence plots, feature importance analysis, and SHAP (SHapley Additive exPlanations) values are used during EDA to assess the influence of individual features on the model's output. For example, in a fraud detection model, EDA might reveal that certain features, such as the time of day or the transaction amount, have a strong influence on the model's predictions. From a data governance perspective, it is important to ensure that these insights are documented and communicated to stakeholders, providing transparency into how the model makes decisions. Governance frameworks should include guidelines for conducting and reporting on interpretability analyses, ensuring that models are not only accurate but also explainable and trustworthy.

Data Cleaning and Transformation

Data cleaning and transformation processes ensure that the raw data collected from various sources is refined and structured in a way that makes it suitable for ML model training. The importance of data cleaning and transformation lies in their ability to identify and correct errors, remove inconsistencies, and standardize data formats, thereby preventing the propagation of faulty or misleading information through the ML pipeline. Effective data governance requires that these tasks are not only performed rigorously but also documented and monitored to ensure transparency and accountability.

Data cleaning involves a range of activities designed to improve the quality of the data. This includes handling missing values, correcting inaccuracies, and removing duplicates, which are common issues in real-world datasets. For example, in a financial dataset, missing transaction amounts or duplicate records could significantly skew the analysis and lead to incorrect predictions if not properly addressed. Data governance frameworks establish the protocols and standards for these cleaning activities, ensuring that the processes are consistent, repeatable, and aligned with the overall goals of the organization. This governance ensures that the data feeding into ML models is reliable and of high quality, which is essential for building models that are both accurate and trustworthy.

Transformation is the process of converting data into a format that is more suitable for analysis and modeling. This can include scaling numerical data, encoding categorical variables, and creating new features from existing data. These transformations are vital for ensuring that the data aligns with the requirements of the specific ML algorithms being used. For instance, some algorithms require that features are on a similar scale, necessitating normalization or standardization. Data governance plays a key role in overseeing these transformations, providing guidelines for how data should be transformed, and ensuring that these steps are performed consistently across different datasets and projects. By governing the data cleaning and transformation processes, organizations can maintain a high standard of data quality, leading to more reliable and interpretable ML models.

Handling Duplicates

Duplicate values occur when the same data record appears multiple times within a dataset. This can happen for various reasons, including data entry errors, system glitches, or the merging of multiple datasets. For instance, in a customer database, a single individual might appear multiple times if their information was entered more than once due to a typo in the name or address. In a transactional dataset, a single transaction might be recorded multiple times due to a system error during data capture. These duplicates, if not handled appropriately, can distort the analysis, leading to overestimation of the frequency or importance of certain events, which can significantly impact the performance of ML models.

One of the key challenges in handling duplicates is identifying them accurately. While exact duplicates, where every field in the record is identical, are relatively straightforward to detect and remove, partial or near duplicates present a more complex problem. Partial duplicates occur when records are nearly identical but have slight differences in one or more fields. For example, two entries for the same customer might differ only by a misspelled name or an abbreviated address. Identifying and resolving such partial duplicates requires sophisticated techniques that go beyond simple exact matching.

From a data governance perspective, the process of identifying duplicates must be systematic and methodical. This involves establishing rules and criteria for what constitutes a duplicate and developing algorithms to detect duplicates based on these criteria. For example, in a dataset containing customer records, duplicates might be identified by comparing key fields such as name, address, and phone number. However, slight variations in these fields, such as different spellings or formatting, can complicate the detection process. To address this, data governance frameworks can incorporate fuzzy matching techniques, which allow for approximate matches between records. Fuzzy matching algorithms calculate the similarity between strings, taking into account potential typographical errors, phonetic similarities, and other common variations. For instance, the Levenshtein distance algorithm measures the number of single-character edits required to change one string into another, providing a useful metric for identifying near duplicates.

Once duplicates have been identified, the next step is to determine how to handle them. In some cases, duplicates can be simply removed, especially when they are exact duplicates and add no new information to the dataset. For example, in a sales dataset where each transaction is recorded with a unique transaction ID, any duplicate records with identical transaction IDs can be safely removed without losing any valuable data. However, when dealing with partial duplicates, the decision on how to handle them becomes more complex. Simply removing one of the duplicates may result in the loss of important information, while retaining both duplicates could skew the analysis.

One approach to handling partial duplicates is to merge them into a single, consolidated record. This process, known as deduplication or record linkage, involves combining the information from the duplicate records to create a comprehensive and accurate representation of the data. For example, if two customer records differ only in the spelling of the name but agree on all other fields, the records can be merged by selecting the most accurate or complete version of the name. In more complex cases, where different fields in the duplicate records contain different information, data governance frameworks can provide guidelines for resolving these conflicts. This might involve establishing rules for prioritizing certain fields, using external reference data to verify the information, or applying machine learning models to predict the most likely correct value based on the available data.

The process of merging duplicates can be particularly challenging when dealing with high-dimensional data, where records contain a large number of fields. In such cases, resolving conflicts between different fields in duplicate records requires a systematic approach that balances the need for accuracy with the complexity of the data. Data governance frameworks can include protocols for handling high-dimensional duplicates, such as the use of weighted scoring systems that assign different levels of importance to different fields based on their relevance to the analysis. For instance, in a financial dataset, fields related to transaction amounts and dates might be given higher priority when resolving duplicates than fields related to customer demographics, which might be less critical to the analysis.

Duplicates in the training data can lead to overfitting, where the model learns to recognize and rely on patterns that are specific to the duplicated data rather than generalizable trends in the overall dataset. For example, if a dataset used to train a fraud detection model contains multiple identical records of fraudulent transactions, the model might overestimate the likelihood of certain types of fraud, leading to an increased rate of false positives. To mitigate this risk, data governance frameworks should include guidelines for detecting and removing duplicates not only in the training data but also in the validation and test datasets. This ensures that the model is trained and evaluated on a clean, representative dataset that accurately reflects the real-world data it will encounter in production.

Handling duplicates also has significant implications for data privacy and security, particularly in regulated industries such as finance, healthcare, and telecommunications. Duplicates can increase the risk of data breaches, as multiple copies of the same sensitive information may be stored in different locations, making it more difficult to secure and monitor. For example, in a healthcare dataset containing patient records, duplicates could result in multiple copies of a patient's medical history being stored in different systems, increasing the risk of unauthorized access or accidental disclosure. Data governance frameworks should include protocols for securely managing and deduplicating sensitive data, ensuring that only a single, consolidated version of each record is stored and that appropriate access controls and encryption methods are applied.

Data governance also plays a crucial role in ensuring compliance with data privacy regulations, such as the General Data Protection Regulation (GDPR) in Europe and the California Consumer Privacy Act (CCPA) in the United States. These regulations impose strict requirements on how personal data is collected, stored, and processed, including provisions for data minimization and the right to be forgotten. Duplicates can complicate compliance with these regulations, as multiple copies of the same data can make it more difficult to identify and delete all instances of a record when requested by the data subject. For example, if a customer requests that their data be deleted under GDPR, the organization must ensure that all duplicates of the customer's data are removed from all systems. Data governance frameworks should include guidelines for identifying and managing duplicates in a way that facilitates compliance with data privacy regulations, including the implementation of automated tools for detecting and removing duplicates across different systems and databases.

Handling duplicates also has implications for data quality and integrity, which are critical components of data governance. Duplicates can introduce inconsistencies and inaccuracies into the dataset, leading to erroneous conclusions and undermining the reliability of the analysis. For example, in a marketing dataset used to analyze customer preferences, duplicates could result in the overrepresentation of certain customer segments, skewing the results and leading to incorrect decisions about marketing strategies. Data governance frameworks should include protocols for regularly auditing datasets for duplicates and ensuring that any duplicates are handled in a way that preserves data quality and integrity. This might involve implementing automated data quality checks that flag potential duplicates for review or using machine learning models to predict the likelihood of duplicates based on patterns in the data.

Organizations often need to integrate data from multiple sources, such as different departments, business units, or external partners. However, integrating data from disparate sources can introduce duplicates, particularly when different systems use different identifiers or naming conventions for the same entities. For example, in a retail organization, customer data from the online store and the physical store might be stored in separate databases with different customer IDs, leading to duplicates when the data is merged. Data governance frameworks should include guidelines for managing duplicates during data integration, such as the use of standard identifiers, cross-referencing with external reference data, or implementing data deduplication tools that can automatically identify and merge duplicates across different systems.

It is important to recognize that in some cases, duplicates may be intentionally introduced into a dataset as part of the data collection or processing workflow. For example, in A/B testing, where different versions of a product or service are tested with different groups of users, duplicates may be introduced to ensure that each group receives the same treatment. In such cases, data governance frameworks should include protocols for documenting and managing these intentional duplicates, ensuring that they are clearly identified and do not interfere with the analysis or lead to incorrect conclusions.

Managing Outliers

Outliers, which are data points that significantly deviate from the majority of observations, present substantial challenges in machine learning (ML). These anomalies can heavily influence the outcomes of models, leading to distorted predictions, diminished accuracy, and potentially faulty decision-making. Within the framework of data governance, the management of outliers is a crucial element of data cleaning and transformation, ensuring that the integrity of data is upheld, that models perform as expected, and that compliance with regulatory standards is maintained.

Outliers may emerge in datasets for a variety of reasons, such as errors during data entry, malfunctioning equipment, instances of fraud, or naturally occurring rare events. For instance, an unusually large transaction in a financial dataset might be due to an error in data entry, or it could signify a fraudulent transaction. In healthcare, an abnormally high or low patient vital sign might indicate a measurement error or an actual critical health event. When outliers are not properly managed, they have the potential to distort the outcomes of statistical analyses and machine learning models. Addressing these anomalies within a structured data governance framework is essential to ensuring data quality, improving model accuracy, and maintaining the integrity of analytical processes.

Identifying outliers is the first critical step in managing them. Simple statistical methods like identifying points beyond three standard deviations from the mean can be effective for normally distributed data, but such methods may not suffice in more complex datasets. Real-world data often do not follow a normal distribution, and outliers may not be immediately evident. For example, in skewed distributions like income data, where a small number of individuals earn substantially more than the majority, traditional methods may fail to identify true outliers. More sophisticated techniques such as the interquartile range (IQR) method or robust statistical methods like Median Absolute Deviation (MAD) can provide more accurate detection, especially in skewed or multidimensional datasets where outliers might exist in combinations of features rather than isolated dimensions.

Deciding how to handle outliers once they are identified involves several potential strategies, each carrying its own benefits and drawbacks. Removing outliers from the dataset is a common method, especially when these data points are suspected to be errors or anomalies that do not represent genuine underlying patterns. For instance, in a retail sales dataset, a record showing an implausibly high number of items sold in a single transaction might be an error and could be removed to avoid skewing the analysis. However, the decision to remove outliers should be made carefully, as doing so can also result in the loss of valuable information, particularly when the outliers represent rare but significant events, such as detecting fraud in financial transactions.

Transforming outliers into less extreme values through winsorization is another approach. This method involves capping the extreme values at a specified percentile, thereby reducing their impact on the analysis. In a dataset of employee salaries, for example, outliers at the high end might be adjusted to a more reasonable maximum salary, mitigating the influence of these extreme values on the model. While winsorization can reduce the distortion caused by outliers, it also changes the original data, a transformation that must be justified and documented within the data governance framework.

Modeling outliers separately is a more nuanced approach that involves treating these data points as a distinct subset requiring separate analysis. Rather than removing or altering outliers, this method involves analyzing them independently to understand their unique characteristics. In a dataset used for predicting customer churn, for example, outliers representing customers with extremely high spending might be modeled separately to better understand their behavior and loyalty. This strategy preserves the information contained in the outliers while preventing them from distorting the primary analysis. However, it necessitates careful consideration and documentation within the data governance framework, as it involves managing and maintaining separate models or analyses for different data subsets.

Outliers can significantly influence model training and validation, often leading to overfitting, where the model becomes too complex and captures noise rather than meaningful patterns. For example, in a housing price prediction model, outliers representing luxury properties might cause the model to overestimate prices for more typical properties. Data governance frameworks should incorporate guidelines to mitigate the risk of overfitting due to outliers. Robust modeling techniques less sensitive to outliers, such as decision trees or ensemble methods like Random Forests or Gradient Boosting Machines, are often effective. Additionally, applying regularization techniques, such as Lasso or Ridge regression, can help penalize extreme coefficients resulting from outliers, thereby reducing their impact on the model.

Anomaly detection techniques are particularly relevant in the context of managing outliers, especially in unsupervised learning settings where labeled data is not available. These algorithms are designed to identify unusual patterns in the data that could indicate anomalies. In a network security application, for instance, anomaly detection might be used to identify irregular network traffic patterns that suggest a cyberattack. Similarly, in a manufacturing context, it could be used to detect equipment failures or production defects by identifying outliers in sensor data. Implementing anomaly detection requires careful calibration and validation to ensure the algorithm effectively identifies true anomalies without generating excessive false positives or negatives. This involves setting appropriate thresholds for what constitutes an anomaly and continuously monitoring the algorithm's performance as the data and environment change.

Data privacy and security concerns are also paramount when managing outliers, especially in the context of sensitive or personal data. Outliers, in some cases, can be used to identify individuals or reveal private information, even in anonymized datasets. For instance, in a dataset containing health records, an outlier representing a rare medical condition could potentially be traced back to a specific individual, thus compromising their privacy. Data governance frameworks must address these risks by implementing measures that protect the privacy and confidentiality of the individuals represented in the data. Techniques such as differential privacy can be employed to ensure that outliers do not inadvertently lead to re-identification. Compliance with relevant data protection regulations, such as the General Data Protection Regulation (GDPR), is essential in these contexts.

Ethical considerations are deeply intertwined with the management of outliers in machine learning. Decisions about whether to exclude or transform outliers can lead to biased or unfair outcomes, particularly if the outliers represent marginalized or minority groups. For example, a hiring algorithm that excludes outliers in employment history or education might disproportionately affect candidates from non-traditional backgrounds, resulting in biased hiring decisions. Data governance frameworks should include protocols for evaluating the ethical implications of managing outliers, ensuring that these practices do not lead to discrimination or unfairness. Bias audits can be conducted to assess the impact of outlier treatment on various demographic groups, and fairness-aware modeling techniques can be implemented to explicitly account for the potential bias introduced by outliers.

The presence of outliers can complicate data integration and interoperability, especially when integrating data from multiple sources that use different standards or measurement units. In global organizations, data from different regions might be recorded in various currencies or units of measurement, leading to outliers when the data is merged. To address this, data governance frameworks should include protocols for managing outliers during data integration. Standardization techniques can harmonize data from different sources, while data validation checks can identify and resolve potential outliers before they impact the analysis.

Outliers, in some cases, provide valuable insights that should be preserved rather than removed or transformed. In financial risk management, for example, outliers representing extreme market events or financial crises may be crucial for understanding and mitigating risk. In scientific research, outliers might signify novel discoveries or unexpected results that challenge existing theories. Data governance frameworks should include guidelines for identifying and preserving valuable outliers, ensuring they are not inadvertently discarded during data cleaning and transformation processes. This might involve creating separate analyses or models for outliers or employing specialized statistical techniques to study the properties and implications of these anomalies in greater depth.

Effective management of outliers is crucial in the machine learning pipeline, significantly impacting data quality, model performance, privacy, compliance, and ethical considerations. From a data governance perspective, addressing outliers requires a comprehensive, systematic approach that goes beyond mere detection and removal. By implementing advanced techniques for handling outliers, ensuring compliance with data privacy regulations, and preserving the integrity and fairness of the analysis, data governance frameworks help organizations build robust, reliable, and trustworthy machine learning models. These models, in turn, deliver accurate and actionable insights while adhering to the highest standards of data quality, compliance, and ethical responsibility.

Processing Textual Data

Textual data, unlike structured numerical data, is inherently messy and varied. It comes from diverse sources such as social media, customer reviews, emails, legal documents, and academic papers, each with its own formatting, language style, and nuances. This variability makes it challenging to apply uniform processing techniques. One significant challenge is the need to standardize the text data. Standardization involves converting the text into a consistent format, which typically includes lowercasing all words, removing punctuation, and handling contractions. For example, converting "I've" to "I have" or "don't" to "do not" ensures consistency in the dataset, which is critical for accurate ML model training. Standardization also helps in reducing the dimensionality of the data, which is particularly important when dealing with large text corpora.

Text normalization is another critical aspect of processing textual data. Normalization involves reducing words to their base or root form, a process known as stemming or lemmatization. Stemming, which removes prefixes and suffixes to reduce words to their root form, can sometimes lead to words that are not linguistically accurate but still useful for certain applications. For instance, "running," "runner," and "ran" might all be reduced to "run" using stemming. Lemmatization, on the other hand, considers the morphological analysis of words and returns a valid word form, such as converting "running" to "run" or "better" to "good." The choice between stemming and lemmatization depends on the specific ML application and the need for linguistic accuracy versus processing speed. From a data governance perspective, it is essential to document and justify the choice of normalization techniques to ensure that they align with the intended use of the data and the goals of the ML project.

Handling stopwords is another crucial step in processing textual data. Stopwords are common words such as "the," "is," "in," and "and," which generally do not contribute significant meaning to the text and can be removed to reduce noise in the data. However, the decision to remove stopwords is not always straightforward. In some contexts, stopwords can carry important semantic information. For example, in sentiment analysis, the word "not" in "not happy" is crucial to understanding the sentiment of the text. Removing such stopwords could lead to a loss of critical information and skew the results of the analysis. Therefore, a data governance framework must establish clear guidelines for handling stopwords, ensuring that their removal is contextually appropriate and does not compromise the integrity of the data.

Text data often contains typos, slang, abbreviations, and other informal language that need to be cleaned and standardized before analysis. This process, known as text cleaning, involves correcting spelling errors, expanding abbreviations, and translating slang into standard language. For instance, "u" might be expanded to "you," or "gr8" to "great." The challenge lies in maintaining the original meaning of the text while cleaning it. Text cleaning is particularly important in datasets derived from social media, where informal language is common. Inaccurate cleaning can lead to misinterpretation of the data, affecting the outcomes of the ML model. Data governance plays a vital role in ensuring that text cleaning processes are rigorous, consistent, and well-documented, with clear rules on how different types of informal language should be handled.

Sentiment analysis is a common application of textual data processing in ML, where the goal is to determine the sentiment or emotional tone behind a body of text. Sentiment analysis relies heavily on the quality and accuracy of the processed textual data. For example, in analyzing customer reviews, accurately identifying and processing phrases such as "not bad" or "hardly disappointing" is critical to correctly assessing the sentiment. Misinterpretation or incorrect processing of such phrases can lead to inaccurate sentiment classification, affecting business decisions based on the analysis. Data governance frameworks should include guidelines for ensuring the accuracy of sentiment analysis models, including the careful handling of negations, sarcasm, and other linguistic nuances that can affect sentiment.

Tokenization is another essential process in preparing textual data for ML. Tokenization involves splitting text into individual words or tokens, which are then analyzed separately. For instance, the sentence "Machine learning is powerful" would be tokenized into ["Machine", "learning", "is", "powerful"]. The choice of tokenization method can significantly impact the performance of the ML model. Word-level tokenization, character-level tokenization, and subword tokenization are different approaches, each with its own advantages and drawbacks. For example, word-level tokenization may struggle with rare or compound words, while character-level tokenization may result in a high-dimensional feature space that complicates model training. Subword tokenization, which breaks down words into meaningful subunits, can address these issues but requires more complex processing. Data governance frameworks should specify the tokenization methods used and ensure that these methods are appropriate for the specific use case.

Named entity recognition (NER) is a specialized task in processing textual data where the goal is to identify and classify proper nouns or entities within the text, such as names of people, organizations, locations, and dates. For example, in a legal document, NER might identify "John Doe," "Acme Corporation," and "January 1, 2020" as distinct entities. This process is particularly important in applications like document classification, information retrieval, and knowledge extraction. However, NER models can be sensitive to the quality of the input data. Inaccurate or inconsistent text processing can lead to misidentification of entities, which can significantly affect the accuracy of the model. From a data governance perspective, it is crucial to ensure that the text processing steps leading up to NER are thoroughly vetted and that the NER process itself is subject to rigorous validation and monitoring.

Dealing with multilingual data is another challenge in processing textual data for ML. In today's globalized world, datasets often contain text in multiple languages, requiring language-specific processing techniques. For instance, tokenization rules, stopword lists, and stemming algorithms vary significantly across languages. Processing English text using rules designed for German or Chinese text would likely result in poor model performance. Data governance frameworks must ensure that appropriate language-specific processing techniques are applied and that the language of the text is correctly identified and tagged before processing. This is especially important in applications like customer service, where multilingual support is often required, and misprocessing can lead to misunderstandings and customer dissatisfaction.

Sentiment and emotion detection models that process textual data must also account for cultural and contextual differences in language use. The same word or phrase can have different connotations in different cultures or contexts. For example, the word "cool" might indicate temperature in one context but express approval or admiration in another. Failing to account for these nuances can lead to incorrect interpretations of the text. Data governance frameworks should include protocols for incorporating cultural and contextual knowledge into the text processing pipeline, ensuring that models are trained on diverse datasets that reflect the intended use case.

Processing textual data also involves handling and preserving context. Words and phrases can have different meanings depending on their context, and this must be carefully managed during the data transformation process. For example, the word "bank" could refer to a financial institution, the side of a river, or even a verb meaning to tilt. Properly handling such ambiguities requires sophisticated context-aware processing techniques, such as word sense disambiguation, which aims to determine the correct meaning of a word based on its surrounding text. Data governance frameworks must ensure that these techniques are applied appropriately and that the transformed data accurately reflects the intended meaning of the text.

The ethical implications of processing textual data cannot be overlooked. Textual data often contains personal, sensitive, or potentially biased information, which must be handled with care to avoid perpetuating stereotypes or violating privacy rights. For example, an ML model trained on biased text data might learn and propagate harmful stereotypes, leading to unfair or discriminatory outcomes. Data governance frameworks should include guidelines for detecting and mitigating bias in textual data, ensuring that data processing techniques do not inadvertently introduce or amplify bias. This might involve implementing fairness-aware preprocessing methods, conducting bias audits on the processed data, and ensuring transparency and accountability throughout the ML pipeline.

Data privacy is another critical concern when processing textual data, especially when dealing with personally identifiable information (PII) or sensitive content. For instance, customer support logs might contain names, addresses, phone numbers, and other sensitive details that need to be anonymized or redacted before being used in an ML model. Failing to adequately protect this information could lead to privacy breaches and legal repercussions, especially under regulations such as the General Data Protection Regulation (GDPR) or the California Consumer Privacy Act (CCPA). Data governance frameworks must include strict protocols for anonymizing or pseudonymizing PII in textual data, ensuring that all legal and ethical requirements are met before the data is processed and analyzed.

Maintaining data lineage and traceability is also essential when processing textual data. Given the complexity and variability of text data, it is crucial to keep detailed records of all transformations applied to the data, from tokenization and normalization to lemmatization and stopword removal. This ensures that the data processing steps are transparent, reproducible, and auditable, allowing data scientists and governance professionals to trace back the origins of any issues that arise in the ML models. Data governance frameworks should mandate comprehensive documentation of all text processing steps, including the rationale for each decision and the specific techniques used. This documentation not only supports compliance with regulatory requirements but also facilitates ongoing monitoring and refinement of the text processing pipeline.

The processing of textual data for machine learning also necessitates careful attention to the management and storage of the processed data. Text data can be voluminous and requires efficient storage solutions that support rapid access and processing. For example, large-scale text corpora used in Natural Language Processing (NLP) applications may need to be stored in distributed systems that allow for parallel processing and scalable storage. Data governance frameworks should ensure that the storage infrastructure is capable of handling the specific demands of textual data processing, including considerations for data security, access control, and scalability. Moreover, the governance framework should address the management of intermediate data products generated during text processing, such as tokenized text or extracted features, ensuring that these products are stored securely and are accessible only to authorized personnel.

Processing textual data often involves the use of pre-trained language models, such as BERT, GPT, or Word2Vec, which are designed to capture the semantic and contextual relationships between words. These models can significantly enhance the performance of ML models by providing rich representations of text data. However, the use of pre-trained models also introduces challenges related to model interpretability, bias, and alignment with the specific use case. Data governance frameworks should include guidelines for the selection, evaluation, and customization of pre-trained language models, ensuring that they are appropriate for the task at hand and do not introduce unwanted biases or inaccuracies. This might involve fine-tuning the models on domain-specific text data, conducting thorough evaluations of model performance, and ensuring that the models are transparent and interpretable to all stakeholders.

Handling Time Series

Time-series data, characterized by observations collected sequentially over time, often includes various anomalies such as missing values, outliers, and non-stationary behavior. These issues must be addressed systematically to ensure that the data is suitable for analysis and modeling. For example, in a financial time-series dataset, missing values might occur due to holidays or technical issues, leading to gaps that can distort trend analysis and forecasting. The presence of such gaps requires careful handling, often involving techniques such as forward or backward filling, interpolation, or even more sophisticated methods like Kalman filtering. These methods aim to preserve the temporal integrity of the data while minimizing the risk of introducing bias or errors. Data governance frameworks must establish clear guidelines for selecting the appropriate method for handling missing values in time-series data, ensuring that the chosen approach aligns with the nature of the data and the objectives of the ML project.

Time-series data often contains outliers, which can arise from data entry errors, sensor malfunctions, or rare events. In financial datasets, an outlier might represent an unusually large transaction due to a one-time event, such as a major investment or market crash. In weather data, an outlier could be an extreme temperature reading caused by an equipment failure. These outliers can significantly impact the accuracy of ML models, particularly in forecasting applications where the model's sensitivity to extreme values may lead to unreliable predictions. Managing outliers in time-series data involves determining whether they should be removed, transformed, or modeled separately. This decision depends on the context of the data and the goals of the analysis. For instance, if the outliers represent genuine events that could recur, such as market crashes, they may need to be retained and modeled explicitly to ensure that the ML model is capable of handling similar events in the future. Data governance frameworks should include protocols for assessing the impact of outliers on time-series data and provide guidelines for their appropriate treatment.

Non-stationarity is another common challenge in time-series data, where the statistical properties of the data, such as mean and variance, change over time. Non-stationary data can complicate the development of ML models, as many algorithms assume stationarity for effective analysis. For example, a time-series dataset of stock prices may exhibit trends, seasonality, or changing volatility, making it non-stationary. To address this, data transformation techniques such as differencing, detrending, or seasonal adjustment may be applied to stabilize the data. These transformations are essential for improving the accuracy and reliability of ML models, but they must be implemented carefully to avoid introducing artifacts or losing important information. Data governance frameworks should ensure that any transformations applied to time-series data are thoroughly documented and that the rationale for these transformations is clearly articulated. This documentation is critical for maintaining the traceability and reproducibility of the data, which are key principles of data governance.

Seasonality, or periodic fluctuations in time-series data, is a critical aspect to consider when handling such data. Many time series exhibit seasonal patterns, such as daily, weekly, or yearly cycles, which can significantly influence the behavior of the data. For example, retail sales data typically shows strong seasonal effects, with spikes during holidays and lulls during off-peak periods. Identifying and adjusting for seasonality is essential for accurate forecasting and analysis. Techniques such as seasonal decomposition, which separates the data into trend, seasonal, and residual components, can be used to isolate and analyze these effects. Data governance frameworks must ensure that the process of identifying and adjusting for seasonality is conducted systematically and that any adjustments made are appropriate for the specific characteristics of the data. This involves not only the technical aspects of seasonal adjustment but also a deep understanding of the domain in which the time-series data is used.

Aggregation is another important process in handling time-series data, particularly when dealing with data collected at different granularities. For instance, data might be collected at minute, hourly, daily, or monthly intervals, depending on the application. Aggregating data to a higher level, such as converting minute-level data to hourly data, can help reduce noise and simplify the analysis. However, aggregation can also lead to a loss of detail, potentially obscuring important patterns or trends in the data. For example, in electricity consumption data, minute-level spikes in usage might be averaged out in hourly data, leading to a loss of information that could be critical for detecting peak demand periods. Data governance frameworks should provide guidance on the appropriate level of aggregation for time-series data, balancing the need for simplicity with the importance of preserving important features of the data. This may involve implementing multiple levels of aggregation and analyzing the data at different granularities to ensure that all relevant patterns are captured.

Time alignment, or the synchronization of time-series data collected from different sources or systems, is another challenge in handling time-series data. In many cases, data from different sources may be recorded with slight differences in timing, leading to misalignment that can affect the accuracy of analysis and modeling. For example, in a smart grid system, electricity usage data from different sensors may be recorded at slightly different times, leading to inconsistencies in the aggregated data. Addressing this issue requires time alignment techniques that adjust the timestamps of the data to ensure that they are properly synchronized. This can involve techniques such as interpolation, time warping, or resampling, depending on the nature of the data and the degree of misalignment. Data governance frameworks should include protocols for time alignment, ensuring that the methods used are appropriate for the specific characteristics of the time-series data and that any adjustments made are carefully documented.

Handling missing time intervals is a critical issue in time-series data, where data points may be missing for certain periods due to various reasons, such as sensor failures, data transmission issues, or deliberate omissions. Missing time intervals can lead to gaps in the data that complicate analysis and modeling. For instance, in a time-series dataset of daily stock prices, missing data for certain days can affect the accuracy of trend analysis and forecasting models. Addressing missing time intervals may involve techniques such as imputation, where the missing values are estimated based on the available data, or more advanced methods such as temporal data augmentation, where synthetic data points are generated to fill in the gaps. Data governance frameworks should provide clear guidelines on how to handle missing time intervals in time-series data, ensuring that the chosen methods are appropriate for the nature of the data and the goals of the analysis.

The handling of time zone differences is another crucial consideration in time-series data, particularly in global applications where data is collected from multiple locations across different time zones. For example, in a global supply chain system, time-series data on shipment times, production schedules, and inventory levels may be recorded in different time zones, leading to inconsistencies and potential errors in analysis. Converting all timestamps to a common time zone, such as Coordinated Universal Time (UTC), can help address this issue, but it also requires careful consideration of the implications for data analysis and interpretation. Data governance frameworks should ensure that time zone conversions are handled consistently and that any adjustments made are clearly documented. This is particularly important in applications where time zone differences can have a significant impact on the results of the analysis, such as in financial markets or international logistics.

Handling time-series data also involves managing data drift, where the statistical properties of the data change over time. Data drift can occur due to changes in the underlying processes that generate the data, such as shifts in consumer behavior, technological advancements, or regulatory changes. For instance, in a time-series dataset of website traffic, data drift might occur due to changes in user preferences or the introduction of new marketing strategies. Detecting and managing data drift is critical for ensuring that ML models remain accurate and reliable over time. This may involve techniques such as rolling window analysis, where the model is periodically retrained on the most recent data, or drift detection algorithms that monitor for changes in the data distribution. Data governance frameworks should include protocols for detecting and managing data drift in time-series data, ensuring that models are regularly updated and validated to account for changes in the data.

The ethical considerations of handling time-series data must also be addressed, particularly in applications where the data represents sensitive or personal information. For example, time-series data on individual health metrics, such as heart rate or activity levels, may contain patterns that reveal personal information or behaviors. Ensuring the privacy and security of this data is critical, especially in light of regulations such as the General Data Protection Regulation (GDPR) or the California Consumer Privacy Act (CCPA). Data governance frameworks should include strict protocols for anonymizing or pseudonymizing time-series data, ensuring that individual identities are protected while maintaining the integrity of the data for analysis and modeling. This may involve techniques such as differential privacy, where noise is added to the data to protect individual privacy while preserving the overall patterns and trends in the data.

The traceability and reproducibility of time-series data processing are essential components of data governance. Given the complexity and variability of time-series data, it is crucial to maintain detailed records of all data cleaning and transformation steps, including the rationale for each decision and the specific methods used. This documentation ensures that the processing steps are transparent, reproducible, and auditable, allowing data scientists and governance professionals to trace back the origins of any issues that arise in the ML models. Data governance frameworks should mandate comprehensive documentation of all time-series data processing steps, ensuring that the data transformation process is fully transparent and that any adjustments made are carefully justified and recorded.

Feature Engineering

Feature engineering begins with the transformation of raw data into a structured format that can be used effectively by machine learning algorithms. This process often involves the creation of new variables or features that capture the underlying patterns in the data. For example, in a dataset containing customer transaction histories, raw data such as transaction dates, amounts, and product categories can be transformed into features like total spend per month, average transaction size, or frequency of purchases in specific categories. These features can provide more relevant and actionable insights for the model, leading to improved predictive performance. However, from a data governance perspective, the creation and use of these features must be carefully managed to ensure that they accurately represent the underlying data and do not introduce biases or errors into the model.

One of the key challenges in feature engineering is ensuring that the features are consistent, reliable, and interpretable. Consistency is particularly important when features are derived from multiple data sources or involve complex calculations. For instance, if a feature is created by combining data from different departments or systems, such as sales and customer service data, it is crucial to ensure that the data from each source is harmonized and that the calculations used to create the feature are consistent across the entire dataset. Any inconsistencies in the data or the feature creation process can lead to misleading results and reduce the reliability of the model. Data governance frameworks should establish protocols for validating and documenting the feature engineering process, ensuring that the features are created in a consistent and transparent manner.

Interpretability is another critical aspect of feature engineering, especially in applications where the decisions made by the ML model have significant consequences, such as in finance, healthcare, or criminal justice. For example, a credit scoring model that uses features derived from customer transaction histories must ensure that these features are interpretable and understandable by human decision-makers. If a feature is too complex or abstract, it may be difficult for stakeholders to understand how it contributes to the model's predictions, leading to a lack of trust in the model. Data governance frameworks should include guidelines for creating interpretable features, ensuring that they are transparent and aligned with the decision-making processes of the organization.

The process of feature selection is also a critical component of feature engineering, involving the identification of the most relevant and informative features for the model. This process can significantly impact the performance of the model, as the inclusion of irrelevant or redundant features can lead to overfitting and reduce the generalizability of the model. For example, in a predictive maintenance model for manufacturing equipment, features related to machine age, usage patterns, and maintenance history might be highly relevant, while features related to employee schedules or weather conditions might be less informative. Selecting the right features requires a deep understanding of the domain and the specific problem being addressed. Data governance frameworks should provide guidelines for feature selection, ensuring that the process is rigorous, transparent, and aligned with the goals of the ML project.

Feature engineering also involves the creation of interaction features, which capture the relationships between different variables in the dataset. For example, in a marketing campaign analysis, an interaction feature might be created by multiplying the number of ad impressions by the number of clicks, capturing the interaction between exposure and engagement. Interaction features can provide valuable insights into the dynamics of the data and improve the predictive power of the model. However, the creation of interaction features must be carefully managed to avoid overcomplicating the model or introducing spurious correlations. Data governance frameworks should include protocols for the creation and validation of interaction features, ensuring that they are based on sound statistical principles and that their inclusion in the model is justified.

Handling categorical features is another important aspect of feature engineering, particularly when dealing with non-numeric data. Categorical features, such as product categories, customer segments, or geographic regions, need to be encoded into a numerical format that can be used by machine learning algorithms. One common method of encoding categorical features is one-hot encoding, where each category is represented by a binary variable. For example, if a dataset contains a feature for customer region with categories such as "North," "South," "East," and "West," one-hot encoding would create four binary variables, one for each region. While one-hot encoding is simple and effective, it can lead to a high-dimensional feature space, especially when dealing with features that have a large number of categories. Data governance frameworks should include guidelines for handling categorical features, ensuring that the chosen encoding method is appropriate for the data and the ML model.

Feature engineering also involves the handling of missing values, which can be a significant challenge in many datasets. Missing data can arise for various reasons, such as data entry errors, incomplete data collection, or system failures. The presence of missing values can significantly impact the performance of the model, leading to biased predictions or reduced accuracy. Several methods can be used to handle missing values during feature engineering, including imputation, where missing values are replaced with estimated values, and deletion, where records with missing values are removed from the dataset. For example, in a customer survey dataset, missing responses might be imputed using the mean or median of the observed data, or the records with missing responses might be removed altogether. The choice of method for handling missing values depends on the specific characteristics of the data and the goals of the ML project. Data governance frameworks should include protocols for handling missing values during feature engineering, ensuring that the chosen method is appropriate, transparent, and well-documented.

Feature engineering for time-series data presents additional challenges, as the temporal aspect of the data must be carefully managed to ensure that the features capture the underlying patterns and trends. Time-series data, such as stock prices, sensor readings, or web traffic, often exhibit trends, seasonality, and autocorrelation, which must be accounted for during feature engineering. For example, in a predictive model for stock prices, features might be created to capture the moving average, volatility, or momentum of the stock over different time periods. These features can provide valuable insights into the temporal dynamics of the data and improve the predictive power of the model. However, the creation of time-series features must be carefully managed to avoid overfitting or introducing spurious correlations. Data governance frameworks should include guidelines for feature engineering with time-series data, ensuring that the features are based on sound statistical principles and that they accurately capture the temporal patterns in the data.

Handling imbalanced data is another critical aspect of feature engineering, particularly in applications where the target variable is imbalanced, such as fraud detection or disease diagnosis. In these cases, the minority class may be underrepresented in the dataset, leading to biased predictions and reduced model accuracy. For example, in a fraud detection model, fraudulent transactions might represent only a small fraction of the total transactions, leading the model to favor the majority class of non-fraudulent transactions. Several techniques can be used to address class imbalance during feature engineering, including resampling methods such as oversampling the minority class or undersampling the majority class, and the creation of synthetic features using techniques such as SMOTE (Synthetic Minority Over-sampling TEchnique). Data governance frameworks should include protocols for handling imbalanced data during feature engineering, ensuring that the chosen method is appropriate, transparent, and well-documented.

Feature engineering for natural language processing (NLP) presents unique challenges, as text data must be transformed into numerical features that can be used by machine learning algorithms. For example, in a sentiment analysis model, features might be created from the frequency of specific words or phrases or from the sentiment score of the text. However, the creation of text features must be carefully managed to ensure that they accurately capture the meaning and context of the text. This might involve the use of advanced techniques such as word embeddings, which represent words as dense vectors in a continuous space, or the use of pre-trained language models such as BERT or GPT. Data governance frameworks should include guidelines for feature engineering with text data, ensuring that the features are based on sound linguistic principles and that they accurately capture the meaning and context of the text.

Feature engineering also involves the management and documentation of the features created during the process. As the number of features increases, it becomes increasingly important to manage and document the features to ensure that they are used correctly and consistently across the ML pipeline. This might involve the use of feature stores, which are centralized repositories for managing and sharing features across different models and projects. Feature stores can help ensure that the features are standardized, validated, and reused consistently across the organization, reducing the risk of errors and improving the efficiency of the ML pipeline. Data governance frameworks should include guidelines for managing and documenting features, ensuring that they are stored, validated, and shared in a way that aligns with organizational standards and best practices.

The use of feature engineering in automated machine learning (AutoML) presents additional challenges, as the process of creating and selecting features is automated and may not be fully transparent or interpretable. For example, in an AutoML platform, the process of feature engineering might be driven by optimization algorithms that select and create features based on their impact on model performance without necessarily considering the interpretability or ethical implications of the features. Data governance frameworks should include guidelines for the use of AutoML, ensuring that the automated feature engineering process is transparent, interpretable, and aligned with the goals and ethical standards of the organization.

The role of feature engineering in model explainability is also critical, particularly in applications where the decisions made by the model have significant consequences. For example, in a credit scoring model, the features used to predict creditworthiness must be interpretable and explainable to human decision-makers, ensuring that they can understand and trust the model's predictions. This might involve the use of Explainable AI techniques, such as SHAP (SHapley Additive exPlanations) or LIME (Local Interpretable Model-agnostic Explanations), which provide insights into the contribution of each feature to the model's predictions. Data governance frameworks should include guidelines for ensuring the explainability of features, ensuring that they are transparent, interpretable, and aligned with the decision-making processes of the organization.

Normalization and Scaling

Normalization involves adjusting the values of features in a dataset so that they fall within a specific range, often between 0 and 1. This process is particularly important in ML because different features in a dataset can have vastly different scales. For instance, in a dataset used for predicting housing prices, the square footage of a house might range from hundreds to thousands, while the number of bedrooms might only range from 1 to 5. If these features are not normalized, the ML model may disproportionately weight the larger values, such as square footage, over smaller but equally important features, like the number of bedrooms. This can lead to biased predictions and reduce the model's overall performance. Data governance frameworks should include guidelines for when and how to apply normalization, ensuring that it is done consistently across datasets and that the choice of normalization method is appropriate for the specific application.

Scaling, on the other hand, involves adjusting the distribution of data so that it has a specific standard deviation, often standardizing data to have a mean of 0 and a standard deviation of 1. This process is crucial for algorithms that are sensitive to the scale of data, such as Support Vector Machines (SVM) and k-nearest neighbors (k-NN). Without proper scaling, these algorithms might perform poorly because they assume that all features are on a similar scale. For example, in a medical dataset where one feature represents the number of hospital visits (a relatively small number) and another represents the cost of treatments (a much larger number), failing to scale these features could lead the model to focus too much on the cost of treatment while ignoring the number of hospital visits, potentially skewing the results. Data governance should ensure that scaling methods are applied systematically and that the rationale behind the choice of scaling technique is well-documented.

One of the challenges in normalization and scaling is the selection of the appropriate method for the specific data and ML model. There are several techniques available, including min–max normalization, z-score standardization, and robust scaling. Min–max normalization transforms the data to fit within a specific range, such as 0 to 1, making it useful when the data needs to be bound within a particular scale. Z-score standardization, on the other hand, adjusts the data based on the mean and standard deviation, making it suitable for datasets with a Gaussian distribution. Robust scaling uses the median and interquartile range, making it more resistant to outliers. Choosing the wrong method can lead to suboptimal model performance or even introduce new biases. Data governance frameworks must provide clear guidelines for selecting the appropriate normalization or scaling technique, considering the nature of the data and the specific requirements of the ML model.

Another critical consideration is the handling of outliers during normalization and scaling. Outliers are extreme values that can significantly skew the results of these processes, leading to distorted feature distributions. For instance, in a financial dataset, outliers in income data, such as extremely high salaries, can cause min–max normalization to compress the range of more typical incomes, reducing the granularity and potentially hiding meaningful differences among middle-income earners. Robust scaling can mitigate this issue by focusing on the interquartile range, but this approach may not always be appropriate, especially if the outliers represent important information rather than noise. Data governance must establish protocols for identifying and handling outliers before applying normalization and scaling, ensuring that the treatment of outliers aligns with the goals of the analysis and the ethical standards of the organization.

Normalization and scaling also have significant implications for the interpretability of ML models. When features are normalized or scaled, their original units and ranges are lost, which can make it more difficult to interpret the model's outputs. For example, if a model is trained on normalized data, the feature coefficients in a linear regression model may not have a clear or intuitive interpretation because they no longer correspond to the original units of measurement. This can be particularly problematic in domains where interpretability is crucial, such as healthcare or finance, where decision-makers need to understand how specific features contribute to the model's predictions. Data governance frameworks should include guidelines for documenting and explaining the effects of normalization and scaling on model interpretability, ensuring that stakeholders can still make sense of the model's predictions even after these transformations have been applied.

The choice of normalization and scaling techniques can also impact the fairness of ML models. In some cases, features related to demographic or socioeconomic factors might require special consideration to ensure that the model does not inadvertently introduce or amplify biases. For instance, normalizing income data in a way that overly compresses the range of low-income individuals could lead to biased predictions in applications like credit scoring or job recruitment, where these features play a significant role. Data governance must ensure that normalization and scaling are applied in a way that promotes fairness, possibly by conducting bias audits before and after these transformations to detect and mitigate any unintended consequences.

Another aspect of normalization and scaling is their impact on data privacy and security. In some cases, the transformed data might reveal patterns or correlations that were not apparent in the raw data, potentially leading to unintended disclosures of sensitive information. For example, in a dataset containing health records, normalizing features related to patient demographics could inadvertently expose patterns that link certain health conditions to specific populations, raising privacy concerns. Data governance should include protocols for assessing the privacy implications of normalization and scaling, ensuring that these transformations do not compromise the confidentiality of the data or violate regulatory requirements such as the General Data Protection Regulation (GDPR) or the Health Insurance Portability and Accountability Act (HIPAA).

Normalization and scaling are also critical in the context of model validation and testing. When datasets are split into training, validation, and test sets, it is important to ensure that normalization and scaling are applied consistently across all sets. Failing to do so can lead to issues where the model performs well on the training data but poorly on the validation or test data because the features are not on the same scale. For example, if min–max normalization is applied separately to the training and test sets, the resulting scales might differ, leading to inconsistent model performance. Data governance frameworks should enforce strict protocols for applying normalization and scaling, ensuring that the same transformations are applied uniformly across all data splits.

The role of normalization and scaling extends beyond individual features to entire datasets, particularly in applications involving data integration or transfer learning. When datasets from different sources or domains are combined, it is crucial to normalize and scale the data to ensure consistency and comparability. For instance, in a transfer learning scenario where a pre-trained model is fine-tuned on a new dataset, the features in the new dataset must be normalized and scaled to match the distributions seen during the initial training. Failure to do so can lead to poor model adaptation and reduced performance. Data governance should provide clear guidelines for managing normalization and scaling in data integration and transfer learning scenarios, ensuring that the data is harmonized and that the model remains effective across different contexts.

The choice of normalization and scaling techniques can also influence the efficiency and scalability of ML models, particularly in large-scale or real-time applications. For instance, certain scaling techniques, such as min–max normalization, might require the entire dataset to be available in memory, which can be impractical for very large datasets. On the other hand, z-score standardization can be more computationally efficient because it only requires the mean and standard deviation, which can be calculated incrementally. In real-time applications, such as online recommendation systems or fraud detection, the normalization and scaling techniques must be efficient enough to handle incoming data streams without introducing significant latency. Data governance should include considerations for the computational efficiency of normalization and scaling, ensuring that the chosen techniques are appropriate for the scale and speed requirements of the application.

Normalization and scaling also play a crucial role in ensuring that ML models remain robust and reliable over time. In many cases, the distributions of features in the data can change over time, a phenomenon known as concept drift. If normalization and scaling are not regularly updated to reflect these changes, the model's performance can degrade. For instance, in a financial application, changes in economic conditions might lead to shifts in income distributions, requiring the normalization parameters to be recalculated. Data governance should enforce protocols for monitoring and updating normalization and scaling parameters, ensuring that the ML model remains accurate and effective in the face of changing data distributions.

Feature engineering is closely tied to normalization and scaling, as the process of creating new features often requires careful consideration of how these features are normalized or scaled. For example, interaction features created by multiplying or dividing existing features might have different scales that need to be aligned through normalization. Similarly, features derived from logarithmic or exponential transformations might require special scaling techniques to ensure that they are on a comparable scale with other features. Data governance should ensure that feature engineering and normalization are integrated processes, with clear guidelines for how new features should be normalized and scaled within the context of the broader ML pipeline.

Normalization and scaling are not one-time processes but ongoing tasks that require continuous monitoring and adjustment. As new data is collected or as the business context evolves, the assumptions underlying the original normalization and scaling techniques might no longer hold. For instance, a normalization strategy that worked well during the initial model development might become less effective as the data evolves, leading to shifts in feature distributions. Data governance should include protocols for regularly reviewing and updating normalization and scaling strategies, ensuring that they remain aligned with the current data landscape and the objectives of the ML project.

The documentation and transparency of normalization and scaling processes are also critical from a data governance perspective. Stakeholders, including data scientists, engineers, and decision-makers, need to understand how these processes have been applied and how they might impact the model's predictions. This is particularly important in regulated industries, where transparency and accountability are paramount. For instance, in healthcare, the normalization and scaling of features related to patient outcomes must be clearly documented to ensure that the model's predictions can be trusted and validated by clinicians. Data governance should enforce strict documentation standards, ensuring that all normalization and scaling steps are thoroughly recorded and that the rationale behind these decisions is transparent and accessible.

Normalization and scaling are also crucial for ensuring that ML models are transferable and interoperable across different systems and platforms. In many organizations, ML models are deployed across multiple environments, such as development, testing, and production, each with its own data characteristics. Ensuring that the data is normalized and scaled consistently across these environments is essential for maintaining model performance and reliability. For instance, a model developed in one country might be deployed in another with different demographic or economic conditions, requiring adjustments to the normalization and scaling strategies to account for these differences. Data governance should provide guidelines for ensuring the transferability and interoperability of ML models, with a focus on maintaining consistency in normalization and scaling across different environments.

Encoding Categorical Variables

Categorical variables come in various forms, including nominal variables, which have no intrinsic order, and ordinal variables, which do have a meaningful order. For instance, a nominal variable could be something like the type of pet (dog, cat, bird), while an ordinal variable might represent customer satisfaction levels (low, medium, high). Converting these categorical variables into a numerical format that preserves their meaning and relevance is essential for ensuring that machine learning models can effectively leverage the information contained within these variables. Data governance frameworks must ensure that the encoding methods used for these variables are appropriate for their nature and that they align with the broader objectives of the ML project.

One common method of encoding categorical variables is one-hot encoding, where each category is transformed into a binary vector. For example, if a dataset contains a feature for the type of vehicle (car, truck, motorcycle), one-hot encoding would create three binary variables, one for each category. While this method is straightforward and effective, it can lead to a significant increase in the dimensionality of the dataset, particularly when dealing with variables that have many categories. This increased dimensionality can pose challenges for model training and performance, as it can lead to overfitting and increased computational complexity. Data governance frameworks should provide guidelines for managing the dimensionality of datasets when using one-hot encoding, ensuring that the chosen approach balances the need for accurate representation of categorical variables with the practical constraints of model performance and scalability.

Label encoding is another method used to convert categorical variables into numerical format. This approach involves assigning a unique integer to each category, such as encoding "dog," "cat," and "bird" as 0, 1, and 2, respectively. While label encoding is simple and efficient, it can introduce unintended ordinal relationships between categories, leading the model to interpret the encoded values as having a natural order, which may not be the case. For example, encoding "car," "truck," and "motorcycle" as 0, 1, and 2 might cause the model to assume that "truck" is somehow greater or less than "car" or "motorcycle," which could skew the model's predictions. Data governance frameworks should ensure that the use of label encoding is carefully considered, particularly for nominal variables where no inherent order exists, to avoid introducing biases or inaccuracies into the model.

Ordinal encoding is particularly relevant when dealing with ordinal variables, where the order of categories is meaningful. For example, customer satisfaction levels (low, medium, high) can be encoded as 0, 1, and 2, preserving the inherent order of the categories. This method is effective in representing the ordinal nature of the data, allowing the model to understand and leverage the relationships between the categories. However, the choice of numerical values assigned to the categories can impact the model's performance and interpretation. For instance, if the difference between "low" and "medium" is not the same as the difference between "medium" and "high," using a simple 0, 1, 2 encoding might not accurately reflect the underlying relationships. Data governance frameworks should provide guidelines for selecting appropriate encoding schemes for ordinal variables, ensuring that the encoded values accurately represent the relationships between categories and align with the objectives of the ML model.

Target encoding is a more advanced technique that involves encoding categorical variables based on the mean of the target variable for each category. For example, in a dataset used for predicting customer churn, a categorical variable representing customer region might be encoded based on the average churn rate for each region. This approach can be particularly powerful when there is a strong relationship between the categorical variable and the target variable. However, target encoding carries the risk of introducing data leakage, where information from the target variable is inadvertently introduced into the model during training, leading to overly optimistic performance estimates. Data governance frameworks should include protocols for mitigating data leakage when using target encoding, such as implementing cross-validation or regularization techniques, and ensuring that the encoded values are used in a way that preserves the integrity and fairness of the model.

Encoding high cardinality categorical variables, where a variable has a large number of unique categories, presents additional challenges. For instance, a categorical variable representing customer IDs or product codes might have thousands or even millions of unique values. One-hot encoding such variables would lead to an impractically large feature space, making it difficult to train the model effectively. Techniques such as hashing or embedding can be used to handle high cardinality variables more efficiently. Hashing involves mapping categories to a fixed number of bins using a hash function, which reduces the dimensionality of the feature space. However, hashing can lead to collisions, where multiple categories are mapped to the same bin, potentially causing information loss. Embeddings, on the other hand, involve learning dense vector representations of categories and capturing the relationships between them in a continuous space. Embeddings are particularly useful in Natural Language Processing (NLP) tasks, where categorical variables such as words or phrases are encoded into vector spaces that capture semantic meaning. Data governance frameworks should provide guidelines for handling high cardinality variables, ensuring that the chosen encoding method is appropriate for the data and the ML model and that it minimizes information loss and computational complexity.

Bias and fairness are critical considerations when encoding categorical variables, particularly when these variables represent sensitive attributes such as race, gender, or socioeconomic status. The way categorical variables are encoded can inadvertently introduce or exacerbate biases in the model, leading to unfair or discriminatory outcomes. For example, if a categorical variable representing gender is encoded in a way that implicitly ranks one gender over another, the model might develop biased predictions that favor one group. Data governance frameworks should include guidelines for assessing and mitigating bias in the encoding of categorical variables, ensuring that the encoding process is transparent, fair, and aligned with ethical standards. This might involve conducting bias audits on the encoded data, using fairness-aware encoding techniques, and ensuring that the encoded features do not introduce unintended biases into the model.

Interpretability is another important aspect of encoding categorical variables, particularly in applications where the decisions made by the ML model have significant consequences. For instance, in a credit scoring model, the encoded categorical variables must be interpretable to human decision-makers, allowing them to understand how these variables contribute to the model's predictions. Complex encoding techniques, such as embeddings, can make it difficult to interpret the encoded variables, as the original categorical values are transformed into abstract numerical representations that may not have a clear or intuitive meaning. Data governance frameworks should include guidelines for ensuring the interpretability of encoded categorical variables, providing clear documentation and explanations of how the encoding process affects the model's predictions and how stakeholders can interpret the encoded features.

Data privacy and security are also critical considerations when encoding categorical variables, particularly when dealing with sensitive or personally identifiable information (PII). For example, a categorical variable representing customer IDs or social security numbers must be encoded in a way that preserves privacy and prevents re-identification. Techniques such as pseudonymization, where sensitive data is replaced with artificial identifiers, can be used to encode categorical variables while protecting privacy. However, the choice of encoding method must be carefully managed to ensure that it does not compromise the integrity or utility of the data. Data governance frameworks should include protocols for encoding sensitive categorical variables, ensuring that the encoding process complies with relevant data protection regulations, such as the General Data Protection Regulation (GDPR) or the California Consumer Privacy Act (CCPA), and that it aligns with the organization's privacy and security standards.

Handling missing values in categorical variables is another important aspect of the encoding process. Missing values can arise for various reasons, such as incomplete data collection or data entry errors, and they must be handled appropriately to ensure that the model can process the categorical variables effectively. Imputation techniques, such as filling in missing values with the most frequent category or using predictive models to estimate the missing values, can be used to handle missing categorical data. However, the choice of imputation method can affect the encoded values and, consequently, the model's performance. For instance, imputing missing values with the most frequent category might lead to overrepresentation of that category, skewing the model's predictions. Data governance frameworks should provide guidelines for handling missing values in categorical variables, ensuring that the chosen imputation method is appropriate for the data and the ML model and that it preserves the integrity and fairness of the encoded features.

Feature interaction is a critical consideration when encoding categorical variables, particularly when these variables interact with other features in the dataset. For example, in a predictive model for customer behavior, the interaction between categorical variables such as customer segment and product category might provide valuable insights into purchasing patterns. Encoding these interactions can involve creating new features that capture the combined effect of multiple categorical variables, such as by multiplying or concatenating the encoded values. However, the creation of interaction features can increase the complexity of the model and the risk of overfitting, particularly when dealing with high cardinality variables. Data governance frameworks should include guidelines for encoding feature interactions, ensuring that the interaction features are created in a way that enhances the model's predictive power while minimizing the risk of overfitting and preserving the interpretability of the model.

The documentation and transparency of the encoding process are critical from a data governance perspective. Stakeholders, including data scientists, engineers, and decision-makers, need to understand how categorical variables have been encoded and how the encoding process might affect the model's predictions. This is particularly important in regulated industries, where transparency and accountability are paramount. For example, in healthcare, the encoding of categorical variables related to patient demographics must be clearly documented to ensure that the model's predictions can be trusted and validated by clinicians. Data governance frameworks should enforce strict documentation standards, ensuring that all encoding steps are thoroughly recorded and that the rationale behind the encoding decisions is transparent and accessible.

The choice of encoding method can also impact the efficiency and scalability of ML models, particularly in large-scale or real-time applications. For instance, one-hot encoding can lead to a significant increase in the dimensionality of the dataset, which can be computationally expensive and time-consuming to process. Hashing or embedding techniques can provide more efficient alternatives, but they come with their own trade-offs, such as potential information loss or reduced interpretability. Data governance should include considerations for the computational efficiency of encoding methods, ensuring that the chosen techniques are appropriate for the scale and speed requirements of the application and that they do not compromise the quality or reliability of the model.

High Cardinality and Dimensionality Reduction

High cardinality refers to features that have a large number of unique values or categories. These features can be challenging to manage in ML models because they can lead to overfitting, increased computational complexity, and difficulties in interpretation. For instance, a categorical variable such as "user ID" or "product code" might have millions of unique values, leading to a massive expansion in the feature space if not handled appropriately. One common method for dealing with high cardinality is to use encoding techniques, such as one-hot encoding or target encoding, which transform categorical variables into a numerical format that can be processed by ML algorithms. However, one-hot encoding, for example, can dramatically increase the dimensionality of the dataset, leading to a sparse matrix that is computationally expensive to process and prone to overfitting. Target encoding, while more efficient, can introduce data leakage if not properly managed, where information from the target variable inadvertently influences the model during training.

Dimensionality reduction is a related process that aims to reduce the number of features in a dataset while retaining as much relevant information as possible. High-dimensional datasets can be problematic for ML models because they increase the risk of overfitting, where the model becomes too complex and captures noise rather than meaningful patterns. Dimensionality reduction techniques such as Principal Component Analysis (PCA), Singular Value Decomposition (SVD), and autoencoders are commonly used to reduce the feature space while preserving the underlying structure of the data. For example, PCA transforms the original features into a set of uncorrelated components, ranked by the amount of variance they explain in the data. This allows the model to focus on the most informative features, reducing the risk of overfitting and improving computational efficiency. However, PCA and similar techniques can lead to a loss of interpretability because the transformed features no longer correspond to the original variables, making it difficult to understand how the model makes its predictions.

The interplay between cardinality and dimensionality reduction is particularly important in the context of machine learning data governance. Managing high cardinality and reducing dimensionality require careful consideration of the trade-offs between model performance, interpretability, and ethical considerations. For instance, in a credit scoring model, reducing the dimensionality of the dataset might improve the model's performance but could also obscure the contributions of specific features that are important for ensuring fairness and transparency. Data governance frameworks should provide clear guidelines for managing high cardinality and dimensionality reduction, ensuring that the chosen techniques are appropriate for the specific application and that they align with the broader goals of the ML project.

One of the key challenges in managing high cardinality is ensuring that the encoded features are interpretable and meaningful. For instance, in a marketing analytics model, encoding a high cardinality variable such as "product SKU" using one-hot encoding might result in thousands of binary variables, each representing a different product. This can make it difficult to interpret the model's predictions, as it becomes unclear how specific products contribute to the model's decisions. Data governance frameworks should include guidelines for ensuring that the encoding of high cardinality variables is done in a way that preserves interpretability, such as by grouping similar categories together or using techniques like embeddings, which can capture the relationships between categories in a continuous space. Embeddings are particularly useful in Natural Language Processing (NLP) tasks, where high cardinality variables such as words or phrases are encoded into dense vector representations that capture their semantic meaning. However, embeddings can also be challenging to interpret, as the resulting vectors do not correspond to specific words or phrases but rather to abstract representations that are learned by the model.

Dimensionality reduction techniques also raise important considerations for data governance, particularly in terms of the trade-offs between model accuracy and interpretability. For example, while PCA can significantly reduce the number of features in a dataset, the resulting components are often difficult to interpret, as they are linear combinations of the original features. This can make it challenging to explain the model's predictions, particularly in regulated industries where transparency and accountability are paramount. In a healthcare application, for instance, using PCA to reduce the dimensionality of a dataset might improve the model's performance, but it could also make it difficult for clinicians to understand how specific patient features contribute to the model's predictions. Data governance frameworks should provide guidelines for balancing the need for dimensionality reduction with the requirement for model interpretability, ensuring that the chosen techniques align with the ethical and regulatory standards of the organization.

High cardinality variables also present challenges in terms of computational efficiency and scalability. For example, in a large-scale e-commerce application, a variable such as "customer ID" might have millions of unique values, making it impractical to use one-hot encoding due to the resulting increase in dimensionality. Hashing techniques can be used to address this issue by mapping categories to a fixed number of bins, reducing the dimensionality of the feature space. However, hashing can lead to collisions, where multiple categories are mapped to the same bin, potentially leading to a loss of information. Data governance frameworks should include guidelines for handling high cardinality variables in a way that balances computational efficiency with the need to preserve the integrity and utility of the data. This might involve using hashing in conjunction with other techniques, such as feature selection or dimensionality reduction, to ensure that the resulting features are both efficient to process and informative for the model.

Dimensionality reduction techniques such as autoencoders, which are a type of neural network used to learn efficient representations of data, also raise important considerations for data governance. Autoencoders can be particularly effective at reducing the dimensionality of high-dimensional datasets, such as images or text, while preserving the underlying structure of the data. For example, in an image recognition application, an autoencoder might be used to reduce the dimensionality of a dataset of images, capturing the most important features of the images in a lower-dimensional space. However, autoencoders can also be challenging to interpret, as the resulting representations are often abstract and do not correspond to specific features or variables. Data governance frameworks should include guidelines for ensuring that the use of autoencoders aligns with the broader goals of the ML project, such as by providing documentation and explanations of how the autoencoder was trained and how the resulting features contribute to the model's predictions.

The ethical implications of managing high cardinality and reducing dimensionality cannot be overlooked, particularly in applications where the features used in the model may lead to biased or unfair outcomes. For example, in a hiring algorithm, the use of high cardinality variables such as "university attended" or "past employers" might lead to biased predictions that favor candidates from certain demographic groups, leading to discriminatory hiring practices. Data governance frameworks should include guidelines for assessing the ethical implications of handling high cardinality and reducing dimensionality, ensuring that the techniques used do not introduce or exacerbate biases in the model. This might involve conducting bias audits on the encoded and reduced features, using fairness-aware algorithms that explicitly account for potential biases, and ensuring that the resulting features are transparent and interpretable.

The role of cardinality and dimensionality reduction in model validation and testing is also critical from a data governance perspective. When datasets are split into training, validation, and test sets, it is important to ensure that the same encoding and dimensionality reduction techniques are applied consistently across all sets. Failing to do so can lead to issues where the model performs well on the training data but poorly on the validation or test data because the features are not on the same scale or because the dimensionality reduction has introduced discrepancies between the sets. For instance, if PCA is applied separately to the training and test sets, the resulting components might differ, leading to inconsistent model performance. Data governance frameworks should enforce strict protocols for applying cardinality and dimensionality reduction techniques, ensuring that the same transformations are applied uniformly across all data splits and that the results are consistently validated and tested.

Managing high cardinality and reducing dimensionality also have significant implications for data privacy and security, particularly when dealing with sensitive or personally identifiable information (PII). For example, a high cardinality variable such as "user ID" might be encoded in a way that inadvertently reveals patterns or correlations that could lead to re-identification of individuals, even in anonymized datasets. Dimensionality reduction techniques such as PCA can also introduce privacy risks, as the resulting components might still contain enough information to identify individuals or reveal sensitive attributes. Data governance frameworks should include protocols for assessing the privacy implications of handling high cardinality and reducing dimensionality, ensuring that these processes comply with relevant data protection regulations, such as the General Data Protection Regulation (GDPR) or the Health Insurance Portability and Accountability Act (HIPAA), and that they align with the organization's privacy and security standards.

The documentation and transparency of cardinality and dimensionality reduction processes are also critical from a data governance perspective. Stakeholders, including data scientists, engineers, and decision-makers, need to understand how these processes have been applied and how they might affect the model's predictions. This is particularly important in regulated industries, where transparency and accountability are paramount. For example, in a financial application, the encoding of high cardinality variables related to customer transactions or the reduction of dimensionality in a credit scoring model must be clearly documented to ensure that the model's predictions can be trusted and validated by auditors or regulators. Data governance frameworks should enforce strict documentation standards, ensuring that all steps in handling high cardinality and reducing dimensionality are thoroughly recorded and that the rationale behind these decisions is transparent and accessible.

The choice of cardinality and dimensionality reduction techniques can also influence the efficiency and scalability of ML models, particularly in large-scale or real-time applications. For instance, one-hot encoding of high cardinality variables might lead to a significant increase in the dimensionality of the dataset, which can be computationally expensive and time-consuming to process. Dimensionality reduction techniques such as PCA or autoencoders can provide more efficient alternatives, but they come with their own trade-offs, such as potential loss of interpretability or increased complexity in model training. Data governance should include considerations for the computational efficiency of these techniques, ensuring that the chosen methods are appropriate for the scale and speed requirements of the application and that they do not compromise the quality or reliability of the model.

Cardinality and dimensionality reduction are also critical for ensuring that ML models remain robust and reliable over time. As new data is collected or as the business context evolves, the assumptions underlying the original encoding and dimensionality reduction techniques might no longer hold. For example, a dimensionality reduction strategy that worked well during the initial model development might become less effective as the data evolves, leading to shifts in feature distributions. Similarly, high cardinality variables that were effectively managed in the past might require new approaches as the dataset grows or as new categories are introduced. Data governance should enforce protocols for regularly reviewing and updating cardinality and dimensionality reduction strategies, ensuring that they remain aligned with the current data landscape and the objectives of the ML project.

The integration of cardinality and dimensionality reduction with other data cleaning and transformation processes is also essential from a data governance perspective. These processes do not occur in isolation; they are part of a broader pipeline that includes feature engineering, normalization, and scaling. For example, the reduction of dimensionality using PCA might need to be followed by normalization to ensure that the resulting components are on a similar scale. Similarly, the encoding of high cardinality variables might need to be integrated with feature selection techniques to ensure that only the most relevant features are included in the model. Data governance should ensure that cardinality and dimensionality reduction are integrated processes, with clear guidelines for how they should be combined with other data cleaning and transformation tasks to build a coherent and effective ML pipeline.

The role of cardinality and dimensionality reduction in feature engineering is also critical, as the process of creating new features often requires careful consideration of how these features are encoded and reduced. For example, interaction features created by multiplying or dividing existing features might have different scales that need to be aligned through dimensionality reduction. Similarly, features derived from categorical variables might require encoding techniques that reduce cardinality while preserving the relationships between categories. Data governance should ensure that feature engineering and cardinality/dimensionality reduction are integrated processes, with clear guidelines for how new features should be managed within the context of the broader ML pipeline.

Data Quality Rules

Data quality rules are designed to validate data against specific criteria before it is processed and used for model training, thus preventing the introduction of errors or inconsistencies that could compromise model performance. In the context of ML data governance, data quality rules encompass a wide range of checks, including but not limited to, validations for data completeness, accuracy, consistency, uniqueness, and conformity to predefined business logic. By establishing and enforcing these rules, organizations can ensure that the data entering the ML pipeline meets the necessary standards for quality, ultimately leading to more trustworthy and effective models.

Implementing data quality rules as part of ML data governance is not just a technical necessity but a strategic imperative. High-quality data is the bedrock of reliable and interpretable ML models, and without stringent data quality controls, the risk of "garbage in, garbage out" scenarios becomes significant. These rules help mitigate this risk by systematically identifying and rectifying data issues during the preprocessing stage, which is crucial for maintaining the integrity of the ML pipeline. Moreover, the rigorous application of data quality rules supports regulatory compliance and ethical standards, ensuring that the data used in ML models is not only technically sound but also aligns with broader organizational and societal expectations.

Range Checks

Range checks are a fundamental aspect of data validation and quality assurance in machine learning (ML) pipelines, ensuring that numerical data adheres to predefined boundaries before it is utilized in model training or decision-making processes. From the perspective of ML data governance, range checks are not just a technical necessity but a critical practice that safeguards the integrity, reliability, and fairness of the data. Properly implemented, range checks can prevent the introduction of erroneous or anomalous data into ML models, which is crucial for maintaining accuracy and trust in the outputs generated by these models.

In the context of data validation, range checks are applied to numerical features to verify that their values fall within an expected and acceptable range. These ranges are typically defined based on domain-specific knowledge, regulatory requirements, or historical data. For instance, in a financial dataset, the value of a transaction might be expected to fall within a certain range based on the type of transaction, the region, or the customer's historical behavior. A transaction amount that significantly deviates from this range might indicate an error, fraud, or an outlier that requires further investigation. By setting these boundaries, range checks serve as an initial filter to catch data points that do not conform to the expected norms, thereby preventing them from skewing the results of ML models.

The importance of range checks becomes even more pronounced in sectors like healthcare, where the data's accuracy directly impacts patient outcomes. For example, in an ML model predicting patient deterioration in an intensive care unit, vital signs such as blood pressure, heart rate, and oxygen saturation must be within plausible physiological ranges. A blood pressure reading of 300/200 mmHg or a heart rate of 15 beats per minute would immediately raise red flags, prompting a review of the data entry process or the functioning of the monitoring equipment. Such outliers, if not caught and corrected, could lead to dangerous model predictions, potentially resulting in inappropriate medical interventions. Therefore, range checks are essential not only for ensuring data quality but also for protecting the lives and well-being of individuals affected by the decisions driven by ML models.

Range checks are also critical in industrial applications, such as predictive maintenance in manufacturing. In this context, sensors constantly monitor equipment to detect signs of wear or failure. The data generated by these sensors, such as temperature, vibration levels, or pressure, must be within specific operational ranges. For example, the temperature of a machine part might typically range from 50 to 150 degrees Celsius under normal conditions. A sudden spike to 300 degrees could indicate a potential failure, warranting immediate maintenance action. Without proper range checks, such an anomaly might go unnoticed, leading to machine breakdowns and costly production delays. Therefore, range checks play a pivotal role in ensuring that the data feeding into predictive maintenance models is accurate and reliable, ultimately contributing to the smooth operation of industrial processes.

The application of range checks is not limited to individual data points but can also extend to ensuring the consistency of relationships between different variables. For instance, in a retail dataset, the price of a product and its discount rate should be logically consistent. A product listed with a price of $100 and a discount of 110% would be an obvious error, as the discount cannot exceed the price of the product. Similarly, in a transportation dataset, the distance traveled by a vehicle and the time taken should have a reasonable correlation. A vehicle covering 500 miles in 5 minutes would indicate a data entry mistake or a malfunctioning GPS device. By implementing range checks that consider the relationships between variables, data governance frameworks can ensure that the dataset not only contains valid individual data points but also maintains logical consistency across the entire dataset.

The implementation of range checks must also account for the potential biases that can be introduced if the boundaries are set based on skewed historical data. For instance, in a hiring dataset, if the range for acceptable salaries is set based on historical data that reflects gender or racial pay gaps, these biases could be perpetuated in the ML model's predictions. A salary that appears as an outlier based on biased historical data might actually be a reflection of equitable pay practices in a more progressive company. Therefore, range checks must be designed with an awareness of the potential for bias, and the parameters used should be regularly reviewed and updated to reflect changes in societal norms or organizational policies. This consideration is crucial for ensuring that ML models do not inadvertently reinforce existing inequalities.

The challenge of setting appropriate ranges becomes more complex when dealing with data from different sources or environments. In a global organization, data may be collected from multiple regions, each with its own economic, cultural, or environmental context. For example, temperature data from manufacturing plants in different countries might have different acceptable ranges based on local climate conditions. A temperature reading that is normal in a plant in Siberia might be considered an anomaly in a plant in the Middle East. Therefore, range checks must be tailored to account for these contextual differences, ensuring that the data is validated appropriately based on its origin.

This requires a nuanced understanding of the data's context and the application of range checks that are flexible enough to accommodate variability while still maintaining rigorous data quality standards.

In real-time data processing environments, such as financial trading or autonomous vehicles, range checks must be implemented with a focus on both speed and accuracy. In these scenarios, data is constantly being generated and fed into ML models, with decisions being made in fractions of a second. For instance, in high-frequency trading, prices and volumes of stocks are monitored in real time, and any data point that falls outside the expected range could indicate market manipulation or a system error. Implementing range checks in such high-speed environments requires the use of efficient algorithms that can validate data points almost instantaneously, without introducing significant latency. This presents a challenge for data governance frameworks, which must balance the need for thorough data validation with the demands of real-time processing.

The evolution of machine learning models over time also necessitates dynamic range checks that can adapt to changing data distributions. In many cases, the distribution of data can shift over time due to changes in market conditions, user behavior, or external factors such as regulations. For example, in an e-commerce platform, the range of acceptable transaction amounts might change as the platform expands into new markets or introduces new products. Static range checks that were set during the initial deployment of the ML model might no longer be relevant, leading to an increase in false positives or negatives during data validation. To address this, data governance frameworks should implement dynamic range checks that can be updated regularly based on new data, ensuring that the ML models remain accurate and reliable even as the underlying data evolves.

Another advanced consideration in range checks is the handling of extreme outliers that may represent valid but rare events. For example, in the financial sector, an extraordinarily large transaction might be flagged as an outlier by a range check, but it could actually be a legitimate transaction, such as the sale of a company or a significant investment. In such cases, it is important to have a data governance framework that allows for manual review or exception handling, where flagged data points can be assessed by domain experts before being excluded from the dataset. This approach ensures that rare but valid data points are not discarded, which could otherwise lead to a loss of valuable information and reduce the generalizability of the ML model.

Range checks are also crucial for ensuring compliance with regulatory requirements, particularly in industries such as finance, healthcare, and environmental monitoring. Regulatory bodies often set specific guidelines for the acceptable ranges of certain data points, and organizations must ensure that their data adheres to these guidelines to avoid legal penalties or sanctions. For example, in the financial industry, regulations may specify the acceptable ranges for transaction amounts, interest rates, or risk scores. Data governance frameworks must incorporate these regulatory requirements into their range checks, ensuring that the data used in ML models is compliant with all relevant laws and regulations. This not only protects the organization from legal risks but also enhances the credibility and trustworthiness of its ML models.

Referential Integrity Checks

The concept of referential integrity is rooted in the relational database model, where it ensures that relationships between tables are maintained correctly. For example, in a database where there is a relationship between customers and orders, each order must be linked to a valid customer record. If an order references a customer ID that does not exist in the customer table, it indicates a breach of referential integrity. Such breaches can lead to inconsistencies in the data, which can significantly impact the performance and accuracy of ML models trained on that data. For instance, if an ML model is being used to predict customer behavior based on order history, any inconsistencies between orders and customer data could result in incorrect predictions, potentially leading to flawed business decisions.

In the context of machine learning, referential integrity checks are not just about maintaining data quality; they also play a critical role in ensuring that the data accurately represents the real-world relationships it is intended to model. For instance, in a healthcare dataset where patient records are linked to treatment records, it is essential to ensure that every treatment record correctly references an existing patient. If a treatment record points to a non-existent patient, it could lead to significant errors in the ML model's predictions about treatment outcomes. This is particularly critical in applications like predictive analytics in healthcare, where the accuracy of the model can directly impact patient care and treatment decisions. Data governance frameworks must therefore establish robust protocols for performing referential integrity checks, ensuring that all data relationships are valid and consistent before the data is used for ML training and analysis.

The challenges of maintaining referential integrity become more pronounced in large-scale, complex datasets where data is integrated from multiple sources. For example, in a supply chain management system that integrates data from various vendors, manufacturers, and retailers, ensuring that all references between these entities are valid can be a daunting task. A purchase order from a retailer might reference a product ID that is sourced from multiple manufacturers, each with its own set of associated data. Ensuring that all these references are consistent and valid is essential for accurate inventory management and forecasting models. If a purchase order references a product ID that does not exist in the product catalog, it could lead to errors in inventory management, such as stockouts or overstocking, ultimately affecting the accuracy of the ML models used for demand forecasting.

Data governance frameworks should include detailed guidelines for performing referential integrity checks in such complex, multi-source datasets. This involves not only validating references within a single dataset but also ensuring that references across different datasets are consistent and accurate. For example, if a Customer Relationship Management (CRM) system is integrated with an e-commerce platform, the customer IDs in the CRM system must match the customer IDs used in the e-commerce platform's order records. Any discrepancies between these systems could lead to significant issues in the ML models used for customer segmentation or personalized marketing, as the models might be trained on incomplete or inaccurate data.

Referential integrity checks also have significant implications for data security and privacy, particularly in contexts where data is subject to regulatory requirements such as the General Data Protection Regulation (GDPR) or the Health Insurance Portability and Accountability Act (HIPAA). For instance, in a dataset containing personally identifiable information (PII), it is crucial to ensure that all references to PII are valid and that any changes to or deletions of PII are consistently reflected across all related datasets. If a customer requests that their data be deleted under GDPR's "right to be forgotten," it is essential to ensure that all references to that customer's data are removed from the system. Failure to maintain referential integrity in such cases could result in data breaches or violations of regulatory requirements, leading to significant legal and financial consequences for the organization. Data governance frameworks must therefore include stringent protocols for maintaining referential integrity in compliance with data protection regulations, ensuring that all references to sensitive data are handled correctly and securely.

The process of performing referential integrity checks often involves the use of automated tools and technologies, particularly in large-scale datasets where manual checks would be impractical. Modern data management platforms typically include built-in tools for enforcing referential integrity, such as foreign key constraints in relational databases. These tools can automatically check that every reference in a dataset points to a valid entity in the related dataset, preventing the insertion of invalid data and flagging any inconsistencies for review. However, the effectiveness of these tools depends on how they are configured and used. For example, in a relational database, foreign key constraints must be properly defined and enforced to ensure that all references between tables are valid. Data governance frameworks should ensure that the technology used for referential integrity checks is correctly configured to meet the specific needs of the organization and that data professionals are adequately trained in its use.

The challenges of maintaining referential integrity are further compounded in dynamic, real-time data environments, where data is continuously updated, and relationships between data entities may change rapidly. For instance, in a real-time financial trading system, trades are executed and recorded continuously, with each trade referencing specific financial instruments and accounts. Ensuring that all these references remain valid in real time is critical for the accuracy of the trading algorithms and risk management models that rely on this data. If a trade references an account that has been closed or a financial instrument that has been delisted, it could lead to significant errors in the ML models used for trading strategies or risk assessments. Data governance frameworks must therefore include protocols for performing referential integrity checks in real-time data environments, ensuring that data relationships are continuously validated and that any discrepancies are quickly identified and resolved.

The role of data lineage in referential integrity checks cannot be overlooked. Data lineage refers to the tracking of data as it moves through different stages of processing, from its source to its final destination. Understanding the lineage of data is essential for maintaining referential integrity, particularly in complex data pipelines where data is transformed, aggregated, and integrated from multiple sources. For example, in a data pipeline that aggregates sales data from multiple regions, it is important to track the lineage of each data point to ensure that references to regional sales offices, product IDs, and customer segments are maintained accurately throughout the pipeline. If the lineage of the data is not properly tracked, it could lead to discrepancies in the final dataset, with references to non-existent or outdated entities. Data governance frameworks should include guidelines for tracking data lineage as part of the referential integrity checking process, ensuring that data relationships are maintained consistently throughout the entire data pipeline.

The implications of referential integrity extend to model interpretability and explainability, particularly in applications where the decisions made by the ML model have significant consequences. For example, in a credit scoring model, the relationship between a borrower's credit history and their current financial status is critical for making accurate predictions. If the dataset used to train the model contains references to outdated or incorrect credit history records, it could lead to biased or inaccurate predictions, potentially resulting in unfair credit decisions. Maintaining referential integrity in such datasets is essential for ensuring that the model's predictions are based on accurate and up-to-date information, thereby improving the interpretability and trustworthiness of the model. Data governance frameworks should include protocols for ensuring referential integrity in datasets used for high-stakes ML applications, ensuring that the relationships between data entities are accurately represented and that the model's predictions can be trusted.

Business Rule Checks

Business rule checks are an essential aspect of data quality assurance within machine learning (ML) pipelines, playing a crucial role in ensuring that the data used in models aligns with the organization's operational guidelines and strategic objectives. From an ML data governance perspective, business rule checks are not merely a technical formality but a vital process that ensures the integrity, accuracy, and relevance of data. These checks are designed to enforce the specific rules and constraints that govern the use of data in the context of a business's unique requirements and are critical for maintaining the trustworthiness of ML models.

The importance of business rule checks stems from the fact that data, before being used in ML models, must adhere to predefined rules that reflect the organization's operational realities and decision-making frameworks. These rules can range from simple constraints, such as ensuring that dates are valid and fall within a specific range, to more complex business logic that governs how data is processed, combined, and interpreted. For example, in a financial institution, a business rule might stipulate that no transaction can be processed if the account balance falls below a certain threshold.

Implementing this rule in the data validation process ensures that ML models trained on this data do not incorporate invalid or erroneous transactions, which could lead to faulty predictions and decisions.

In retail, a business rule might require that any order exceeding a certain value must be flagged for additional review. This rule ensures that high-value transactions are scrutinized for potential fraud or errors, thereby protecting the business from financial losses. When applied to the data used in ML models, this rule ensures that outlier transactions, which could disproportionately influence model predictions, are appropriately handled. Without such business rule checks, the ML model might overemphasize these outliers, leading to skewed insights that do not accurately reflect the typical customer behavior or business operations.

Business rule checks are particularly crucial when dealing with data that crosses different functional areas within an organization. For instance, in a supply chain management system, business rules might dictate that inventory levels must always be positive and that the lead time for restocking must fall within a certain range. When data from various departments, such as sales, procurement, and logistics, is integrated, these rules ensure that the combined dataset accurately reflects the operational constraints of the business. This integration is especially important in ML models that predict supply chain outcomes, as any violation of these business rules could lead to incorrect forecasts, resulting in stockouts or overstock situations that disrupt the business.

One of the key challenges in implementing business rule checks within ML data governance is ensuring that the rules are consistently applied across all relevant datasets and processes. This consistency is particularly challenging in large organizations where data is sourced from multiple systems, each with its own data structures and formats. For example, a multinational company might have separate sales systems for different regions, each with its own set of business rules. Ensuring that a global ML model adheres to all these regional rules requires a robust governance framework that can reconcile these differences and enforce consistent business rule checks across the entire dataset.

The implementation of business rule checks also involves dealing with the temporal aspects of data. Many business rules are time-sensitive, meaning that they are only applicable during certain periods or under specific conditions. For instance, a retail business might have a rule that discounts are only applicable during promotional periods and that any transaction outside this period should not apply the discount. In an ML context, this requires ensuring that the data used for model training accurately reflects the temporal conditions under which the business rules apply. This can be complex when dealing with historical data, where the relevant time periods may have already passed, requiring careful validation to ensure that the data remains relevant and accurate for the model's purpose.

Data governance frameworks must also address the issue of exceptions to business rules, which are common in many business environments. Exceptions occur when a specific data point or scenario does not adhere to the standard business rules but is still valid due to special circumstances. For instance, a business rule might state that customer orders cannot be shipped if the delivery address is incomplete. However, there may be cases where orders are shipped to repeat customers with known addresses, even if the current order form is incomplete. Handling these exceptions requires a nuanced approach in data validation, where the business rule checks are flexible enough to accommodate valid exceptions without compromising the overall quality and integrity of the data.

The automation of business rule checks is a critical consideration in ML data governance, particularly in environments where data is generated and processed in real time. Automated rule checks can ensure that data entering the ML pipeline is instantly validated against the business rules, preventing invalid or erroneous data from contaminating the model training process. For example, in a financial trading system, real-time data on market transactions must be instantly validated against business rules that govern trading limits, transaction types, and compliance requirements. Any data that fails these checks should be immediately flagged and excluded from the ML models to prevent erroneous predictions that could lead to significant financial risks.

Another critical aspect of business rule checks is their role in ensuring regulatory compliance, which is a key concern in many industries, such as finance, healthcare, and telecommunications. Regulatory requirements often impose specific rules on how data should be handled, processed, and reported. For example, financial institutions must comply with anti-money laundering (AML) regulations, which require that certain types of transactions be flagged and reported. Business rule checks ensure that the data used in ML models adheres to these regulatory requirements, preventing the organization from inadvertently violating the law. In healthcare, business rule checks might enforce compliance with patient privacy regulations, ensuring that ML models do not use or expose sensitive patient information in ways that contravene regulations such as HIPAA.

The effectiveness of business rule checks in ML data governance also depends on the ability to adapt and update the rules as the business environment evolves. Business rules are not static; they must evolve in response to changes in the market, technology, and regulatory landscape. For example, a retail business might adjust its pricing and discount rules in response to competitive pressures or changes in consumer behavior. Ensuring that these updated rules are reflected in the data validation process is crucial for maintaining the relevance and accuracy of ML models. This requires a data governance framework that supports the continuous monitoring and updating of business rules, ensuring that the ML models remain aligned with the current business objectives and constraints.

Business rule checks play a crucial role in maintaining the interpretability and transparency of ML models, which is increasingly important in an era where explainability is a key concern. When business rules are clearly defined and consistently applied, the resulting ML models are more likely to produce outcomes that align with the business's operational logic and decision-making frameworks. This makes it easier for stakeholders to understand and trust the model's predictions, as they can see how the model's outputs are grounded in the business's established rules and practices. For example, a credit scoring model that incorporates business rule checks to enforce lending criteria will produce scores that are more easily explainable to both internal stakeholders and external regulators, ensuring that the model's decisions can be trusted and validated.

Data Enrichment and Integration

Data enrichment involves enhancing the existing datasets with additional information, which could come from external sources, third-party data, or by deriving new features from the existing data. This process is essential for providing a more comprehensive and nuanced dataset that can improve the accuracy and robustness of ML models. However, from a governance perspective, data enrichment must be carefully managed to ensure that the added data is reliable, relevant, and compliant with regulatory standards. This includes verifying the accuracy of the external data, ensuring that it aligns with the original dataset, and maintaining transparency about the sources and methods used for enrichment. Proper governance ensures that enriched data contributes positively to model performance without introducing biases, inaccuracies, or ethical concerns.

Data integration, on the other hand, involves combining data from different sources or systems into a unified dataset that can be used for ML model training and analysis. This process is crucial for organizations that rely on data from various departments, databases, or external partners, as it allows them to create a cohesive view of the information. Effective data integration requires robust governance practices to manage the challenges of data consistency, compatibility, and quality across diverse sources. This includes establishing standards for data formats, ensuring that data is harmonized correctly, and resolving any conflicts or discrepancies that arise during the integration process. By implementing strong governance practices around data integration, organizations can ensure that their ML models are built on a solid foundation of high-quality, well-integrated data, which is critical for generating reliable and actionable insights.

Merging External Data

Merging external data during data enrichment and integration is a complex but essential aspect of machine learning (ML) data governance. This process involves incorporating data from outside sources, such as third-party vendors, public datasets, or industry-specific repositories, into an organization's internal datasets. When done effectively, this integration can significantly enhance the richness and utility of the data available for ML model training and analysis. However, merging external data also presents numerous challenges that require careful governance to ensure that the resulting dataset maintains high standards of quality, integrity, and compliance with regulatory requirements.

The first challenge in merging external data is assessing the quality and reliability of the data sources. External datasets can vary widely in terms of accuracy, completeness, and timeliness, and integrating data from unreliable sources can introduce errors and biases into the ML models. For example, an organization may purchase demographic data from a third-party vendor to enhance its customer segmentation models. If this external data is outdated or collected using inconsistent methodologies, it could lead to inaccurate customer profiles and flawed marketing strategies. To mitigate this risk, ML data governance frameworks must establish rigorous criteria for evaluating external data sources before they are integrated. This might include verifying the data's provenance, assessing its alignment with the organization's internal data standards, and conducting preliminary analyses to detect any obvious errors or inconsistencies.

Once the quality of the external data is assured, the next step in the integration process involves aligning this data with the existing internal datasets. This alignment is often challenging due to differences in data formats, structures, and definitions. For instance, an external dataset might use different naming conventions for similar attributes, or it might categorize data in a way that is not directly comparable with the organization's internal data. These discrepancies can create significant obstacles in merging the datasets effectively. For example, if an external dataset categorizes income levels into broad ranges, while the internal data contains precise income figures, merging these datasets could lead to a loss of granularity and accuracy in the final dataset. To address these challenges, organizations must implement data governance protocols that standardize the data formats and definitions before integration. This process may involve mapping external data attributes to the corresponding internal attributes, harmonizing categories, and transforming data into a consistent format that aligns with the organization's existing data infrastructure.

Another critical aspect of merging external data during data enrichment and integration is ensuring data privacy and security. External data sources, especially those that involve personal or sensitive information, can pose significant risks if not handled correctly. For example, integrating customer data from an external vendor with internal CRM data could expose the organization to potential data breaches or violations of privacy regulations like the General Data Protection Regulation (GDPR). If the external data includes personal identifiers, such as names or addresses, it is crucial to implement robust data anonymization techniques before merging the datasets. Data governance frameworks must define and enforce protocols for data anonymization, encryption, and access control to protect sensitive information and comply with legal requirements. These protocols should also ensure that any data sharing agreements with external vendors include clear provisions for data security and compliance, thereby minimizing the risk of unauthorized access or data misuse.

Merging external data also raises important questions about data ownership and intellectual property rights. When integrating third-party data, organizations must ensure that they have the appropriate rights to use the data for their intended purposes, including ML model development and analysis. For instance, if an organization purchases social media data from an external vendor to enhance its sentiment analysis models, it must ensure that the data is licensed for use in ML applications. Failure to secure the necessary rights could lead to legal disputes and reputational damage. Data governance frameworks must establish clear guidelines for verifying data ownership and usage rights before integrating external data. This includes reviewing the terms of data licenses, negotiating data usage agreements with vendors, and maintaining detailed records of data acquisition and usage.

One of the more advanced challenges in merging external data involves dealing with potential biases that the external data might introduce into the ML models. External datasets often reflect the biases inherent in the methods and contexts in which they were collected. For example, if an external dataset used for customer profiling is biased toward certain demographic groups, integrating this data with internal datasets could lead to biased ML models that unfairly favor or disadvantage particular groups. This is particularly problematic in applications like credit scoring or hiring, where biased models can have significant ethical and legal implications. To address these concerns, data governance frameworks must include robust bias detection and mitigation strategies as part of the data integration process. This might involve conducting bias audits on the external data, comparing the distribution of key attributes in the external data to the internal data, and applying techniques such as reweighting or resampling to correct for any detected biases.

Another advanced consideration in merging external data is the impact on data lineage and traceability. Data lineage refers to the ability to trace the origins and transformations of data throughout its lifecycle. When external data is integrated into an organization's internal datasets, it is crucial to maintain clear and detailed records of the data's source, how it was transformed, and how it was integrated. This ensures that any issues or errors that arise in the ML models can be traced back to their source, facilitating debugging and model refinement. For example, if an ML model begins producing unexpected predictions, the ability to trace the data lineage can help data scientists determine whether the issue stems from the external data or from the integration process. Data governance frameworks must ensure that all data transformations and integrations are thoroughly documented, with clear records of the data's lineage maintained throughout the ML pipeline.

Scalability is another key issue when merging external data during data enrichment and integration. As organizations increasingly rely on external data sources to enhance their ML models, the volume and complexity of the data being integrated can grow exponentially. This can create significant challenges in terms of data storage, processing, and management. For instance, integrating large-scale geospatial data from external providers to enhance location-based services can strain an organization's data infrastructure, leading to performance bottlenecks and increased costs. Data governance frameworks must address scalability by implementing efficient data integration architectures that can handle large volumes of external data without compromising performance. This might involve leveraging cloud-based storage solutions, implementing distributed processing frameworks, or using data compression techniques to manage the increased data load.

The process of merging external data also has implications for data quality monitoring and continuous improvement. Once external data has been integrated, it is essential to continuously monitor the quality of the combined dataset to ensure that it remains accurate, relevant, and aligned with the organization's objectives. For example, if an organization integrates real-time data feeds from external sensors into its predictive maintenance models, it must regularly assess the quality of the sensor data to ensure that it remains accurate and reliable over time. Data governance frameworks should include provisions for ongoing data quality monitoring, with specific metrics and thresholds defined for assessing the quality of the integrated dataset. This continuous monitoring is critical for detecting and addressing any issues that might arise from changes in the external data sources or the integration process itself.

The integration of external data should be aligned with the organization's overall data strategy and business objectives. Merging external data is not just a technical exercise; it is a strategic decision that can significantly impact the organization's ability to achieve its goals. For instance, an organization that integrates external market data to enhance its competitive analysis models must ensure that the data is aligned with its strategic priorities, such as expanding into new markets or improving customer retention. Data governance frameworks must ensure that decisions about merging external data are made in the context of the organization's broader data strategy, with clear alignment between the data being integrated and the organization's business objectives. This alignment helps ensure that the integration of external data delivers tangible value and supports the organization's long-term success.

Merging external data during data enrichment and integration is a complex process that requires careful consideration of multiple factors, including data quality, privacy, ownership, bias, lineage, scalability, and alignment with business objectives. From an ML data governance perspective, this process must be meticulously managed to ensure that the integrated data enhances the organization's ML models while maintaining high standards of quality, compliance, and ethical responsibility. By addressing these challenges through robust data governance frameworks, organizations can effectively leverage external data to build more accurate, reliable, and valuable ML models that drive business success.

Entity Resolution

Entity resolution is a critical process during data enrichment and integration in machine learning (ML) pipelines, particularly from a data governance perspective. It involves identifying and merging records that refer to the same real-world entity across different datasets. This process is crucial for ensuring the accuracy, consistency, and reliability of data used in ML models, especially when integrating data from multiple sources. Entity resolution is not merely a technical exercise; it is deeply intertwined with the governance of data quality, privacy, compliance, and ethical considerations within an organization's ML operations.

Entity resolution is essential when combining data from diverse sources, such as internal databases, third-party vendors, or public repositories. These sources often contain overlapping or redundant information about the same entities, such as customers, products, or locations, but may refer to them in different ways. For example, a customer might be represented in one dataset by their full name and in another by their email address. Without proper entity resolution, these records could be treated as separate entities, leading to fragmented, inconsistent, and inaccurate data. This can have significant implications for ML models, as it can result in duplicated or conflicting records that skew the model's predictions, reduce its accuracy, and undermine its trustworthiness.

The complexity of entity resolution arises from the variations in how entities are represented across different datasets. These variations can include differences in naming conventions, typographical errors, and inconsistent formatting. For instance, a person might be listed as "John Doe" in one dataset, "J. Doe" in another, and "Jonathan Doe" in a third. These discrepancies make it challenging to determine whether these records refer to the same individual. The process of resolving these variations requires sophisticated algorithms that can match records based on fuzzy logic, probabilistic matching, or machine learning techniques. These algorithms must be carefully calibrated to balance the need for accuracy with the risk of false positives (incorrectly merging different entities) and false negatives (failing to merge records that refer to the same entity).

From a data governance perspective, it is crucial to establish clear guidelines and protocols for entity resolution to ensure that the process is consistent, transparent, and aligned with the organization's data quality standards. This involves defining the criteria for matching records, selecting the appropriate algorithms, and setting thresholds for deciding when records should be merged. For example, in a Customer

Relationship Management (CRM) system, the entity resolution process might involve matching records based on a combination of attributes, such as name, address, and phone number, with specific rules for handling discrepancies or missing values. These rules must be documented and consistently applied across all datasets to maintain the integrity of the entity resolution process.

Entity resolution also has significant implications for data privacy and security. When merging records from different sources, it is essential to ensure that sensitive information is not inadvertently exposed or misused. For instance, if records from a healthcare database are merged with records from a marketing database, there is a risk that personal health information could be linked to identifiable individuals in the marketing dataset, potentially violating privacy regulations such as the Health Insurance Portability and Accountability Act (HIPAA) in the United States or the General Data Protection Regulation (GDPR) in Europe. Data governance frameworks must include protocols for anonymizing or pseudonymizing sensitive data before performing entity resolution, ensuring that the process complies with relevant privacy laws and regulations.

The issue of bias is another critical consideration in entity resolution. Bias can be introduced at various stages of the entity resolution process, from the design of matching algorithms to the selection of data attributes used for matching. For example, if an entity resolution algorithm disproportionately relies on certain attributes, such as zip codes or surnames, it may inadvertently introduce demographic biases into the data, leading to skewed ML models that unfairly favor or disadvantage certain groups. Data governance must address these risks by implementing bias detection and mitigation strategies, such as auditing the entity resolution process, using diverse and representative datasets for training matching algorithms, and ensuring that the process is transparent and accountable.

Entity resolution also plays a crucial role in maintaining data lineage and traceability, which are key aspects of data governance. Data lineage refers to the ability to track the origins, transformations, and movements of data throughout its lifecycle. In the context of entity resolution, this means maintaining detailed records of how and why records were merged, which data sources were involved, and what algorithms or rules were applied. This traceability is essential for auditing the entity resolution process, troubleshooting issues, and ensuring that the merged data can be trusted. For example, if an error is discovered in the merged dataset, the ability to trace the data lineage can help identify the source of the error, whether it was due to incorrect matching, data entry errors, or other issues. Data governance frameworks must ensure that entity resolution processes are thoroughly documented and that data lineage is preserved at every stage of the integration process.

Scalability is another significant challenge in entity resolution, especially as organizations increasingly rely on large-scale, real-time data integration. As the volume and variety of data grow, the computational resources required to perform entity resolution also increase, potentially leading to performance bottlenecks and delays in data processing. For example, in a global financial institution that integrates customer data from multiple regions and systems, the entity resolution process must be able to handle millions of records in real time, without compromising accuracy or speed. Data governance must address these scalability challenges by implementing efficient, scalable algorithms and architectures that can perform entity resolution at scale, while also ensuring that the process remains aligned with the organization's data quality and governance standards.

The integration of external data sources during entity resolution also presents unique challenges, particularly in terms of data compatibility and harmonization. External data sources often use different standards, formats, and definitions for the same entities, making it difficult to achieve accurate and consistent entity resolution. For example, an organization might acquire customer data from a third-party vendor that uses a different schema for customer records, such as different formats for names, addresses, or phone numbers. Harmonizing these records with the organization's internal data requires careful preprocessing, including standardizing data formats, resolving discrepancies, and ensuring that the merged records accurately reflect the real-world entities they represent. Data governance must establish protocols for harmonizing external data during entity resolution, ensuring that the process is consistent with the organization's data standards and that the integrated data is reliable and accurate.

Entity resolution also plays a critical role in ensuring the ethical use of data in ML models. When records are merged, it is essential to consider the potential ethical implications of combining data from different sources, particularly when the data involves sensitive or personal information. For example, merging social media data with financial records could create detailed profiles of individuals that raise ethical concerns about surveillance, privacy, and consent. Data governance frameworks must include ethical guidelines for entity resolution, ensuring that the process respects individuals' privacy rights, that data is used in a manner consistent with the organization's ethical principles, and that the potential risks and benefits of merging data are carefully considered and balanced.

The success of entity resolution depends on the active involvement of key stakeholders, including data scientists, data engineers, legal and compliance teams, and business leaders. These stakeholders must work together to define the rules, criteria, and objectives of the entity resolution process, ensuring that it aligns with the organization's overall data strategy and business goals. For example, in a healthcare organization, data scientists and data engineers might work together to develop and implement entity resolution algorithms, while legal and compliance teams ensure that the process complies with patient privacy regulations. Data governance frameworks must facilitate this collaboration by establishing clear roles and responsibilities, providing the necessary tools and resources, and fostering a culture of accountability and transparency.

Data Warehousing

A data warehouse serves as a centralized repository where data from various sources is consolidated, processed, and made available for analysis. This centralized nature is particularly important in ML, where models often require diverse datasets—ranging from transactional data, customer behavior analytics, to external market trends—integrated seamlessly to generate accurate predictions. The process of data enrichment within a data warehouse involves aggregating data from multiple sources, transforming it into a consistent format, and enhancing it with additional context or information that can improve the utility and relevance of the data for ML models. For instance, in the retail industry, a data warehouse might integrate point-of-sale data, customer demographics, and social media sentiment analysis to provide a comprehensive view of consumer behavior, which can be used to develop personalized marketing strategies or demand forecasting models.

The design and architecture of a data warehouse are critical to its effectiveness in supporting ML processes. Data warehouses are typically structured to handle large volumes of historical data, making them ideal for training ML models that require extensive datasets for pattern recognition and prediction. However, the architecture must also support the scalability and flexibility needed to accommodate the continuous inflow of new data. For example, in a financial institution, a data warehouse might be designed to ingest and store transactional data in near real time, allowing ML models to detect fraud or assess credit risk with the most current information available. Ensuring that the data warehouse can scale to meet the demands of growing data volumes while maintaining performance and data integrity is a key concern in ML data governance.

Data warehouses often employ a schema-on-read or schema-on-write approach to organize data, each with its implications for ML data governance. Schema-on-write, where data is structured according to a predefined schema before being stored, ensures that the data entering the warehouse is clean, consistent, and ready for analysis. This method is advantageous for maintaining data quality but can be rigid, requiring significant upfront effort to define the schema and process the data. On the other hand, schema-on-read allows for greater flexibility by storing raw data and applying the schema at the time of analysis. While this approach supports diverse and evolving data sources, it poses challenges for data governance, as it requires robust data validation and transformation processes to ensure data quality at the point of use. In an ML context, where the accuracy and reliability of models depend heavily on the quality of input data, organizations must carefully consider which approach aligns best with their data governance strategy and operational needs.

The process of data integration within a data warehouse is also a critical component of ML data governance. Data integration involves consolidating data from disparate sources, such as databases, cloud services, and third-party vendors, into a cohesive dataset that can be used for analysis and model training. This process requires the implementation of extract, transform, and load (ETL) pipelines, which are essential for cleaning, normalizing, and transforming data before it is loaded into the warehouse. For example, in the healthcare industry, integrating patient records, clinical trial data, and genomic information into a data warehouse allows for the development of predictive models that can improve patient outcomes. However, ensuring that the data is accurately transformed and integrated without introducing errors or biases is a significant governance challenge. Data governance frameworks must include stringent controls and validation procedures to monitor the ETL process, ensuring that the data warehouse remains a reliable and trustworthy source for ML applications.

Security and privacy are paramount concerns in the management of data warehouses, particularly in industries handling sensitive information such as healthcare, finance, or government. Data warehouses store vast amounts of data, including personally identifiable information (PII) and confidential business data, making them attractive targets for cyberattacks. From an ML data governance perspective, protecting the data within a warehouse is crucial to maintaining the integrity of ML models and ensuring compliance with data protection regulations such as the General Data Protection Regulation (GDPR) or the Health Insurance Portability and Accountability Act (HIPAA). Data governance frameworks must enforce robust security protocols, including encryption, access controls, and auditing, to safeguard data within the warehouse. For instance, Role-Based Access Control (RBAC) can be implemented to ensure that only authorized personnel have access to sensitive data, while encryption can protect data at rest and in transit. Regular security audits and vulnerability assessments are also necessary to identify and mitigate potential risks, ensuring that the data warehouse remains secure over time.

The issue of data lineage and traceability is another critical aspect of data warehousing from a data governance perspective. Data lineage refers to the ability to track the origin, movement, and transformation of data as it flows through the warehouse. This capability is essential for ensuring data quality, as it allows organizations to trace errors or inconsistencies back to their source and rectify them promptly. For example, if an ML model produces unexpected results, data lineage can help identify whether the issue originated from a specific data source, an error in the ETL process, or a misconfiguration within the warehouse. Traceability is also crucial for compliance with regulatory requirements that mandate transparency in data processing and reporting. Data governance frameworks must ensure that comprehensive data lineage is maintained within the warehouse, supported by tools and processes that document every step of the data lifecycle, from ingestion to analysis.

The governance of data warehousing also involves managing the lifecycle of the data stored within the warehouse. Data lifecycle management (DLM) encompasses the policies and practices that govern the retention, archiving, and deletion of data, ensuring that data is managed efficiently and in compliance with legal and regulatory requirements. For instance, data retention policies might specify how long different types of data must be stored, while archiving policies determine when and how data is moved to long-term storage. In an ML context, where historical data is often critical for model training and validation, organizations must balance the need to retain data for analysis with the requirements to delete or anonymize data after a certain period to comply with privacy regulations. Data governance frameworks must establish clear DLM policies that align with both operational needs and regulatory mandates, ensuring that the data warehouse supports the long-term goals of the organization while minimizing legal and compliance risks.

Data warehousing also plays a significant role in supporting real-time analytics and ML model deployment. As organizations increasingly seek to leverage real-time data for decision-making, data warehouses must be capable of processing and delivering data with minimal latency. This requires the implementation of streaming data pipelines, which can ingest, process, and store data in near real time, allowing ML models to operate on the most current information available. For example, in the retail sector, real-time data on customer interactions can be used to update personalization models on the fly, delivering tailored recommendations as customers browse an online store. However, integrating real-time data into a warehouse presents challenges in terms of data consistency, latency, and processing power. Data governance frameworks must ensure that real-time data pipelines are designed to maintain data quality and consistency, with mechanisms in place to handle data discrepancies, synchronization issues, and processing delays.

The issue of data democratization is also relevant in the context of data warehousing and ML data governance. Data democratization refers to making data accessible to a broader range of users within an organization, enabling them to leverage data for decision-making without requiring advanced technical skills. A well-governed data warehouse can facilitate data democratization by providing a centralized, consistent, and user-friendly platform for accessing and analyzing data. For instance, business analysts and data scientists can use the data warehouse to run queries, generate reports, and build ML models without needing to access or integrate data from multiple disparate sources. However, data democratization must be balanced with the need to protect sensitive data and maintain data quality. Data governance frameworks must establish clear guidelines for data access and usage within the warehouse, ensuring that users have the tools and permissions they need to work with data while safeguarding the integrity and security of the warehouse.

Data Versioning and Lineage Tracking

Data versioning and lineage processes ensure that every iteration and transformation of data throughout its lifecycle is meticulously recorded and traceable, providing a clear historical record of how data has been altered or updated. Versioning allows organizations to maintain multiple versions of datasets, enabling them to track changes over time, roll back to previous versions if necessary, and compare different iterations to understand the impact of data modifications on ML model performance. This is particularly important in regulated industries, where maintaining accurate records of data usage and transformation is required for compliance and auditing purposes. By implementing robust data versioning practices, organizations can ensure the integrity and consistency of the data used in their ML pipelines, thereby enhancing the reliability and reproducibility of their models.

Lineage tracking, on the other hand, provides a detailed map of the data's journey from its origin through various processing stages to its final use in ML models. This transparency is crucial for understanding the context and provenance of data, which directly influences the trustworthiness and accountability of ML models. By maintaining comprehensive lineage records, organizations can identify the sources of data, the transformations it has undergone, and how it has been integrated with other datasets. This capability not only supports troubleshooting and debugging efforts when models produce unexpected results but also facilitates informed decision-making by providing clear insights into how data-related decisions were made. Together, data versioning and lineage tracking form the backbone of a robust data governance framework, ensuring that data quality and integrity are maintained throughout the ML lifecycle.

Reproducibility

Reproducibility in machine learning (ML) is a critical aspect that ensures models can be consistently retrained, validated, and deployed using the same data, algorithms, and processing techniques. From an ML data governance perspective, reproducibility is tightly linked to data versioning and lineage tracking, which are essential for maintaining the integrity and traceability of the data used throughout the ML lifecycle. The ability to reproduce ML experiments and results is not only a technical necessity but also a cornerstone of building trust in ML models, ensuring compliance with regulatory standards, and facilitating collaboration across teams.

Data versioning plays a crucial role in enabling reproducibility by maintaining a complete history of all changes made to datasets. Each version of a dataset is uniquely identified and stored, allowing data scientists to access specific versions at any point in time. For example, if a model trained on a particular dataset version needs to be retrained or validated, data versioning ensures that the exact same data can be retrieved and used, ensuring consistent results. This is particularly important in situations where models are deployed in production environments and need to be periodically retrained with updated data. Without robust versioning practices, it would be challenging to guarantee that the same data is used in subsequent training sessions, potentially leading to discrepancies in model performance and outcomes.

The importance of data versioning extends to scenarios where multiple teams or individuals are collaborating on a project. In large organizations, it is common for different teams to work on different aspects of an ML project simultaneously. Data versioning allows these teams to work with consistent datasets, even as changes are made. For instance, one team might be responsible for feature engineering, while another focuses on model training. By using versioned datasets, both teams can ensure that their work is based on the same data, preventing conflicts and inconsistencies. Moreover, if a particular version of the dataset leads to unexpected model behavior, data scientists can roll back to a previous version to diagnose and address the issue, further reinforcing the reproducibility of the model.

Lineage tracking complements data versioning by providing a detailed map of the data's journey through various stages of preprocessing, transformation, and integration. This transparency is crucial for understanding how data has been manipulated and how those changes impact model outcomes. For example, in a predictive maintenance application, sensor data might undergo multiple transformations, such as normalization, aggregation, and feature extraction, before being used to train an ML model. Lineage tracking ensures that every step of this process is documented, allowing data scientists to trace the origin of specific features and understand the impact of each transformation on the final model. If the model produces unexpected results, lineage tracking provides a clear path to investigate whether any of the transformations introduced errors or biases, enabling data scientists to reproduce and refine their experiments.

The combination of data versioning and lineage tracking is particularly valuable in regulated industries where compliance with data governance standards is mandatory. For example, in the healthcare sector, where patient data is used to develop diagnostic models, regulators may require that all data transformations and model training processes be fully documented and reproducible. Data versioning ensures that the exact datasets used in model development can be retrieved and audited, while lineage tracking provides a transparent record of how the data was processed. This level of traceability is essential for demonstrating compliance with regulations such as the Health Insurance Portability and Accountability Act (HIPAA) in the United States or the General Data Protection Regulation (GDPR) in Europe. By ensuring that every step of the data pipeline is reproducible, organizations can protect themselves against legal and regulatory risks.

Reproducibility also plays a critical role in addressing biases and ethical concerns in ML models. Data versioning and lineage tracking allow organizations to scrutinize the data and transformations that were used to train models, ensuring that biases are identified and mitigated. For example, if an ML model used in hiring decisions is found to be biased against certain demographic groups, data scientists can use lineage tracking to trace the origin of the bias back to specific datasets or transformations. By revisiting the data version and transformations used, they can modify the pipeline to address the bias and retrain the model, ensuring that the results are both fair and reproducible. This capability is essential for building trust in ML models, particularly in high-stakes applications where fairness and transparency are paramount.

Advanced ML workflows often involve complex data pipelines that integrate multiple data sources, transformations, and models. In such environments, ensuring reproducibility requires sophisticated versioning and lineage tracking mechanisms that can handle the complexity of the data flow. For example, in a financial institution using ML for credit risk assessment, data from various sources, such as customer transactions, credit scores, and external economic indicators, may be combined and processed through multiple steps before being used in a model. Data versioning ensures that each source dataset and intermediate dataset is stored and retrievable, while lineage tracking documents the entire process, from raw data ingestion to final model predictions. This level of detail is crucial for reproducing the entire pipeline, especially when fine-tuning or updating models based on new data.

Version control systems designed for code, such as Git, have inspired similar approaches in data management, leading to the development of tools and platforms that support data versioning and lineage tracking. These tools allow data scientists to manage datasets in much the same way they manage code, with the ability to branch, merge, and revert to previous versions as needed. This integration of version control into data management enhances reproducibility by providing a structured and auditable process for managing data changes. For example, a data scientist working on a machine learning project might use a platform that automatically tracks all versions of the dataset, records every change made during data preprocessing, and generates a lineage graph that visualizes the data flow. Such tools not only improve the reproducibility of the model but also foster collaboration by enabling teams to work on different branches of the data pipeline and merge their changes in a controlled manner.

Reproducibility is also essential for the continuous improvement and optimization of ML models. As new data becomes available or as models are retrained with updated data, being able to reproduce previous experiments and compare them with new ones is key to understanding the impact of changes. Data versioning allows data scientists to compare the performance of models trained on different versions of the data, while lineage tracking provides insights into how data transformations and feature engineering steps influence the results. For example, if a model's performance improves after adding a new data source or modifying a feature, lineage tracking can help identify which changes were responsible for the improvement. This iterative process of refining models based on reproducible experiments is crucial for achieving the best possible performance and ensuring that models remain relevant as the data landscape evolves.

Reproducibility in the context of data versioning and lineage tracking is a foundational aspect of ML data governance that ensures models are trustworthy, compliant, and capable of delivering consistent results. By meticulously recording every change to the data and documenting the entire data pipeline, organizations can reproduce their experiments, address biases, comply with regulations, and continuously improve their models. This capability is essential for building robust and reliable ML systems that can adapt to changing data and business environments while maintaining the highest standards of data quality and governance.

Disaster Recovery

Data versioning is fundamental to disaster recovery because it allows organizations to maintain historical records of datasets. Each version of a dataset is stored with a unique identifier, enabling precise tracking of changes over time. In the event of a disaster, such as a data breach, system failure, or accidental deletion, data versioning allows an organization to restore the affected datasets to their exact state before the incident occurred. For example, consider a scenario where an ML model used for financial forecasting is corrupted due to a system crash. Without data versioning, it would be nearly impossible to recover the exact dataset used to train the model, leading to potential discrepancies in the model's outputs once it is retrained. Data versioning ensures that the correct version of the dataset can be retrieved, allowing the model to be accurately reconstructed, thus preserving the integrity of the forecasting process.

Lineage tracking complements data versioning by providing a detailed map of the data's journey through various stages of processing and transformation. In a disaster recovery scenario, lineage tracking becomes critical for ensuring that not only is the correct version of the data restored, but also that all subsequent transformations and processing steps are accurately reapplied. For instance, in an ML system that processes customer data for personalized marketing, the data might go through multiple stages, including cleaning, normalization, and feature engineering. If a disaster disrupts this pipeline, simply restoring the raw data is insufficient. Lineage tracking allows the organization to reconstruct the entire pipeline, ensuring that all data transformations are correctly reapplied. This capability is crucial for maintaining the consistency and reliability of the model's predictions once operations resume.

The importance of data versioning and lineage tracking in disaster recovery extends to compliance with regulatory requirements, which often mandate the ability to recover data and demonstrate the continuity of operations. In highly regulated industries such as healthcare, finance, and telecommunications, organizations must be able to prove that their data handling processes are robust enough to withstand disruptions. For example, the General Data Protection Regulation (GDPR) in Europe requires organizations to implement measures that protect against data loss and to ensure the availability and access to personal data in the event of a technical incident. Data versioning enables the recovery of the exact data that was in use before the incident, while lineage tracking ensures that the organization can demonstrate how the data was processed, even after a disaster. This traceability is essential for audits and compliance reviews, where organizations must provide evidence of their data governance practices and disaster recovery capabilities.

One of the advanced considerations in disaster recovery is the integration of automated backup and recovery processes that leverage data versioning and lineage tracking. Automated systems can regularly snapshot datasets and track changes, storing these versions in secure, offsite locations to protect against local failures. In the event of a disaster, these systems can automatically trigger recovery processes that retrieve the latest versions of the data and reapply all necessary transformations based on the recorded lineage. For example, in an e-commerce platform that relies on ML models to recommend products, the data and model pipelines might be backed up every few hours. If a disaster strikes, the system can automatically restore the latest version of the data and the corresponding model, ensuring minimal downtime and continuity in customer recommendations. This level of automation reduces the potential for human error during recovery and ensures that the restored data and models are consistent with pre-disaster operations.

Disaster recovery planning also involves testing and validating the recovery processes to ensure they function correctly when needed. Data versioning and lineage tracking provide the framework for these tests by allowing organizations to simulate disasters and assess their ability to recover. For instance, an organization might simulate a scenario where a critical dataset is lost and then attempt to restore it using versioning and lineage records. The success of this recovery would depend on how well the versioning system captures changes and how accurately the lineage tracking documents the data's transformation journey. By regularly testing these recovery processes, organizations can identify potential gaps in their disaster recovery plans and refine their versioning and lineage tracking practices to ensure greater resilience.

Data versioning and lineage tracking also play a crucial role in protecting against data corruption, which can be particularly insidious because it might not be immediately obvious. Corrupted data can propagate through an ML pipeline, leading to flawed models and erroneous predictions. With data versioning, organizations can revert to an earlier, uncorrupted version of the data, thereby preventing the spread of corruption. Lineage tracking enables them to identify the point at which the corruption occurred, allowing for targeted interventions that correct the issue without the need to redo the entire data processing pipeline. For example, in a predictive maintenance system for industrial machinery, corrupted sensor data could lead to incorrect maintenance schedules, potentially resulting in equipment failures. By tracing the lineage of the data, the organization can pinpoint the stage where the corruption occurred, correct it, and restore the system to its original state using the versioned data, thus preventing costly downtime.

From a security perspective, data versioning and lineage tracking also contribute to disaster recovery by enabling organizations to respond to cyberattacks that compromise data integrity. In the event of a ransomware attack, where data is encrypted and held hostage, having versioned backups stored in secure, immutable locations can be the difference between paying a ransom and restoring operations independently. Lineage tracking allows organizations to verify the integrity of their data post-recovery, ensuring that no unauthorized changes were made during the attack. For instance, if a financial institution's transaction data is targeted by ransomware, the ability to restore the data to a pre-attack version and verify its integrity through lineage tracking is crucial for resuming operations without compromising customer trust or regulatory compliance.

The strategic implementation of data versioning and lineage tracking within disaster recovery plans also supports the broader goals of business continuity. By ensuring that data and models can be swiftly and accurately restored, these practices minimize downtime and reduce the financial impact of disasters. For example, in the context of a supply chain ML system, a disaster that disrupts inventory data could halt production and distribution. With versioning and lineage tracking, the organization can quickly recover the necessary data, reestablish accurate inventory levels, and resume operations with minimal disruption. This capability is particularly important in industries where even brief interruptions can have significant downstream effects, such as in manufacturing or logistics.

Mitigating Bias

Mitigating bias in machine learning (ML) is a critical concern, particularly when data versioning and lineage tracking are employed as part of an ML data governance strategy. Bias in ML models can lead to unfair outcomes, perpetuate existing inequalities, and undermine the trustworthiness of the models. From a data governance perspective, versioning and lineage tracking are not just technical tools but integral components in the ongoing effort to detect, measure, and mitigate bias throughout the ML lifecycle, particularly during data preprocessing. These processes help ensure that the data used in model training is accurate, consistent, and free from systemic biases that could affect the fairness and reliability of the models.

Bias can enter an ML model at various stages, but it often originates from the data itself. Data versioning plays a crucial role in managing and mitigating bias by allowing data scientists to track the evolution of datasets over time. When a dataset is versioned, each iteration of the data is stored with a unique identifier, making it possible to compare different versions and identify changes that might introduce or exacerbate bias. For instance, in a dataset used to train an employment recommendation system, early versions of the data might include biased samples that overrepresent certain demographics, leading to biased recommendations. By comparing these early versions with later, more refined versions that include a more balanced demographic representation, data scientists can assess the impact of data adjustments on model fairness. Data versioning ensures that these iterations are preserved, allowing for a systematic examination of how changes to the dataset influence bias in the model's outputs.

Lineage tracking complements data versioning by providing a detailed map of how data is transformed and processed through various stages before being used in an ML model. This transparency is vital for detecting and mitigating bias during data preprocessing. For example, consider a credit scoring model where customer data undergoes multiple preprocessing steps, including normalization, imputation, and feature engineering. If the data lineage reveals that certain demographic features were disproportionately normalized or that missing values were imputed using biased assumptions, these transformations could introduce or reinforce bias in the model. By examining the lineage of the data, data scientists can identify specific preprocessing steps that contribute to biased outcomes and take corrective action. Lineage tracking also enables the documentation of these corrective actions, ensuring that future audits can verify the steps taken to mitigate bias.

Detecting bias during data preprocessing requires a systematic approach that leverages both data versioning and lineage tracking. One of the key challenges in detecting bias is identifying when and where it occurs in the data pipeline. For instance, in a healthcare ML model designed to predict patient outcomes, bias might be introduced during the feature selection process if certain demographic features are given undue weight or if the data is skewed toward a particular patient population. By using data versioning, data scientists can compare different versions of the feature selection process, evaluating how changes in feature importance or selection criteria impact the model's predictions. Lineage tracking provides the context needed to understand why certain features were selected or omitted, offering insights into potential sources of bias.

Measuring bias during data preprocessing involves quantifying the extent to which certain groups are underrepresented, overrepresented, or unfairly treated by the data. One common method is to calculate fairness metrics, such as demographic parity, equalized odds, or disparate impact, which assess the relative treatment of different groups within the dataset. Data versioning is crucial in this context because it allows these metrics to be tracked over time, providing a historical record of how the fairness of the data has evolved. For example, in an ML model used for loan approval, data versioning can help track whether changes to the dataset, such as the inclusion of more diverse income groups, improve fairness metrics related to racial or gender parity. If biases are detected, lineage tracking can reveal the preprocessing steps that contributed to the biased outcomes, enabling data scientists to adjust the process and measure the impact of those adjustments.

Mitigating bias during data preprocessing often requires iterative refinement of both the data and the preprocessing techniques. Data versioning supports this iterative process by ensuring that each step is documented and that previous versions of the data can be revisited if necessary. For instance, if a bias audit reveals that a particular version of a dataset leads to unfair outcomes in an ML model, data scientists can revert to an earlier version, apply different preprocessing techniques, and compare the results. This process of iterative refinement is essential for fine-tuning the model and ensuring that bias is minimized. Lineage tracking adds an additional layer of transparency by documenting the specific transformations and decisions made at each step, allowing for a thorough examination of how these changes affect bias in the model's predictions.

An advanced topic in mitigating bias is the use of debiasing techniques during data preprocessing. These techniques can include reweighting, resampling, or the application of fairness constraints to the data. Data versioning is critical in this context because it allows data scientists to experiment with different debiasing techniques while preserving the ability to compare the outcomes of each approach. For example, in a hiring model, reweighting the data to give more importance to underrepresented groups can help mitigate bias, but it is essential to track how this reweighting impacts the model's overall performance. By versioning the dataset before and after reweighting, data scientists can evaluate whether the debiasing technique successfully reduces bias without introducing new issues. Lineage tracking ensures that the process of applying these techniques is well-documented, providing a clear record of how debiasing methods were implemented and how they influenced the final model.

The interplay between data versioning and lineage tracking is particularly important when addressing complex biases that arise from interactions between multiple features. For instance, in a financial ML model, bias might not only stem from individual features like income or education level but also from the interaction between these features. Data versioning allows for the tracking of different interaction terms and their impact on model predictions, while lineage tracking provides insights into how these interactions were identified and processed during data preprocessing. If a particular interaction term is found to contribute to biased outcomes, data scientists can modify or remove the interaction, version the dataset, and use lineage tracking to document the change. This process helps ensure that the model's fairness is systematically improved while maintaining transparency and accountability.

Data versioning and lineage tracking also play a crucial role in ensuring that bias mitigation efforts are sustainable and scalable across different ML models and datasets. In large organizations, multiple ML models might be developed simultaneously, each with its own data requirements and potential biases. Data versioning allows for the standardization of bias detection and mitigation practices across these models by providing a consistent framework for tracking changes to datasets. For example, if a particular bias mitigation technique proves effective in one model, the versioned dataset can serve as a reference for applying similar techniques to other models. Lineage tracking further supports this standardization by documenting the specific preprocessing steps that contributed to bias reduction, allowing these steps to be replicated or adapted for other models.

Another consideration is the role of human oversight in mitigating bias, particularly when using automated tools for data preprocessing. While automation can enhance efficiency, it also risks perpetuating biases if not carefully managed. Data versioning and lineage tracking facilitate human oversight by providing the tools needed to review and audit the data preprocessing process. For example, if an automated feature

198

selection tool is used, data versioning can track the different sets of features selected over time, while lineage tracking can document the criteria used by the tool to make these selections. If a bias is detected, human reviewers can use this information to adjust the tool's parameters, ensuring that the model remains fair and unbiased. This integration of human oversight into the data governance process is essential for maintaining the ethical integrity of ML models.

Bias audits are another critical aspect of mitigating bias during data preprocessing, and data versioning and lineage tracking are essential tools for conducting these audits effectively. A bias audit involves a systematic review of the data and preprocessing steps to identify potential sources of bias and evaluate the effectiveness of mitigation efforts. Data versioning provides the historical context needed for these audits, allowing auditors to trace the evolution of the dataset and assess whether bias has been reduced or exacerbated over time. Lineage tracking offers the detailed documentation required to understand the specific preprocessing decisions that contributed to bias, enabling auditors to make informed recommendations for further improvements. This auditability is crucial for ensuring that bias mitigation efforts are not only effective in the short term but also sustainable over the long term.

Data versioning and lineage tracking are also essential for addressing the ethical and legal implications of bias in ML models. As regulations around AI and ML continue to evolve, organizations are increasingly required to demonstrate that their models are fair and unbiased. Data versioning ensures that organizations can provide a complete history of the datasets used in model training, showing that they have taken steps to address bias at each stage of the data pipeline. Lineage tracking provides the necessary transparency to demonstrate that these steps were based on sound reasoning and that the resulting models are compliant with ethical and legal standards. For example, in the context of GDPR, organizations must be able to explain how personal data was processed and how decisions were made. Data versioning and lineage tracking provide the tools needed to meet these requirements, ensuring that bias mitigation efforts are both effective and compliant with regulatory expectations.

Regulatory Compliance

Data versioning is essential for regulatory compliance as it allows organizations to maintain a comprehensive history of all datasets used in ML models. This capability is particularly important when regulations require that specific versions of data be preserved for auditing purposes. For example, the General Data Protection Regulation

(GDPR) mandates that organizations must be able to demonstrate how personal data has been processed, stored, and used. Data versioning provides a detailed record of every change made to a dataset, including additions, deletions, and modifications. This historical record is invaluable when responding to regulatory inquiries, as it allows organizations to show exactly which version of a dataset was used at any point in time. For instance, if a customer requests information about how their data was used to generate a particular decision, the organization can retrieve the specific version of the dataset that was in use when that decision was made, ensuring transparency and accountability.

Lineage tracking complements data versioning by mapping the entire lifecycle of data from its source to its final use in an ML model. This traceability is critical for demonstrating compliance with regulations that require detailed documentation of data processing activities. In the context of GDPR, organizations must not only keep track of the data they hold but also provide clear documentation of how that data has been processed and transformed. Lineage tracking offers this transparency by documenting every step of the data pipeline, from initial data collection to preprocessing, feature engineering, and model training. For example, if an ML model uses customer data to predict credit risk, lineage tracking can document how the data was cleaned, normalized, and transformed before being fed into the model. This level of detail is crucial for compliance, as it ensures that the organization can demonstrate that the data was handled in accordance with legal requirements.

Detecting bias during data preprocessing is a critical component of regulatory compliance, particularly in jurisdictions that enforce fairness and non-discrimination laws. Data versioning and lineage tracking provide the tools needed to systematically identify and address bias before it becomes embedded in an ML model. For example, in the United States, the Fair Housing Act prohibits discrimination in housing-related decisions based on race, gender, and other protected characteristics. If an ML model used for mortgage approvals is trained on biased data, it could inadvertently perpetuate discriminatory practices. Data versioning allows organizations to compare different versions of the dataset, identifying changes that might introduce or exacerbate bias. Lineage tracking provides the context needed to understand how data preprocessing steps, such as feature selection or normalization, might contribute to biased outcomes. By tracing the lineage of the data, organizations can pinpoint the source of bias and take corrective action, such as adjusting the preprocessing techniques or reweighting certain features to ensure a fairer model.

Advanced regulatory requirements often mandate that organizations not only mitigate bias but also continuously monitor their models to ensure ongoing compliance. Data versioning and lineage tracking are key to this ongoing monitoring, as they provide the tools needed to audit and update models as new data becomes available or as regulations evolve. For example, an organization might develop a model for predicting loan defaults that complies with current regulations, but as new fairness standards are introduced, the model might need to be retrained on a more diverse dataset. Data versioning allows the organization to access previous datasets and compare them with newer, more inclusive versions, ensuring that the model is updated in line with regulatory expectations. Lineage tracking provides the necessary documentation to show how the model was updated and how data preprocessing was adjusted to meet the new standards.

The role of automation in regulatory compliance is another important consideration, particularly when dealing with large-scale datasets and complex ML pipelines. Automated tools for data versioning and lineage tracking can enhance the efficiency and accuracy of compliance efforts, reducing the risk of human error. For example, automated versioning systems can track every change made to a dataset in real time, ensuring that no version is lost or overwritten. Automated lineage tracking can document every step of the data pipeline, from initial ingestion to final model deployment, providing a complete and accurate record of data processing activities. This automation is particularly valuable in highly regulated industries, such as finance or healthcare, where the volume and complexity of data can make manual tracking impractical. By integrating automated versioning and lineage tracking into their data governance frameworks, organizations can ensure that they remain compliant with regulatory requirements while also improving the efficiency and accuracy of their ML workflows.

Best Practices for Data Quality and Preprocessing

Establishing robust data quality and preprocessing practices is crucial for the success of machine learning (ML) projects. In the following sections, we will explore key considerations and strategies for ensuring high-quality data.

Defining Data Quality Standards

One of the foundational elements of data quality is defining standards that align with the specific requirements of the project and domain.

Domain-Specific Relevance

First of all, data quality standards must be tailored to the specific requirements of the ML project and the domain in which they are being applied. For instance, in the financial industry, data accuracy is paramount due to the high stakes involved in decision-making, such as in credit scoring or fraud detection. A financial ML model might require precision down to the last decimal point, necessitating stringent accuracy checks and regular updates to data inputs. In contrast, an ML model used for sentiment analysis on social media data might tolerate more variability, as the nature of social media content is inherently noisy and less structured. In this context, relevance might be more important than precision, focusing on capturing the nuances of public opinion rather than ensuring exact numerical accuracy.

Defining these standards requires a deep understanding of the intended use of the data. For example, in healthcare, data quality standards must account for the critical nature of patient outcomes, where even minor inaccuracies can lead to significant consequences. Data might need to be cross-referenced with multiple sources, such as electronic health records (EHRs) and medical imaging databases, to ensure its reliability. Understanding the domain-specific relevance of data quality helps in establishing appropriate benchmarks and criteria for evaluating the data used in ML projects.

Data Fitness for Purpose

Ensuring that data is "fit for purpose" is central to data quality. This concept involves verifying that the data captures all necessary information and characteristics required to address the specific problem the ML model aims to solve. For instance, in a supply chain optimization model, the data must accurately reflect real-time inventory levels, lead times, and demand forecasts to generate reliable predictions. If the data does not meet these criteria, the model may produce suboptimal recommendations, leading to stockouts or overstocking.

Assessing the fitness for purpose often involves using techniques such as feature importance analysis or calculating correlation coefficients to identify the most relevant features. For example, in a predictive maintenance model, features like machine age, usage patterns, and historical maintenance records may be crucial. If these features are not well-represented or are of poor quality, the model's predictive power will be compromised. Ensuring that data is fit for its intended purpose requires continuous monitoring and adjustment of data quality standards to align with the evolving goals of the ML project.

Data Profiling as a Baseline

Initiating any ML project with comprehensive data profiling is essential to establish a baseline understanding of data quality. Data profiling involves analyzing the statistical properties of the data, including data types, distributions, outliers, and the presence of missing values. For example, profiling might reveal that a dataset contains a significant number of outliers in numerical fields, which could indicate data entry errors or anomalies that need to be addressed before model training.

Data profiling also helps in identifying potential data quality issues early in the process. For instance, if profiling reveals a large proportion of missing values in critical fields, such as customer transaction histories in a banking application, this might necessitate a data cleaning strategy focused on imputation or interpolation. By providing an initial assessment of data quality, profiling informs the subsequent steps of data cleaning and preprocessing, ensuring that the data is well-prepared for model training.

Data Quality Assessment

Automated Data Validation Checks

Automated data validation checks are crucial for maintaining data quality throughout the ML pipeline. These checks can be designed to identify discrepancies, errors, and inconsistencies early in the data processing stages. For example, in an e-commerce ML model predicting customer churn, automated validation might involve comparing customer transaction data against source systems to ensure accuracy. If discrepancies are detected, such as mismatched customer IDs or incorrect transaction amounts, these issues can be flagged for further investigation.

Automated validation checks are particularly important in large-scale ML projects where data is continuously ingested from multiple sources. For instance, in a real-time fraud detection system, data from credit card transactions, customer profiles, and external threat intelligence feeds must be validated to ensure that the inputs to the model are accurate and up-to-date. Implementing automated checks reduces the risk of errors propagating through the pipeline and ensures that only high-quality data is used for model training.

Data Quality Scorecards

Data quality scorecards offer a structured way to monitor and assess key data quality metrics over time. These scorecards might track metrics such as data completeness, accuracy, and the number of errors detected during validation. For example, in a healthcare ML project aimed at predicting patient readmissions, a scorecard might track the completeness of patient demographic data, the accuracy of diagnostic codes, and the frequency of missing lab results. These metrics provide a snapshot of data quality at any given point, allowing for quick identification of areas needing improvement.

Scorecards are particularly useful in regulated industries where data quality must be documented and reported to comply with legal standards. For instance, in the financial sector, regulators may require organizations to maintain detailed records of data quality assessments as part of their compliance obligations. Data quality scorecards facilitate this by providing a clear, auditable trail of the organization's data quality efforts, ensuring that the data used in ML models meets regulatory requirements.

Data Quality Dashboards

Data quality dashboards provide a visual representation of key metrics and trends related to data quality, enabling easy monitoring and identification of potential issues. For instance, a dashboard might display trends in data accuracy, highlighting periods when data quality dipped due to external factors, such as system outages or data source changes. In an ML project focused on personalized marketing, a dashboard could reveal fluctuations in customer data completeness, indicating that certain data sources are not consistently providing the necessary information.

Dashboards can be integrated into existing data governance platforms, providing a centralized view of data quality across the organization. This integration allows data scientists, data engineers, and business stakeholders to collaborate more effectively by sharing a common understanding of data quality issues. For example, in a large-scale ML deployment involving multiple teams, a centralized dashboard could help coordinate efforts to address data quality problems that impact the entire organization, such as inconsistencies in product catalog data across different regions.

Data Preprocessing

Data preprocessing is a fundamental step in preparing data for machine learning models. This stage involves a series of tasks designed to ensure that the data is accurate, relevant, and ready for analysis. One of the crucial aspects of preprocessing is data cleaning, which sets the stage for all subsequent data preparation activities.

Data Cleaning with Version Control

Data cleaning is a critical step in the preprocessing pipeline, involving the identification and correction of errors, inconsistencies, and inaccuracies in the data. For instance, in a customer segmentation model, data cleaning might involve correcting misspelled customer names, standardizing address formats, or resolving duplicate entries. These steps ensure that the data is accurate and consistent, reducing the risk of errors in the model's predictions.

Maintaining version control throughout the data cleaning process is essential for tracking changes made to the data and enabling rollbacks if necessary. For example, in a sales forecasting model, if data cleaning involves removing outliers or imputing missing values, version control allows data scientists to compare different versions of the cleaned data to assess the impact of these changes on model performance. If a particular cleaning step leads to a decline in model accuracy, version control enables a quick rollback to a previous version, allowing the data scientists to refine their cleaning strategy without starting from scratch.

Version control also facilitates collaboration among teams, as it provides a clear record of who made changes to the data and when. This transparency is particularly important in large organizations where multiple teams may be working on the same dataset. For instance, in a healthcare ML project involving data from multiple hospitals, version control ensures that all teams are working with the same version of the data, reducing the risk of inconsistencies and improving the overall quality of the model.

Feature Engineering with Documentation

Feature engineering involves creating new features from existing data to enhance the predictive power of ML models. For example, in a churn prediction model, feature engineering might involve creating a new feature that captures the number of customer support interactions within the past month. This feature could provide valuable insights into customer behavior that are not captured by the raw data alone.

Documenting the rationale behind feature engineering choices is critical for maintaining transparency and reproducibility. For instance, if a new feature significantly improves model performance, data scientists should document why this feature was created, how it was engineered, and its impact on the model. This documentation is essential for future model updates, as it allows data scientists to understand the reasoning behind past decisions and assess whether those decisions are still valid in light of new data or changes in the business environment.

Feature engineering documentation also supports collaboration and knowledge sharing within the organization. For example, in a retail ML project, documenting feature engineering techniques could help other teams working on similar models, such as those focused on inventory optimization or demand forecasting. By sharing this knowledge, organizations can build on past successes and avoid repeating mistakes, leading to more effective and efficient ML models.

Standardized Preprocessing Pipelines

Developing standardized preprocessing pipelines is a best practice that promotes consistency and reduces the risk of errors during data preparation. These pipelines encapsulate the sequence of preprocessing steps applied to the data, such as data cleaning, feature engineering, and normalization. For instance, in a credit scoring model, a standardized pipeline might include steps for handling missing values, encoding categorical variables, and scaling numerical features. By standardizing these steps, organizations ensure that all models are built on a consistent foundation, improving the reliability of their predictions.

Standardized preprocessing pipelines also facilitate scalability and repeatability in ML projects. For example, if an organization develops a successful model for predicting customer churn in one region, the same preprocessing pipeline can be applied to other regions, ensuring that the models are comparable and consistent across the organization. Standardization also reduces the risk of human error, as data scientists can rely on tested and validated pipelines rather than creating new ones from scratch for each project.

Versioning these pipelines is important for ensuring traceability and enabling rollbacks if necessary. For instance, if a new version of a preprocessing pipeline is found to degrade model performance, version control allows data scientists to revert to the previous version and investigate the issue without disrupting the entire project.

This capability is particularly important in production environments where model performance directly impacts business outcomes, such as in real-time fraud detection or personalized marketing campaigns.

Data Lineage Tracking

To ensure that data transformations are well-documented and transparent, data lineage tracking becomes crucial. This process involves two key components: documenting data transformations and mapping data lineage.

Documenting Data Transformations

Thorough documentation of all data transformations applied during preprocessing is essential for maintaining transparency and auditability in the ML lifecycle. For instance, in a healthcare ML model, documenting how patient data is normalized, how missing values are imputed, and how features are engineered ensures that the model's predictions can be traced back to their origins. This documentation is particularly important in regulated industries, where organizations must demonstrate compliance with data protection and privacy regulations.

Data transformation documentation should include details on the techniques used, the rationale behind the choices, and the impact of these transformations on the data. For example, in a predictive maintenance model, data scientists might document how sensor data is aggregated and how outliers are handled to ensure that the model accurately predicts equipment failures. This documentation not only supports compliance efforts but also facilitates model validation and debugging by providing a clear record of how the data was processed.

Data Lineage Mapping

Data lineage mapping tools provide a visual representation of the data's journey from its source to its final use in the ML model. For example, in a financial ML model, data lineage mapping might show how raw transaction data is transformed through various stages, such as data cleaning, feature engineering, and model training. This transparency is crucial for identifying potential sources of bias or error and ensuring that the model is based on accurate and reliable data.

Data lineage mapping enhances accountability by allowing organizations to pinpoint where data quality issues might have originated and assess their impact on the model. For instance, if a financial institution detects a bias in its credit scoring model, lineage mapping can help trace the bias back to specific data transformations or preprocessing steps. This capability is essential for mitigating bias and ensuring that the model complies with fairness and non-discrimination regulations.

Data lineage mapping also supports model governance by providing a clear record of how data is handled throughout the ML lifecycle. This record is invaluable for audits, as it allows organizations to demonstrate that they have followed best practices for data quality and preprocessing. For example, in a healthcare ML project, lineage mapping can show how patient data is anonymized and processed to ensure compliance with data protection regulations, such as the General Data Protection Regulation (GDPR).

Collaboration and Communication

Effective collaboration and communication are essential for maintaining high data quality and ensuring successful ML projects. This involves several key practices and roles that contribute to a well-coordinated approach to data management.

Data Stewards and Ownership

Clear ownership of data quality within the data governance framework is essential for ensuring that data quality issues are addressed promptly and effectively. For instance, in a retail ML project, a designated data steward might be responsible for overseeing the quality of sales and customer data. This role involves collaborating with data scientists and data engineers to identify data quality issues, such as missing values or inconsistencies, and coordinating efforts to resolve them.

Data stewards play a critical role in bridging the gap between business stakeholders and technical teams, ensuring that data quality standards are aligned with the organization's goals. For example, in a financial institution, a data steward might work with risk management teams to ensure that the data used in credit scoring models is accurate and up-to-date. This collaboration helps ensure that the models are reliable and that the organization can make informed decisions based on high-quality data.

Cross-Functional Training

Cross-functional training programs are essential for fostering a shared understanding of data quality concepts and best practices across the organization. For instance, training data scientists on data governance principles can help them appreciate the importance of data quality and encourage them to prioritize data validation and cleaning in their workflows. Similarly, training business stakeholders on the technical aspects of data quality can help them better understand the impact of data quality on model performance and decision-making.

Cross-functional training also promotes collaboration by enabling different teams to speak a common language when discussing data quality issues. For example, in a healthcare organization, training programs might bring together data scientists, clinicians, and IT professionals to develop a shared understanding of the importance of accurate and consistent patient data. This collaboration is essential for ensuring that data quality issues are addressed holistically and that the ML models built on this data are reliable and trustworthy.

Data Quality Champions

Identifying and cultivating "data quality champions" within the organization is a best practice for promoting awareness and continuous improvement of data quality practices. These individuals act as advocates for data quality, raising awareness of data quality issues and encouraging their colleagues to prioritize data quality in their workflows. For example, in a large enterprise with multiple ML projects, data quality champions might lead initiatives to standardize data validation checks or promote the use of automated data cleaning tools.

Data quality champions also play a critical role in driving cultural change within the organization, fostering a mindset that values data quality as a key enabler of ML success. For instance, in a tech company developing AI products, data quality champions might work with product managers to ensure that data quality considerations are integrated into the product development lifecycle. This proactive approach helps ensure that data quality is not an afterthought but an integral part of the organization's ML strategy.

Continuous Improvement

To ensure the ongoing effectiveness of data quality management, organizations must embrace a mindset of continuous improvement.

Regular Data Quality Reviews

Scheduling regular data quality reviews is essential for assessing the effectiveness of existing data quality practices and identifying areas for improvement. For example, in an e-commerce company, data quality reviews might involve evaluating the accuracy of customer and product data, identifying any trends in data errors, and assessing the impact of these errors on ML models used for recommendation systems. These reviews provide valuable insights into the strengths and weaknesses of the organization's data quality efforts and inform the development of strategies for continuous improvement.

Regular data quality reviews also help organizations stay aligned with evolving industry standards and best practices. For example, as new data quality tools and techniques emerge, organizations can incorporate these innovations into their data quality frameworks, ensuring that they remain at the forefront of data quality management. This continuous improvement mindset is essential for maintaining the reliability and effectiveness of ML models in a rapidly changing technological landscape.

Lessons Learned Documentation

Documenting lessons learned from data quality challenges is a best practice that fosters continuous improvement and knowledge sharing within the organization. For instance, if an ML project encounters data quality issues that impact model performance, documenting the root causes of these issues, the mitigation strategies employed, and the outcomes of these strategies provides valuable insights for future projects. This documentation serves as a reference for data scientists, data engineers, and business stakeholders, helping them avoid repeating the same mistakes and build on past successes.

Lessons learned documentation also supports organizational learning by capturing the collective experience of the team and making it accessible to others. For example, in a financial institution, lessons learned from a credit scoring project might be shared with teams working on other risk management models, helping them anticipate and address similar data quality challenges. This knowledge sharing fosters a culture of continuous improvement and ensures that the organization's data quality practices evolve in response to new challenges and opportunities.

Benchmarking and Industry Standards

Benchmarking data quality practices against industry standards is a best practice that ensures organizations remain competitive and compliant with evolving expectations. For instance, in the healthcare industry, organizations might benchmark their data quality practices against standards set by regulatory bodies, such as the Health Insurance Portability and Accountability Act (HIPAA). This benchmarking helps identify areas where data quality management can be optimized and ensures that the organization's practices align with industry best practices.

Benchmarking also provides a basis for setting data quality goals and measuring progress. For example, in a technology company developing AI products, benchmarking against industry leaders might involve comparing data quality metrics, such as data accuracy, completeness, and timeliness. By setting ambitious but achievable goals based on these benchmarks, organizations can drive continuous improvement in their data quality practices and ensure that their ML models are built on a solid foundation.

Integration with Broader Data Governance Framework

Integrating data quality and preprocessing practices into a broader data governance framework is crucial for ensuring that these efforts are aligned with the organization's overall data management strategies.The following sections delve into how various aspects of data governance intersect with data quality, highlighting the importance of alignment and integration for effective data management.

Alignment with Data Security Policies

Aligning data quality and preprocessing procedures with data security policies is essential for ensuring that data quality efforts are implemented while adhering to best practices for data protection. For example, in a healthcare organization, data anonymization or pseudonymization techniques might be employed during data preprocessing to protect patient privacy while maintaining data quality. These techniques ensure that sensitive data is protected in accordance with regulations, such as GDPR, while still being suitable for use in ML models.

Aligning data quality practices with data security policies also helps organizations manage the risks associated with data breaches or unauthorized access. For instance, in a financial institution, ensuring that data quality checks are performed on encrypted data can prevent unauthorized access to sensitive information while maintaining the accuracy and reliability of the data used in ML models. This alignment ensures that data quality and security are not treated as separate concerns but as complementary aspects of a holistic data governance framework.

Data Quality Impact Assessments

Conducting data quality impact assessments is a best practice that evaluates the potential impact of data quality issues on model performance and decision-making. For example, in a marketing ML model, an impact assessment might evaluate how data quality issues, such as missing customer demographic information or incorrect transaction data, affect the accuracy of customer segmentation and targeting. These assessments help prioritize data cleaning efforts and allocate resources toward addressing the most critical data quality problems.

Data quality impact assessments also support risk management by identifying potential areas where data quality issues could lead to significant negative outcomes. For instance, in a financial institution, an impact assessment might reveal that inaccuracies in credit scoring data could lead to incorrect loan approvals or rejections, exposing the organization to financial and reputational risks. By identifying these risks early, organizations can take proactive steps to mitigate them, ensuring that their ML models are reliable and trustworthy.

Data Governance Tools and Technologies

Leveraging data governance tools and technologies is essential for automating data quality checks, streamlining data lineage tracking, and facilitating data quality reporting. For instance, in a large enterprise with multiple ML projects, data governance platforms can automate the validation of data sources, track data transformations, and generate real-time data quality reports. These tools enhance the efficiency and effectiveness of data quality management efforts, ensuring that data quality is maintained throughout the ML lifecycle.

Data governance tools also support scalability by enabling organizations to manage data quality across large, complex datasets and multiple ML projects. For example, in a global technology company, data governance platforms might be used to manage data quality for ML models developed in different regions, ensuring that all models adhere to the same high standards of data quality. By integrating these tools into their data governance frameworks, organizations can ensure that their ML models are built on a foundation of reliable, high-quality data, enabling them to achieve their strategic goals and maintain a competitive edge in their respective industries.

Policies and Procedures for Data Quality and Preprocessing

Within a robust data governance framework for ML, well-defined policies and procedures for data quality and preprocessing are crucial for ensuring data reliability, consistency, and suitability for model development.

Define Data Quality Standards

To ensure the efficacy and reliability of data used in machine learning models, policies must define precise standards for various aspects of data quality. These standards serve as the foundation for assessing data across different dimensions, each crucial for maintaining the integrity and usefulness of the data.

Data Relevance

Policies must define clear criteria for assessing the relevance of data, which may include evaluating whether the data captures the necessary information and characteristics to address the intended problem. For example, in a natural language processing (NLP) project aimed at sentiment analysis, policies should mandate the use of recent and contextually relevant textual data, such as social media posts or customer reviews, rather than outdated or unrelated texts. Metrics such as feature importance analysis or correlation coefficients between features and the target variable can be employed to assess the relevance of different data attributes. This ensures that only the most impactful and contextually appropriate data is used for model training, thereby enhancing the model's ability to generalize and perform effectively.

Data Accuracy

Accuracy is a cornerstone of data quality, and policies must establish rigorous procedures for verifying the accuracy of data used in ML models. This involves comparing the data against reliable source systems, detecting outliers, and implementing data cleansing techniques to correct inconsistencies or errors. For instance, in a predictive maintenance model for industrial equipment, data accuracy can be ensured by cross-referencing sensor readings with maintenance logs and technician reports. Any discrepancies, such as a sensor reporting abnormal temperature readings that are not reflected in the maintenance logs, should be flagged for investigation. Policies should also mandate regular audits of data sources to ensure that they continue to provide accurate information. In cases where data inaccuracies are detected, predefined protocols should guide the process of rectification, such as recalibrating sensors, revising data entry procedures, or applying statistical correction methods.

Data Completeness

Policies should clearly define acceptable levels of missing data for different attributes and outline strategies for handling such cases. For example, in a clinical trial dataset, the policy might specify that no more than 5% of critical patient attributes, such as age or medical history, can be missing. Strategies for addressing missing data, such as imputation or removal, should be tailored to the specific context and documented accordingly. In a retail recommendation system, for instance, missing transactional data might be imputed using historical purchasing patterns, whereas in a high-stakes financial model, it might be more appropriate to exclude records with missing critical variables altogether. The chosen approach should be justified and documented as part of the overall data governance strategy, ensuring transparency and reproducibility.

Data Consistency

Consistency across data formats and representations is vital for ensuring that data from multiple sources can be integrated and analyzed effectively. Policies should enforce the use of standardized data dictionaries, data types, and validation rules to maintain consistency. For example, in a global supply chain ML model, policies should mandate the use of a consistent unit of measurement for all inventory data, such as converting all quantities to metric units. Inconsistencies, such as mixing metric and imperial units, can lead to significant errors in the model's predictions. Similarly, policies should ensure that categorical variables, such as product categories or customer segments, are represented consistently across different datasets, avoiding issues like case sensitivity or inconsistent labeling. Implementing these policies requires collaboration between data engineers, data scientists, and domain experts to define and enforce consistency standards across all data sources and stages of preprocessing.

Data Timeliness

For time-series data or any data where the temporal aspect is critical, policies must specify acceptable data freshness levels to ensure that the ML model is trained and deployed using the most relevant and up-to-date information. In a stock market prediction model, for example, policies might dictate that data used for training should be no more than 24 hours old to reflect the most current market conditions. Stale data can lead to models that are out of sync with current trends, resulting in poor performance and potentially costly decisions. Policies should also outline procedures for regular data refresh cycles, ensuring that the data pipeline is continuously updated to incorporate the latest information. In an IoT-based smart city traffic management system, for instance, policies might require real-time data feeds from traffic sensors to be updated every minute, with any delays or data gaps flagged for immediate attention.

Data Quality Assessment

Assessing data quality is fundamental to ensuring that the data used in machine learning (ML) models is reliable, accurate, and suitable for the intended analysis. This process involves several key activities, including data profiling, validation, monitoring, and reporting.

Data Profiling

Data profiling should be a mandated activity at the outset of any ML project, providing a baseline understanding of the data's quality and characteristics. Policies should require comprehensive profiling to analyze statistical properties, such as data types, distributions, the presence of outliers, and missing values. For example, in a fraud detection model for financial transactions, data profiling might reveal skewed distributions of transaction amounts, indicating potential anomalies or outliers that need to be addressed during preprocessing. Profiling also helps identify patterns in missing data, such as certain fields being consistently incomplete, which might suggest systemic issues in data collection. This baseline understanding informs subsequent actions, such as data cleaning and transformation, and helps ensure that the data is suitable for the intended ML task.

Data Validation

Policies should define a rigorous data validation process to verify the accuracy, consistency, and integrity of data against predefined standards. This process can involve automated data validation checks, such as ensuring that numerical fields fall within expected ranges or that categorical fields match a predefined set of allowable values. For instance, in an insurance claims processing ML model, validation checks might include verifying that claim amounts are within the expected range for the type of claim and that all required fields, such as policy number and claim date, are populated correctly. Manual reviews might also be necessary for specific data subsets, particularly in cases where automated checks cannot capture the nuances of the data. For example, in a sentiment analysis model, manual validation might involve reviewing a sample of text data to ensure that sentiment labels are correctly applied. By enforcing strict validation procedures, organizations can reduce the risk of errors and inconsistencies in the data used for ML model development.

Data Monitoring

Ongoing data monitoring is essential for maintaining data quality throughout the ML lifecycle. Policies should establish procedures for continuous monitoring to detect changes in data quality, such as the emergence of new biases, shifts in data distributions, or the introduction of inconsistencies. For example, in a dynamic pricing model for e-commerce, monitoring might reveal that the average price sensitivity of customers has changed due to seasonal trends or economic factors. Such changes could necessitate adjustments to the model's features or retraining with updated data. Automated data quality checks, such as anomaly detection algorithms, can be implemented to flag potential issues in real time, allowing for prompt corrective action. Periodic manual assessments might also be necessary to review data quality metrics and ensure that the monitoring processes are functioning effectively. This proactive approach to data quality monitoring helps maintain the reliability and performance of ML models over time.

Data Quality Reporting

Policies should mandate the generation of comprehensive data quality reports that document the findings of data profiling, validation, and monitoring activities. These reports should provide a detailed account of identified data quality issues, their potential impact on the ML model, and the proposed solutions for addressing them. For example, a data quality report for a customer churn prediction model might highlight high levels of missing data in customer interaction records, suggest imputation techniques, and estimate the potential impact on model accuracy. Such reports are essential for ensuring transparency and accountability throughout the data quality management process. They also serve as valuable documentation for audits, providing evidence that the organization is actively managing data quality risks in accordance with its governance policies.

Data Preprocessing Procedures

The following sections outline key areas of data preprocessing, each with specific policies and procedures designed to optimize data quality and model outcomes.

Data Cleaning

Effective data cleaning is crucial for addressing errors, inconsistencies, and missing values in the dataset, ensuring that the data is suitable for model training. Policies should outline specific strategies for data cleaning, such as data imputation, outlier removal, and normalization of skewed data distributions. For instance, in a loan approval ML model, data cleaning might involve imputing missing income data based on similar applicants or removing outliers in credit score data that are unlikely to represent typical applicants. The chosen cleaning techniques should be documented, along with the rationale behind their selection, to ensure that the process is transparent and reproducible. This documentation is particularly important in regulated industries, where organizations must demonstrate that their data processing methods comply with legal and ethical standards.

Data Transformation

Data transformation involves modifying features to enhance their suitability for model training, and policies should establish clear procedures for this process. For example, in an image recognition model, data transformation might involve resizing images to a standard resolution, applying data augmentation techniques, or normalizing pixel values to ensure consistent input to the model. In a text-based ML model, transformation might involve tokenizing text, removing stop words, and applying stemming or lemmatization to reduce words to their root forms. These transformations help improve the model's ability to learn from the data by ensuring that the features are represented

217

in a consistent and meaningful way. Policies should also mandate the documentation of all transformation steps, detailing the techniques used and their impact on the data. This ensures that the transformations can be replicated and that their effects on model performance can be evaluated.

Feature Engineering

Policies should provide guidelines for feature engineering, including the criteria for selecting and creating features, as well as the documentation of these processes. For example, in a predictive maintenance model, feature engineering might involve creating a new feature that captures the rate of change in sensor readings over time, which could be a strong predictor of equipment failure. The rationale behind feature engineering choices should be thoroughly documented, including how the new features are expected to enhance the model's performance and the results of any tests conducted to validate their effectiveness. This documentation is crucial for ensuring transparency and enabling future teams to understand and replicate the feature engineering process.

Data Versioning

Data versioning is essential for tracking changes made to the data during preprocessing, allowing for the rollback to previous versions if necessary, and facilitating data lineage tracking throughout the ML development process. Policies should establish clear versioning practices, including the use of version control systems that log every modification to the dataset. For example, in a financial forecasting model, versioning might involve keeping track of different versions of the dataset that reflect various stages of preprocessing, such as raw data, cleaned data, and data with engineered features. This versioning allows data scientists to compare the performance of models trained on different versions of the data, helping them identify the preprocessing steps that lead to the best outcomes. It also supports auditability, ensuring that all data transformations are documented and can be traced back to their source.

Documentation

Thorough documentation of all data preprocessing steps is essential for ensuring transparency and reproducibility in ML projects. Policies should mandate the detailed documentation of techniques used, the rationale behind their selection, and the impact of preprocessing on the data. For example, in a recommendation system for online retail, documentation might include details on how user interaction data was cleaned, how purchase histories were transformed into features, and how categorical data, such as product categories, was encoded. This documentation should be maintained in a central repository, accessible to all team members involved in the ML project, and updated regularly as new preprocessing steps are applied. Comprehensive documentation not only supports transparency but also facilitates collaboration, knowledge sharing, and the replication of successful ML models in future projects.

Governance Mechanisms and Enforcement

This section explores the key components of governance that ensure data quality management is systematically integrated into organizational practices and adhered to rigorously.

Data Quality Management Roles

Clear roles and responsibilities for data quality management are essential within the data governance framework. Policies should define specific roles, such as data stewards, who are responsible for overseeing data quality within particular domains. For example, in a healthcare organization, a data steward might be assigned to manage the quality of patient records, ensuring that data is accurate, complete, and consistent across all systems. Data stewards work closely with data scientists, engineers, and other stakeholders to identify data quality issues, implement corrective actions, and ensure that data quality standards are met throughout the ML lifecycle. By clearly defining these roles, organizations can ensure that data quality management is prioritized and that accountability is maintained.

Data Quality Training

Training programs are critical for equipping relevant personnel, such as data scientists and engineers, with the skills needed to assess, monitor, and maintain data quality. Policies should mandate regular training sessions that cover key topics, such as data quality metrics, data cleansing techniques, and best practices for data preprocessing within the ML context. For example, a training program for a financial institution might focus on techniques for detecting and correcting errors in transaction data, or methods for ensuring that customer data is consistently formatted and validated. Training programs should also emphasize the importance of data quality in ML outcomes, helping participants understand how poor data quality can lead to biased or inaccurate models. By investing in ongoing training, organizations can build a culture of data quality awareness and ensure that their teams have the knowledge and tools needed to maintain high standards.

Data Quality Audit and Review

Regular audits and reviews are essential for ensuring that data quality practices are effective and aligned with the organization's goals. Policies should establish procedures for conducting data quality audits, which might involve external data quality experts or internal data governance teams assessing data quality practices and identifying areas for improvement. For example, an audit of a retail ML model might involve reviewing the data cleaning procedures applied to sales data, evaluating the effectiveness of feature engineering techniques, and assessing the consistency of data across different sources. Audits should be conducted periodically, with the findings documented and shared with relevant stakeholders. These audits provide valuable insights into the strengths and weaknesses of the organization's data quality efforts and help ensure that best practices are continually refined and updated.

Data Quality Metrics and KPIs

Relevant data quality metrics and key performance indicators (KPIs) are essential for measuring the effectiveness of data quality initiatives. Policies should define specific metrics to track, such as data completeness rates, the number of data errors detected, or the time spent on data cleaning activities. For example, in an ML model for customer churn prediction, metrics might include the percentage of complete customer profiles, the accuracy of transaction data, or the number of outliers identified and corrected. Tracking these metrics over time provides a clear picture of data quality trends and helps organizations identify areas where data quality efforts can be optimized. KPIs should be aligned with the organization's overall goals, ensuring that data quality initiatives contribute to the success of ML projects and the broader business objectives.

Data Quality Improvement Processes

Continuous improvement processes are vital for adapting data quality practices to the evolving requirements of ML projects. Policies should outline a structured approach for incorporating feedback from data quality audits and reviews, implementing new data quality tools and technologies, and refining data quality practices as needed. For example, if an audit reveals that a specific data validation check is consistently missing errors, the policy might mandate the implementation of a new validation tool or the revision of the existing check. Continuous improvement processes ensure that data quality practices remain effective and responsive to changes in the data environment, ultimately supporting the long-term success of ML initiatives.

Integration with Broader Data Governance Framework

It is essential to integrate data quality and preprocessing procedures with broader organizational policies and frameworks. This section explores how aligning these procedures with data security policies, tracking data lineage, and implementing standardization and best practices ensures a comprehensive approach to managing data quality within the ML lifecycle.

Alignment with Data Security Policies

Data quality and preprocessing procedures must be aligned with data security policies to ensure that data quality efforts do not compromise data security. For example, in a healthcare organization, policies might require that data quality checks are performed on encrypted patient data, ensuring that sensitive information is protected while still allowing for the detection and correction of errors. Data anonymization or pseudonymization techniques might also be employed during preprocessing to balance data quality needs with data privacy considerations. These techniques ensure that data can be cleaned and transformed without exposing personally identifiable information (PII) or violating data protection regulations. By aligning data quality and security policies, organizations can ensure that their data governance efforts support both the accuracy and integrity of their data and the protection of sensitive information.

Data Lineage Tracking

Data lineage tracking is a critical component of data governance, ensuring that the origin, transformations, and usage of data are documented throughout the ML lifecycle. Policies should mandate the use of data lineage tools that provide a clear, auditable trail of how data is processed and transformed from its source to its final use in the ML model. For example, in a financial institution, lineage tracking might document how transaction data is ingested, cleaned, transformed, and used to train a fraud detection model. This transparency is essential for maintaining data quality, as it allows organizations to trace errors or inconsistencies back to their source and assess the impact of data quality issues on the model. Lineage tracking also supports compliance with regulatory requirements, as it provides evidence that the organization is following best practices for data governance and ensuring that data quality is maintained throughout the data lifecycle.

Standardization and Best Practices

Data governance frameworks can promote the development and adoption of standardized approaches for data quality assessment and preprocessing within the context of ML. This can ensure consistency and effectiveness across different ML projects within the organization. Policies should promote the development and adoption of standardized approaches that can be applied across various domains and use cases. For example, in a global enterprise with multiple ML initiatives, standardizing data validation checks, data cleaning techniques, and feature engineering practices ensures that all models are built on a consistent foundation, regardless of the specific application or region. Standardization also facilitates the sharing of best practices across teams, enabling organizations to leverage their collective knowledge and experience to continually improve data quality and preprocessing efforts. By establishing and enforcing standardized procedures, organizations can ensure that their ML models are built on high-quality, well-prepared data, ultimately leading to more reliable and effective outcomes.

Summary

This chapter explores the essential components required for effective machine learning (ML) data governance, with a particular emphasis on data quality and preprocessing. It underscores the critical role these processes play in ensuring that ML models are reliable, accurate, and trustworthy. A thorough examination of how data quality standards must be rigorously defined, assessed, and maintained to support the development of robust ML models is provided. The discussion covers the implementation of standardized preprocessing procedures, such as data cleaning, transformation, and feature engineering, which are crucial for preparing data for model training.

The importance of documenting and versioning each step in the data preprocessing pipeline is highlighted as a way to enhance transparency and reproducibility. These practices ensure that data handling processes align with regulatory compliance requirements, making them resilient to audits and legal scrutiny. The narrative also emphasizes the need for continuous monitoring and improvement of data quality practices, recognizing the importance of adaptability as new data is integrated and ML models evolve. By addressing these elements, a comprehensive guide is provided for establishing a robust data governance framework that supports successful ML initiatives.

CHAPTER 4

Data Privacy and Security Considerations

Vast amounts of information are collected, analyzed, and used to train algorithms that can identify patterns, make predictions, and automate tasks. However, this data often contains sensitive information about individuals, organizations, or financial transactions. Balancing the need for data to fuel machine learning innovation with the imperative to protect privacy and security is a critical challenge in our data-driven world. This chapter delves into the essential considerations for data privacy and security within the context of machine learning.

Introduction to Data Privacy and Security

Machine learning (ML) thrives on data, but this data often contains personal information, raising concerns about privacy and security. Data governance in ML becomes incomplete without robust frameworks that address these critical aspects. Data privacy ensures individuals have control over their personal data and how it's used, while data security safeguards that data from unauthorized access, modification, or loss. Both are fundamental for building trust and mitigating risks associated with ML deployments.

One example illustrating the challenges of data privacy in ML is the Netflix Prize competition, which aimed to improve movie recommendations. Netflix publicly released anonymized datasets of user ratings, believing they had sufficiently protected users' identities. However, researchers were able to de-anonymize a portion of the data by cross-referencing it with publicly available IMDb ratings, revealing sensitive information about users' viewing habits. This incident highlighted the risks of re-identification in seemingly anonymized data and underscored the need for more sophisticated privacy-preserving techniques in ML.

225

A. Nandan Prasad, *Introduction to Data Governance for Machine Learning Systems*, https://doi.org/10.1007/979-8-8688-1023-7_4

In another example, consider the case of an ML model used by a large retailer to predict customer pregnancy status based on purchasing patterns. The model identified a young woman as pregnant before she had informed her family, leading to targeted marketing materials being sent to her home. This breach of privacy not only caused personal distress but also raised ethical concerns about the extent of data collection and the potential misuse of sensitive information. Such scenarios demonstrate the importance of implementing data privacy measures that respect individual autonomy and avoid invasive practices.

Data breaches and privacy violations can have severe consequences for organizations and individuals alike. For instance, the 2017 Equifax data breach, which exposed the personal information of over 147 million people, serves as a stark reminder of the potential fallout from inadequate data security measures. The breach resulted in identity theft, financial losses, and a $700 million settlement with the Federal Trade Commission (FTC). In the context of ML, such breaches could expose training data containing sensitive information, leading to significant legal, financial, and reputational damage.

Privacy violations can erode user trust and lead to regulatory fines. A notable example is the British Airways data breach in 2018, where the personal and payment information of around 500,000 customers was compromised. This breach not only led to a substantial fine under GDPR but also damaged the airline's reputation, highlighting the critical need for robust data governance frameworks that include stringent security measures like encryption, access controls, and intrusion detection systems.

Beyond compliance, organizations must strive to build trust with their stakeholders. For example, Apple has emphasized privacy as a core value, implementing features like on-device processing and differential privacy to minimize data collection and protect user information. This approach has helped Apple build a reputation for strong data privacy practices, distinguishing it from competitors and fostering customer loyalty.

Data privacy and security regulations are constantly evolving, requiring data governance frameworks to be adaptable. The rise of federated learning, where models are trained across decentralized data sources without sharing raw data, presents new privacy challenges. For instance, while federated learning can enhance privacy by keeping data local, it also introduces risks such as model inversion attacks, where adversaries attempt to reconstruct training data from the model's outputs. To address these challenges, organizations must continuously evolve their data governance strategies, incorporating cutting-edge techniques like secure multi-party computation and homomorphic encryption to protect user privacy in ML environments.

Data Privacy Regulations and Compliance

Data privacy regulations are emerging worldwide, dictating how organizations collect, manage, and utilize data in the digital age. Within the framework of data governance for ML, understanding and adhering to these regulations is crucial for ensuring responsible data practices and fostering trust with users.

The landscape of data privacy regulations is not a monolithic entity. Different regions have enacted their own frameworks to safeguard individual privacy. Prominent examples include the General Data Protection Regulation (GDPR) in the European Union, the California Consumer Privacy Act (CCPA) in the United States, and the Personal Information Protection and Electronic Documents Act (PIPEDA) in Canada. These regulations often share core principles like the right to access, rectify, and erase personal data, but they may differ in specific details and enforcement mechanisms. Data governance frameworks for ML need to be comprehensive, accounting for the regulations applicable to the organization's location and the data it utilizes.

Compliance with data privacy regulations shouldn't be viewed as a mere checkbox exercise. Organizations need to establish a culture of data privacy within their ML projects. This involves embedding data privacy principles throughout the ML development lifecycle. Data governance frameworks should outline clear data collection processes, with specific consent mechanisms for individual data use. Additionally, data minimization practices, where only the data necessary for a specific ML application is collected, should be prioritized.

In the context of ML, transparency around data usage becomes paramount. Data governance frameworks need to foster open communication about how data is used for model training and decision-making. This could involve providing users with clear descriptions of the data collected, the purpose for its use, and the potential risks associated with ML deployments. Additionally, providing avenues for individuals to exercise their data privacy rights, such as opting out of data collection or requesting data deletion, strengthens trust and empowers individuals to control their data footprint within the ML ecosystem.

Data privacy regulations are not static; they constantly evolve to address new technologies and societal concerns. Data governance frameworks need to be adaptable to accommodate these changes. For instance, regulations might be updated to address the unique data privacy challenges associated with emerging technologies like facial recognition or artificial intelligence-powered recruitment tools. By remaining informed about regulatory updates and proactively adjusting data governance practices, organizations can ensure their ML endeavors remain compliant and trustworthy.

Ultimately, data privacy compliance is not just about adhering to regulations; it's about building a sustainable future for ML. By prioritizing data privacy and fostering trust with users, organizations can empower the responsible development and deployment of ML models. This trust translates into user adoption and ultimately fuels the continued innovation and positive impact of ML across various sectors. As the field of ML continues to grow, robust data governance practices that address data privacy regulations will be fundamental for harnessing the potential of ML for good.

Data Anonymization and Pseudonymization

Data anonymization and pseudonymization are data transformation techniques employed within data governance for ML to protect individual privacy. Anonymization aims to permanently remove or obfuscate personal identifiers (PII) within the data, rendering it impossible to re-identify individuals from the anonymized dataset. Here, the focus is on creating a truly anonymous experience, akin to a mask that completely conceals the wearer's identity. Common anonymization techniques include Suppression, Generalization, Aggregation and Perturbation.

Suppression involves removing sensitive attributes like names, addresses, or social security numbers entirely, effectively erasing any direct links back to individuals. However, a limitation of suppression is that it can lead to data loss, reducing the utility of the dataset for ML tasks. For example, in the healthcare industry, suppressing specific patient details can hinder the model's ability to predict patient outcomes accurately.

Generalization replaces specific data points with broader categories. For instance, a specific zip code might be replaced with a city or region, offering a less granular view of the data. While generalization can help protect privacy, it also risks oversimplifying data, which can reduce the precision of ML models. A practical example is in credit risk modeling, where generalizing financial data may lead to less accurate risk assessments, potentially impacting both the lender and borrower.

Aggregation is used to combine data points into groups, diminishing the risk of identifying individuals within the group. Imagine anonymizing purchase history data, where individual transactions are aggregated into broader categories like total spending per month for a specific demographic. However, aggregation can also obscure significant outliers or unique patterns in the data, which might be critical for detecting fraud or other anomalies in financial services.

Perturbation introduces intentional noise or modifications to the data while preserving its statistical properties relevant for ML tasks. Think of adding slight variations to dates or locations in a dataset to obscure individual identities while retaining the overall trends and patterns within the data. Although perturbation can be effective, it comes with the risk that the introduced noise might distort the data to the extent that it negatively impacts the model's accuracy. For instance, in location-based services, perturbing geographic data might lead to incorrect recommendations or navigation errors.

Despite these techniques, the risk of re-identification remains a significant challenge. For instance, in the famous case of the AOL search data release in 2006, anonymized search queries were made public. However, journalists were able to re-identify individuals by cross-referencing the anonymized data with other publicly available information, leading to severe privacy breaches. This case study highlights that anonymization is not foolproof, especially when datasets are rich in detail or can be linked with external data sources.

Genomic data presents a significant re-identification risk, even when anonymized. In one case, researchers were able to re-identify individuals who had participated in a study by cross-referencing their anonymized DNA sequences with publicly available genealogy databases. For instance, using Y-chromosome data and surname inference, researchers identified individuals who had provided their DNA for scientific research. This case underlined the challenges of protecting privacy in datasets where the data itself (like DNA) is inherently unique to the individual.

In a widely discussed case, researchers were able to re-identify individuals from anonymized health records by linking them with publicly available data. For example, in Massachusetts, then-governor William Weld's health records were re-identified by linking anonymized hospital discharge data with voter registration lists using only a few data points, such as zip code, birth date, and gender. This demonstrated that anonymization techniques like suppression or generalization could still leave enough information for re-identification when combined with other datasets.

Anonymization aims to permanently remove or obfuscate personal identifiers (PI) to ensure that individuals cannot be re-identified within a dataset. However, the process of achieving truly irreversible anonymization is far more complex than it might initially appear. One of the primary challenges is that data is rarely isolated. In modern data environments, datasets are often interconnected, either directly or indirectly, with other data sources. This interconnectedness creates opportunities for re-identification, even when anonymization techniques have been rigorously applied. For instance, removing or generalizing direct identifiers like names, social security numbers, or addresses might seem sufficient. However, indirect identifiers—or quasi-identifiers—such as birth dates, zip codes, or even patterns of behavior, can still uniquely identify individuals when combined with external data sources.

Quasi-identifiers are pieces of information that, on their own, may not reveal an individual's identity, but when combined with other quasi-identifiers or datasets, can uniquely identify someone. For example, in the famous study involving Massachusetts Governor William Weld, researchers used just three quasi-identifiers—zip code, birth date, and gender—to re-identify his supposedly anonymized health records. This case illustrates that even seemingly innocuous data points can lead to re-identification when linked with other datasets, a phenomenon that becomes more pronounced as datasets grow in size and diversity.

As datasets increase in dimensionality—that is, the number of attributes or variables included—the challenge of anonymization becomes more pronounced. Each additional attribute increases the uniqueness of data points within the dataset, making it easier to differentiate one individual from another. For instance, in large retail datasets, where numerous attributes like purchase history, time stamps, and location data are recorded, anonymizing this data while preserving its utility for machine learning is incredibly difficult. The more attributes a dataset has, the higher the likelihood that a combination of them can uniquely identify an individual, even if all direct identifiers have been removed.

The increasing availability of external data sources, such as social media, public records, and commercial databases, exacerbates the challenge of anonymization. Anonymized data can often be linked to these external sources to re-identify individuals. For example, anonymized location data from a mobile app can be cross-referenced with social media check-ins to identify where a person lives or works. The more datasets that exist and the more they can be linked, the more difficult it becomes to ensure that anonymization is truly irreversible.

Pseudonymization, on the other hand, adopts a different approach. It replaces PII with pseudonyms, artificial identifiers that don't directly reveal an individual's identity. Unlike anonymization, pseudonymization allows for the data to be potentially re-identified under specific circumstances, with the appropriate access controls and safeguards in place. This approach can be likened to a masquerade mask that allows for some level of recognition if the mask-wearer is known, but still provides a degree of anonymity within the broader crowd. Common pseudonymization techniques include tokenization and hashing.

Tokenization involves replacing PII with unique, random tokens. Imagine substituting a customer name with a randomly generated alphanumeric code, severing the direct link between the data and the individual. While tokenization is effective in reducing re-identification risks, it relies heavily on the security of the tokenization system. If the tokens and the original data are not stored separately and securely, the system becomes vulnerable to attacks, as was demonstrated in a breach of a major payment processing company, where poorly implemented tokenization led to the exposure of millions of credit card details.

Hashing involves applying a one-way cryptographic function to PII, generating a unique identifier (hash) that cannot be easily reversed to reveal the original data. This approach is akin to creating a fingerprint for the data, but without the ability to reconstruct the original details from the fingerprint alone. However, hashing is not entirely secure, as demonstrated by the LinkedIn data breach in 2012. Hackers were able to reverse-engineer hashed passwords using brute-force attacks, exposing millions of user accounts. This incident illustrates that while hashing can obscure data, it is not infallible, especially if the hash function used is weak or if attackers have significant computational resources.

While both anonymization and pseudonymization offer privacy benefits, they present challenges for data governance in ML. The risk of re-identification, the potential loss of data utility, and the vulnerabilities in pseudonymization techniques underscore the need for continuous assessment and improvement of privacy-preserving methods. In real-world scenarios, the effectiveness of these techniques often depends on the context in which they are applied and the robustness of the overall data governance framework. As the field of ML continues to evolve, so too must the strategies for safeguarding privacy, requiring ongoing innovation and vigilance to protect individuals' data in increasingly complex data environments.

Data Utility vs. Privacy

Extensive anonymization techniques can significantly alter the data, potentially compromising its quality and reducing its usefulness for ML models. Striking a balance between protecting privacy and maintaining data utility is crucial, yet this balance is not a one-size-fits-all proposition. Different industries face unique challenges and have varied approaches to managing the trade-offs between data privacy and utility. Data governance frameworks need to assess the specific privacy risks associated with the data and select anonymization or pseudonymization techniques that minimize the impact on data quality. However, the real-world implications of these decisions are often complex and carry significant consequences.

In healthcare, the protection of patient privacy is paramount due to the sensitive nature of medical data. Anonymization techniques like data masking, generalization, and k-anonymity are commonly used to de-identify patient records before sharing them for research or public health purposes. However, these techniques can sometimes strip away critical details needed for accurate medical research, such as the exact sequence of treatment events or rare conditions that might only appear in highly granular data. The trade-off here often leans more heavily toward privacy, particularly in regions governed by strict regulations like HIPAA in the United States, but this can come at the cost of potentially slower or less precise medical advancements.

In a case where a dataset was anonymized to protect patient identities, researchers found that it was impossible to track the effectiveness of specific treatments for rare diseases due to the loss of detailed temporal information. This resulted in a delay in identifying effective treatment protocols, illustrating the tension between patient privacy and the utility of medical data.

In the retail sector, detailed customer data is crucial for personalization, targeted marketing, and customer experience optimization. However, anonymization techniques that remove too much detail can diminish the effectiveness of these marketing efforts. Retailers often face a trade-off between respecting customer privacy and leveraging detailed purchasing behavior data to drive sales. Unlike healthcare or finance, where the consequences of reduced data utility can be more severe, the retail industry might be more willing to accept some level of risk in favor of maintaining data utility.

A major retail chain implemented a pseudonymization strategy that replaced customer identifiers with pseudonyms to protect privacy while still allowing for targeted marketing. However, this approach led to less effective personalization because the pseudonyms could not fully capture the nuances of individual customer behaviors, resulting in lower engagement rates and reduced sales.

The choice between prioritizing privacy or data utility has real-world implications that extend beyond just the technical quality of the data. These decisions can affect organizational outcomes, regulatory compliance, and public trust.

Impact on Innovation

In industries like healthcare and finance, where innovation relies heavily on data-driven insights, overly aggressive anonymization can stifle innovation. If critical details are lost in the anonymization process, organizations may struggle to develop new products, services, or treatments that rely on precise data.

Regulatory Compliance vs. Operational Efficiency

Strict privacy regulations often compel organizations to prioritize privacy over utility, but this can lead to inefficiencies. For example, compliance with GDPR may require extensive anonymization that makes data less useful for operational purposes like customer service optimization or fraud prevention. Organizations must navigate this balance carefully to avoid compromising their ability to operate effectively while still meeting regulatory requirements.

Customer Trust and Brand Reputation

The way organizations handle the balance between privacy and utility also impacts customer trust and brand reputation. Consumers are increasingly aware of privacy issues, and organizations that fail to protect personal data can face backlash. On the other hand, if data is anonymized to the point where services become less personalized or effective, customers might lose interest, impacting loyalty and sales.

Re-identification Risks

Anonymization techniques, while essential for protecting privacy, are increasingly being challenged by the sophistication of re-identification attacks. The growing availability of external data sources, combined with the power of advanced technologies like artificial intelligence (AI) and machine learning (ML), has made it easier for malicious actors to re-identify individuals from anonymized datasets. This reality exposes the limitations of traditional anonymization methods and underscores the need for more robust and adaptive data governance frameworks.

Re-identification attacks have evolved significantly in recent years, driven by advancements in AI and ML. These technologies can analyze vast amounts of data, recognize patterns, and correlate information across multiple datasets with unprecedented efficiency. For example, AI algorithms can rapidly process and cross-reference anonymized data with publicly available information, social media profiles, or commercial databases, making it easier to pinpoint an individual's identity even when direct identifiers have been removed.

In a well-publicized case, researchers used a machine learning algorithm to re-identify individuals in an anonymized genomic dataset by cross-referencing it with genealogical databases. Despite the removal of direct identifiers like names and addresses, the unique patterns in the genetic data, combined with external genealogical information, enabled the identification of individuals and their relatives. This example highlights how AI can overcome traditional anonymization efforts, posing significant risks to privacy, particularly in sensitive domains like genomics.

The risk of re-identification is further amplified in large, interconnected datasets. As organizations increasingly aggregate and analyze data from various sources, the complexity of these datasets grows, creating more opportunities for re-identification. Even seemingly innocuous data points, when combined with other datasets, can reveal unique patterns that lead to re-identification.

Consider the anonymized location data shared by a mobile app developer. Although the dataset had removed specific user identifiers, AI-powered analysis can reveal that patterns in the location data, such as frequent visits to certain addresses at specific times, could be linked to individual identities by cross-referencing with social media check-ins. This kind of re-identification attack illustrates the vulnerability of anonymized data in the age of big data and AI.

AI is not only used for re-identification but is also challenging the very foundations of anonymization. Advanced AI techniques, such as deep learning, can infer missing information in anonymized datasets, predict likely data points, or even reconstruct original datasets from anonymized versions. This predictive capability makes traditional anonymization methods increasingly inadequate.

Deep learning models have been used to reverse-engineer anonymized image datasets. In one study, researchers trained a neural network to predict the original features of faces that had been blurred or pixelated, effectively de-anonymizing the images. This technology demonstrates that anonymization techniques that were once considered secure can be rendered obsolete by AI, as these models can fill in the gaps left by anonymization processes.

Given the increasing sophistication of re-identification attacks, data governance frameworks must evolve to address these new challenges. Merely relying on traditional anonymization techniques is no longer sufficient. Organizations must implement additional safeguards to mitigate re-identification risks effectively.

Advanced Anonymization Techniques

Data governance frameworks should incorporate advanced anonymization techniques such as differential privacy, which introduces mathematical noise to the data in a way that maintains overall data utility while reducing the likelihood of re-identification. Differential privacy is particularly effective against AI-driven attacks because it limits the amount of information any one individual contributes to the dataset, making it more difficult for AI to accurately reconstruct original data. The U.S. Census Bureau has adopted differential privacy techniques for the 2020 census data to protect individual privacy while still providing useful data for policymakers and researchers. This approach reduces the risk of re-identification from AI-driven analysis while maintaining the overall statistical integrity of the data.

Data Access Controls and Monitoring

Implementing strict data access controls and continuous monitoring is essential. By limiting who can access sensitive data and under what conditions, organizations can reduce the risk of re-identification. Additionally, monitoring access patterns and data usage can help detect and prevent potential re-identification attempts. A financial institution might restrict access to anonymized transaction data, only allowing access to authorized personnel under specific conditions. Furthermore, by implementing real-time monitoring of data access, the institution can quickly identify and respond to any suspicious activities that might indicate an attempt to re-identify individuals.

Anonymization Reversal Procedures

In cases where anonymization techniques are applied, organizations should have procedures in place to reverse anonymization if necessary. These procedures should be tightly controlled and audited, ensuring that re-identification can only occur under legally sanctioned circumstances, such as during a criminal investigation. In the healthcare industry, a hospital might anonymize patient data for research purposes but retain the ability to re-identify patients if new information becomes available that could significantly impact their treatment. Such reversal procedures would be governed by strict ethical and legal guidelines to protect patient privacy.

Regular Audits and Updates

Data governance frameworks should include regular audits of anonymization techniques and data protection measures. As AI and re-identification technologies evolve, so too must the strategies to counteract them. Continuous updates to anonymization methods and the adoption of new technologies are critical to staying ahead of potential threats. A tech company might conduct annual audits of its anonymization techniques, employing third-party experts to assess the effectiveness of its privacy measures against the latest AI-driven re-identification methods. Based on the audit results, the company can update its practices to enhance data protection.

Pseudonymization and Reversibility

Pseudonymization offers the possibility of re-identification, which creates its own set of challenges and risks. While it is a valuable technique for protecting privacy, it is crucial to recognize that pseudonymization does not eliminate the risk of re-identification but rather mitigates it by separating identifiable information from the dataset. This separation, however, introduces new vulnerabilities, particularly around key management and the potential for re-identification through data breaches or insider threats.

The effectiveness of pseudonymization largely depends on the secure management of the pseudonymization keys—those unique identifiers that can link the pseudonymized data back to the original personal information. If these keys are not adequately protected, the entire pseudonymization process can be compromised, leading to a breach of privacy. Consider a healthcare provider that uses pseudonymization to protect patient records. The pseudonymization keys are stored in a secure database, separate from the pseudonymized records. However, if the key management system is not adequately secured, a malicious actor who gains access to these keys can re-identify the patient data, effectively undoing the pseudonymization. This scenario illustrates that even with pseudonymization in place, poor key management can nullify the privacy protections, exposing sensitive health information.

Another significant risk associated with pseudonymization is the potential for re-identification through insider threats or data breaches. Unlike fully anonymized data, pseudonymized data can be re-identified if someone with access to the pseudonymization keys decides to misuse their privileges or if these keys are exposed in a data breach. This vulnerability is particularly concerning in environments where multiple stakeholders or systems have access to the pseudonymized data and the corresponding keys. In a financial institution, pseudonymization is used to protect customer transaction data during analysis. However, if an insider with access to both the pseudonymized data and the keys decides to misuse this access—perhaps by selling the re-identified data on the dark Web—the privacy of thousands of customers could be compromised. This risk is exacerbated if the organization does not have strong monitoring and auditing processes in place to detect and prevent unauthorized access to the pseudonymization keys.

Challenges in Key Management and Access Controls

Effective key management is critical to maintaining the integrity of pseudonymization, yet it poses several challenges:

- **Key Storage Security:** Pseudonymization keys must be stored in a highly secure environment, often involving encryption and access control mechanisms. However, if the storage environment itself is compromised, such as through a data breach, the keys—and thus the pseudonymized data—become vulnerable to re-identification. A government agency stores pseudonymization keys in an encrypted database, but the database is compromised during a cyberattack. Despite the encryption, if the attackers manage to decrypt the keys, they can re-identify sensitive personal data from previously pseudonymized government records, leading to potential misuse of that data.

- **Access Control and Role-Based Permissions:** Robust access controls are necessary to ensure that only authorized personnel have access to pseudonymization keys. Implementing role-based access control (RBAC) can help limit who can access, manage, or use these keys, reducing the risk of insider threats. However, managing these permissions across large organizations can be complex and prone to errors. In a large corporation, access to pseudonymization keys is restricted to a small team of data engineers. However, due to a misconfiguration in the RBAC system, an unauthorized employee gains access to these keys and inadvertently re-identifies a dataset containing employee performance reviews, leading to a significant breach of privacy within the organization.

- **Key Rotation and Expiry:** Regularly rotating pseudonymization keys and setting expiration dates can help mitigate the risk of long-term exposure in the event of a breach. However, this requires careful management to ensure that key rotation does not disrupt ongoing operations or analysis that relies on consistent pseudonymization. A retail company implements a policy of rotating pseudonymization keys every six months. During one rotation, a miscommunication between the IT and data analytics teams leads to the loss of access to historical pseudonymized data, which disrupts critical analytics projects. This incident highlights the operational challenges associated with key rotation and the need for coordinated management processes.

Mitigating the Risks of Re-identification in Pseudonymization

Given these risks, it is essential for data governance frameworks to implement comprehensive strategies for managing and protecting pseudonymization keys:

1. **Separation of Duties:** Implementing a separation of duties principle ensures that no single individual has access to both the pseudonymized data and the keys required for re-identification. This reduces the risk of insider threats and ensures that multiple parties must be involved in any re-identification process. For example, in a pharmaceutical company, the data management team is responsible for maintaining the pseudonymized data, while the IT security team manages the pseudonymization keys. Any request for re-identification requires the approval and involvement of both teams, ensuring that no single party can compromise the pseudonymization process.

2. **Strong Encryption and Multi-factor Authentication (MFA):** Encrypting pseudonymization keys and requiring multi-factor authentication for access adds an additional layer of security, making it significantly more difficult for unauthorized users to gain access to these critical assets. A financial institution may use multi-factor authentication for any access to the pseudonymization key database, coupled with strong

238

encryption. Even if a hacker gains access to one factor (such as a password), they cannot access the keys without the second factor, significantly reducing the risk of re-identification through a breach.

3. **Continuous Monitoring and Auditing:** Regular monitoring and auditing of access to pseudonymization keys are essential for detecting any unauthorized attempts to access or misuse these keys. Automated alerts and audit logs can help organizations respond quickly to potential threats. An e-commerce company could implement continuous monitoring of its key management system, with automated alerts set up to notify the security team of any unusual access patterns or failed login attempts. This proactive approach helps the company quickly detect and respond to potential breaches, ensuring the continued security of pseudonymized customer data.

4. **Incident Response Plans:** In the event of a breach, having a well-defined incident response plan that includes specific procedures for addressing the exposure of pseudonymization keys is crucial. This plan should outline the steps to be taken to mitigate the impact of the breach, such as key revocation, re-encryption of data, and notification of affected individuals. A healthcare provider experiences a data breach that exposes pseudonymization keys. The provider's incident response plan includes immediate key revocation and re-encryption of the affected datasets, as well as notifying patients whose data may have been compromised. This swift response helps limit the damage and maintains trust with patients.

In conclusion, while pseudonymization is a powerful tool for protecting privacy, it introduces its own set of challenges, particularly around key management and the risks of re-identification through data breaches or insider threats. Data governance frameworks must address these challenges by implementing robust key management practices, strong access controls, continuous monitoring, and comprehensive incident response plans to ensure the effectiveness of pseudonymization in safeguarding sensitive data.

Key Strategies

To ensure data anonymization and pseudonymization contribute effectively to data privacy within data governance for ML, organizations can implement the several strategies.

Data Classification and Risk Assessment

Classify data based on its sensitivity and the potential privacy risks associated with its disclosure. Data governance frameworks can then determine the appropriate level of anonymization or pseudonymization required for different data categories. For instance, anonymizing purchase history data might be less critical compared to anonymizing medical records. Some of the steps that organizations can take are as follows:

> **Develop a Data Classification Schema:** Create a detailed schema that categorizes data based on sensitivity, such as public, internal, confidential, and highly sensitive. This schema should be aligned with industry standards and regulatory requirements.

> **Conduct a Privacy Risk Assessment:** For each data category, perform a risk assessment that evaluates the potential impact of data breaches or re-identification. This assessment should consider factors like data volume, the likelihood of re-identification, and the potential harm to individuals.

> **Map Data Flows:** Document how data moves through your systems, identifying points where sensitive data is accessed, stored, or processed. This will help in determining where anonymization or pseudonymization is most needed.

> **Define Anonymization and Pseudonymization Policies:** Based on the classification and risk assessment, establish clear policies that dictate the level of anonymization or pseudonymization required for each data category. For example, medical records might require strong anonymization, while internal business data might only need pseudonymization.

Privacy-Enhancing Technologies (PETs)

Leverage PETs like differential privacy, which adds controlled noise to data to protect individual privacy while preserving its statistical properties for ML algorithms. Imagine adding slight variations to purchase amounts within a dataset – this injects noise that protects individual spending habits while still allowing the model to learn overall spending patterns. Below are the some of the guidelines for PETs:

> **Evaluate and Select PETs:** Assess the available PETs such as differential privacy, homomorphic encryption, and secure multi-party computation. Choose technologies that align with your specific ML use cases and data sensitivity levels.

> **Integrate PETs into Data Pipelines:** Embed the selected PETs directly into your data pipelines. For instance, apply differential privacy at the data ingestion stage to ensure that all data entering the ML pipeline has the necessary privacy protections.

> **Test PETs for Effectiveness:** Regularly test the PETs to ensure they provide the intended level of privacy without excessively degrading data utility. This can be done through simulations or real-world testing with anonymized data.

> **Monitor and Update PETs:** As new PETs emerge or as regulatory requirements evolve, continuously monitor the effectiveness of your existing technologies and update them as necessary to maintain robust data privacy.

Purpose-Based Data Transformation

Anonymize or pseudonymize data only to the extent necessary for the specific ML application. This helps to minimize the impact on data utility while still achieving the desired level of privacy protection. There might be situations where a less rigorous anonymization technique suffices, depending on the specific needs of the ML model. Below are some of the actions that organization can take to implement Purpose-Based Data Transformation.

Define Specific Use Cases: Clearly define the purpose of data usage within each ML project. This should include the specific outcomes, the necessary data granularity, and the acceptable trade-offs between privacy and utility.

Tailor Anonymization to Purpose: Apply anonymization or pseudonymization techniques that are appropriate for the specific ML use case. For example, if the ML model only needs aggregated data, generalization or aggregation can be applied to reduce privacy risks without significantly impacting data utility.

Document Transformation Decisions: Maintain detailed documentation of the data transformation decisions, including the rationale for choosing specific techniques. This documentation should be reviewed periodically to ensure it remains relevant and effective.

Data Sandboxes and Secure Enclaves

Create secure environments where anonymized or pseudonymized data can be accessed and analyzed for ML purposes. This restricts access to the data and reduces the risk of unauthorized disclosure. Imagine creating a virtual vault for the anonymized data, accessible only to authorized personnel for ML development. Implementing Data Sandboxes require some specific actions like

Set Up Data Sandboxes: Create isolated environments where anonymized or pseudonymized data can be accessed securely for development and testing purposes. These sandboxes should be designed to prevent data from being exported or accessed by unauthorized users.

Implement Secure Enclaves: For highly sensitive data, use secure enclaves that leverage hardware-based security to protect data during processing. These enclaves ensure that even if the environment is compromised, the data remains protected.

Control Access to Sandboxes and Enclaves: Implement strict access controls for data sandboxes and secure enclaves. Access should be granted based on the principle of least privilege, ensuring that users only have access to the data necessary for their tasks.

Audit Access Logs: Regularly audit access logs to detect any unauthorized access attempts or unusual activity within these secure environments. Any anomalies should be investigated promptly to mitigate potential risks.

Data Governance Training and Awareness

Educate data scientists, data engineers, and other stakeholders involved in ML projects about data anonymization and pseudonymization techniques, their limitations, and the importance of data governance practices in ensuring responsible data use. By fostering a culture of data privacy awareness, organizations can encourage responsible use of anonymization and pseudonymization techniques throughout the ML development lifecycle.

Develop a Comprehensive Training Program: Create a training program for all employees involved in ML projects, focusing on data anonymization, pseudonymization, and overall data governance practices. This program should be tailored to different roles, from data scientists to IT staff.

Regularly Update Training Content: Ensure that the training content is regularly updated to reflect the latest privacy regulations, technologies, and best practices. Include real-world case studies to illustrate the potential consequences of poor data governance.

Foster a Culture of Privacy: Encourage a culture of privacy by integrating data governance discussions into regular team meetings, promoting transparency about data handling practices, and recognizing employees who demonstrate strong privacy practices.

Conduct Periodic Assessments: Periodically assess the effectiveness of the training program through quizzes, simulations, and feedback from participants. Use these assessments to identify areas for improvement and to reinforce key concepts.

Publish Data Anonymization Policies: Publicly share your organization's data anonymization and pseudonymization policies, explaining the measures taken to protect user data. This builds trust with customers and stakeholders by demonstrating your commitment to privacy.

Establish Clear Accountability Mechanisms: Define roles and responsibilities within the organization for managing data privacy, including who is accountable for breaches or failures in anonymization efforts. This should include a clear chain of command for incident response.

Implement a Breach Notification Protocol: Develop a protocol for notifying affected individuals and regulators in the event of a data breach involving anonymized or pseudonymized data. This protocol should outline the steps to be taken, including investigation, mitigation, and communication.

Regularly Review Governance Practices: Conduct regular reviews of your data governance practices to ensure they remain effective in protecting privacy. These reviews should involve key stakeholders from across the organization and should result in actionable recommendations for improvement.

Key Future Trends

As the field of machine learning (ML) continues to evolve, new technologies and methodologies are emerging that promise to enhance the effectiveness and ethical considerations of data handling. This section explores some of the most promising trends that are shaping the future of data privacy and security in ML.

Homomorphic Encryption

This emerging technology allows computations to be performed on encrypted data without decrypting it. This could enable ML models to analyze data while it remains anonymized, potentially offering a more robust approach to privacy-preserving ML. Imagine training an ML model on encrypted medical data, allowing the model to learn patterns from the data without ever revealing the identities of the patients involved.

Federated Learning

This approach trains ML models on decentralized datasets without directly sharing the data. This can minimize the need for data anonymization or pseudonymization, as the raw data never leaves its original location. However, data governance frameworks need to adapt to address privacy challenges associated with federated learning environments. Federated learning offers a promising approach for privacy-preserving ML, but data governance needs to ensure robust security measures are in place to protect data privacy across decentralized locations.

Privacy by Design

Incorporating privacy considerations throughout the entire ML development lifecycle, from data collection to model deployment, will be crucial for fostering responsible data practices and building trust with users. By prioritizing privacy from the outset, organizations can ensure data anonymization and pseudonymization techniques are implemented effectively and ethically within their ML projects.

Data anonymization and pseudonymization are valuable tools in the data governance toolkit for ML. They enable organizations to leverage data for valuable insights while safeguarding individual privacy. However, it's a continuous process – balancing the need for robust privacy protections with the utility of the data for ML models. By implementing well-defined data governance practices, embracing emerging technologies, and fostering a culture of transparency and accountability, organizations can ensure data privacy remains a priority in the exciting world of machine learning. Ultimately, this responsible approach paves the way for building trustworthy ML models that empower innovation and address real-world challenges while safeguarding individual privacy, leading to a more positive future for all.

Data Encryption and Access Controls

Data encryption scrambles data using cryptographic algorithms, rendering it unreadable to anyone without the decryption key. This ensures even if unauthorized individuals gain access to the data, they cannot decipher its meaning. Data governance frameworks for ML leverage various encryption techniques to protect data at rest, in transit, and in use.

Data Encryption at Rest

This technique encrypts data when it is stored on devices like hard drives or cloud storage platforms. Encryption algorithms like Advanced Encryption Standard (AES) transform the data into an unreadable format, requiring a decryption key to unlock its true meaning. This safeguards data from unauthorized access even if storage systems are compromised.

Data Encryption in Transit

When data is transferred between systems, such as during data collection or model training, it becomes vulnerable to interception. Data encryption in transit protects the data during these transfers. Techniques like Secure Sockets Layer (SSL)/Transport Layer Security (TLS) protocols encrypt data communication channels, ensuring only authorized systems can access the data in its readable form.

Data Encryption in Use

While less common, data encryption in use scrambles data even while it is being processed for ML tasks. This technique offers an additional layer of security, especially for highly sensitive data. Homomorphic encryption, a developing technology, allows computations to be performed on encrypted data without decryption, potentially offering a more robust approach for privacy-preserving ML.

Data encryption alone is not sufficient. Data governance frameworks need robust access controls to determine who can access the data, what actions they can perform, and what data they are authorized to see.

User Authentication and Authorization

User accounts are established within the data governance framework, and each user is assigned a unique identifier and credentials. Authentication verifies user identities, while authorization determines the specific permissions each user has for accessing, modifying, or deleting data. For instance, data scientists working on a specific ML project might be granted access to relevant datasets, but not to other sensitive data used for different ML applications.

Role-Based Access Control (RBAC)

This approach assigns permissions based on user roles within the organization. Data governance frameworks define roles like data scientist, data analyst, and data administrator, with each role having pre-defined access levels to specific data sets and functionalities. RBAC simplifies access management and reduces the risk of unauthorized access by ensuring users only have the level of access necessary for their specific roles.

Attribute-Based Access Control (ABAC)

This more granular approach considers various attributes beyond user roles when determining access. These attributes could include the type of data being accessed, the purpose of access, or the user's location. Imagine an ABAC system granting access to a specific dataset only if the user is a data scientist working on a specific ML project and accessing the data from a secure company network.

Data Governance Challenges

While encryption protects data at rest and in transit, it also poses challenges for model training. Traditional encryption methods render data unusable for algorithms, necessitating decryption before processing. This creates a security vulnerability as decrypted data becomes susceptible to unauthorized access. Homomorphic encryption, a promising solution, enables computations on encrypted data without decryption, but its computational overhead often limits its practical application in large-scale machine learning scenarios.

Another significant challenge lies in balancing data privacy with model performance. Differential privacy, a technique that adds noise to data to protect individual privacy, can impact model accuracy. Striking the right balance between privacy and utility is crucial, requiring careful consideration of data sensitivity and model requirements.

Access controls are equally complex in the machine learning context. Determining appropriate access levels for data scientists, engineers, and other stakeholders is challenging. Overly restrictive access can hinder collaboration and innovation, while overly permissive access increases the risk of data breaches. Role-based access controls (RBAC) can be effective, but they must be carefully defined and enforced to ensure appropriate data access.

Moreover, the dynamic nature of machine learning introduces additional complexities. As models evolve and new data is incorporated, access controls must be continuously evaluated and adjusted. This requires robust monitoring and auditing mechanisms to detect and address potential security breaches.

The increasing use of cloud-based platforms for machine learning introduces further challenges. While cloud providers offer various security features, organizations must carefully assess their security posture and implement additional controls to protect sensitive data. Data sovereignty and compliance with data residency regulations also become important considerations.

The rise of federated learning, where models are trained on decentralized data, exacerbates privacy and security concerns. Ensuring data privacy while enabling model collaboration is a complex challenge that requires innovative solutions.

Overcoming Challenges

Navigating the complexities of machine learning (ML) involves addressing various challenges related to data security, model performance, and privacy concerns. This section explores key issues such as balancing encryption with model performance and managing the trade-offs between privacy and accuracy.

Balancing Encryption and Model Performance

When integrating encryption technologies into ML workflows, it's essential to balance the need for data security with the impact on model performance. Encryption methods, such as homomorphic encryption and differential privacy, offer robust privacy protections but can introduce significant computational overheads and affect the accuracy of ML models.

Honest Assessment of Homomorphic Encryption

Homomorphic encryption (HE) is a promising technique that allows computations to be performed on encrypted data without requiring decryption, thus protecting data privacy even during processing. However, it is important to acknowledge the significant limitations of this technology, particularly its computational cost and the slow adoption rate in the industry. Current implementations of HE are computationally intensive and can drastically slow down the processing time, making it impractical for many real-time applications or large-scale machine learning (ML) models.

Actionable Strategies

To address the challenges associated with homomorphic encryption, consider the following strategies:

Selective Application: To mitigate the computational burden, consider applying homomorphic encryption selectively to the most sensitive data subsets, such as financial transactions or personal health records, while using less computationally expensive techniques like tokenization or pseudonymization for other data. This approach helps preserve data utility while managing the performance trade-offs.

Hybrid Encryption Models: Combine homomorphic encryption with traditional encryption methods. For example, use symmetric encryption for large data sets and apply homomorphic encryption only when computation on encrypted data is necessary. This hybrid approach can optimize performance while still benefiting from the privacy guarantees of HE.

Performance Optimization: Collaborate with technology vendors and research institutions to explore cutting-edge optimizations for homomorphic encryption. This might include exploring more efficient encryption schemes or leveraging hardware acceleration (such as GPUs or FPGAs) to improve computational efficiency.

Differential Privacy and Model Accuracy Trade-offs

Differential privacy (DP) introduces noise to datasets to protect individual privacy, but this can degrade model accuracy. The challenge lies in balancing privacy with the need for accurate ML models. There are a few actionable strategies worth considering:

> **Feature-Specific Differential Privacy:** Apply differential privacy selectively to sensitive features within a dataset, such as personal identifiers or medical history, while allowing less sensitive data to remain unaltered. This approach minimizes the impact on model accuracy by protecting only the most critical information.

> **Local Differential Privacy (LDP):** Implement LDP, which applies differential privacy at the data source rather than in a centralized manner. This method is particularly useful in distributed systems or when collecting data from multiple users. By protecting data before it is aggregated, LDP reduces the overall privacy-accuracy trade-off.

> **Regular Calibration:** Continuously calibrate the amount of noise added through differential privacy based on the specific use case and the acceptable accuracy threshold for the ML model. This calibration should be data-driven, taking into account the sensitivity of the data and the model's performance requirements.

Refining Access Controls

Traditional role-based access controls (RBAC) can be restrictive and may not accommodate the dynamic needs of modern data-driven environments. To address this, organizations should consider more advanced and flexible access control models.

Innovative Approaches to Granular Access Controls

> **Attribute-Based Access Control (ABAC):** Enhance RBAC with ABAC, which takes into account additional attributes such as the data's sensitivity, the user's clearance level, and environmental factors (e.g., time of access, location). ABAC allows for more granular and context-aware access policies.

Dynamic Access Controls: Implement dynamic access controls that adapt based on user behavior and real-time data usage patterns. These controls can adjust permissions dynamically, increasing or restricting access as needed based on detected behaviors or anomalies.

Anomaly Detection and Automated Responses: Integrate anomaly detection systems with access controls. These systems can monitor user activity in real time, identifying unusual access patterns that may indicate a security threat. Upon detection, automated responses such as temporary access suspension or alerts can be triggered to mitigate risks.

Zero-Trust Security Model: Adopt a zero-trust security model where every access request is verified based on multiple factors, regardless of the user's location within the network. This model assumes that threats can originate from inside and outside the organization, thus every access request is treated with skepticism and requires re-authentication.

Data Governance Framework and Culture

A comprehensive data governance framework is essential for addressing encryption and access control challenges. Clear policies, procedures, and standards should be established, encompassing data classification, encryption, access controls, and incident response.

Regular security audits and vulnerability assessments are indispensable. These evaluations should focus on identifying weaknesses in encryption, access controls, and data handling practices. Employee training is crucial to foster a security-conscious culture. Data privacy and security awareness programs should be mandatory for all employees, emphasizing the importance of protecting sensitive information.

Actionable Strategies

Develop Clear Data Governance Policies: Establish detailed policies that cover all aspects of data governance, including data classification, encryption protocols, access control mechanisms, and incident response procedures. These policies should be aligned with regulatory requirements and industry best practices.

251

Regular Security Audits: Conduct frequent security audits and vulnerability assessments to identify and address weaknesses in encryption and access controls. These audits should be comprehensive, covering both technical implementations and procedural adherence.

Employee Training Programs: Create mandatory, role-specific training programs for all employees, particularly those involved in data handling and ML development. Training should emphasize the importance of data security, the specific encryption and access control practices in place, and the role each employee plays in maintaining data integrity.

Collaboration and Partnerships

Collaboration between data scientists, security experts, and legal teams is essential. Data scientists can provide insights into data requirements for model development, while security experts can assess risks and recommend appropriate safeguards. Legal teams can ensure compliance with regulations and provide guidance on data privacy. Partnerships with technology providers can accelerate the adoption of advanced security solutions. Consider collaborating with specialized encryption, privacy, and security vendors to leverage their expertise.

Actionable Strategies

Cross-Functional Teams: Form cross-functional teams that include data scientists, security experts, legal advisors, and compliance officers. These teams should regularly collaborate to assess the risks associated with data usage and develop integrated security solutions.

Partner with Specialized Vendors: Collaborate with technology vendors that specialize in encryption, privacy, and security. These partnerships can provide access to the latest innovations and expertise, helping organizations stay ahead of emerging threats.

> **Industry Consortia and Knowledge Sharing:** Join industry consortia or working groups focused on data security and privacy. These forums can provide valuable insights, share best practices, and facilitate the adoption of new technologies through collective learning.

Continuous Evaluation and Improvement

Data governance is an ongoing process. Regularly review and update encryption and access control policies to align with evolving threats and business requirements. Conduct privacy impact assessments to evaluate the potential impact of new technologies and data practices.

Actionable Strategies

> **Regular Policy Reviews:** Schedule periodic reviews of encryption and access control policies to ensure they remain aligned with the latest threat landscapes and business needs. These reviews should be informed by ongoing risk assessments and feedback from stakeholders.

> **Privacy Impact Assessments (PIAs):** Conduct PIAs whenever new technologies, data practices, or ML models are introduced. These assessments evaluate the potential privacy risks and recommend appropriate safeguards.

> **Iterative Improvement Cycles:** Implement an iterative cycle of testing, feedback, and improvement for all data security practices. This ensures that encryption and access control mechanisms are continuously optimized based on real-world performance and emerging threats.

Data Masking and Obfuscation

Data masking and obfuscation are data transformation techniques that alter sensitive information within a dataset while aiming to maintain its statistical properties relevant for ML models. Data masking replaces sensitive data with fictitious values, while data obfuscation aims to distort the data in a way that renders it unusable for re-identification purposes. Data governance frameworks can leverage various techniques to achieve these goals.

Data masking techniques offer a way to protect sensitive information within datasets while still allowing for their use in machine learning models. Substitution, permutation, and aggregation are three common masking approaches, each with its own strengths and considerations.

Substitution involves replacing sensitive data points with entirely different, but realistic, values. Imagine a dataset containing customer names. Substitution might replace real names with fictitious names generated based on real name patterns. This preserves the overall structure of the data (e.g., name length, distribution of first letter) while obscuring the identities of individual customers. However, the effectiveness of substitution depends on the type of data being masked. For highly sensitive data like social security numbers, simply replacing the numbers with random digits might not be sufficient to prevent re-identification, especially if additional data sources are available.

Permutation, on the other hand, focuses on rearranging real data points within a dataset. Imagine anonymizing purchase history data. Permutation might shuffle zip codes within the dataset, altering the specific locations where purchases occurred, while maintaining the overall distribution of purchases across different geographic regions. This technique can be effective for anonymizing location data or other categorical variables. However, permutation might not be suitable for numerical data, as rearranging values could significantly alter statistical properties relevant for ML models.

Aggregation involves combining data points into broader categories that mask individual identities. Imagine anonymizing financial data. Aggregation might combine income ranges into monthly spending brackets, obscuring individual spending habits while preserving overall spending trends. This technique is particularly useful for numerical data, allowing for the creation of anonymized datasets that still retain valuable statistical information for model training. However, the level of aggregation needs to be carefully considered. Overly broad aggregation can render the data unusable for ML tasks that require more granular insights.

Data obfuscation techniques take a different approach to data masking, aiming to distort the data itself to make re-identification difficult. Character swapping, adding noise, and k-anonymity are three common obfuscation methods, each offering varying levels of protection and introducing different considerations.

Character swapping involves a seemingly simple yet effective tactic: exchanging specific characters within data points. Imagine anonymizing patient names. Character swapping might replace the first and last letters of names. This approach maintains a similar name structure while obscuring the original identities. However, the effectiveness of character swapping depends on the data type. For short data points like names, it might be easier to reverse the obfuscation, especially with access to additional information. Additionally, for numerical data, character swapping could significantly alter the data's meaning, rendering it unusable for ML tasks.

Character swapping is a basic yet widely used technique in data masking and obfuscation. While the example of anonymizing patient names through character swapping is a simple illustration, real-world scenarios, especially in sensitive fields like healthcare and finance, require more complex implementations and considerations.

In healthcare, data masking is critical for protecting patient privacy, particularly when sharing medical records for research, clinical trials, or compliance purposes. Consider a scenario where a hospital needs to share patient data with a third-party research organization. The dataset includes sensitive information such as patient names, dates of birth, medical conditions, and treatment histories. Simply swapping characters in patient names (e.g., changing "John Doe" to "Dohn Joe") is insufficient in a healthcare context. While this may obscure the identity on the surface, it does little to protect against re-identification, especially when the dataset contains other quasi-identifiers like dates of birth or ZIP codes. A more sophisticated approach would involve combining character swapping with other masking techniques such as hashing or tokenization to create a layered defense.

When applying character swapping to numerical data such as patient IDs or medical codes (e.g., ICD codes), the potential for rendering data unusable is significant. In medical datasets, even minor alterations to codes can lead to incorrect diagnoses, treatment errors, or misinterpretation of data by ML models. To mitigate this, character swapping might be selectively applied only to non-critical fields, while critical fields are masked using more appropriate techniques like encryption or format-preserving tokenization.

Adding noise introduces controlled variations or random modifications to sensitive data points. Imagine anonymizing birthdates. Adding noise might involve slightly shifting birth years or dates within a permissible range. This technique injects a layer of uncertainty that makes it difficult to re-identify individuals based on their anonymized data. However, the amount of noise added needs to be carefully balanced. Excessive noise can distort the data and reduce its statistical value for ML models. Finding the right balance between privacy protection and data utility is crucial for this approach.

K-anonymity offers a more robust approach to obfuscation. It ensures that every combination of k identifying attributes appears at least k times within the anonymized dataset. Imagine anonymizing medical data. K-anonymity might achieve this by grouping patients based on age, gender, and zip code – ensuring there are always at least k patients with the same combination of these anonymized attributes. This significantly reduces the risk of re-identification by making it difficult to pinpoint a specific individual within the anonymized data. However, achieving k-anonymity often requires significant data modification or aggregation, which can impact the data's usability for ML tasks, especially for smaller datasets.

While data masking and obfuscation offer privacy benefits, they present challenges for data governance in ML.

Data Utility vs. Privacy

Extensive masking or obfuscation techniques can significantly alter the data, potentially compromising its quality and reducing its usefulness for ML models. Striking a balance between protecting privacy and maintaining data utility is crucial. Data governance frameworks need to assess the specific privacy risks associated with the data and select masking or obfuscation techniques that minimize the impact on data quality. Imagine obfuscating a dataset to such an extent that income ranges become meaningless for training an ML model focused on predicting loan defaults.

Re-identification Risks

Masking techniques might not be perfect, and with access to additional data sources, it might be possible to re-identify individuals from masked datasets. Data governance frameworks need to consider the potential re-identification risks associated with different masking techniques and implement additional safeguards like data access controls and anonymization reversal procedures (where applicable) to further mitigate these risks. For instance, masking medical data by replacing patient names might still allow for re-identification if the data is combined with other datasets containing patient location information.

Data Skew and Bias

Masking or obfuscation techniques can inadvertently introduce bias or skew into the data, potentially impacting the performance or fairness of ML models. Data governance frameworks need to be vigilant about monitoring for unintended bias introduced during data masking or obfuscation processes and implement strategies to mitigate these biases. Imagine masking income data in a way that disproportionately affects a specific demographic group, potentially leading to biased outcomes in an ML model used for loan approvals.

Data governance frameworks can address these challenges by implementing the several strategies.

Data Classification and Risk Assessment

Classify data based on its sensitivity and the potential privacy risks associated with its disclosure. Data governance frameworks can then determine the appropriate level of masking or obfuscation required for different data categories. High-risk data like medical records might require more rigorous masking techniques compared to anonymizing purchase history data.

Privacy-Preserving Machine Learning (PPML)

Explore PPML techniques that allow for training ML models directly on masked or obfuscated data. This eliminates the need for complete data de-identification while still enabling the development of privacy-preserving ML models.

Differential Privacy

This technique adds controlled noise to data to protect individual privacy while preserving its statistical properties for ML algorithms. Imagine adding slight variations to purchase amounts within a dataset – this injects noise that protects individual spending habits while still allowing the model to learn overall spending patterns.

Data Governance Training and Awareness

Educate data scientists, data engineers, and other stakeholders involved in ML projects about data masking and obfuscation techniques, their limitations, and the importance of data governance practices in ensuring responsible data use. Fostering a culture of data privacy awareness can encourage responsible use of masking and obfuscation techniques throughout the ML development lifecycle.

Data masking and obfuscation offer valuable tools for data governance in ML. However, they are not a silver bullet. Transparency and accountability remain essential principles. Organizations need to be transparent about their data masking and obfuscation practices, informing users about the measures taken to protect their privacy. Users have the right to understand how their data is being used and to what extent it is being masked or obfuscated.

Data governance frameworks need to establish clear accountability mechanisms for data breaches or misuse of masked or obfuscated data. This ensures appropriate responses and mitigation strategies are implemented in case of incidents. Even with masked data, a data breach could still have reputational consequences for the organization, highlighting the importance of robust data governance practices.

Data masking and obfuscation are valuable tools in the data governance toolkit for ML. They enable organizations to leverage data for valuable insights while safeguarding individual privacy. However, it's an ongoing process, akin to a magician perfecting their illusions. Data governance frameworks need to continuously assess the balance between data utility and privacy, while mitigating the risks of re-identification and bias. By implementing well-defined data governance practices, embracing emerging technologies, and fostering a culture of transparency and accountability, organizations can ensure data masking and obfuscation contribute to responsible data governance in the exciting world of machine learning. Ultimately, this approach paves the way for building trustworthy ML models that empower innovation and address real-world challenges while safeguarding the privacy of individuals in the ever-evolving digital landscape.

Data Retention and Disposal Policies

Data retention policies directly influence the quality and relevance of data available for machine learning (ML) model development and maintenance. Retaining historical data for extended periods can provide valuable insights into trends and patterns, which are crucial for model accuracy and predictive power. However, the balance between retaining valuable data and managing data volume is delicate. If outdated or irrelevant data is kept too long, it can degrade model performance by introducing noise or biases that no longer reflect current realities.

In the financial services industry, data retention policies are crucial for the development and ongoing refinement of credit scoring models. Retaining extensive historical data allows institutions to train models that can accurately predict default risks across different economic cycles. However, retaining outdated financial data—such as credit behavior from a decade ago—may introduce biases that do not reflect current consumer behavior or economic conditions. This can result in models that either overestimate or underestimate credit risk, leading to poor lending decisions.

To address this, financial institutions may implement a tiered data retention policy. This policy could involve keeping detailed transaction data for a shorter period (e.g., five years) while retaining aggregated or anonymized historical data for longer durations to capture broader trends without the risk of outdated data impacting model accuracy.

Improper data disposal can have severe legal and regulatory consequences, particularly in industries governed by strict data protection laws such as the General Data Protection Regulation (GDPR) in Europe or the Health Insurance Portability and Accountability Act (HIPAA) in the United States. Failure to properly dispose of data, especially personal or sensitive information, can lead to data breaches, regulatory fines, and legal action from affected individuals or organizations.

In healthcare, improper disposal of patient records—whether physical or digital—can result in significant legal repercussions under HIPAA. For example, if a healthcare provider fails to securely delete electronic health records (EHRs) after the mandated retention period, and these records are later accessed in a data breach, the organization could face substantial fines and legal liability. Beyond financial penalties, the provider's reputation could suffer, leading to a loss of patient trust.

To mitigate these risks, healthcare organizations must implement stringent data disposal policies, including the use of certified data destruction services for physical records and secure deletion methods for digital records. These methods might include data wiping, degaussing, or even physical destruction of storage devices to ensure that no recoverable data remains.

Techniques for Data Retention and Disposal

This section explores various methods and considerations for handling inactive data and the importance of proper data management techniques.

Data Retention

Different types of storage solutions and archival techniques come into play depending on the data's relevance and usage requirements.

Archiving Inactive Data

Archiving is not merely about moving inactive data to long-term storage; it requires careful consideration of how this data might be used in the future. For example, in regulatory investigations, archived data may be required to demonstrate compliance or to provide evidence in legal proceedings. In the pharmaceutical industry, data from clinical trials must be retained for several years post-study to comply with regulatory requirements. Archived trial data needs to be stored in a way that is both secure and accessible, should it be needed for future research or regulatory review.

Data Warehouses and Data Lakes

Data warehouses often serve as the backbone for historical analysis and trend identification. However, retaining data indefinitely can lead to excessive storage costs and management complexity. Conversely, data lakes, which store raw data in its original format, can accumulate outdated or irrelevant data over time, necessitating periodic cleansing. In a retail setting, a data warehouse might retain sales data for a decade to analyze long-term customer trends, while a data lake could store raw transaction logs for a shorter period. Implementing retention policies that differentiate between these storage types helps manage costs and maintain data relevance.

Data Disposal

When data is no longer needed, it must be disposed of securely to prevent unauthorized access or recovery. Effective disposal techniques are critical.

Permanently Erasing Data

Simply deleting data from a system does not guarantee it is irretrievable. Incomplete deletion can leave data fragments recoverable, especially in cloud environments where multiple backups might exist. Secure deletion practices, including data wiping or cryptographic erasure, ensure that data cannot be recovered. In the financial sector, improper disposal of customer data, such as credit card numbers, can lead to severe breaches. To prevent this, financial institutions often use secure deletion methods, including overwriting data multiple times, to ensure it cannot be retrieved.

Anonymization and Pseudonymization

While anonymization and pseudonymization are effective for protecting privacy, they must be applied in a way that balances the need for privacy with the potential future utility of the data. For example, pseudonymized data may still be valuable for longitudinal studies in healthcare, where patient outcomes need to be tracked over time. In the telecom industry, customer call records might be pseudonymized to allow for service quality analysis while protecting individual privacy. However, after the analysis is complete and the retention period has expired, these records must be securely deleted to comply with privacy regulations.

Challenges and Mitigations

Storing data beyond its useful life consumes resources and increases the risk of data breaches. Data governance frameworks need to establish clear criteria for determining data retention periods based on legal requirements, business needs, and potential future use cases. Implement automated data retention policies that trigger data deletion or archiving once data reaches the end of its useful life. Regularly audit these policies to ensure compliance.

Premature data disposal can lead to the loss of valuable information for legal compliance, auditing purposes, or future ML projects. Data governance frameworks need to ensure data retention policies consider regulatory compliance timelines and potential future research needs. Establish minimum retention periods based on legal and business needs, and only dispose of data after these periods have lapsed. Include safeguards that require multiple approvals before data can be permanently deleted.

Locating and identifying all data relevant to a specific retention or disposal policy can be difficult, especially within complex data ecosystems. Data governance frameworks should promote data lineage practices to track the movement and transformation of data throughout its lifecycle. Use data lineage tools to map the flow and transformation of data across systems, ensuring that all relevant data is accounted for in retention and disposal policies.

Ensuring complete data deletion across all storage locations can be challenging. Data governance frameworks need to implement robust deletion procedures that account for potential backups or archived copies of the data. Maintain regular backups and implement version control systems that allow for data recovery in case of accidental deletion. Use access controls to limit who can delete data.

Accidental deletion of critical data can significantly disrupt operations. Data governance frameworks need to establish data backup procedures and implement safeguards to prevent accidental data deletion. Failure to comply with data retention requirements mandated by regulations or legal proceedings can result in penalties or legal action. Data governance frameworks need to ensure data retention policies align with relevant legal and regulatory obligations.

Data retention practices can perpetuate biases present in the original data. Data governance frameworks need to consider potential biases and implement strategies to mitigate them, such as periodic data cleansing or incorporating diverse data sources over time. Periodically review and cleanse historical data, incorporating more diverse datasets over time to mitigate biases in ML models.

Regulations like the General Data Protection Regulation (GDPR) grant individuals the right to request the erasure of their personal data. Data governance frameworks need to establish procedures for handling such requests and ensuring compliance with relevant privacy regulations.

Data Retention and Disposal Policies

Effective data retention and disposal policies are crucial components of a robust data governance framework for ML. Several key elements need to be considered while preparing such policies.

Alignment with Business Needs

The data retention period should be determined by the business value of the data and its potential future use cases. A cost-benefit analysis can be helpful in defining appropriate retention periods, balancing the need for data availability with storage costs and privacy concerns.

Legal and Regulatory Compliance

Data retention policies must adhere to relevant laws and regulations governing data privacy and security. This might include regulations like the GDPR, HIPAA (Health Insurance Portability and Accountability Act) in the healthcare sector, or industry-specific data retention requirements. Data governance frameworks need to stay updated on evolving legal landscapes and ensure policies adapt accordingly.

Data Classification

Implementing a data classification system allows for differentiated retention and disposal approaches based on data sensitivity. Highly sensitive data might require stricter anonymization or shorter retention periods compared to less sensitive data. Data governance frameworks should define clear classification criteria and corresponding retention and disposal practices.

Transparency and Accountability

Data retention and disposal policies should be transparent and readily accessible to all stakeholders. Individuals should be informed about how long their data will be retained and how it will be disposed of. Data governance frameworks need to establish clear accountability mechanisms for ensuring adherence to these policies.

Regular Review and Updates

Data retention and disposal policies are not static documents. They need to be reviewed and updated periodically to reflect evolving business needs, technological advancements, and changes in legal and regulatory requirements. Data governance frameworks should establish a regular review process for these policies.

Data Subject Requests

Data governance frameworks need to incorporate procedures for handling data subject requests related to data retention and disposal. This might include requests for access to personal data, rectification of inaccurate data, or deletion requests under regulations like the GDPR. Clear processes for handling such requests are essential to ensure transparency and compliance.

Integration with Broader Data Governance Framework

Data retention and disposal policies should not exist in isolation. They need to be integrated seamlessly with the broader data governance framework, aligning with data security, access control, and data privacy principles.

By considering these elements, data governance frameworks can establish effective data retention and disposal policies. These policies safeguard privacy, minimize legal and regulatory risks, and ensure the responsible management of data throughout its lifecycle within ML projects. Ultimately, this fosters trust and transparency in the use of data for machine learning, paving the way for ethical and responsible advancements in the field.

Privacy by Design and Data Protection Impact Assessments (DPIAs)

Privacy by Design (PbD) is a proactive approach that integrates privacy considerations throughout the entire ML development lifecycle. It emphasizes building privacy safeguards directly into the design and development of ML models, rather than attempting to address privacy issues as an afterthought.

PbD emphasizes collecting and utilizing only the minimum amount of data necessary for the intended ML task. This minimizes the potential privacy risks associated with collecting and storing vast quantities of data. Data governance frameworks should encourage data minimization practices and establish clear guidelines for determining the minimum data requirements for specific ML projects.

Data Anonymization and Pseudonymization techniques modify data to remove personally identifiable information (PII) or replace it with non-identifiable pseudonyms. PbD advocates for incorporating anonymization or pseudonymization techniques early in the data collection and processing stages, reducing the risks associated with storing and processing PII. Data governance frameworks should define clear guidelines for when and how anonymization or pseudonymization techniques should be implemented.

Privacy-enhancing technologies (PETs) encompass a range of tools and techniques designed to protect privacy while enabling data analysis and utilization. PbD encourages the exploration and integration of PETs, such as homomorphic encryption, which allows computations on encrypted data without decryption. Data governance frameworks can promote awareness and adoption of PETs within ML projects.

PbD emphasizes building ML models that are transparent and explainable. This allows individuals to understand how their data is used in the model and the rationale behind its decisions. Data governance frameworks should encourage the development of explainable AI techniques and promote transparency in how data is utilized within ML projects.

Data Protection Impact Assessments (DPIAs) are systematic processes designed to identify, assess, and mitigate potential privacy risks associated with data processing activities, including those within ML projects. DPIAs play a crucial role in ensuring compliance with data privacy regulations like the General Data Protection Regulation (GDPR).

DPIAs are mandatory for processing activities that pose a high risk to individuals' privacy. Data governance frameworks should establish clear criteria for identifying high-risk ML projects based on factors like the type of data collected, the purpose of the ML model, and the potential impact on individuals.

DPIAs involve a comprehensive analysis of potential privacy risks associated with the ML project. This includes factors like data breaches, unauthorized data access, or discriminatory outcomes produced by the model. Data governance frameworks should provide resources and guidance for conducting thorough risk assessments within the context of ML projects.

DPIAs culminate in the development of mitigation strategies designed to address identified privacy risks. This might involve implementing additional privacy safeguards, such as data access control mechanisms or anonymization techniques. Data governance frameworks should ensure that data controllers (organizations using data for ML) have effective processes for implementing mitigation strategies identified through DPIAs.

Privacy by Design (PbD) and Data Protection Impact Assessments (DPIAs) play a critical role in ensuring responsible data governance for machine learning (ML) projects. PbD emphasizes building privacy safeguards directly into the design and development of ML models, rather than addressing privacy concerns as an afterthought. This proactive approach minimizes privacy risks throughout the ML lifecycle.

PbD encourages practices like data minimization (using only the minimum data needed), anonymization or pseudonymization (removing or replacing personal identifiers), and utilizing privacy-enhancing technologies (PETs) to protect data while enabling analysis. Additionally, PbD promotes transparency and explainability in ML models, allowing individuals to understand how their data is used.

DPIAs are systematic evaluations that identify, assess, and mitigate potential privacy risks associated with processing data in ML projects. They are mandatory for high-risk processing activities defined by data privacy regulations like GDPR.

DPIAs help organizations develop strategies to address identified privacy risks. This might involve implementing additional data security measures, adopting alternative data minimization techniques, or incorporating fairness and bias detection mechanisms within the ML model itself.

By integrating privacy considerations from the outset and conducting regular risk assessments, PbD and DPIAs minimize the potential for privacy breaches and misuse of data in ML projects. DPIAs ensure compliance with data privacy regulations, protecting organizations from legal and reputational risks.

PbD and DPIAs foster a culture of responsible data practices within organizations, building trust with individuals whose data is used for ML projects. PbD principles encourage transparency in how data is used and how ML models arrive at decisions, giving individuals greater control over their information.

PbD and DPIAs are essential tools for ensuring that ML projects leverage data responsibly and ethically, safeguarding individual privacy while enabling the development of innovative and beneficial applications.

Data Security Monitoring and Incident Response

Machine learning models, by their nature, are particularly susceptible to a range of data security threats due to the sensitive information they often process, such as customer records, financial transactions, or personally identifiable information (PII).

The complexity of these models further complicates the detection of security breaches, making continuous data security monitoring not just important, but essential.

For instance, a financial institution using an ML model to detect fraudulent transactions might inadvertently expose sensitive customer data during the training process if proper security measures are not in place. Data breaches in this context can lead to unauthorized access to training datasets or the models themselves, potentially compromising both the integrity of the data and the security of the institution's systems. Monitoring tools like Security Information and Event Management (SIEM) systems, such as Splunk or IBM QRadar, are critical in such scenarios. These tools collect and analyze log data from various points in the ML pipeline, enabling real-time detection of unauthorized data access or suspicious activity. For example, if an unexpected surge in data access requests is detected from a user account during non-business hours, the SIEM system can flag this as a potential security incident, prompting immediate investigation.

In cloud-based ML environments, where data and models are distributed across various locations, the challenges multiply. The dynamic nature of cloud infrastructures, with data constantly moving between different storage systems and processing units, makes it difficult to maintain visibility and control. Cloud Security Posture Management (CSPM) tools, like Prisma Cloud, can help by continuously scanning cloud environments for misconfigurations or vulnerabilities. For instance, CSPM might identify an improperly configured storage bucket that inadvertently exposes sensitive training data to the public internet, a situation that could lead to a massive data breach if not quickly rectified. This proactive identification of vulnerabilities allows organizations to address potential threats before they are exploited.

Another critical aspect of data security monitoring in ML is the detection of data poisoning attacks, where malicious actors inject corrupt data into the training datasets. This type of attack can subtly manipulate the behavior of the ML model, leading to biased or inaccurate outputs that can have far-reaching consequences. For example, in a healthcare application, a poisoned dataset might cause an ML model to misdiagnose diseases, potentially endangering patients' lives. To counter this, anomaly detection algorithms within a User and Entity Behavior Analytics (UEBA) system can monitor data inputs for unusual patterns that deviate from the norm, such as unexpected spikes in certain types of input data that could indicate an ongoing poisoning attempt. These systems, like Exabeam, can provide alerts when anomalous data behaviors are detected, allowing data scientists to investigate and remove compromised data before it adversely affects the model.

Model theft is another significant threat, particularly when considering the intellectual property value of sophisticated ML models. In industries like finance or pharmaceuticals, where ML models can represent a significant competitive advantage, the theft of a model could allow an adversary to replicate or undermine the original organization's capabilities. Monitoring tools that track model access and usage, such as Endpoint Detection and Response (EDR) systems like CrowdStrike Falcon, play a vital role here. These tools can detect unauthorized attempts to access or extract the model, whether it's deployed on-premises or in the cloud. For example, if an EDR system detects that a model file is being accessed by an unfamiliar IP address, it can immediately restrict access and alert the security team, preventing potential theft.

Privacy violations also represent a critical concern, especially as ML models increasingly rely on large datasets that might include sensitive personal information. Regulations like GDPR impose strict requirements on how such data can be handled, and failure to comply can result in severe penalties. Continuous monitoring ensures that data anonymization and pseudonymization techniques are correctly applied throughout the ML pipeline. For example, a Data Loss Prevention (DLP) system might monitor outgoing data streams to ensure that no raw PII leaves the organization's secure environment, automatically blocking or encrypting such data if it detects a compliance violation.

Moreover, the integration of monitoring tools with robust incident response protocols is key to mitigating the impact of security incidents. When a breach is detected, the immediate availability of detailed logs and real-time activity records facilitates rapid identification of the breach's origin, enabling a more precise and effective response. For example, if a monitoring system detects an unauthorized model download, a Security Orchestration, Automation, and Response (SOAR) platform could automatically isolate the affected system, initiate a data wipe, and notify all relevant stakeholders, thus containing the threat and preventing further damage.

In a distributed ML environment, where models are trained and deployed across multiple cloud services, maintaining this level of visibility and control is challenging. Multi-cloud security management tools offer a solution by providing a unified security interface across different cloud platforms. This allows organizations to enforce consistent security policies and monitor data access across all environments. For instance, a multi-cloud security tool might detect that sensitive data is being transferred from one cloud service to another without the appropriate encryption, triggering an automatic halt to the transfer and notifying the security team to investigate.

Continuous data security monitoring is crucial for detecting and responding to threats promptly. By proactively monitoring data access, usage, and anomalies, organizations can identify potential security incidents before they escalate into major breaches. It not only enhances an organization's ability to detect and respond to threats but also supports compliance with data privacy regulations and strengthens overall risk management strategies. By identifying and addressing vulnerabilities in real time, organizations can prevent breaches that might otherwise compromise the integrity of their ML models. This proactive approach to security fosters a culture of vigilance, ensuring that data protection remains a priority throughout the ML lifecycle. Integrating these practices into the broader data governance framework sends a clear message to all stakeholders about the critical importance of data security in the era of machine learning.

Data security monitoring empowers proactive threat detection. By continuously analyzing data access patterns, usage trends, and system activity, organizations can identify suspicious activity early on. Anomaly detection algorithms can flag unusual access attempts, data modifications, or potential breaches, allowing for swift intervention before significant damage occurs. This proactive approach minimizes potential losses and fosters a more secure environment for data-driven initiatives.

Data security monitoring strengthens incident response capabilities. When a security breach does occur, the insights gleaned from monitoring can significantly improve the response process. Detailed logs and activity records facilitate pinpointing the source of the incident, enabling containment efforts to be targeted effectively. Having a clear understanding of the attack vector allows for faster eradication and minimizes the potential for further compromise. This translates to reduced downtime, cost savings, and the ability to restore trust in the integrity of the ML system.

Data security monitoring fosters better compliance with data privacy regulations. As the use of sensitive data in ML models grows, adhering to stringent data privacy regulations becomes paramount. Continuous monitoring allows organizations to demonstrate compliance by providing a clear audit trail of data access and usage. This can involve logging user activity, tracking data transformations, and ensuring anonymization techniques are implemented effectively. By showcasing a commitment to data security through robust monitoring practices, organizations can build trust with regulators and stakeholders.

Data Security Monitoring Strategies

The first strategy focuses on **access control and user activity monitoring**. This involves tracking all attempts to access data used for ML projects. Information such as user identities, timestamps, and the type of data accessed is meticulously logged. Anomaly detection algorithms can be utilized to identify unusual access patterns, such as attempts to access data outside of normal working hours or from unauthorized locations. This continuous vigilance helps to identify potential insider threats or unauthorized attempts to access sensitive data.

Another crucial strategy involves **data usage monitoring**. This goes beyond simply tracking who accesses the data and focuses on how the data is used within the ML lifecycle. Monitoring techniques track the data transformations applied during preprocessing, the data used for model training, and the data utilized for model predictions. By scrutinizing data usage patterns, organizations can ensure data is employed only for authorized purposes and prevent unauthorized modification or manipulation of data used for training ML models.

Data integrity monitoring plays a vital role in safeguarding data security. This strategy involves regularly assessing the integrity of data used for ML projects. Techniques like data hashing and digital signatures can be employed to detect inconsistencies, errors, or signs of tampering within the data. Regular data integrity checks ensure that the data used to train and deploy ML models remains reliable and trustworthy, ultimately leading to more accurate and robust models.

Moving beyond data itself, **model monitoring** is another critical strategy. This involves keeping a watchful eye on the behavior of deployed ML models in production environments. By tracking model performance metrics, organizations can identify unexpected changes in the model's output, which might indicate potential security vulnerabilities or data poisoning attempts. Furthermore, model monitoring can detect the emergence of biases over time, allowing for corrective actions to be taken to ensure the fairness and ethical use of ML models.

Finally, a comprehensive data security monitoring strategy embraces **log management and analysis**. All data access attempts, data usage patterns, and system activity logs should be centralized and analyzed for potential security threats. Advanced analytics tools can be utilized to identify correlations between seemingly disparate events, potentially uncovering sophisticated cyberattacks targeting the ML environment. By meticulously analyzing logs, organizations can gain valuable insights into potential security vulnerabilities and take proactive steps to mitigate them. These data security monitoring strategies, when implemented within the broader ML data governance framework, create a robust and secure foundation for building and deploying responsible ML applications.

Implementing an Incident Response (IR) Plan

As organizations increasingly rely on machine learning (ML) models to drive business decisions, the need for robust data security and incident response (IR) strategies has become paramount. The very nature of ML, which often involves the processing of vast amounts of sensitive data, introduces unique security challenges that traditional cybersecurity measures may not adequately address. An effective incident response plan tailored specifically for ML data governance must encompass the full spectrum of potential security threats, from data breaches and poisoning attacks to adversarial manipulations and model theft. By preparing for these scenarios and integrating advanced monitoring techniques, organizations can not only protect their valuable data assets but also ensure the integrity and reliability of their ML models.

At the heart of any robust incident response plan for ML is the understanding that the models themselves are assets, much like the data they process. In many cases, these models are the culmination of extensive research, development, and training on proprietary datasets, making them both valuable and vulnerable. For instance, consider a financial institution that has developed a sophisticated ML model to predict stock market movements based on real-time financial data. The model, trained on vast amounts of historical market data, represents a significant competitive advantage. If an attacker were to gain unauthorized access to this model, they could either steal the model for their own use or, more insidiously, manipulate its predictions to benefit from market fluctuations. An effective IR plan must anticipate such scenarios and include specific protocols for monitoring model access, detecting suspicious activities, and responding to incidents of model theft or manipulation.

Data poisoning is another critical threat that organizations must prepare for in their incident response strategies. In data poisoning attacks, adversaries inject malicious data into the training datasets used to develop ML models. The goal of such attacks is to corrupt the model's learning process, leading to biased or inaccurate outputs. Consider a healthcare organization that uses an ML model to diagnose medical conditions based on patient data. If an attacker were to successfully poison the training data with false or misleading medical records, the resulting model could produce incorrect diagnoses, potentially endangering patients' lives. Detecting data poisoning requires sophisticated monitoring tools capable of analyzing training data in real time for anomalies that might indicate an ongoing attack. Once a poisoning attempt is detected, the IR plan should include immediate steps to halt the training process, isolate the compromised data, and retrain the model on a verified dataset.

Adversarial attacks present another complex challenge for incident response in the context of ML. Unlike data poisoning, where the attack targets the training data, adversarial attacks focus on the model itself, manipulating its inputs to produce incorrect or unexpected outputs. These attacks often involve subtle perturbations to the input data that are imperceptible to human observers but cause the ML model to misclassify the data. For example, in a scenario where an ML model is used to identify objects in video surveillance footage, an adversarial attack could involve slightly altering the pixels of an image so that a stop sign is incorrectly classified as a yield sign. Such an attack could have serious consequences, particularly in autonomous driving systems that rely on accurate object detection. An incident response plan must include provisions for detecting and mitigating adversarial attacks, which may involve real-time monitoring of model predictions for signs of manipulation and the deployment of adversarial training techniques to harden models against such threats.

Model theft, while often overshadowed by more direct attacks like data poisoning or adversarial manipulation, is a significant risk in its own right. As ML models become increasingly sophisticated and valuable, they become prime targets for theft. This could involve the outright copying of a model's code and weights, or more subtly, the extraction of the model's intellectual property through API queries. Consider a company that has developed a proprietary ML model for predicting consumer behavior. If a competitor were to gain access to this model, they could reverse-engineer it to replicate its functionality, undermining the original developer's competitive advantage. An effective IR plan must address this risk by implementing strict access controls, monitoring API usage for signs of unusual activity, and ensuring that any detected breach is responded to swiftly and decisively.

When an organization's ML model is compromised, the response must be swift and comprehensive to minimize damage and restore trust. The first step is containment, which involves isolating the affected systems to prevent further unauthorized access or damage. For instance, if a model deployed in a cloud environment is found to be compromised, the organization might immediately revoke all access tokens, disable the relevant APIs, and shift operations to a backup model if available. Simultaneously, incident response teams should begin a thorough investigation to determine the extent of the compromise, including how the attacker gained access, what data or models were affected, and whether any sensitive information was exfiltrated.

The next phase of the response involves eradication, where the organization works to remove the attacker's access and repair any vulnerabilities that were exploited. This might include patching software, updating security protocols, or even reconfiguring the network to prevent future breaches. In the case of a compromised ML model, this could involve retraining the model using a clean dataset, applying additional security measures such as encryption, and conducting a comprehensive audit of the model's deployment environment to identify and rectify any security gaps.

Recovery is the process of restoring normal operations after a security incident. For ML models, this might involve re-deploying the model once it has been verified as secure, re-establishing secure data pipelines, and communicating with stakeholders about the steps taken to address the breach. It's crucial during this phase to conduct a post-incident analysis to learn from the incident and strengthen the organization's security posture. This might involve updating the incident response plan to address any shortcomings revealed during the breach, implementing additional monitoring tools, or enhancing employee training on security best practices.

The importance of ongoing monitoring in the context of ML data governance cannot be overstated. Continuous data security monitoring provides the visibility needed to detect threats early, whether they involve unauthorized data access, data poisoning, adversarial attacks, or model theft. Advanced tools such as Security Information and Event Management (SIEM) systems, User and Entity Behavior Analytics (UEBA), and anomaly detection algorithms are critical components of this monitoring infrastructure. These tools can detect patterns that deviate from the norm, such as unexpected spikes in data access or unusual model outputs, which might indicate an ongoing attack.

For example, if a SIEM system detects an unusual number of requests to a model API from an unfamiliar IP address, this could indicate an attempt to steal the model. The system could trigger an automatic response, such as blocking the IP address or escalating the alert to the security team for further investigation. Similarly, UEBA tools can analyze user behavior to detect insider threats, such as an employee accessing sensitive datasets outside of their typical work hours. By correlating these activities with other security data, the organization can quickly identify and respond to potential breaches.

In distributed and cloud-based environments, monitoring poses additional challenges due to the complexity and scale of these systems. Data and models are often spread across multiple locations, making it difficult to maintain consistent security controls. Cloud Security Posture Management (CSPM) tools are essential for maintaining visibility in these environments. They continuously scan cloud configurations for vulnerabilities and misconfigurations that could be exploited by attackers. For instance, a CSPM tool might detect that a cloud storage bucket containing training data is publicly accessible, which could lead to data exposure. By identifying and addressing these issues before they are exploited, organizations can significantly reduce their risk of a breach.

Adversarial machine learning is a rapidly evolving field that presents unique challenges for incident response. Unlike traditional cybersecurity threats, adversarial attacks exploit the very mechanics of ML models, using specially crafted inputs to deceive the model into making incorrect predictions. These attacks can be subtle and difficult to detect, as the adversarial examples often appear normal to human observers. For instance, in a facial recognition system, an adversarial attack might involve adding imperceptible noise to an image so that the model misidentifies the person. To defend against these attacks, organizations need to incorporate adversarial robustness into their incident response plans.

One approach is to use adversarial training, where the model is trained on both normal and adversarial examples to improve its resilience. However, this technique is not foolproof, as new adversarial strategies are constantly being developed. Monitoring tools that can detect when a model's predictions deviate significantly from the expected norm are crucial for identifying these attacks in real time. For example, if a facial recognition model suddenly starts misidentifying a large number of individuals after processing a particular batch of images, this could indicate an adversarial attack. The incident response plan should include protocols for immediately investigating such anomalies, which might involve reverting to a previous model version or temporarily suspending the model's use until the issue is resolved.

When an adversarial attack is detected, the response must be swift and thorough. This might involve analyzing the inputs that triggered the attack to understand how the adversarial example was crafted, updating the model to resist similar attacks, and enhancing monitoring tools to detect future adversarial attempts. The organization should also communicate with stakeholders, including customers and regulators, to explain the incident and the steps taken to address it.

The legal and regulatory implications of data security incidents in ML cannot be ignored. Data breaches that involve personally identifiable information (PII) are subject to stringent reporting requirements under regulations such as the General Data

Protection Regulation (GDPR) in Europe or the California Consumer Privacy Act (CCPA) in the United States. Failure to comply with these regulations can result in substantial fines and damage to the organization's reputation. An incident response plan must include clear protocols for reporting data breaches to the appropriate authorities within the required timeframes.

For example, if an ML model used in a healthcare application is compromised and patient data is exposed, the organization must notify the relevant regulatory bodies and affected individuals. The incident response team should work closely with legal and compliance teams to ensure that all reporting requirements are met and that the organization's response is aligned with legal obligations. This might involve providing detailed reports on the nature of the breach, the data affected, and the steps taken to mitigate the impact.

In the aftermath of a data security incident, it is critical to conduct a thorough post-incident review to identify the root cause of the breach and to improve the organization's defenses. This review should include an analysis of how the incident was detected, the effectiveness of the response, and any gaps in the organization's security posture. For instance, if the breach was due to a misconfiguration in a cloud environment, the review might recommend changes to cloud security policies or additional training for staff responsible for managing cloud resources. The insights gained from this review should be used to update the incident response plan, ensuring that the organization is better prepared for future incidents.

Effective communication is also a key component of incident response. During and after a security incident, it is important to keep all stakeholders informed, including employees, customers, partners, and regulators. Transparent communication helps to build trust and can mitigate the reputational damage caused by a breach. For example, if an ML model used in a financial services application is compromised, the organization might issue a public statement explaining the incident, the steps taken to protect customer data, and the measures being implemented to prevent future breaches. Internally, the incident response team should provide regular updates to senior management, ensuring that they are aware of the situation and can make informed decisions.

In a complex and rapidly evolving threat landscape, the ability to adapt and respond to new challenges is crucial. This is particularly true in the context of ML, where the threat landscape is constantly changing as new attack vectors emerge. Organizations must be proactive in their approach to data security, continuously updating their incident response plans to address new risks. This might involve adopting new technologies, such as advanced threat detection tools or AI-driven security analytics, to enhance their ability to detect and respond to threats in real time.

For example, as adversarial attacks become more sophisticated, organizations may need to invest in specialized tools that can simulate adversarial scenarios and test the resilience of their ML models. These tools can help organizations identify potential weaknesses in their models before they are exploited by attackers. By incorporating these simulations into the incident response process, organizations can stay ahead of emerging threats and ensure that their models remain secure.

Fostering a culture of security awareness within the organization is essential for effective incident response. Employees at all levels must understand the importance of data security and be trained to recognize and respond to potential threats. This includes not only technical staff, such as data scientists and IT professionals, but also non-technical employees who may have access to sensitive data or systems. Regular security training and awareness programs can help to ensure that everyone in the organization is prepared to respond to a security incident, reducing the risk of human error and enhancing the overall effectiveness of the incident response plan.

Implementing an incident response plan for data security monitoring and incident response in the context of ML requires a comprehensive and proactive approach. Organizations must be prepared to address a wide range of security threats, from data breaches and poisoning attacks to adversarial manipulations and model theft. By integrating advanced monitoring tools, developing robust response protocols, and fostering a culture of security awareness, organizations can protect their valuable data assets and ensure the integrity and reliability of their ML models. As the threat landscape continues to evolve, ongoing vigilance and adaptation will be key to maintaining a strong security posture and minimizing the impact of security incidents.

Integrating Data Security Monitoring and IR with Data Governance

Data security monitoring and IR should be seamlessly integrated with the broader data governance framework for effective ML project management. This integration creates a secure environment for building and deploying ML models, fostering trust in their reliability and ethical use. Let's explore seven key strategies for achieving successful integration.

The foundation for effective integration lies in establishing clear **data security policies and procedures**. These policies should align with broader data governance principles, outlining a comprehensive approach to data security within the ML lifecycle. They should encompass data access controls, defining who can access sensitive data and under what circumstances. Furthermore, the policies should detail usage limitations, specifying how data can be employed for model training and predictions. Additionally, data anonymization techniques should be outlined to minimize privacy risks associated with using sensitive data in ML models. Finally, these policies should mandate data security awareness training for personnel involved in ML projects, ensuring everyone understands their role in data protection.

Next comes the allocation of clear **data governance roles and responsibilities**. The data governance framework should designate specific roles responsible for data security. This might involve assigning data security officers who oversee data security practices and incident response. Their responsibilities could include managing access control systems, conducting regular risk assessments, and ensuring compliance with data security regulations. Furthermore, data scientists and engineers should be trained to recognize and report suspicious activity related to data access or model behavior. Clearly defining roles and responsibilities fosters a culture of data security accountability within the organization.

Data security tools and technologies play a crucial role in facilitating data security monitoring and IR. A variety of tools can be leveraged to achieve comprehensive security coverage. Security information and event management (SIEM) systems serve as a centralized hub for log aggregation and analysis. By consolidating logs from diverse sources, including data access attempts, system activity, and model behavior, SIEM systems enable the detection of anomalies that might indicate potential security threats. Data loss prevention (DLP) tools provide another layer of protection by preventing unauthorized data exfiltration. DLP solutions can monitor data transfers and identify attempts to move sensitive data outside of authorized channels. Additionally, encryption technologies safeguard data at rest and in transit, minimizing the risk of unauthorized access even if a breach occurs. Furthermore, Machine Learning-based threat detection models can be employed to identify sophisticated cyberattacks specifically targeting the vulnerabilities of ML systems.

Data lineage tracking is another crucial element in integrating data security monitoring and IR. This involves implementing mechanisms to trace the origin, transformations, and usage of data throughout the ML lifecycle. Tracking data lineage provides valuable insights during security incidents. If a security breach occurs, it allows investigators to identify the source of compromised data and assess the potential impact on deployed ML models. Furthermore, data lineage tracking facilitates impact assessments in case of data quality issues, allowing for the identification of affected models and the implementation of corrective actions. Robust data lineage tracking ensures a comprehensive understanding of data flow within the ML environment, aiding in effective data security monitoring and IR efforts.

Continuous improvement is an essential aspect of maintaining a secure ML ecosystem. The data governance framework should promote a culture of continuous improvement in data security practices. This involves regularly reviewing and updating data security monitoring and IR procedures based on several factors. Evolving cyberthreats necessitate adapting security measures to address new vulnerabilities. Furthermore, industry best practices for data security in ML are constantly evolving, prompting organizations to stay current with these advancements. Finally, lessons learned from past incidents should be incorporated into the data governance framework. Conducting vulnerability assessments and penetration testing at regular intervals helps to identify and address potential security weaknesses within the ML environment. By fostering a culture of continuous improvement, organizations can proactively address evolving security threats and maintain a robust environment for developing trustworthy ML models.

Integrating data security monitoring and IR with the broader data governance framework creates a secure foundation for ML projects. By implementing a combination of data security policies, designated roles and responsibilities, appropriate security tools and technologies, data lineage tracking, and continuous improvement practices, organizations can build confidence in the reliability and fairness of their ML models. Ultimately, a secure and responsible approach to data security empowers organizations to unlock the full potential of ML while minimizing risks and fostering trust in this transformative technology.

Best Practices for Data Security Monitoring and Incident Response

As machine learning (ML) systems become increasingly integral to the functioning of modern enterprises, the importance of implementing robust data security monitoring and incident response (IR) strategies within ML data governance frameworks has become more critical than ever. The complexities inherent in ML environments—ranging from the vast amounts of data processed to the advanced algorithms employed—create unique security challenges that necessitate meticulous planning and execution of data security practices. Best practices in this area encompass granular access control, rigorous data usage monitoring, comprehensive data integrity checks, and the deployment of advanced security tools, all aimed at safeguarding sensitive data and ensuring the reliability of ML models.

Data Access Control and User Activity Monitoring

Traditional data governance practices often rely on Role-Based Access Control (RBAC) to manage permissions, where users are granted access based on predefined roles within an organization. While this method has proven effective in many scenarios, it can be too rigid when applied to the dynamic and complex environments characteristic of ML projects. ML pipelines typically involve various stages—data collection, preprocessing, model training, evaluation, and deployment—each requiring different levels of access to data. Granular access controls provide a more flexible and nuanced approach, enabling organizations to assign specific permissions tailored to the exact needs of each user or group.

Consider a scenario in which an organization's ML team includes data scientists, model developers, and data engineers, each playing a distinct role in the development lifecycle. A data scientist engaged in exploratory data analysis might need read-only access to a subset of anonymized data, allowing them to derive insights without risking data integrity. In contrast, a model developer responsible for refining and training ML models might require both read and write access to data and model outputs, as they iteratively improve model performance. By employing granular access controls, the organization can fine-tune access levels to match these requirements, thereby reducing the risk of data breaches or unauthorized modifications.

Multi-factor authentication (MFA) is another critical security measure that complements granular access controls by adding an additional layer of verification before access is granted. MFA typically requires users to authenticate their identity using multiple methods, such as a password combined with a one-time code sent to their mobile device, a biometric scan, or a hardware token. For example, if a data engineer needs to access a cloud-based storage system containing sensitive training data, MFA would require them to provide both their password and a biometric fingerprint scan, ensuring that even if their password is compromised, unauthorized access is prevented. The implementation of MFA significantly enhances security by reducing the likelihood of unauthorized access, particularly in environments where data is highly sensitive or regulated.

Monitoring user activity is another cornerstone of secure data governance in ML environments. By systematically logging all data access attempts, organizations can maintain a comprehensive audit trail that details who accessed what data, when, and for what purpose. This audit trail is invaluable for both real-time security monitoring and post-incident investigations. For example, in the event of a suspected data leak, security teams can analyze these logs to identify any unauthorized access attempts or unusual patterns of behavior, such as a user downloading large datasets at odd hours or accessing data from an unfamiliar location.

User Behavior Analytics (UBA) enhances the effectiveness of user activity monitoring by applying machine learning algorithms to detect anomalies in user behavior. UBA systems can establish a baseline of normal activity for each user, such as typical login times, locations, and data access patterns. Any deviation from this baseline—such as an employee accessing a sensitive database from a foreign IP address—can be flagged as suspicious. For instance, if a data scientist who usually accesses data during office hours suddenly starts accessing data late at night from a remote location, the UBA system could trigger an alert, prompting a security investigation. This proactive approach enables organizations to detect and mitigate potential insider threats or unauthorized access attempts before they escalate into more severe security incidents.

Data Usage Monitoring

Securing the ML data lifecycle requires vigilant monitoring of how data is utilized across different stages of the ML pipeline. Data usage monitoring involves not only tracking access to data but also keeping a close eye on how data is processed, transformed, and ultimately used in model training and deployment. One key aspect of data usage monitoring is the meticulous tracking of data transformations during the preprocessing stage. Data transformations—such as normalization, feature scaling, and encoding—are crucial for preparing raw data for model training. However, deviations from standard data processing techniques or unexpected alterations to feature engineering processes could indicate unauthorized data manipulation.

For example, consider a financial institution that uses an ML model to detect fraudulent transactions. During the preprocessing stage, transaction amounts might be normalized to ensure that the model treats them consistently, regardless of the currency. If an unauthorized party gains access to the preprocessing pipeline and introduces a subtle modification—such as scaling transaction amounts differently for certain regions—this could skew the model's predictions, leading to either an increase in false positives (legitimate transactions flagged as fraudulent) or false negatives (fraudulent transactions going undetected). By monitoring data transformations, organizations can detect such unauthorized modifications early and take corrective action before they impact model performance.

Data usage monitoring also plays a critical role in detecting data poisoning attacks, where adversaries deliberately inject corrupted data into the training datasets with the intent of manipulating the model's behavior. Such attacks can have far-reaching consequences, particularly in scenarios where ML models are used to make high-stakes decisions. For instance, an attacker might target an ML model used in autonomous vehicles by poisoning the training data with examples that cause the model to misinterpret stop signs as yield signs. The consequences of such an attack could be catastrophic, leading to accidents and loss of life. By monitoring the input data for anomalies, such as unexpected distributions or patterns that deviate from historical norms, organizations can detect and mitigate data poisoning attempts before they compromise the integrity of the ML model.

Integrating data lineage tracking with data usage monitoring further strengthens the security framework by providing a comprehensive view of the data's journey through the ML pipeline. Data lineage tracking enables organizations to trace the origin, transformations, and final usage of data, offering visibility into every step of the data's lifecycle. This capability is particularly valuable in the event of a security breach, as it allows security teams to identify the source of compromised data and assess the impact on downstream processes. For example, if a data breach leads to the corruption of a dataset used in model training, data lineage tracking can help pinpoint when and where the breach occurred, allowing for a targeted and efficient response.

Data lineage tracking also facilitates impact assessments by identifying which models or processes were potentially affected by data quality issues. For instance, if an organization discovers that a dataset used for training a customer segmentation model was tampered with, data lineage tracking can help determine whether other models or processes that relied on the same dataset were also compromised. This information is crucial for prioritizing remediation efforts and ensuring that all affected models are retrained or updated with clean, verified data. By combining data usage monitoring with data lineage tracking, organizations gain a holistic understanding of how data flows within their ML environment, enabling them to proactively address security threats and ensure the integrity of data used for model training and deployment.

Data Integrity Monitoring

The foundation of trustworthy ML models lies in the integrity of the data used for training and deployment. Data integrity monitoring within the ML data governance framework ensures that the data remains accurate, consistent, and reliable throughout its lifecycle. Implementing robust data quality checks is a best practice for maintaining data integrity. These checks involve validating the completeness, consistency, and accuracy of datasets before they are used in model training.

For example, consider an e-commerce platform that uses an ML model to predict customer preferences based on purchase history. Before feeding the data into the model, the organization should perform data quality checks to ensure that all purchase records are complete, consistent, and free from errors. This might involve checking for missing values, duplicates, and outliers that could skew the model's predictions. Statistical techniques and data visualization tools can be employed to identify anomalies in the data, such as unusually high purchase amounts that could indicate fraudulent transactions. By addressing these issues early in the data pipeline, organizations can prevent them from compromising the model's performance and ensure that the predictions are based on accurate and reliable data.

Data hashing and digital signatures provide additional layers of security to ensure data integrity. Data hashing involves generating a unique mathematical fingerprint, or hash value, for a dataset. Any alteration to the data, whether intentional or accidental, will result in a different hash value, immediately alerting security teams to potential tampering. For instance, if an organization suspects that a dataset used for training a fraud detection model has been compromised, they can compare the current hash value with the original hash to determine whether any changes have occurred. This quick verification process helps maintain data integrity throughout the ML pipeline, ensuring that the model is trained on unaltered, trustworthy data.

Digital signatures act like electronic seals that verify the authenticity and origin of the data. By employing cryptographic techniques, digital signatures ensure that the data has not been modified since it was signed by an authorized source. For example, in a healthcare setting, a digital signature might be used to authenticate patient records before they are used in an ML model designed to predict disease outcomes. If the signature verification fails, it indicates that the data has been tampered with, prompting further investigation. Digital signatures provide a strong assurance of data integrity, particularly in scenarios where data is shared between different organizations or stakeholders.

Anomaly detection algorithms offer a proactive approach to data integrity monitoring by continuously analyzing data characteristics and identifying unusual patterns that might indicate data tampering. These algorithms are particularly useful in detecting subtle changes that might not be immediately apparent through manual inspection. For example, an anomaly detection algorithm might identify a sudden spike in specific data points or unexpected changes in the distribution of data values, signaling an attempt to manipulate the data. In a financial services context, such an anomaly might involve a sudden increase in high-value transactions, which could indicate money laundering activities. By continuously monitoring for such anomalies, organizations can promptly identify and address attempts to manipulate data, preventing them from compromising the integrity and performance of ML models.

Model Monitoring

Model monitoring is a critical facet of ensuring the security and reliability of ML models within the data governance framework. This practice involves continuously tracking key performance metrics to detect potential security threats early and take corrective action before they impact the model's outputs. One of the fundamental aspects of model monitoring is tracking performance metrics such as accuracy, precision, recall, and F1 score. These metrics provide a snapshot of the model's effectiveness and help identify unexplained deviations that might indicate issues with the underlying data or the model itself.

For instance, consider a credit scoring model used by a bank to assess loan applications. If the model's accuracy suddenly drops without any changes to the data or the model architecture, it could be a sign of a data poisoning attack. Malicious actors might inject corrupted data into the training process to manipulate model behavior, causing the model to produce biased or inaccurate outputs. By continuously monitoring these metrics, organizations can quickly identify and address such issues before they escalate, ensuring that the model continues to make accurate and fair predictions.

Monitoring fairness metrics like disparity scores is also crucial, particularly in scenarios where ML models are used to make decisions that affect individuals, such as loan approvals, hiring recommendations, or criminal sentencing. Sudden changes in fairness metrics could indicate the introduction of bias, potentially stemming from data breaches that expose sensitive information or manipulation attempts targeting specific demographics within the data. For example, if a loan approval model begins to exhibit unexpected bias against a particular demographic group, it could be a sign that sensitive data was inadvertently leaked during a security incident, affecting the training data and influencing the model's decision-making process. Detecting such biases early is critical for maintaining the fairness and ethical integrity of ML models.

Leveraging model explainability techniques empowers security teams to delve deeper into the inner workings of ML models and understand how they arrive at their predictions. Explainability techniques, such as SHAP (SHapley Additive exPlanations) values or LIME (Local Interpretable Model-agnostic Explanations), provide insights into the factors that most strongly influence a model's decisions. For example, if a hiring model starts favoring candidates from a specific region, explainability techniques can help determine whether this bias was introduced due to tampered data or an attack on the model itself. Understanding the root cause of such biases is essential for implementing corrective measures and ensuring that the model remains fair and unbiased.

Model drift is another critical aspect of model monitoring that requires careful attention. Model drift occurs when a model's performance gradually deteriorates over time due to changes in the underlying data distribution. This phenomenon is particularly common in dynamic environments where the data landscape evolves rapidly, such as in e-commerce, finance, or social media. Security incidents, such as data exfiltration or unauthorized data modifications, can significantly alter the data landscape, leading to model drift and impacting the reliability of predictions. For example, if a customer segmentation model is trained on data that is later altered due to a security breach, the model may no longer accurately reflect the customer base, leading to incorrect segmentation and targeted marketing efforts.

To mitigate model drift, organizations should regularly refresh training data with high-quality sources and implement drift detection algorithms that proactively identify performance deviations. Drift detection algorithms can monitor the consistency of key performance metrics over time and trigger alerts when significant deviations are detected. For instance, if a recommendation system for an online retailer starts suggesting irrelevant products due to shifts in customer behavior patterns, the drift detection algorithm would flag this issue, prompting the organization to retrain the model with updated data. By addressing model drift early, organizations can maintain the accuracy and reliability of their ML models in the face of potential security threats and changing data environments.

Threat Intelligence and Threat Modeling

Threat intelligence and threat modeling are essential components of a proactive security strategy within machine learning (ML) data governance. Threat intelligence involves the continuous gathering and analysis of information about emerging security threats, vulnerabilities, and attack vectors that could potentially target ML systems. In the context of ML, this might include tracking the latest developments in adversarial attacks, data poisoning techniques, and other sophisticated methods that malicious actors use to compromise ML models and data pipelines. For instance, an organization that develops and deploys ML models in the healthcare sector might monitor threat intelligence feeds for information on new types of attacks that target patient data, such as ransomware that encrypts sensitive training datasets. By staying informed about these emerging threats, the organization can implement preemptive measures to safeguard its ML systems and protect sensitive data.

Threat intelligence feeds are typically sourced from a combination of public databases, private security firms, and industry-specific security consortiums. These feeds provide real-time updates on known vulnerabilities, malicious IP addresses, malware signatures, and other indicators of compromise (IOCs) relevant to the organization's ML environment. For example, if a new exploit targeting a specific ML framework is discovered, threat intelligence feeds can alert the organization to the risk, allowing security teams to patch the vulnerability before it is exploited. In this way, threat intelligence serves as an early warning system, enabling organizations to stay ahead of potential attackers and strengthen their security posture.

Threat modeling, on the other hand, is a systematic process used to identify, analyze, and prioritize potential security threats specific to an organization's ML systems. The goal of threat modeling is to understand the potential attack surface of an ML pipeline, including identifying the assets that need protection, the pathways through which these assets could be compromised, and the threat actors most likely to target the organization. For instance, a financial institution using ML models for fraud detection might conduct threat modeling to identify potential threats such as data poisoning, where attackers inject fraudulent data to skew the model's outputs, or model inversion attacks, where adversaries attempt to reconstruct sensitive training data by querying the model.

The threat modeling process typically involves creating detailed diagrams of the ML pipeline, highlighting critical components such as data ingestion points, preprocessing stages, model training environments, and deployment interfaces. These diagrams help security teams visualize potential attack vectors and identify weak points where security controls may be insufficient. For example, if the threat model reveals that an ML model's API endpoint is exposed to the internet without proper authentication, the organization can take immediate steps to secure the endpoint by implementing robust access controls and monitoring for suspicious activity. By systematically identifying and addressing potential vulnerabilities, threat modeling helps organizations mitigate the risks associated with deploying ML systems in increasingly hostile environments.

Integrating threat intelligence with threat modeling enhances the effectiveness of both strategies. Threat intelligence provides the up-to-date information necessary to inform the threat modeling process, ensuring that the models reflect the latest threats and vulnerabilities. Conversely, the insights gained from threat modeling can guide the organization's threat intelligence efforts by identifying the specific types of threats that are most relevant to the ML systems in use. For example, if threat modeling reveals that the organization's primary concern is adversarial attacks against its image recognition models, the threat intelligence team can focus on gathering information related to the latest adversarial techniques, tools, and defenses. This targeted approach ensures that the organization's security efforts are both relevant and effective, minimizing the risk of costly breaches or model failures.

Together, threat intelligence and threat modeling provide a comprehensive approach to securing ML systems. While threat intelligence keeps organizations informed about the ever-evolving threat landscape, threat modeling ensures that security measures are strategically applied to protect the most critical assets and pathways. By continuously updating both threat intelligence and threat models, organizations can adapt to new challenges, ensuring that their ML systems remain secure and resilient in the face of increasingly sophisticated cyberthreats.

Explainable AI (XAI) for Security

Explainable AI (XAI) is an emerging field that focuses on making machine learning (ML) models more transparent, interpretable, and understandable by humans. In the context of data security, XAI plays a crucial role by enabling security teams to gain insights into how ML models make decisions, identify potential biases, and detect security threats that might otherwise go unnoticed. For instance, in a scenario where an ML model used for loan approvals suddenly begins to deny loans disproportionately to a particular demographic, XAI techniques can help security teams understand whether the model's behavior is due to biased training data, a data poisoning attack, or an adversarial manipulation.

XAI techniques such as SHAP (SHapley Additive exPlanations), LIME (Local Interpretable Model-agnostic Explanations), and model-specific interpretability methods provide detailed explanations of the factors that influence a model's predictions. For example, SHAP values can be used to determine the contribution of each feature to a specific prediction, allowing security teams to identify whether certain features are being manipulated to alter the model's output. In the context of adversarial attacks, XAI can reveal how small perturbations to input data are leading to significant changes in the model's predictions, helping to diagnose and mitigate these attacks. By providing a clearer understanding of the inner workings of ML models, XAI enables security teams to take targeted actions to protect against specific threats.

One of the key benefits of XAI in security is its ability to identify unintended consequences of security incidents that might not be immediately apparent. For example, if a data breach exposes sensitive demographic data that is later used in model training, XAI can help determine whether the breach has introduced biases into the model's decision-making process. This is particularly important in sectors such as healthcare, finance, and criminal justice, where biased ML models can have serious ethical and legal implications. By using XAI to assess the impact of data breaches and other security incidents on model fairness and reliability, organizations can take proactive steps to address these issues and maintain the integrity of their ML systems.

XAI also plays a critical role in enhancing anomaly detection within ML systems. Anomaly detection models are often used to identify unusual patterns or behaviors that may indicate a security threat, such as a sudden spike in data access or an unexpected shift in user behavior. However, the black-box nature of many ML models can make it difficult to understand why certain anomalies are flagged and others are not. XAI provides the transparency needed to interpret the outputs of anomaly detection models, helping security teams to distinguish between true threats and false positives. For example, if an anomaly detection model flags a spike in data access as suspicious, XAI can reveal whether the spike was due to a legitimate business need or an unauthorized attempt to exfiltrate data.

Another important application of XAI in security is in the development of robust defenses against adversarial attacks. Adversarial attacks involve deliberately manipulating input data to deceive an ML model into making incorrect predictions. These attacks can be highly sophisticated, making them difficult to detect and defend against. However, XAI techniques can be used to analyze how adversarial examples are affecting the model's decision-making process. By understanding the model's vulnerabilities, security teams can develop more effective defenses, such as adversarial training, which involves retraining the model on adversarial examples to improve its resilience. For instance, if an adversarial attack is found to exploit a specific feature of the input data, XAI can guide the modification of that feature's handling in the model to reduce the attack's effectiveness.

Incorporating XAI into the broader data security monitoring and incident response framework offers significant advantages for organizations seeking to protect their ML systems from emerging threats. By making ML models more interpretable, XAI not only enhances security teams' ability to detect and respond to attacks but also helps build trust in the models' outputs. This is particularly important in regulated industries, where organizations must demonstrate that their ML models are both secure and fair. As the field of XAI continues to evolve, its integration into security practices will become increasingly essential for organizations looking to maintain the integrity, fairness, and reliability of their ML systems in a rapidly changing threat landscape.

Security Tools and Technologies

The effective implementation of security tools and technologies is paramount for robust data security monitoring and incident response in ML data governance. One of the most critical tools in this context is Security Information and Event Management (SIEM) systems. SIEM systems aggregate logs from various sources within the ML environment, including user activity logs, system logs, and application logs. By centralizing this data, SIEM facilitates real-time analysis and anomaly detection, enabling security teams to identify suspicious activity that might indicate a security threat.

For example, if a SIEM system detects a sudden spike in data transfer activity from a particular user account, it could indicate an attempt to exfiltrate sensitive data from the ML pipeline. The SIEM system can automatically flag this activity for further investigation and, if necessary, trigger automated responses, such as temporarily suspending the user account or blocking the data transfer. By providing real-time insights into potential security threats, SIEM systems play a critical role in protecting ML environments from data breaches and other malicious activities.

Data Loss Prevention (DLP) tools act as a critical line of defense against data exfiltration in ML environments. DLP solutions monitor data transfers across the network and can identify attempts to move sensitive data outside of authorized channels. For instance, if an employee tries to upload a dataset containing personally identifiable information (PII) to an unauthorized cloud storage platform, the DLP system can detect the attempt and block the upload, preventing the data from being exposed. By configuring DLP policies to align with data classification schemes, organizations can ensure that sensitive data used for ML projects is protected from unauthorized access and transfer.

Encryption is another essential component of securing data within ML environments. Encrypting data at rest and in transit ensures that even if attackers gain access to data storage systems or intercept network traffic, the data remains unreadable without the decryption key. For example, when sensitive customer data is being transmitted between different components of the ML pipeline, such as from the data warehouse to the model training environment, encryption ensures that the data cannot be accessed or altered by unauthorized parties. Implementing strong encryption algorithms, such as AES-256, and securely managing encryption keys are crucial for maintaining data confidentiality and integrity in ML projects.

Machine learning-based threat detection tools offer an additional layer of security by utilizing advanced algorithms to analyze security data and identify sophisticated cyberattacks specifically targeting vulnerabilities within ML environments. These tools can detect complex attack patterns that might not be apparent through traditional security measures. For example, an ML-based threat detection tool could analyze network traffic patterns to identify a slow, stealthy data exfiltration attempt, where an attacker is gradually transferring small amounts of data over time to avoid detection. By leveraging these tools, organizations can proactively detect and respond to emerging threats, safeguarding their ML models and the sensitive data used for training and deployment.

Recovery and Remediation

Swift and effective recovery following a security incident in an ML environment is crucial for minimizing the impact of the breach and restoring normal operations. A robust recovery plan should outline clear steps for incident containment, system restoration, and data recovery, ensuring that the organization can quickly return to normal operations while minimizing data loss and disruption.

Incident containment involves isolating the affected systems to prevent the attacker from causing further damage. For instance, if a data breach is detected in the ML pipeline, the organization might immediately revoke all access tokens, disable the relevant APIs, and shift operations to a backup model if available. Containment measures should be implemented as quickly as possible to limit the spread of the breach and protect the organization's sensitive data and ML models.

System restoration involves restoring the affected systems to their pre-incident state. This might involve rolling back to a previous backup, reinstalling software, or reconfiguring security settings. For example, if an ML model's training environment is compromised, the organization might restore the environment from a recent backup, ensuring that the model is retrained using clean, uncorrupted data. System restoration should be carefully planned and executed to ensure that all components of the ML pipeline are fully functional and secure before resuming normal operations.

Data recovery is a critical aspect of the recovery process, particularly in cases where data has been lost or corrupted due to the security incident. Organizations should have robust data backup procedures in place to ensure that critical data can be restored quickly and with minimal loss. For instance, if an attacker corrupts a dataset used for model training, the organization can restore the dataset from a recent backup, ensuring that the model can be retrained using clean data. Data recovery efforts should be closely coordinated with the overall incident response process to ensure that data is restored in a timely and secure manner.

Vulnerability remediation strategies are vital for preventing future security breaches. After investigating the incident and conducting a root cause analysis, organizations should implement measures to address the exploited weaknesses. This might involve patching software vulnerabilities, updating security configurations, or implementing additional security controls within the ML pipeline. For instance, if the incident was caused by a misconfiguration in the cloud environment, the organization might update its cloud security policies and conduct additional training for employees to prevent similar issues in the future. By addressing the root cause of the security breach, organizations can significantly reduce the risk of future incidents and strengthen their overall security posture.

Continuous Improvement

Maintaining robust data security practices within the ML data governance framework requires continuous improvement. Regular reviews of data security monitoring and incident response practices are essential for identifying potential gaps and areas for enhancement. These reviews should assess the effectiveness of existing security measures, identify new threats, and evaluate the need for updated mitigation strategies.

For instance, an organization might conduct quarterly reviews of its SIEM system to ensure that it is effectively capturing and analyzing logs from all relevant sources within the ML environment. If the review identifies gaps in the logging coverage, such as missing logs from a newly deployed ML model, the organization can update the SIEM configuration to address the issue. Regular reviews help ensure that the organization's security practices remain effective and up-to-date in the face of evolving threats.

Vulnerability assessments and penetration testing specifically tailored for ML environments further strengthen the organization's security posture. These assessments involve simulating cyberattacks to identify vulnerabilities within the ML pipeline. For example, a penetration test might attempt to exploit a known vulnerability in the model's API to gain unauthorized access to the model's predictions or training data. By conducting these activities regularly, organizations can proactively identify and address weaknesses before malicious actors have the opportunity to exploit them.

Post-incident reviews are a valuable tool for continuous improvement. After a security incident has been contained and remediated, conducting a thorough post-incident review is crucial for analyzing lessons learned and identifying potential improvements to data security practices. For example, if an organization experiences a data breach due to a phishing attack, the post-incident review might recommend additional employee training on recognizing and avoiding phishing emails. By incorporating these insights into the organization's security practices, organizations can strengthen their ML data governance framework and effectively mitigate future security threats.

Building upon Established Best Practices

Building upon established best practices, advanced approaches elevate data security monitoring and incident response within the ML data governance framework. These strategies equip organizations with enhanced capabilities to identify and mitigate sophisticated security threats targeting their ML environments.

Proactive security measures, such as threat intelligence and threat modeling, are key to staying ahead of potential attackers. Threat intelligence involves gathering and analyzing information about emerging threats specific to ML environments. For instance, by subscribing to threat intelligence feeds that focus on ML-related vulnerabilities, an organization can stay informed about the latest attack vectors and vulnerabilities. This knowledge informs threat modeling, a process that identifies potential security weaknesses within the ML pipeline. For example, an organization might simulate an attack scenario where an adversary attempts to inject malicious data into the model's training dataset. By simulating these scenarios, the organization can pinpoint critical assets and vulnerabilities, allowing for the implementation of targeted security controls to mitigate potential risks.

Deception technologies and honeypots provide an additional layer of defense against sophisticated attackers. Deception technologies deploy strategically crafted lures and fake data to mislead attackers. For example, an organization might deploy a honeypot that mimics a real ML environment but contains no sensitive data. If an attacker attempts to gain access to the honeypot, the organization can analyze their behavior to gain valuable insights into their tactics and techniques. By deploying honeypots and analyzing attacker behavior, organizations can strengthen their overall security posture and gain valuable intelligence on potential threats.

Cloud security considerations are essential when leveraging cloud platforms for ML workloads. Implementing best practices for data security and access control in the cloud is critical for protecting sensitive data and ML models. For instance, an organization might use granular access controls and identity and access management (IAM) tools to ensure that only authorized users can access sensitive data stored in the cloud. Additionally, organizations should leverage cloud-native security features and services, such as encryption and intrusion detection systems (IDS), to protect data stored and processed in the cloud environment. Regularly monitoring cloud security posture and addressing potential vulnerabilities specific to cloud-based ML deployments is essential for maintaining a strong security posture.

Explainable AI (XAI) techniques play a crucial role in securing ML models by providing transparency into how models arrive at their predictions. For example, if an organization suspects that a data breach has introduced biases into the model, XAI can help pinpoint the specific data points that contributed to the bias. By understanding how security incidents might influence model behavior, organizations can proactively mitigate potential risks and ensure that their ML models remain fair and reliable.

Policies and Procedures for Data Privacy and Security

Data Access Control

Data access control is the cornerstone of data privacy and security in ML environments. While Role-Based Access Control (RBAC) remains a foundational approach, providing users with access based on their roles within an organization, the dynamic nature of ML workflows necessitates a more granular approach. In an ML setting, access control must be adaptable, taking into account the varying levels of data sensitivity and the specific needs of different users throughout the ML pipeline. For instance, a data scientist who is conducting exploratory analysis may only require read-only access to anonymized datasets, whereas a model developer might need full access to datasets to train and validate models. By tailoring data access to the specific needs of each user, organizations can minimize the potential for data breaches and reduce the overall attack surface.

Granular access control is not just about assigning the correct permissions; it is also about establishing clear, well-documented procedures for granting and revoking access. User provisioning should be tightly controlled, ensuring that new users are granted only the access they need to perform their tasks. This often involves implementing Multi-factor authentication (MFA) to add an additional layer of security, requiring users to authenticate their identities through multiple channels before accessing sensitive data. For example, a user might need to enter a password and then confirm their identity via a code sent to their mobile device or through a biometric scan. This reduces the likelihood of unauthorized access, even if a user's password is compromised.

De-provisioning, or the process of revoking access when a user no longer requires it, is equally important. In many organizations, users accumulate permissions over time as they move between projects, leading to excessive access that could be exploited if their account is compromised. Automated de-provisioning tools can help manage this by regularly reviewing user access and revoking permissions that are no longer necessary. For instance, if a data scientist moves from one ML project to another, their access to the datasets and resources from the previous project should be revoked automatically to prevent any lingering access that could be misused.

Training and awareness are critical components of effective data access control. Regular security training programs should be implemented to educate all personnel involved in ML projects about the importance of data security, the risks of unauthorized access, and the specific controls in place to protect data. These programs should cover a wide range of topics, from the basics of data classification and encryption to the specific security threats that are prevalent in ML environments, such as data poisoning and adversarial attacks. By fostering a culture of security awareness, organizations can reduce the risk of human error and ensure that all users understand the importance of following established data access protocols.

The principle of least privilege is central to the design of any access control policy. In the context of ML data governance, this principle dictates that users should be granted the minimum level of access required to perform their tasks. For example, a junior data analyst working on data cleaning should not have access to the full production dataset but rather to a subset that is sufficient for their work. This approach limits the potential damage that can be caused by compromised accounts or accidental misuse of data, ensuring that the most sensitive data is only accessible to those who absolutely need it.

Data Security

Data security in ML environments involves a multifaceted approach that includes data classification, encryption, and privacy-preserving techniques, all of which must be carefully tailored to the specific needs of the organization and the sensitivity of the data involved. Datasets used in ML projects should be classified based on their sensitivity, with clear labeling and access controls that reflect the importance of protecting highly sensitive data. For example, datasets containing personally identifiable information (PII) or proprietary business information should be classified as high sensitivity, requiring strict access controls and advanced encryption techniques to ensure their protection.

Encryption is one of the most critical tools in the data security arsenal, particularly in ML environments where data is often stored and processed across various systems and networks. Implementing robust encryption practices is essential for safeguarding data both at rest and in transit. Data at rest refers to data stored on servers, databases, or other storage devices, while data in transit refers to data being transferred across networks.

For example, sensitive training data stored in a cloud database should be encrypted using strong encryption algorithms such as AES-256, ensuring that even if attackers gain access to the storage system, they cannot read the data without the decryption key. Similarly, data being transferred between different components of the ML pipeline, such as between a data warehouse and a model training environment, should be encrypted during transmission to protect it from interception.

Privacy-preserving techniques such as data anonymization and pseudonymization play a crucial role in enabling organizations to use sensitive data in ML models while minimizing privacy risks. Anonymization involves removing all PII from the dataset, making it impossible to trace the data back to specific individuals. For example, before using a dataset containing patient health records to train an ML model, an organization might remove all direct identifiers such as names, addresses, and social security numbers. Pseudonymization, on the other hand, replaces PII with fictitious identifiers, allowing the data to be linked to an individual only by those with access to the pseudonymization key. This technique is particularly useful in scenarios where it is necessary to maintain some level of linkage between records for analysis, such as in longitudinal studies where patient data is tracked over time. By applying these techniques, organizations can train ML models on sensitive data without compromising individual privacy.

Data Loss Prevention (DLP) policies and tools are essential for preventing the unauthorized transfer of sensitive data outside of the organization's secure environment. DLP systems monitor data transfers across networks and can identify and block unauthorized attempts to move sensitive data, such as when an employee tries to upload a confidential dataset to an external cloud storage service. These tools can be configured to alert security teams to suspicious activities, such as large-scale data exports or attempts to send encrypted data through unsecured channels. For instance, if an employee with access to sensitive financial data attempts to email the data to an external recipient, the DLP system could automatically block the email and notify the security team, preventing a potential data breach.

Data Usage

Meticulously tracking how data is used throughout the ML pipeline is critical for ensuring data integrity, preventing unauthorized modifications, and maintaining a clear audit trail. Data usage monitoring involves keeping detailed records of all data transformations that occur during the preprocessing stage, where raw data is cleaned, normalized, and prepared for model training. These records should include information about who performed the transformation, when it was done, and what specific changes were made to the data. For example, if a dataset containing customer purchase history is being cleaned to remove outliers, the organization should document the criteria used for identifying and removing these outliers, along with the identities of the personnel involved in the process. This ensures that any changes to the data are transparent and can be audited if necessary.

Comprehensive logging of data usage throughout the ML lifecycle is essential for detecting and responding to potential security incidents. These logs should capture a wide range of information, including timestamps, user identities, data accessed, and the specific actions performed, such as downloading, modifying, or deleting data. For example, if an organization suspects that an insider is attempting to exfiltrate sensitive data, security teams can analyze these logs to identify any unusual patterns of data access, such as a user downloading large volumes of data at an unusual time or accessing datasets that are not typically required for their role. By maintaining detailed logs of data usage, organizations can quickly identify suspicious activities and take appropriate action to prevent data breaches or other security incidents.

Validating model input and output data is a critical practice for preventing data poisoning attacks, where malicious actors inject corrupted data into the training process to manipulate model behavior. These attacks can have severe consequences, particularly in high-stakes environments where ML models are used to make critical decisions. For example, an attacker might target an ML model used in a financial trading system by injecting fraudulent data that causes the model to make incorrect predictions, leading to significant financial losses. To prevent such attacks, organizations should implement rigorous validation techniques to check for inconsistencies, outliers, and unexpected changes in data distribution between input and output data. This might involve comparing the distribution of input features to expected norms or analyzing the outputs of the model for unusual patterns that could indicate a manipulation attempt.

Data lineage tracking is another essential tool for maintaining data integrity and security in ML environments. By tracing the origin, transformations, and final usage of data used in ML projects, organizations can gain a comprehensive view of the data lifecycle and quickly identify the source of any compromised data. For instance, if a data breach occurs and a dataset used in model training is found to be corrupted, data lineage tracking can help pinpoint when and where the corruption occurred, allowing security teams to take swift action to mitigate the risks. This might involve retraining affected models with clean data, updating data transformation processes to prevent future incidents, and conducting a thorough investigation to identify the root cause of the breach.

Incident Response

Developing a robust incident response (IR) plan is essential for effectively managing and mitigating the impact of data security incidents in ML environments. A clear classification system for data security incidents is crucial for ensuring that the appropriate response is initiated based on the severity and potential impact of the incident. This classification system should categorize incidents into different levels, such as low, medium, and high severity, based on factors such as the sensitivity of the data involved, the extent of the compromise, and the potential consequences for the organization. For example, a low-severity incident might involve a minor data access violation with no immediate impact, while a high-severity incident could involve a significant data breach affecting sensitive customer information, potentially leading to regulatory penalties and reputational damage.

Escalation procedures are a critical component of the IR plan, ensuring that incidents are promptly reported to the appropriate personnel and that the response is coordinated effectively. For high-impact incidents, it is essential that the incident is immediately escalated to senior management, the legal department, and the security operations center (SOC) to ensure a swift and coordinated response. For example, if an ML model used in a healthcare application is compromised and patient data is exposed, the IR team must quickly notify senior management and the legal department to assess the potential regulatory implications and coordinate the response with external stakeholders, such as healthcare regulators and affected patients.

Establishing a dedicated data security IR team is essential for handling security incidents effectively. This team should consist of personnel with expertise in data security, ML, and incident response best practices. The team's responsibilities include detecting incidents, conducting investigations, containing the incident to prevent further damage, eradicating the threat, and recovering affected systems. For example, if a data poisoning attack is detected in an ML pipeline, the IR team might work to isolate the affected model, analyze the input data to identify the source of the attack, and implement measures to prevent future occurrences. The team should also coordinate the recovery process, which might involve retraining the model on clean data, restoring affected systems from backups, and conducting a thorough post-incident review to identify any gaps in the organization's security posture.

Clear communication protocols are vital during an ML security incident, ensuring that all stakeholders are informed in a timely and transparent manner. These protocols should define who needs to be notified during an incident, such as management, legal counsel, affected users, and external regulators, as well as the timeframe for communication. For example, if an organization experiences a data breach affecting customer information, the communication protocol should specify that affected customers must be notified within a certain timeframe, as required by data breach notification laws such as the General Data Protection Regulation (GDPR) or the California Consumer Privacy Act (CCPA). Timely and transparent communication helps build trust with stakeholders and ensures that everyone involved understands the scope of the incident and the steps being taken to address it.

Organizations must comply with relevant data breach notification regulations to minimize legal risks and maintain customer trust. Regulations such as GDPR and CCPA require organizations to notify affected individuals within a specific timeframe in the event of a data breach. Developing data breach notification procedures is essential for ensuring compliance with these regulations. For instance, if an organization's ML model is compromised, resulting in the unauthorized access of customer data, the IR team must work closely with the legal department to ensure that breach notifications are sent out in accordance with regulatory requirements. This includes providing affected individuals with details about the breach, the data involved, the potential risks, and the steps being taken to mitigate the impact.

Privacy Considerations

Privacy by Design principles should be embedded throughout the ML development lifecycle, ensuring that data privacy is prioritized from the outset of any project. This proactive approach involves considering privacy at every stage of the ML process, from data collection and preprocessing to model training, evaluation, and deployment. For example, when designing an ML model for personalized marketing, Privacy by Design would involve minimizing data collection to only the data necessary for model training, applying data anonymization techniques to protect customer identities, and implementing robust data governance frameworks to manage data access and usage. By embedding privacy into the design of ML systems, organizations can reduce the risk of privacy breaches and ensure compliance with data protection regulations.

Compliance with data subject rights regulations, such as GDPR, is critical for maintaining transparency and trust with individuals whose data is used in ML projects. GDPR grants individuals the right to access, rectify, erase, and restrict the processing of their personal data, and organizations must develop procedures for managing these data subject rights requests effectively. For instance, if a customer requests the erasure of their data from an ML model used for credit scoring, the organization must have processes in place to identify all instances of the customer's data within the system, remove it from the dataset, and ensure that the model is retrained without the customer's data. Failure to comply with data subject rights can result in significant regulatory penalties and damage to the organization's reputation.

Differential privacy is a powerful technique for training ML models on sensitive data while minimizing privacy risks. This technique involves introducing controlled noise into the data, making it difficult to identify information about specific individuals within the dataset. For example, when training a model on location data to predict traffic patterns, differential privacy can be used to add random noise to the data points, ensuring that the model's predictions are accurate at the aggregate level without compromising the privacy of individual users. By applying differential privacy, organizations can balance the need for accurate ML models with the imperative to protect individual privacy.

Model explainability techniques are essential for understanding how ML models arrive at their predictions and identifying potential privacy biases that may have been introduced during the model development process. Explainability is particularly important in regulated industries, such as finance and healthcare, where decisions made by ML models can have significant impacts on individuals. For instance, if an ML model used for loan approvals is found to exhibit bias against a particular demographic group, explainability techniques can help identify the features and data points that are contributing to this bias. This allows organizations to take corrective action, such as reweighting features, adjusting the model's parameters, or retraining the model on a more representative dataset, ensuring that the model's predictions are fair and non-discriminatory.

Embedding explainability into the ML pipeline also supports transparency and accountability, which are critical for maintaining trust with stakeholders. By providing insights into how decisions are made by ML models, organizations can demonstrate that their models are not only accurate and effective but also fair and compliant with privacy regulations. This is particularly important in the context of data breaches or security incidents, where understanding the model's decision-making process can help identify whether any privacy biases were introduced as a result of the incident. By addressing these issues proactively, organizations can mitigate the impact of privacy breaches and maintain the trust of their customers and regulators.

Security Tools and Technologies

Implementing advanced security tools and technologies is essential for safeguarding data and models within the ML environment. Security Information and Event Management (SIEM) systems are a cornerstone of this approach, providing a centralized platform for aggregating and analyzing logs from various sources within the ML pipeline. By consolidating data from user activity logs, system logs, and application logs, SIEM systems enable real-time monitoring and analysis of potential security threats. For example, if a SIEM system detects an unusual spike in data transfer activity from a specific user account, it could indicate a potential data exfiltration attempt. The system can automatically trigger an alert, allowing security teams to investigate and respond before the data is compromised.

Machine learning-based threat detection tools are increasingly being leveraged to enhance security within ML environments. These tools utilize advanced ML algorithms to analyze large volumes of security data, identifying patterns and anomalies that may indicate a sophisticated cyberattack. For example, an ML-based threat detection tool might analyze network traffic to identify a slow and stealthy data exfiltration attempt, where an attacker is gradually transferring small amounts of data over an extended period to avoid detection. By identifying these subtle patterns, ML-based threat detection tools can help organizations detect and mitigate threats that might otherwise go unnoticed.

When deploying ML workloads in cloud environments, it is crucial to implement best practices for data security and access control. Cloud providers offer a range of security features designed to protect data stored and processed in the cloud, such as encryption, intrusion detection systems (IDS), and vulnerability scanning. For instance, when storing sensitive data in a cloud-based data lake, organizations should ensure that data is encrypted both at rest and in transit using strong encryption algorithms. Additionally, regular vulnerability assessments should be conducted to identify and address any security gaps in the cloud infrastructure. By leveraging these cloud-native security features and maintaining a vigilant monitoring posture, organizations can secure their ML deployments against potential threats.

Regular vulnerability assessments and penetration testing are essential for identifying and addressing security weaknesses within the ML pipeline. Vulnerability assessments involve systematically scanning the ML environment to identify potential security gaps, such as outdated software, misconfigured systems, or exposed endpoints. For example, if a vulnerability assessment reveals that a model's API endpoint is accessible without proper authentication, immediate action can be taken to secure the endpoint and prevent unauthorized access. Penetration testing goes a step further by simulating real-world attacks against the ML environment, testing the effectiveness of existing security controls and identifying any weaknesses that could be exploited by attackers. For instance, a penetration test might attempt to exploit a known vulnerability in the model's deployment environment to gain unauthorized access to training data or manipulate the model's predictions. By regularly conducting these assessments, organizations can proactively address security risks and strengthen their overall security posture.

Security tools and technologies should be continuously evaluated and updated to keep pace with the evolving threat landscape. As new vulnerabilities are discovered and new attack vectors emerge, organizations must adapt their security strategies to protect their ML environments. This might involve deploying new security tools, updating existing configurations, or adopting new security frameworks that are specifically designed for ML applications. By staying ahead of the curve, organizations can ensure that their data and models remain secure and resilient against even the most sophisticated cyberthreats.

Recordkeeping and Audit

Recordkeeping and audit processes are integral to effective data privacy and security management in ML environments. These processes ensure that all data access, usage, and security activities are thoroughly documented, providing a comprehensive audit trail that can be used for compliance, investigation, and continuous improvement purposes. Developing policies for data access logs and user activity monitoring record retention is crucial for balancing the need for accountability with privacy considerations. For example, an organization might establish a policy that mandates the retention of access logs for a minimum of five years, ensuring that all data access activities are recorded and available for review in the event of a security incident.

Thorough documentation of security incidents is essential for organizational learning and improvement. When a security incident occurs, it is important to document all aspects of the incident, including the timeline of events, the root cause analysis, the actions taken to mitigate the incident, and the lessons learned. For instance, if an ML model is compromised due to a data poisoning attack, the organization should document the steps taken to identify the attack, isolate the affected model, and retrain the model on clean data. This documentation not only helps improve the organization's incident response capabilities but also serves as a valuable resource for future security planning and risk management.

Regular audits are necessary to ensure that data security practices within the ML environment are functioning effectively and in compliance with internal policies and external regulations. These audits can be conducted internally by the organization's security team or externally by third-party security professionals. For example, an internal audit might assess the effectiveness of data access controls, ensuring that only authorized users have access to sensitive datasets and that access logs are being properly maintained. An external audit, on the other hand, might evaluate the organization's overall security posture, identifying any gaps or weaknesses that need to be addressed. By regularly auditing their data security practices, organizations can identify and mitigate potential risks, ensuring that their ML environments remain secure and compliant.

Regular audits also provide an opportunity for organizations to assess the effectiveness of their incident response procedures. For instance, an audit might review the organization's response to a recent security incident, evaluating how well the incident was detected, contained, and mitigated. The audit findings can then be used to refine and improve the organization's incident response plan, ensuring that it is better equipped to handle future incidents. By continuously improving their incident response capabilities, organizations can reduce the impact of security incidents and minimize the risk of data breaches and other security threats.

Recordkeeping and audit processes also play a critical role in demonstrating compliance with regulatory requirements. For organizations operating in highly regulated industries, such as finance or healthcare, it is essential to maintain detailed records of all data access, usage, and security activities. These records can be used to demonstrate compliance with regulations such as GDPR, CCPA, and HIPAA, ensuring that the organization meets its legal obligations and avoids regulatory penalties. For example, in the event of a data breach, an organization might be required to provide regulators with detailed records of all data access and usage activities, as well as the steps taken to secure the affected data and mitigate the impact of the breach. By maintaining comprehensive records and conducting regular audits, organizations can ensure that they are prepared to meet these compliance requirements and protect their reputation.

Continuous Improvement

The landscape of data privacy and security is constantly evolving, and organizations must regularly review and update their policies and procedures to remain aligned with best practices and address emerging threats. Continuous improvement is a key component of effective ML data governance, ensuring that the organization's security posture remains robust and resilient in the face of new challenges. For example, an organization might conduct an annual review of its data security policies, incorporating lessons learned from recent security incidents and industry advancements into its data governance framework. This might involve updating data classification schemes, revising access control protocols, or adopting new security technologies that offer enhanced protection against emerging threats.

Following a security incident, conducting a thorough post-incident review is crucial for identifying the root cause of the incident and determining how to prevent similar incidents in the future. This review should analyze all aspects of the incident, including the effectiveness of the incident response procedures, the actions taken to mitigate the incident, and the lessons learned. For instance, if an organization experiences a data breach due to a phishing attack, the post-incident review might reveal that additional training is needed to help employees recognize and avoid phishing attempts. By identifying these lessons learned and taking corrective actions, organizations can continuously improve their data security practices and build more resilient ML environments.

Empowering personnel involved in ML projects with data privacy and security knowledge is essential for minimizing human error and enhancing data protection throughout the ML lifecycle. Regular training programs should be implemented to educate participants on data classification, access control protocols, potential security threats in the ML environment, and best practices for handling sensitive data. For example, a training program might cover topics such as how to recognize and report suspicious activities, how to securely transfer sensitive data, and how to apply privacy-preserving techniques when working with personal data. By fostering a culture of data security awareness, organizations can ensure that all personnel are equipped with the knowledge and skills needed to protect sensitive data and respond effectively to security incidents.

Continuous improvement also involves staying informed about the latest developments in data privacy and security, including emerging threats, new technologies, and evolving regulatory requirements. For example, an organization might subscribe to industry-specific threat intelligence feeds, attend security conferences, or participate in security forums to stay up to date on the latest trends and best practices. By staying informed and continuously adapting their security strategies, organizations can ensure that they are prepared to address new challenges and protect their ML environments against even the most sophisticated cyberthreats.

Incorporating feedback from regular audits, training programs, and post-incident reviews into the organization's data governance framework is essential for continuous improvement. This feedback provides valuable insights into the effectiveness of existing security measures and identifies areas where improvements are needed. For example, if an audit reveals that data access controls are not being consistently applied across all departments, the organization might implement additional controls or provide targeted training to ensure compliance. By continuously refining and improving their data privacy and security policies and procedures, organizations can build a more secure and resilient ML environment that is capable of withstanding the ever-changing threat landscape.

Summary

In this chapter, we dove into the vital aspects of data privacy and security within the realm of machine learning (ML) and data governance. With the growing use of ML across various industries, the benefits are immense, but so are the challenges, especially when it comes to protecting sensitive information. We explored how to strike the right balance between harnessing data for innovation and keeping it secure and private.

Starting with data access control, we looked at how granular permissions and multi-factor authentication (MFA) are key to safeguarding sensitive data in ML projects. It's crucial for organizations to implement strong security measures to prevent unauthorized access and ensure that data is managed responsibly throughout its entire lifecycle.

We also tackled the complexities of data security, particularly the challenges of encrypting data at rest, in transit, and in use. While encryption is a must, it needs to be part of a broader strategy that includes robust access controls and ongoing monitoring to guard against potential breaches. We touched on the tricky balance between encryption and model performance, especially when dealing with advanced techniques like homomorphic encryption and differential privacy.

The importance of continuous data security monitoring and incident response was another major focus. By putting advanced monitoring tools and strategies in place, organizations can stay ahead of security threats, ensuring that data remains secure and compliant with ever-changing regulations. This proactive approach helps mitigate risks and builds trust with stakeholders.

Finally, we discussed the evolving landscape of data privacy and security regulations and the need for adaptable governance frameworks. With new technologies like federated learning on the rise, it's crucial for organizations to stay updated on emerging threats, regularly refresh their security practices, and foster a culture that prioritizes data privacy.

CHAPTER 5

Ethical Implications and Bias Mitigation

Machine learning (ML) has revolutionized various sectors by providing predictive power and automation, transforming how organizations make decisions and deliver services. However, the use of machine learning systems comes with significant ethical implications, particularly regarding data governance and bias. Ensuring ethical practices in ML data governance is critical to building systems that are fair, transparent, and accountable. This chapter delves into the ethical implications of ML data governance and strategies for bias mitigation, providing a comprehensive discussion aimed at practitioners, researchers, and policymakers involved in ML and AI development.

Principles of Ethical AI

As machine learning (ML) continues to reshape various aspects of our lives, the need for ethical considerations becomes paramount. These principles provide a framework for developing and deploying AI in a responsible and trustworthy manner.

A core principle is ensuring fairness and non-discrimination in AI systems. This involves mitigating biases that can creep into ML models during data collection, training, and deployment. For instance, consider an organization developing a credit scoring model. If the training data includes historical biases, such as those linked to socioeconomic status, the model might unfairly penalize certain demographic groups, perpetuating systemic inequality. To combat this, organizations must implement rigorous processes for data selection and preprocessing, ensuring diverse representation across all demographics. In a real-world scenario, a financial institution may employ techniques like reweighting or adversarial debiasing to correct imbalances, regularly monitoring the model's performance across different subgroups to identify and address any emerging disparities.

Transparency in AI development fosters trust and accountability. This principle encourages the creation of models with explainable decision-making processes. For example, in healthcare, where AI models assist in diagnosing diseases, it is critical for medical professionals to understand how these systems arrive at their conclusions. By employing techniques such as LIME (Local Interpretable Model-agnostic Explanations) or SHAP (SHapley Additive exPlanations), organizations can provide insights into the decision-making process of complex models, helping doctors to trust the recommendations made by AI while ensuring that they align with clinical reasoning. In practice, a hospital might use these tools to break down the AI's decisions into understandable components, allowing healthcare providers to validate the model's logic and make informed decisions.

The protection of individual privacy is essential in the age of AI. This principle advocates for responsible data collection practices, robust data security measures, and user control over personal information used in AI systems. In a retail environment, where customer data is often leveraged for personalized marketing, organizations must ensure that such data is collected and stored securely. Techniques like differential privacy can be implemented to allow data analysis while safeguarding individual identities. A real-world example could involve an online retailer using differential privacy to analyze shopping trends without exposing individual customer data, thereby maintaining privacy while still gaining valuable insights. Moreover, organizations must regularly update their security protocols to guard against new threats, such as those posed by evolving cyberattack strategies that target personal data.

Despite advancements in AI, human oversight remains crucial. This principle emphasizes the importance of human control over AI systems and the need for clear accountability mechanisms. For instance, in autonomous driving technology, while AI plays a significant role in vehicle navigation, human operators must remain in control to intervene during unexpected situations. A practical example is a ride-sharing company testing self-driving cars, where human drivers are required to monitor the AI's decisions, ready to take over if the system fails to recognize a pedestrian or misinterprets a traffic signal. In such scenarios, clear lines of accountability are essential to ensure that any incidents involving AI systems are appropriately addressed, with responsibility clearly delineated between the human operators and the AI developers.

The development and deployment of AI should promote societal well-being and environmental sustainability. This principle encourages the use of AI for positive social impact and environmental responsibility. For example, AI can be employed to optimize energy consumption in smart grids, reducing overall environmental impact. A tech company might develop AI algorithms that predict energy demand, adjusting supply in real time to reduce waste and lower carbon emissions. By integrating these practices into their operational models, organizations can contribute to a more sustainable future, ensuring that their AI innovations align with broader societal and environmental goals. Through such efforts, organizations can build trustworthy AI systems that not only drive business success but also foster positive outcomes for society and the planet.

Ethical Implications of Machine Learning Data Governance

The ethical implications of machine learning (ML) data governance are profound, cutting across issues of fairness, transparency, privacy, and accountability. As ML continues to be integrated into various sectors, the ethical challenges associated with its data governance become increasingly complex and critical. Unlike traditional software, ML models learn from vast datasets, which may inadvertently encode societal biases and prejudices. These biases, when not properly addressed, can be magnified during the training process, leading to models that perpetuate discrimination and unfair treatment of certain groups.

A case that exemplifies the ethical pitfalls of biased data in ML is the widely discussed COMPAS algorithm, used in the U.S. criminal justice system to assess the likelihood of a defendant reoffending. Despite being intended as a tool for fair decision-making, studies have shown that the COMPAS algorithm disproportionately assigned higher risk scores to Black defendants compared to white defendants, even when controlling for similar criminal histories. This led to harsher sentencing recommendations for Black individuals, revealing how deeply embedded biases in the training data—reflecting historical inequities in the justice system—can result in discriminatory outcomes when not adequately addressed. The ethical failure here lies not only in the biased outcomes but also in the lack of transparency about how these risk scores were calculated, making it nearly impossible for defendants to challenge their assessments.

Transparency, or the lack thereof, in ML models poses another significant ethical concern. Many advanced ML models, particularly those based on deep learning, function as "black boxes," where the decision-making process is not easily interpretable even by their creators. This opacity raises critical questions about accountability.

Consider the case of ML models used in healthcare, where an algorithm might recommend a particular treatment plan for a patient. If the healthcare provider cannot understand how the algorithm arrived at its recommendation, they may struggle to trust or effectively communicate these decisions to patients, potentially compromising patient care. A real-world example is IBM Watson's AI for Oncology, which, despite being highly publicized, faced criticism for offering treatment recommendations that were sometimes unsafe or unsupported by clinical evidence. The lack of transparency in the model's decision-making process made it difficult for medical professionals to validate or understand these recommendations, leading to a broader mistrust in AI-driven healthcare solutions.

Privacy concerns are another area where the ethical implications of ML data governance come to the forefront. The reliance on large datasets, often containing sensitive personal information, presents significant risks if data is not collected, stored, and used ethically. A notorious example of unethical data use is the Cambridge Analytica scandal, where personal data from millions of Facebook users was harvested without consent and used to influence voter behavior in the 2016 U.S. presidential election. This case highlighted how ML models, when fed unethical data, can be used to manipulate public opinion, posing grave risks to democratic processes. The scandal not only led to a significant loss of trust in social media platforms but also sparked global discussions on data privacy regulations, illustrating the far-reaching consequences of neglecting ethical data governance.

Different industries face unique ethical challenges based on the nature of their data and use cases. In the financial sector, for instance, the use of ML models for credit scoring can lead to significant ethical dilemmas. Historically, financial data has been used to develop models that assess creditworthiness, but if these models are trained on biased data, they can reinforce existing economic disparities. A real-world example is the case of Apple's credit card algorithm, which was accused of offering lower credit limits to women compared to men with similar financial profiles. This discrepancy, which was brought to light by tech entrepreneur David Heinemeier Hansson, sparked a broader conversation about gender bias in financial algorithms. The opaque nature of the algorithm, coupled with its significant impact on individuals' financial opportunities, underscored the ethical necessity of transparency and fairness in financial ML models.

In contrast, the healthcare industry grapples with the ethical implications of data privacy and the potential for bias in diagnostic models. As electronic health records (EHRs) become more prevalent, ML models trained on this data have the potential to revolutionize patient care by providing personalized treatment recommendations. However, if these models are not governed ethically, they can also exacerbate health disparities. For example, an ML model trained primarily on data from urban hospitals might not perform well in rural settings, leading to suboptimal care for rural populations. Moreover, the sensitive nature of health data requires stringent privacy protections. The ethical imperative in healthcare ML data governance is not only to ensure that models are fair and unbiased but also to protect patient privacy rigorously. A failure in this domain could lead to significant harm, both to individuals' health and their trust in the healthcare system.

The ethical implications of ML data governance also extend to the realm of law enforcement, where predictive policing models are increasingly being used to allocate resources and identify potential criminal activity. These models are typically trained on historical crime data, which is often biased against marginalized communities due to long-standing systemic inequities in law enforcement practices. For instance, predictive policing tools like PredPol have been criticized for disproportionately targeting minority neighborhoods, reinforcing discriminatory policing practices. The ethical concern here is twofold: not only do these models risk perpetuating racial profiling, but they also lack transparency in how predictions are generated, making it difficult for communities to challenge or understand these decisions. The use of such models raises critical questions about the balance between public safety and the potential for reinforcing systemic bias.

In the realm of employment, ML models are increasingly used in hiring processes to screen resumes and predict candidate success. While these models offer the promise of efficiency, they also introduce significant ethical risks. For example, an AI-powered hiring tool used by Amazon was found to be biased against women, as it favored resumes that included male-dominated terms and experiences. This bias was likely a reflection of the company's historical hiring practices, which the model had been trained on. The ethical failure here was not only the model's discriminatory outcomes but also the lack of accountability and transparency in how the model made its decisions. This case highlights the importance of ethical considerations in the development and deployment of ML models in the workplace, particularly in ensuring that these models do not perpetuate existing biases or create new forms of discrimination.

The ethical implications of ML data governance are particularly pronounced in industries that deal with vulnerable populations. For example, in the insurance industry, ML models are increasingly used to assess risk and determine premiums. While these models can improve accuracy and efficiency, they also raise significant ethical concerns.

If an insurance company uses an ML model that discriminates against individuals based on factors like genetics or socioeconomic status, it could lead to unfairly high premiums for certain groups. This not only exacerbates existing inequalities but also raises questions about the ethics of using sensitive personal data in risk assessment. The challenge for the insurance industry is to develop ML models that are both fair and transparent while also protecting the privacy of individuals.

The ethical implications of ML data governance are not limited to individual industries but also have broader societal impacts. The rise of ML and AI technologies has led to increasing concerns about the potential for these technologies to reinforce existing power imbalances and create new forms of inequality. For example, as ML models are increasingly used to automate decision-making in areas like finance, healthcare, and criminal justice, there is a risk that these models will disproportionately harm marginalized communities. This is particularly concerning given the lack of transparency and accountability in many ML models, which makes it difficult to identify and address biases and other ethical issues.

To mitigate these risks, it is essential for organizations to adopt ethical data governance practices that prioritize fairness, transparency, and accountability. This includes implementing processes for regularly auditing ML models for bias, ensuring that data is collected and used ethically, and providing clear explanations for how models make decisions. In practice, this might involve using tools like fairness-aware ML algorithms, which are designed to minimize bias in model predictions, or explainable AI techniques that make it easier to understand and interpret model decisions. Organizations should also consider the broader societal impacts of their ML models and take steps to ensure that these models are used in ways that promote equity and social justice.

One of the key challenges in ethical ML data governance is balancing the need for innovation with the need for ethical oversight. On one hand, ML technologies have the potential to drive significant advancements in a wide range of industries, from healthcare to finance to criminal justice. On the other hand, these technologies also raise significant ethical concerns, particularly when it comes to issues like bias, transparency, and privacy. To address this challenge, organizations need to develop governance frameworks that allow for both innovation and ethical oversight. This might involve creating interdisciplinary teams that include ethicists, data scientists, and legal experts to oversee the development and deployment of ML models, or establishing ethical review boards that can assess the potential risks and benefits of new ML technologies.

It is important to recognize that the ethical implications of ML data governance are not static but evolve over time as new technologies and use cases emerge. As such, organizations need to adopt a flexible and adaptive approach to data governance that can respond to changing ethical challenges. This might involve regularly updating governance frameworks to reflect new ethical considerations, engaging with stakeholders to identify emerging ethical issues, and continuously monitoring and evaluating the impact of ML models on society. By adopting such an approach, organizations can ensure that their ML models are not only effective but also ethically sound, promoting fairness, transparency, and accountability in all their operations.

Data Privacy and Consent

One of the foremost ethical considerations in ML data governance is data privacy. With the vast amount of data required to train machine learning models, the protection of personal and sensitive information is paramount. Data privacy concerns include unauthorized access, data breaches, and the misuse of personal data. Organizations must ensure that they have robust data privacy policies in place, complying with regulations such as the General Data Protection Regulation (GDPR) and the California Consumer Privacy Act (CCPA).

Consent is another critical aspect. Individuals whose data is being used must provide informed consent, understanding how their data will be used, stored, and protected. Transparent consent mechanisms and clear communication about data use are necessary to uphold ethical standards. However, in practice, ensuring that consent is genuinely informed and meaningful poses significant challenges, particularly as companies seek to balance user consent with the need for large, diverse datasets to drive their data-driven initiatives.

One of the primary challenges in this area is the sheer volume of data being collected, often across multiple platforms and for a wide range of purposes. For example, social media platforms collect vast amounts of personal data, including user interactions, location data, and even browsing habits across the Web. To obtain informed consent, these platforms must clearly communicate how each type of data will be used, which can be a daunting task given the complexity of their operations. Users are often presented with lengthy, jargon-filled privacy policies that are difficult to understand, leading to consent that is more nominal than truly informed. This raises ethical concerns about whether users are genuinely aware of what they are agreeing to when they consent to data collection.

The practical challenges of managing consent also extend to the need for flexibility and scalability in large-scale data operations. For instance, a company that collects data for improving its services might later decide to use the same data for developing new products or for sharing with third-party partners. Ideally, this would require obtaining renewed consent from users for the new purposes, but in practice, this can be logistically challenging, particularly when dealing with millions of users. This creates a tension between the ethical requirement for ongoing, informed consent and the operational need for data versatility.

A real-world example of these challenges can be seen in the healthcare sector, where informed consent is both legally mandated and ethically critical. In clinical trials or research studies, participants must be fully informed about how their data will be used, including any future uses that might not be immediately apparent. However, as medical research increasingly relies on large datasets to identify trends and develop new treatments, the need for broad consent—that is, consent that allows for future, unspecified research uses—becomes more prevalent. While broad consent can facilitate important medical advancements, it also raises ethical concerns about whether participants can truly understand and consent to all potential future uses of their data.

Companies also face the challenge of balancing user consent with the need to ensure the diversity and representativeness of their datasets. For example, an AI company developing a facial recognition system might require a large, diverse dataset to ensure its model works effectively across different demographic groups. However, obtaining meaningful consent from all individuals represented in the dataset can be difficult, particularly when dealing with publicly available images or data collected indirectly through third-party sources. This raises ethical questions about whether it is possible to obtain informed consent in such scenarios and what alternative measures might be necessary to protect user privacy.

The evolving nature of data use presents ongoing challenges for managing consent in a meaningful way. Companies must navigate the complexities of obtaining and maintaining consent while also respecting user autonomy and privacy. This requires not only transparent communication but also the implementation of robust consent management systems that can adapt to changes in data use over time. As the landscape of data privacy continues to evolve, the importance of informed, flexible, and scalable consent mechanisms cannot be overstated, particularly as companies seek to balance the ethical imperative of protecting user privacy with the practical demands of large-scale data operations.

Data Ownership and Intellectual Property

Determining data ownership is fundamental to ethical machine learning (ML) data governance, yet it is fraught with complexity due to the diverse sources from which data is collected and the various stakeholders involved. Organizations often aggregate data from individuals, other businesses, and public datasets, creating a mosaic of information that is both valuable and legally sensitive. Establishing clear ownership rights over this data is essential not only to avoid legal and ethical disputes but also to ensure that data usage aligns with the expectations and rights of all parties involved.

One challenge in establishing data ownership arises from the fact that data is often collected from individuals who may not fully understand or appreciate the value of the information they are providing. For example, social media platforms collect vast amounts of personal data, which they use to drive advertising revenues. While users may have agreed to terms of service that grant the platform rights to use their data, questions remain about whether they truly understand what they have consented to and whether the platform's use of their data respects their ownership rights. This issue is further complicated when data is sold to or shared with third parties, raising concerns about whether the original data providers have relinquished their ownership rights or if those rights continue to exist in some form.

The issue of data ownership becomes even more complex when data is aggregated or anonymized. For instance, a healthcare provider might anonymize patient data to use it for research purposes or to sell it to a third-party company developing new medical treatments. While anonymization is intended to protect patient privacy, it also raises questions about whether the individuals who originally provided the data still have any ownership claims over the anonymized dataset. This becomes particularly contentious if the data is used to develop profitable products or services, leading to debates over whether the individuals whose data made these innovations possible should be compensated or acknowledged in some way.

In the realm of intellectual property (IP), the rights associated with data and algorithms are equally crucial. Organizations that develop proprietary algorithms using vast datasets must navigate the legal landscape to protect their IP while also respecting the rights of the data owners. For example, an AI company might use publicly available data to train a machine learning model that it then patents or licenses to others. The question of who owns the resulting model—the company that developed it or the individuals who generated the data used to train it—can lead to complex legal disputes. This is especially true when the data is sourced from public repositories or platforms where the original data providers may not have explicitly relinquished their IP rights.

The situation is further complicated in cases where multiple entities contribute data to a single ML project. For instance, a consortium of companies might pool their data to develop an industry-wide AI solution, raising questions about how ownership of the resulting data and algorithms is to be divided. In such scenarios, it is essential to establish clear agreements at the outset regarding who owns the data, who has rights to the algorithms developed, and how profits or benefits derived from the project will be shared. Failing to address these issues can lead to conflicts that undermine the collaboration and the value of the project itself.

The legal landscape surrounding data ownership and intellectual property is still evolving, with many jurisdictions grappling with how to apply traditional IP laws to the digital age. For example, the European Union's General Data Protection Regulation (GDPR) provides individuals with certain rights over their data, such as the right to access and delete personal information. However, these rights do not necessarily confer ownership in the traditional sense, leading to ambiguity over how data should be treated in terms of property rights. This ambiguity can be particularly problematic when data is transferred across borders, as different countries have different legal frameworks regarding data ownership and IP.

As organizations continue to harness the power of data and machine learning, the need for clear and enforceable data ownership and IP rights will only grow. By addressing these issues proactively, companies can not only protect their own interests but also build trust with their customers and partners, ensuring that the benefits of data-driven innovation are shared fairly and ethically.

Transparency and Accountability

Transparency and accountability are fundamental pillars of ethical machine learning (ML) governance, particularly as these models become more complex and their decision-making processes more opaque. To achieve meaningful transparency, organizations must go beyond simply disclosing how data is collected, processed, and used. They need to implement specific strategies and frameworks that demystify the inner workings of ML models, ensuring that stakeholders, including those whose data is being used, can comprehend and trust these systems.

One effective strategy for enhancing transparency in ML models is the use of model interpretability tools. These tools, such as LIME (Local Interpretable Model-agnostic Explanations) and SHAP (SHapley Additive exPlanations), allow organizations to explain how their models arrive at specific decisions. For instance, in a financial institution using ML for credit scoring, SHAP values can be employed to illustrate how different factors, such as income, credit history, and debt levels, contribute to the final credit score. This not only helps the institution explain decisions to customers but also enables the identification and correction of any unintended biases that might be influencing the model. By providing clear, understandable explanations, these tools promote transparency and help build trust with both customers and regulators.

Another essential practice is the development of transparent data governance frameworks that document every step of the data lifecycle. This includes detailed records of data sources, the methods used for data collection, preprocessing steps, and the rationale behind data selection. For example, in the healthcare industry, where patient data is used to train diagnostic models, maintaining a transparent data governance framework can help ensure that the data used is representative and ethically sourced. By keeping detailed logs and documentation, healthcare providers can demonstrate that their models are built on high-quality data, free from biases that could lead to incorrect diagnoses. Such transparency not only supports accountability but also provides a foundation for auditing and validating the model's performance over time.

Organizations can also enhance transparency through the adoption of "white-box" models, which are inherently more interpretable than "black-box" models. While black-box models like deep neural networks offer high accuracy, they are often criticized for their lack of interpretability. White-box models, such as decision trees or linear regression, may offer a lower level of accuracy but provide a clear, logical structure that is easy to understand. For example, a company using ML to optimize supply chain logistics might opt for a decision tree model that clearly outlines the decision-making process, showing how different factors like inventory levels, demand forecasts, and supplier reliability are weighted. By prioritizing transparency over sheer predictive power, the company can ensure that stakeholders understand how decisions are made, fostering trust and enabling more informed decision-making.

Accountability in ML systems requires clearly defined roles and responsibilities throughout the development, deployment, and maintenance processes. One effective framework for ensuring accountability is the RACI matrix, which stands for Responsible, Accountable, Consulted, and Informed. This framework helps organizations delineate

317

who is responsible for each aspect of the ML lifecycle. For instance, in an organization deploying an ML-based hiring tool, the data scientists might be responsible for model development, while the HR department is accountable for the ethical implications of the hiring decisions. Legal and compliance teams could be consulted on regulatory matters, and senior management would be informed about the overall performance and risks associated with the tool. By clearly defining these roles, the organization ensures that there is accountability at every stage, from data collection to model deployment, and that ethical concerns are addressed proactively.

Another crucial aspect of accountability is the establishment of mechanisms for addressing errors, biases, or harms caused by ML systems. This can include the implementation of an AI ethics review board, which regularly audits ML models for compliance with ethical standards. For example, a tech company might establish a review board tasked with evaluating the fairness of its facial recognition technology, ensuring that the model does not disproportionately affect certain demographic groups. In cases where the model is found to cause harm, the review board could recommend adjustments or the suspension of the technology's use until the issues are resolved. Such mechanisms not only help organizations respond to ethical issues as they arise but also signal a commitment to responsible AI practices.

To ensure accountability, organizations must also provide avenues for redress when ML systems cause harm. This could involve establishing a formal complaint process where individuals affected by an ML decision can challenge the outcome and seek rectification. For instance, in a scenario where an ML model erroneously denies someone a loan, the affected individual should have a clear pathway to dispute the decision and have their case reviewed by a human decision-maker. By integrating such processes into their governance frameworks, organizations can mitigate the risks of ML deployment and uphold their ethical obligations to the people they serve.

Through the implementation of these strategies—model interpretability tools, transparent data governance frameworks, white-box models, accountability frameworks like the RACI matrix, AI ethics review boards, and avenues for redress—organizations can effectively navigate the challenges of transparency and accountability in ML. These practices not only build trust and foster ethical governance but also ensure that ML systems are used responsibly, with a clear commitment to minimizing harm and upholding the rights of all stakeholders involved.

Bias and Fairness

Bias in ML systems is a significant ethical concern. ML models can inadvertently learn and perpetuate biases present in the training data, leading to unfair and discriminatory outcomes. Bias can manifest in various forms, such as gender, racial, or socioeconomic biases, and can have serious implications, especially in critical applications like hiring, lending, and law enforcement.

Ensuring fairness in ML systems involves identifying and mitigating biases during the data collection, model training, and deployment phases. Continuous monitoring and evaluation are necessary to ensure that models remain fair and do not perpetuate or exacerbate existing inequalities.

Bias Mitigation in Machine Learning

Bias in machine learning can arise from several sources, including biased data, biased model selection, and biased evaluation metrics. It is essential to understand these sources to effectively mitigate bias. **Data bias** occurs when the training data is not representative of the real-world population. This can happen due to historical biases, sampling biases, or data collection methods that exclude certain groups. **Model bias** occurs when the algorithms used in ML favor certain outcomes or groups over others. This can be due to the inherent properties of the algorithms or the choices made during model development. **Evaluation bias** happens when the metrics used to assess model performance do not adequately capture fairness or overlook the impacts on certain groups.

Bias Detection

Bias detection in machine learning (ML) is a critical aspect of data governance that requires constant vigilance, particularly as models are deployed and used in real time. The potential for biased outcomes in ML models poses significant ethical, legal, and operational risks. Organizations must implement sophisticated tools and techniques to identify and mitigate bias throughout the model lifecycle, from training to real-time deployment. Bias detection tools have evolved to meet these challenges, offering a range of methods to ensure that models operate fairly and equitably across diverse populations.

Detecting bias in ML models begins with a thorough examination of the data used for training. Data often carries historical biases that can be perpetuated or even amplified by the model if not carefully managed. Tools like Fairness Indicators allow data scientists to evaluate how different subgroups perform within a model, highlighting disparities in outcomes that may indicate bias. For instance, in a hiring model, Fairness Indicators might reveal that candidates from certain demographic groups consistently receive lower scores, signaling potential bias in the data or the model itself. By integrating such tools into the data preprocessing stage, organizations can proactively address biases before they influence the model's predictions.

Once a model is deployed, the challenge shifts to real-time bias detection. Real-time monitoring tools are essential for identifying and mitigating biases as they emerge during model operation. For example, tools like IBM Watson OpenScale provide continuous monitoring of deployed models, tracking their performance across various metrics, including fairness. OpenScale can automatically detect when a model's predictions begin to skew in favor of or against certain groups, allowing organizations to intervene before the bias causes significant harm. For example, in a credit scoring application, OpenScale might alert the organization if the model starts disproportionately denying loans to applicants from specific demographics, prompting an immediate review and adjustment of the model.

Another advanced technique for real-time bias detection involves adversarial debiasing. This approach integrates an adversarial network with the primary model, designed to detect and correct for biases during the training process. The adversarial network continuously challenges the primary model by trying to predict sensitive attributes such as race or gender based on the model's outputs. If the adversarial network can predict these attributes with high accuracy, it indicates that the primary model is likely biased. The primary model is then adjusted to reduce this predictability, thereby mitigating bias. This method has been particularly effective in applications like facial recognition, where bias detection is crucial for ensuring that the technology works equitably across different skin tones and facial features.

Bias detection tools also extend to the interpretability of ML models, especially those that function as black boxes, such as deep neural networks. Techniques like SHAP (SHapley Additive exPlanations) and LIME (Local Interpretable Model-agnostic Explanations) provide insights into how different features influence the model's predictions. These tools are vital for detecting bias in models where the decision-making process is not transparent. For instance, SHAP can be used in a healthcare model to determine whether race or socioeconomic status is unduly influencing the likelihood of receiving certain treatments. By understanding how these features contribute to the model's decisions, data scientists can identify and address potential biases that might otherwise go unnoticed.

In real-time environments, bias detection also relies on continuous feedback loops. Models must be regularly retrained and updated to reflect new data and changing conditions, which requires tools capable of detecting shifts in data distributions that might introduce bias. Drift detection tools, such as those integrated into platforms like TensorFlow Extended (TFX), monitor the input data and model outputs for significant changes over time. If a drift is detected, indicating that the model's predictions are becoming less reliable or biased, the system can trigger an automatic retraining process or alert data scientists to investigate. For instance, an e-commerce recommendation system might detect that its suggestions are increasingly favoring a certain category of products, potentially due to a recent marketing campaign. Drift detection would flag this shift, allowing the organization to adjust the model to ensure it remains balanced and fair in its recommendations.

Beyond technical tools, real-time bias detection also involves the implementation of ethical AI frameworks that guide the ongoing evaluation of models. These frameworks ensure that bias detection is not solely a technical exercise but is embedded within the organization's broader governance practices. For example, Google's AI Principles emphasize the importance of fairness, accountability, and transparency in all AI applications. By adhering to such principles, organizations can ensure that their bias detection efforts are aligned with ethical standards and societal expectations. Regular audits and reviews, conducted by cross-functional teams that include ethicists, data scientists, and legal experts, help maintain the integrity of the models over time, ensuring that any biases are promptly identified and addressed.

In real-time deployment scenarios, it is also crucial to consider the role of counterfactual fairness. This method evaluates whether a model's predictions would change if certain sensitive attributes were altered while keeping all other factors constant. Counterfactual fairness is particularly useful in dynamic environments where the impact of bias can evolve rapidly. For instance, in a real-time ad-serving platform, counterfactual fairness might be used to assess whether changing a user's gender or ethnicity would lead to different ad recommendations. If the model's outputs differ significantly based on these changes, it indicates a potential bias that needs to be corrected. This approach ensures that the model remains fair and consistent, even as it encounters new and diverse data inputs.

Another critical aspect of real-time bias detection is the deployment of fairness constraints during model training. These constraints are rules or conditions that ensure the model's predictions adhere to predefined fairness criteria. For example, a model might be trained with a fairness constraint that requires equal false positive rates across different demographic groups. This ensures that no group is unfairly penalized by the model's predictions, even in real-time scenarios. Fairness constraints can be enforced through tools like Microsoft's Fairlearn, which allows data scientists to specify fairness goals and integrate them directly into the model development process. This approach has been successfully applied in areas such as loan approval and hiring, where ensuring equitable outcomes is critical to avoiding legal and ethical pitfalls.

The challenge of bias detection in real time also extends to the evaluation of intersectional biases—where multiple forms of bias intersect and amplify each other. Detecting intersectional biases requires tools that can analyze the combined effects of various attributes, such as race, gender, and age, on the model's predictions. Techniques like intersectional auditing involve testing the model across different combinations of these attributes to identify any compounded biases. For example, a real-time hiring platform might use intersectional auditing to evaluate whether women of color are being systematically disadvantaged compared to other candidates. By identifying and addressing these nuanced biases, organizations can ensure that their models operate fairly across all demographic intersections.

Real-time bias detection also benefits from the integration of user feedback mechanisms. Allowing users to report biased or unfair outcomes in real time provides valuable insights that might not be captured by automated tools. For instance, on a social media platform, users might flag content recommendations that seem biased or discriminatory. This feedback can be used to refine the model and improve its fairness over time. Incorporating user feedback into the bias detection process ensures that the model remains responsive to the needs and concerns of its users, fostering greater trust and accountability.

The role of explainable AI (XAI) in real-time bias detection cannot be overstated. XAI tools help bridge the gap between complex ML models and human understanding, making it easier to detect and address biases as they occur. For example, in a real-time financial trading system, XAI tools can provide transparency into the decision-making process, allowing traders and regulators to understand how the model is weighing different factors. If the model starts to favor certain types of trades that could lead to market manipulation or unfair practices, the transparency provided by XAI tools can help detect this bias early and prevent significant financial harm.

As organizations increasingly rely on ML models for critical decision-making, the importance of real-time bias detection tools and techniques continues to grow. By integrating these tools into their data governance frameworks, organizations can ensure that their models remain fair, transparent, and accountable throughout their lifecycle. This not only protects against the risks associated with biased outcomes but also builds trust with users, regulators, and other stakeholders. The ongoing evolution of bias detection tools, combined with ethical AI frameworks, positions organizations to navigate the complex landscape of ML deployment while upholding the highest standards of fairness and equity.

Strategies for Bias Mitigation

Bias mitigation in machine learning (ML) is a critical concern for ensuring fairness and accuracy in model predictions. This section delves into various strategies to address and reduce bias, focusing on techniques that improve the quality of data and the fairness of the models developed from it.

Data-Centric Approach

A data-centric approach to bias mitigation in machine learning (ML) emphasizes the critical role that data plays in shaping model outcomes. By focusing on the quality, diversity, and representativeness of the data used to train ML models, organizations can address and reduce biases that might otherwise lead to unfair or discriminatory decisions. This approach not only involves careful data collection and preprocessing but also the ongoing evaluation and adjustment of data throughout the lifecycle of the ML model.

Diverse and representative data collection is foundational to mitigating bias in ML models. When data is sourced from a narrow or homogeneous population, the resulting model is likely to exhibit biases that reflect the limitations of the training data. For instance, if an ML model used by a bank for loan approvals is primarily trained on data from affluent neighborhoods, it may underperform for applicants from lower-income areas, leading to unfair rejections. To counteract this, organizations must actively seek out data that encompasses a wide range of income levels, ethnicities, and genders, ensuring that the model is reflective of the entire population it is intended to serve. This might involve targeted data collection strategies, such as conducting surveys or acquiring datasets from regions or demographics that are underrepresented in the original data.

The complexity of achieving truly representative data is exemplified in the healthcare industry, where ML models are increasingly used to assist in diagnosing diseases and recommending treatments. If a model is trained primarily on data from urban hospitals that predominantly serve certain racial or socioeconomic groups, it may not perform well in rural or minority populations. For example, a model designed to predict the risk of heart disease might fail to accurately assess the risk for African American patients if the training data lacks sufficient representation from this group. To address this, healthcare organizations might collaborate with rural clinics or community health programs to collect data that better represents these underserved populations. By ensuring that the training data is diverse and representative, the model is more likely to produce accurate and equitable outcomes across all patient groups.

In addition to targeted data collection, organizations can enhance the diversity of their data by leveraging synthetic data generation techniques. Synthetic data, created by algorithms that mimic the statistical properties of real-world data, can be used to supplement datasets that are lacking in diversity. For instance, in a facial recognition system that struggles to accurately identify individuals from certain ethnic backgrounds, synthetic data can be generated to include more examples of those underrepresented groups, thereby improving the model's performance. This approach can be particularly useful in scenarios where collecting real-world data is challenging or where privacy concerns limit access to sensitive information. However, care must be taken to ensure that synthetic data accurately reflects the diversity and complexity of the real-world population, as poorly generated synthetic data can introduce new biases rather than mitigate existing ones.

Preprocessing techniques are another crucial aspect of a data-centric approach to bias mitigation. Data preprocessing involves cleaning and transforming data before it is used to train an ML model. This stage presents an opportunity to identify and correct biases that might be embedded in the data. For example, in a hiring algorithm, the historical data might show a preference for candidates with certain educational backgrounds, reflecting biases in the past hiring practices of the company. By carefully analyzing the data and applying techniques such as reweighting or resampling, data scientists can adjust the dataset to ensure that it is more representative of the desired diversity in the workforce. This might involve giving more weight to candidates from underrepresented educational backgrounds or ensuring that the training data includes a balanced mix of applicants from various demographics.

Addressing intersectional bias is a more advanced aspect of the data-centric approach, requiring an understanding of how different forms of bias can intersect and compound each other. For example, in the context of loan approvals, a model might exhibit bias not just based on race or gender alone, but specifically against women of color, who face unique challenges that differ from those faced by either white women or men of color. To mitigate such biases, the data used to train the model must be carefully examined for these intersectional effects. This might involve segmenting the data to analyze how different combinations of attributes (e.g., race, gender, income level) impact the model's predictions. By doing so, organizations can ensure that their models are fair and equitable across all intersections of identity, rather than merely addressing single-dimensional biases.

Regular data audits and bias detection are essential components of a data-centric approach, ensuring that the data remains representative and unbiased throughout the model's lifecycle. These audits involve continuously monitoring the data and model performance to detect any emerging biases that may arise as new data is collected or as the model is exposed to different populations. For instance, an e-commerce platform using an ML model to recommend products might notice that over time, the recommendations increasingly favor products targeted at a specific demographic, due to changes in user behavior or market trends. By conducting regular audits, the platform can detect this bias early and take corrective action, such as retraining the model with more balanced data or adjusting the recommendation algorithm to ensure fairness across all user groups.

A data-centric approach also necessitates a deep understanding of the cultural and societal contexts in which the data is collected and the model is deployed. Data that is collected in one cultural context may not be appropriate or relevant when applied in another, leading to biased or inaccurate outcomes. For instance, a sentiment analysis model trained on English-language social media posts might not perform well when applied to posts in other languages or cultural contexts, where the same words or phrases may carry different connotations. To address this, organizations must ensure that their data collection methods are culturally sensitive and that the data used to train their models is appropriately localized. This might involve working with local communities or experts to collect and annotate data in a way that accurately reflects the cultural nuances of the target population.

The success of a data-centric approach to bias mitigation depends on the active involvement of a diverse team of data scientists, ethicists, and domain experts. A diverse team brings different perspectives to the data collection and preprocessing processes, helping to identify and address biases that might be overlooked by a more homogeneous group. For example, in the development of a healthcare model, a diverse team might include clinicians from various specialties, ethicists with expertise in healthcare disparities, and data scientists with experience in working with diverse datasets. This collaborative approach ensures that the data used to train the model is scrutinized from multiple angles, reducing the likelihood of bias and improving the overall fairness and accuracy of the model.

By actively seeking out diverse and representative data, leveraging advanced data collection and preprocessing techniques, and maintaining a vigilant approach to bias detection, organizations can significantly mitigate the risks of bias in their ML models. This data-centric approach is not just about ensuring fairness and compliance with ethical standards, but also about building models that are robust, reliable, and effective in delivering accurate outcomes across all segments of the population. As ML continues to play an increasingly prominent role in decision-making across various sectors, adopting a data-centric approach to bias mitigation is essential for creating models that truly serve the diverse needs of society.

Identifying and Mitigating Bias in Sampling Techniques

Stratified sampling is a method where the data is divided into different subgroups or strata, and samples are drawn from each subgroup in proportion to their prevalence in the population. This technique is particularly effective in ensuring that all relevant demographics are adequately represented in the training data. For instance, in the context of a recidivism prediction model used in the criminal justice system, stratified sampling can help ensure that the training data reflects the racial and socioeconomic disparities present within the prison population. Without such careful sampling, the model might overrepresent data from wealthier neighborhoods with lower crime rates, leading to biased predictions that unfairly target individuals from low-income communities.

The importance of stratified sampling becomes even more evident when considering the intersectionality of different demographic factors. For example, in a healthcare setting, an ML model designed to predict patient outcomes might perform well for the general population but poorly for specific subgroups, such as elderly patients from rural areas or women from certain ethnic backgrounds. Stratified sampling can be employed to ensure that these subgroups are adequately represented in the training data, allowing the model to learn the nuances associated with each group. This is particularly important in medical research, where underrepresented groups have historically been excluded from clinical trials, leading to treatments that are less effective or even harmful to those populations.

While stratified sampling is a powerful tool for bias mitigation, it is not without its challenges. One significant challenge is the availability of sufficient data for each subgroup. In some cases, certain subgroups may be underrepresented or have limited data available, making it difficult to achieve proportional representation in the training dataset. This is where techniques such as data augmentation or synthetic data generation come into play. By creating additional synthetic examples for underrepresented groups, data scientists can bolster the training data and ensure that the model has enough information to learn from. For example, in a facial recognition system that lacks sufficient images of individuals with darker skin tones, synthetic data can be generated to fill this gap, thereby reducing the likelihood of biased predictions.

Another advanced technique related to sampling is the use of active learning, where the model is trained iteratively, and at each step, it identifies the most informative samples to add to the training set. This approach can be particularly effective in addressing biases that arise from skewed or imbalanced datasets. For example, in a fraud detection system, where fraudulent transactions are rare compared to legitimate ones, active learning can help the model focus on learning from those rare, but highly informative, fraudulent cases. By carefully selecting samples that the model finds most challenging, active learning ensures that the model is exposed to a more balanced and representative dataset over time, leading to more accurate and unbiased predictions.

Mitigating bias through sampling techniques also requires ongoing monitoring and validation of the model's performance across different subgroups. Even with careful sampling, biases can emerge during the model's operation due to shifts in the underlying data distribution or changes in the population being served. To address this, organizations must implement continuous monitoring systems that track the model's performance across various demographic groups. For instance, in a loan approval model, continuous monitoring might reveal that the model's predictions have become biased against certain income levels or racial groups over time. This could occur due to changes in economic conditions or shifts in the applicant pool that were not fully captured by the original training data. By detecting these biases early, organizations can take corrective action, such as retraining the model with updated data or adjusting the sampling strategy to better reflect the current population.

The ethical implications of sampling techniques in ML cannot be overstated. When biases in sampling lead to unfair outcomes, they can perpetuate systemic inequalities and undermine public trust in AI-driven decision-making. For example, if an ML model used in hiring consistently disadvantages candidates from certain racial or ethnic backgrounds due to biased sampling, it not only harms those individuals but also reinforces existing disparities in employment opportunities. Addressing these issues requires a commitment to ethical data governance, where the selection and use of training data are guided by principles of fairness, transparency, and accountability.

One emerging area of research in bias mitigation through sampling is the development of fairness-aware sampling algorithms. These algorithms are designed to optimize the selection of training data by balancing the trade-offs between model accuracy and fairness. For instance, a fairness-aware sampling algorithm might prioritize the inclusion of samples from historically marginalized groups, even if those samples are less prevalent or more challenging for the model to learn from. This approach ensures that the model is not only accurate but also equitable in its predictions, providing fair outcomes across all demographic groups.

It is important to recognize that bias mitigation through sampling is not a one-time process but an ongoing effort. As ML models are deployed in dynamic environments, the populations they serve may evolve, and new biases may emerge. To effectively mitigate bias, organizations must continuously revisit their sampling strategies, update their training data, and refine their models to reflect the changing realities of the world they operate in. This iterative approach ensures that ML models remain fair and effective over time, adapting to new challenges and serving the diverse needs of society.

Identifying and mitigating bias in sampling techniques is a critical component of a data-centric approach to bias mitigation in machine learning. By employing advanced techniques such as stratified sampling, data augmentation, active learning, and fairness-aware sampling algorithms, organizations can ensure that their models are trained on diverse and representative data, leading to more accurate and equitable outcomes. Continuous monitoring and validation of model performance across different subgroups further help in identifying and addressing biases that may arise during deployment. As the field of ML continues to evolve, the importance of ethical data governance and the ongoing refinement of sampling strategies will remain central to the development of fair and trustworthy AI systems.

Data Cleaning and Anomaly Detection

Data cleaning involves systematically identifying and rectifying errors or inconsistencies within datasets. This process can include the removal of duplicates, correction of inaccuracies, and handling of missing values, all of which can introduce bias if not properly managed. For example, in a healthcare ML model designed to predict patient outcomes, missing or incorrect data could disproportionately affect predictions for certain demographic groups. If data on a specific ethnic group is incomplete or has higher error rates, the model might underperform for that group, leading to inequitable healthcare outcomes. By thoroughly cleaning the data and ensuring that it accurately represents all relevant populations, data scientists can help mitigate such biases before they affect the model's performance.

Anomaly detection is a more nuanced aspect of data cleaning, involving the identification of data points that significantly deviate from expected patterns. These anomalies can be indicative of errors, rare events, or outliers that, if left unaddressed, could skew the model's predictions. For instance, in an ML model used for credit scoring, an anomaly detection algorithm might flag a small number of loan applications with unusually high credit scores that do not align with the applicant's financial history. Upon investigation, it might be discovered that these anomalies are due to data entry errors or fraudulent applications. Correcting these anomalies is crucial because their inclusion in the training data could lead to a model that either overestimates or underestimates creditworthiness, depending on how these anomalies are handled.

In some cases, anomalies might reflect biases embedded in the data collection process itself. For example, consider an ML model used for targeted advertising, where anomaly detection reveals that a particular product category shows unusually low click-through rates for a specific age group. Further analysis might uncover that the product descriptions or advertising materials inadvertently contain age-based biases, such as using language or imagery that does not resonate with older users. This kind of unconscious bias can significantly impact the effectiveness of the advertising campaign, leading to underrepresentation of certain demographics in the model's predictions. By identifying and correcting these anomalies, the model can be adjusted to ensure that it performs equitably across all age groups, leading to more inclusive and effective advertising strategies.

Advanced anomaly detection techniques can also be employed to identify more subtle forms of bias that might not be immediately apparent. For instance, in a facial recognition system, an anomaly detection algorithm could be used to identify instances where the system's confidence levels are significantly lower for certain demographic groups, such as individuals with darker skin tones. These anomalies might not be errors in the traditional sense but could indicate a deeper issue with the training data or model architecture. By investigating these anomalies, data scientists can uncover and address the underlying biases that lead to these disparities, potentially retraining the model with more diverse and representative data or adjusting the algorithm to better handle the variations in skin tones.

The challenge of data cleaning and anomaly detection becomes even more complex in dynamic environments where data is continuously collected and updated. In such cases, anomalies might not just be static outliers but could represent shifts in the underlying data distribution over time. For example, in an e-commerce platform, anomaly detection might reveal that customer behavior is changing in response to a new marketing campaign or economic event. These shifts could lead to biases in the model's predictions if not properly accounted for, as the model might continue to rely on outdated assumptions about customer preferences. To address this, organizations need to implement ongoing anomaly detection and data-cleaning processes that adapt to changes in the data, ensuring that the model remains accurate and unbiased as new data is incorporated.

Another advanced topic in anomaly detection is the use of unsupervised learning techniques to identify biases that are not associated with any known labels or categories. In traditional supervised learning, models are trained on labeled data where the correct outcome is known, making it easier to identify and correct biases. However, in unsupervised learning, where the model must discover patterns in the data without explicit labels, detecting biases becomes more challenging. Anomaly detection algorithms, such as clustering or density estimation techniques, can be employed to identify groups of data points that do not fit the expected distribution. For example, in a social media analysis tool, unsupervised anomaly detection might reveal that certain topics or keywords are systematically underrepresented in discussions among specific demographic groups, indicating a potential bias in the data collection process. By identifying these anomalies, data scientists can take steps to ensure that the model is trained on a more balanced and representative dataset.

Data cleaning and anomaly detection are not just technical processes but also involve ethical considerations, particularly when it comes to deciding how to handle anomalies. In some cases, removing anomalies might inadvertently erase important variations or minority perspectives from the data. For example, in a dataset of student performance, outliers might represent students who excel in unconventional ways or who face unique challenges that are not captured by the majority of the data. Simply removing these outliers could lead to a model that overlooks the needs of these students, reinforcing biases in the educational system. Instead, data scientists must carefully consider the context and implications of each anomaly, deciding whether it represents an error that should be corrected or a legitimate variation that should be preserved and understood.

Moreover, the process of data cleaning and anomaly detection must be transparent and well-documented, allowing for accountability and reproducibility in ML models. Stakeholders, including those affected by the model's predictions, should have visibility into how data is being cleaned and how anomalies are being handled. This transparency is crucial for building trust in the model's outcomes and ensuring that the bias mitigation efforts are both effective and ethically sound. For example, in a legal context where an ML model is used to inform sentencing decisions, transparency in the data-cleaning process is essential to ensure that defendants from all demographic groups are treated fairly and that any potential biases are identified and addressed before they can impact the model's predictions.

De-biasing Techniques

De-biasing techniques play a critical role in a data-centric approach to bias mitigation, focusing on the manipulation and enhancement of datasets to address imbalances and improve the fairness of machine learning (ML) models. These techniques aim to rectify biases that emerge from skewed or unrepresentative data, ensuring that ML models are trained on datasets that accurately reflect the diversity of the populations they serve. By employing strategies such as data augmentation, oversampling, and undersampling, data scientists can create more equitable and effective models capable of making fair decisions across various demographic groups.

Data augmentation is a powerful de-biasing technique that involves artificially generating new data points to increase the size and diversity of the training dataset. This approach is particularly beneficial in scenarios where certain groups are underrepresented in the original data. For example, in the development of facial recognition software, there is often a bias toward recognizing faces with lighter skin tones because the training datasets predominantly consist of images from such demographics. This bias can lead to poor performance when the model encounters faces with darker skin tones, potentially causing significant inaccuracies in real-world applications. To address this issue, data augmentation techniques can be employed to generate synthetic images that represent a broader range of skin tones. Techniques such as generative adversarial networks (GANs) can be used to create realistic images of individuals with diverse facial features, ensuring that the model is trained on a dataset that encompasses a wide variety of appearances. By enhancing the diversity of the training data, the model becomes better equipped to recognize and accurately classify faces from all racial and ethnic backgrounds, reducing the likelihood of biased outcomes in practice.

In addition to data augmentation, oversampling and undersampling are essential techniques for balancing the representation of different classes within a dataset. These methods are particularly useful in addressing class imbalance, a common issue in ML where certain classes are significantly more prevalent than others. For instance, consider an ML model designed for spam filtering, where the training dataset contains a much higher proportion of spam emails compared to legitimate emails. This imbalance can lead the model to overfit on the spam class, resulting in a tendency to classify all emails as spam, which undermines its utility. To mitigate this bias, oversampling can be used to increase the number of legitimate emails in the training data, giving the model more opportunities to learn the characteristics of non-spam content. Techniques such as Synthetic Minority Over-sampling Technique (SMOTE) can be applied, where new legitimate email examples are generated by interpolating between existing samples. This helps to create a more balanced dataset, allowing the model to better differentiate between spam and legitimate emails without disproportionately favoring one class over the other.

Conversely, undersampling can be applied to reduce the number of examples from the majority class, which in the spam filtering example would involve selectively removing some spam emails from the training data. This approach forces the model to pay more attention to the legitimate emails, thereby improving its ability to correctly identify non-spam content. However, undersampling must be applied carefully to avoid discarding valuable information from the majority class that could contribute to the model's overall performance. A nuanced application of both oversampling and undersampling, potentially in combination with more sophisticated techniques like ensemble learning, can help achieve a balanced and unbiased model.

The challenges of de-biasing are not limited to class imbalances but also extend to the subtle ways in which data can encode societal biases. For instance, in natural language processing (NLP) models, biases can be deeply embedded in the text data, reflecting cultural stereotypes and prejudices. A chatbot trained on biased text data might, for example, develop a tendency to associate certain professions with specific genders or ethnicities, leading to biased responses. To mitigate such biases, data augmentation can be used to introduce counterexamples that challenge these stereotypes. For example, by augmenting the dataset with text that associates women with leadership roles or men with caregiving responsibilities, the model can learn a more balanced representation of societal roles. Additionally, techniques like bias-aware word embeddings, where the vectors representing words are adjusted to reduce biased associations, can be employed to further de-bias NLP models.

Another advanced de-biasing technique involves reweighting the data during the training process. Reweighting assigns different weights to data points based on their importance or representation in the training set. For example, in an ML model used for predicting job performance, if the training data includes fewer examples from minority groups, those examples can be given higher weights during training. This approach ensures that the model places more emphasis on learning from these underrepresented groups, leading to predictions that are more accurate and fair across all demographics. Reweighting can be particularly effective in combination with other de-biasing techniques, such as data augmentation or oversampling, creating a robust strategy for addressing multiple forms of bias simultaneously.

The implementation of de-biasing techniques also requires careful consideration of the ethical implications and potential trade-offs involved. For example, while data augmentation and oversampling can improve the representation of minority groups in the training data, there is a risk of overcompensating, leading to models that are overly sensitive to certain features. This could result in unintended consequences, such as reverse discrimination, where the model becomes biased in favor of the previously underrepresented group. To mitigate this risk, it is essential to validate and test the model across different scenarios and demographics, ensuring that it performs fairly and consistently for all users. This might involve using cross-validation techniques that specifically evaluate the model's performance on different subgroups, as well as conducting thorough bias audits to identify and address any remaining disparities.

De-biasing techniques are not a one-time fix but part of an ongoing process of model refinement and improvement. As ML models are deployed and interact with real-world data, new biases may emerge, necessitating continuous monitoring and adjustment. For instance, a facial recognition system that initially performs well across diverse populations might begin to exhibit bias if the demographic composition of its user base changes over time. Regularly updating the training data with new examples that reflect these changes, and reapplying de-biasing techniques as necessary, is crucial for maintaining the model's fairness and effectiveness. This ongoing commitment to de-biasing ensures that ML models remain aligned with ethical standards and societal values, even as the contexts in which they are used evolve.

The importance of a data-centric approach to bias mitigation is further underscored by the need for transparency and accountability in the deployment of ML models. By clearly documenting the de-biasing techniques used and the rationale behind them, organizations can provide transparency into their decision-making processes and build trust with users and stakeholders. For example, in regulatory contexts where ML models are used to make critical decisions, such as in finance or healthcare, providing detailed documentation of the de-biasing methods employed can help demonstrate compliance with fairness and non-discrimination standards. This transparency is not only a legal and ethical requirement but also a key factor in fostering public confidence in the use of AI technologies.

In summary, de-biasing techniques such as data augmentation, oversampling, undersampling, reweighting, and bias-aware word embeddings are essential tools in the data-centric approach to bias mitigation. These techniques enable data scientists to address the inherent biases present in datasets and create ML models that are fair, inclusive, and reliable. By continuously refining these methods and adapting them to the specific challenges of different applications, organizations can ensure that their ML models contribute positively to society and uphold the principles of fairness and equity in every decision they make.

Statistical Analysis and Visualization Techniques

Statistical analysis and visualization techniques are integral components of a data-centric approach to bias mitigation in machine learning (ML). By analyzing the distribution of data across various demographic groups, organizations can uncover hidden biases that may lead to unfair or discriminatory outcomes in ML models. These techniques provide a quantitative foundation for understanding how different groups are represented within a dataset and allow data scientists to take corrective actions to ensure that the model performs equitably across all segments of the population.

A key aspect of this approach is the use of statistical tools to examine the distribution of data across different demographic groups. For instance, when training an ML model for automated resume screening, it is crucial to analyze the distribution of resumes by gender, ethnicity, age, and other relevant factors. If the training data contains a significantly higher proportion of resumes from male candidates, the model might inadvertently learn to associate certain keywords or writing styles more strongly with male applicants. This bias could result in the model favoring male candidates over equally qualified female candidates, perpetuating gender disparities in hiring practices. By conducting a statistical analysis of keyword usage and writing styles across genders, data scientists can identify patterns that might indicate bias in the training data.

Visualization techniques such as boxplots and histograms are particularly effective for identifying disparities in data distribution. A boxplot, for example, can visually represent the distribution of a specific feature, such as years of experience or salary expectations, across different demographic groups. By comparing the boxplots for male and female candidates, data scientists can quickly see if there are significant differences in the data that could lead to biased model predictions. For instance, if the boxplot for male candidates shows a wider range and higher median salary expectations compared to female candidates, this could indicate that the model might learn to associate higher salaries with male candidates, potentially disadvantaging female applicants who may have faced systemic pay inequities in the past.

Histograms offer another powerful visualization tool for detecting bias in data distribution. By plotting the frequency of specific keywords or phrases across different demographic groups, histograms can reveal whether certain terms are disproportionately associated with one group over another. For example, in a resume screening model, a histogram might show that leadership-related keywords are more frequently used by male applicants. This could lead the model to prioritize male candidates for leadership positions, reinforcing gender biases in hiring. Identifying such disparities through visualization allows data scientists to take proactive steps, such as rebalancing the dataset or refining the model to ensure that it does not unfairly favor one group over another.

Advanced statistical techniques, such as t-tests or chi-square tests, can be employed to quantitatively assess the significance of observed disparities in data distribution. For example, in the context of a healthcare ML model predicting disease risk, a t-test could be used to compare the average risk scores assigned to patients from different ethnic groups. If the test reveals a statistically significant difference in risk scores between groups, this might indicate that the model is biased due to imbalances in the training data. Such statistical evidence provides a strong basis for revisiting the model's development process, potentially leading to the collection of additional data or the application of de-biasing techniques to address the identified issues.

The use of statistical analysis and visualization extends beyond identifying biases in individual features to examining the overall structure of the dataset. For example, Principal Component Analysis (PCA) is a technique that can be used to reduce the dimensionality of the data while preserving as much variability as possible. By visualizing the principal components, data scientists can explore how different demographic groups are represented in the dataset. If certain groups cluster together or are separated in the principal component space, this could indicate underlying biases in the way the data is structured. For instance, in a model used for loan approval, PCA might reveal that applicants from certain socioeconomic backgrounds are consistently grouped together, suggesting that the model may be overfitting to features associated with those backgrounds and potentially leading to biased decisions.

Heatmaps are another effective visualization tool for examining correlations between features and demographic attributes. A heatmap can display the strength of correlations between various features, such as education level, job title, and years of experience, with demographic attributes like gender or ethnicity. If the heatmap reveals strong correlations between certain features and a specific demographic group, this could indicate that the model is learning to associate those features with that group, potentially leading to biased outcomes. For example, in an employment model, a heatmap might show a strong correlation between high-level job titles and male applicants, suggesting that the model might be biased in favor of men for senior positions. By identifying these correlations, data scientists can adjust the model or the data preprocessing steps to mitigate such biases.

An essential consideration in using statistical analysis and visualization for bias detection is the interpretability of the results. Visualizations must be clear and accessible to all stakeholders, including those without a technical background, to ensure that the findings can be effectively communicated and acted upon. For instance, in a model used for college admissions, the admissions committee might need to understand how certain features, such as test scores or extracurricular activities, are influencing the model's predictions across different demographic groups. By presenting the data through intuitive visualizations, such as scatter plots or bar charts, data scientists can help the committee identify potential biases and make informed decisions about how to address them.

To fully leverage statistical analysis and visualization for bias mitigation, it is important to integrate these techniques into the model development and evaluation pipeline. This involves not only analyzing the training data before the model is built but also continuously monitoring the model's performance across different demographic groups once it is deployed. For example, a real-time monitoring system could use statistical analysis to track the model's predictions for different groups over time, alerting data scientists if the model begins to exhibit biased behavior. This proactive approach ensures that biases are detected and corrected as soon as they emerge, preventing them from causing harm in the real world.

In addition to detecting bias, statistical analysis and visualization can also be used to evaluate the effectiveness of bias mitigation strategies. For instance, after applying a de-biasing technique, such as reweighting or data augmentation, data scientists can use the same statistical tools to assess whether the disparities in data distribution have been reduced. Boxplots, histograms, and other visualizations can be compared before and after the intervention to determine if the changes have led to a more balanced and fair dataset. This iterative process of analysis, intervention, and evaluation is key to achieving robust bias mitigation in machine learning models.

Statistical analysis and visualization techniques are invaluable tools in the data-centric approach to bias mitigation. By providing insights into how different demographic groups are represented in the data and how the model is likely to behave across these groups, these techniques enable data scientists to identify and address biases at every stage of the ML pipeline. As the use of machine learning continues to expand into critical areas such as healthcare, finance, and criminal justice, the importance of these tools in ensuring fairness and equity in AI systems cannot be overstated.

Model-Centric Approach

A model-centric approach to bias mitigation within machine learning (ML) data governance focuses on refining the intrinsic properties of the model itself to ensure fairness and equity. This approach goes beyond just cleaning and balancing data; it dives deep into the model's architecture, algorithms, and training processes to identify and rectify biases that might emerge from the model's design or its learning patterns. By targeting the core of the ML pipeline, this strategy ensures that the models not only perform well but also operate in a manner that is fair across all demographic groups.

A critical tool in this approach is the fairness matrix, which serves as a structured framework to assess and diagnose potential biases in an ML model's decision-making process. The fairness matrix operates by offering a tabular representation of various fairness metrics across different sensitive attributes, such as gender, race, age, or socioeconomic status. The matrix's value lies in its ability to provide a comprehensive overview of the model's behavior, enabling organizations to quantify and address biases systematically.

Integrating the fairness matrix into existing workflows requires a practical, step-by-step approach that allows organizations to identify, measure, and mitigate biases effectively. The process begins with defining the relevant sensitive attributes and selecting appropriate fairness metrics that align with the specific context of the ML model. For example, in a model used for credit scoring, sensitive attributes might include race, gender, and income level, while relevant fairness metrics could include demographic parity, equalized odds, and predictive parity. Once these are defined, the organization can move on to calculating these metrics for each subgroup.

The first step in utilizing a fairness matrix involves the identification of sensitive attributes relevant to the application. For a hiring algorithm, attributes like gender, ethnicity, and educational background might be critical, whereas for a healthcare prediction model, attributes such as age, race, and socioeconomic status could be more pertinent. After identifying these attributes, the next step is to calculate fairness metrics such as demographic parity, which evaluates whether the model's predicted outcomes are evenly distributed across different groups. For instance, if a model predicts the likelihood of job success, demographic parity would require that the likelihood of positive outcomes is similar for male and female applicants.

Equalized odds is another metric to consider, ensuring that both the true positive rates and false positive rates are consistent across different demographic groups. For example, in a model used for recidivism prediction, equalized odds would require that the model's accuracy in predicting repeat offenses is similar for all racial groups. If one group experiences higher false positive rates—being incorrectly flagged as high-risk more often than others—this would indicate a bias that needs to be addressed.

Predictive parity, another crucial metric, assesses whether the model's positive predictive value is consistent across different groups. This is particularly important in contexts like healthcare, where a model might predict the success of a particular treatment. Predictive parity would ensure that the predicted success rate of the treatment is equally reliable across different demographic groups, such as men and women or different ethnicities.

Once these metrics are calculated and presented in the fairness matrix, the matrix serves as a diagnostic tool that allows data scientists to pinpoint specific areas where the model's fairness might be compromised. For example, if the fairness matrix reveals a significant disparity in false positive rates for a particular demographic group in a criminal justice algorithm, the focus can shift toward reducing false positives for that group. This might involve adjusting the model's decision thresholds, retraining the model with more balanced data, or applying algorithmic debiasing techniques specifically aimed at reducing these disparities.

The fairness matrix does not merely stop at identifying biases; it also plays a strategic role in guiding bias mitigation efforts. By clearly showing where fairness issues lie, the matrix helps prioritize interventions, ensuring that resources are directed toward the most pressing fairness concerns without compromising overall model performance. For example, in a loan approval model, if the matrix indicates that female applicants are more likely to be incorrectly denied loans compared to male applicants, the organization can focus on refining the model's criteria or using reweighting techniques to ensure that the model evaluates all applicants more equitably.

The integration of a fairness matrix into an organization's workflow requires a structured approach that aligns with the model development lifecycle. After identifying sensitive attributes and calculating fairness metrics, the next step is to interpret the results and develop a plan of action. This might involve setting up a cross-functional team that includes data scientists, ethicists, and domain experts to review the fairness matrix and make decisions about the necessary interventions. For example, if a healthcare prediction model shows bias against older patients, the team might decide to collect additional data for this group, apply reweighting to give more importance to older patient data, or even redesign parts of the model to ensure better performance for this demographic.

Organizations can also use the fairness matrix as a tool for continuous monitoring and improvement. As models are deployed and start interacting with real-world data, it is crucial to regularly update the fairness matrix to track how the model's performance changes over time. This ongoing assessment is particularly important in dynamic environments where the data distribution might shift, potentially introducing new biases. For instance, a hiring algorithm might initially perform well across all demographic groups, but as the applicant pool changes, new biases could emerge. Regularly recalculating the fairness metrics and updating the fairness matrix allows the organization to catch and address these biases before they become ingrained in the model's decision-making process.

Incorporating the fairness matrix into existing workflows also involves educating and training the relevant teams on how to use this tool effectively. Data scientists need to be proficient in calculating and interpreting fairness metrics, while managers and decision-makers should understand how to use the matrix to guide policy and strategic decisions. For instance, in a company using ML models for hiring, HR professionals should be trained to understand the implications of the fairness matrix results and how they can influence hiring practices to promote diversity and inclusion.

Organizations should also establish clear guidelines and procedures for using the fairness matrix. This includes defining how often the matrix should be updated, how the results should be communicated to stakeholders, and how the findings should be integrated into the decision-making process. For example, an organization might decide to review the fairness matrix quarterly, with results presented to a governance board that oversees the ethical use of AI. This board would then make recommendations on whether to adjust the model, collect additional data, or implement new bias mitigation strategies.

The fairness matrix can also be integrated into automated ML pipelines, allowing for real-time bias detection and mitigation. By embedding fairness checks directly into the model training and evaluation process, organizations can ensure that biases are addressed at every stage of the model lifecycle. For instance, during model development, an automated system could calculate fairness metrics after each training iteration and adjust the training process if any metric falls below a certain threshold. This proactive approach not only helps prevent biases from becoming entrenched but also ensures that the final model is as fair and equitable as possible.

Integrating the fairness matrix into existing workflows also requires a cultural shift within the organization. It necessitates a commitment to ethical AI practices and a recognition of the importance of fairness in ML models. This cultural shift might involve creating new roles or teams dedicated to ethical AI, establishing policies that prioritize fairness, and fostering an environment where fairness is considered as important as model accuracy or efficiency. For example, a company might create a Chief Ethics Officer role responsible for overseeing the integration of fairness tools like the fairness matrix into all ML projects.

In practice, integrating the fairness matrix into an organization's ML workflows involves several steps: defining sensitive attributes and selecting fairness metrics, calculating these metrics and populating the fairness matrix, interpreting the results and developing a plan of action, continuously monitoring and updating the matrix, and embedding fairness checks into automated ML pipelines. Each step requires careful consideration and collaboration across different teams within the organization to ensure that the model is not only effective but also fair and equitable.

The fairness matrix, when used effectively, becomes a powerful tool for guiding bias mitigation efforts in a model-centric approach to ML. It enables organizations to quantify fairness, prioritize interventions, and continuously monitor and improve the fairness of their models. By integrating the fairness matrix into existing workflows, organizations can build ML models that are not only accurate and efficient but also aligned with ethical standards and societal values.

Leveraging Explainable AI (XAI) Techniques

Leveraging Explainable AI (XAI) techniques is essential for mitigating bias in machine learning (ML) models, particularly as models become more complex and their decisions more difficult to interpret. XAI techniques like LIME (Local Interpretable Model-agnostic Explanations) and SHAP (SHapley Additive exPlanations) allow practitioners to gain insights into how specific features influence a model's predictions. This transparency is crucial for identifying and addressing biases that might be embedded within the model, thereby promoting fairness and accountability in ML systems.

For instance, consider a loan approval model used by a financial institution to predict whether an applicant is likely to repay a loan. XAI techniques can be applied to deconstruct the model's decision-making process and reveal which features are most heavily weighted in its predictions. If SHAP analysis shows that the model places significant emphasis on an applicant's zip code, it might indicate that the model is implicitly associating certain areas with higher default risks. This correlation could be rooted in historical biases in lending practices, such as redlining, where certain neighborhoods—often those with a high concentration of minority populations— were unfairly labeled as high-risk. Even though the model is not explicitly designed to discriminate, the use of biased data can lead to outcomes that perpetuate these inequalities. Identifying this through XAI techniques allows organizations to critically evaluate the model's reliance on such features and consider alternative, more equitable criteria for assessing creditworthiness.

To effectively integrate XAI into existing workflows, organizations need to establish a structured approach that incorporates XAI techniques at various stages of the ML model lifecycle. This integration begins with the development phase, where XAI can be used to scrutinize model prototypes before they are fully deployed. By applying LIME or SHAP early in the development process, data scientists can identify potential biases and adjust the model's architecture or training data accordingly. For example, if LIME reveals that a facial recognition model is disproportionately influenced by skin tone, this could prompt the team to revisit the training data and include a more diverse set of images, ensuring that the model performs well across all demographic groups.

Once a model has been developed and validated, XAI techniques should be employed continuously as part of the model's monitoring and maintenance. This ongoing application of XAI is particularly important in dynamic environments where the data landscape can shift, potentially introducing new biases. For instance, in an e-commerce platform that uses an ML model to recommend products, SHAP analysis might be run periodically to check if the model's recommendations are becoming skewed toward certain demographics. If the analysis reveals that the model increasingly favors products popular among a specific age group, the organization might need to recalibrate the model or adjust the recommendation algorithm to ensure that it remains balanced and fair to all users.

Organizations also need to establish clear guidelines for interpreting the results generated by XAI techniques. The outputs of LIME or SHAP are often complex, requiring careful analysis to draw meaningful conclusions about a model's behavior. Data scientists must be trained to interpret these results correctly, understanding that the presence of a strong feature influence does not necessarily imply causation, but rather indicates a correlation that warrants further investigation. For example, if SHAP shows that educational background is a significant predictor in a hiring model, this does not automatically mean the model is biased against candidates with less prestigious degrees. However, it does suggest that the organization should examine whether the model is unduly favoring certain educational institutions and whether this aligns with the job's actual requirements.

Incorporating XAI into an organization's ML workflows can also enhance collaboration between data scientists and other stakeholders, such as business leaders, ethicists, and legal teams. By providing transparent explanations for model decisions, XAI bridges the gap between technical and non-technical audiences, enabling more informed discussions about the model's fairness and ethical implications. For instance, in the context of a healthcare application, SHAP could be used to explain why an ML model is recommending a particular treatment plan over others. Clinicians can then combine this information with their medical expertise to ensure that the recommendation is not only based on sound data but also aligns with best practices and ethical standards. This collaborative approach helps ensure that the model's outputs are not just technically accurate but also ethically sound and aligned with the organization's values.

To systematically integrate XAI into existing workflows, organizations can follow a step-by-step guide that embeds XAI techniques at key points in the ML lifecycle. The first step involves the initial model development, where XAI tools are used to identify and mitigate potential biases in the early stages. For example, as soon as a prototype of an ML model is ready, LIME can be applied to analyze individual predictions, revealing which features the model is relying on most heavily. This early application of XAI allows the team to detect biases before the model is finalized, providing an opportunity to make necessary adjustments, such as refining the feature set or retraining the model on a more diverse dataset.

After the model has been developed and is ready for deployment, XAI should be incorporated into the validation and testing phases. During these stages, SHAP can be used to conduct a comprehensive analysis of the model's decision-making process across different scenarios and demographic groups. For example, in a hiring model, SHAP can help the organization understand whether the model's predictions are consistent across candidates of different genders, ethnicities, and educational backgrounds. If discrepancies are found, the organization can take corrective actions, such as implementing fairness constraints or collecting additional data to address the underrepresented groups.

Once the model is deployed, XAI techniques must be used for ongoing monitoring and bias detection. Organizations can establish a monitoring framework that includes regular checks using SHAP or LIME to ensure that the model continues to perform fairly over time. For example, in a credit scoring system, periodic SHAP analyses could reveal whether the model's reliance on certain features shifts over time, potentially introducing new biases as the economic landscape changes. If such shifts are detected, the organization can proactively adjust the model to maintain fairness, such as by rebalancing the training data or refining the feature selection process.

To facilitate the integration of XAI into these workflows, organizations should establish clear roles and responsibilities for interpreting and acting on XAI insights. Data scientists should be tasked with conducting the technical analysis, while business leaders and domain experts should be involved in interpreting the results and making decisions about how to address any identified biases. For example, if SHAP analysis reveals that an ML model used in insurance pricing is disproportionately affecting a certain demographic, the data science team would work with actuaries and business leaders to understand the implications and decide on the appropriate mitigation strategies. This might involve adjusting the model to ensure that pricing decisions are based on relevant risk factors rather than demographic proxies that could lead to unfair outcomes.

Another critical aspect of integrating XAI into workflows is ensuring that the explanations generated by XAI tools are accessible and understandable to all stakeholders. This requires developing user-friendly interfaces and visualization tools that can present the results of SHAP or LIME in a clear and intuitive manner. For example, interactive dashboards could be created to allow stakeholders to explore how different features influence model predictions and to simulate how changes to the model or data might affect outcomes. These tools can help bridge the gap between the technical complexity of XAI and the practical needs of decision-makers, ensuring that everyone involved has a clear understanding of the model's behavior and the steps needed to mitigate bias.

Organizations should institutionalize the use of XAI by incorporating it into their governance frameworks and ethical guidelines. This involves establishing policies that require the use of XAI in all ML projects, particularly those that involve high-stakes decisions, such as in finance, healthcare, or criminal justice. For example, an organization might implement a policy that mandates SHAP analysis for any model used in credit scoring, with the results reviewed by a cross-functional team that includes data scientists, compliance officers, and ethical advisors. By embedding XAI into the organization's governance structures, leaders can ensure that bias mitigation is not just an afterthought but a fundamental part of the model development process.

Incorporating XAI into existing workflows requires organizations to think critically about how these techniques can be used to not only interpret models but also to drive actionable changes. For instance, if a SHAP analysis in a recruitment model shows that certain skills or experiences are being undervalued in candidates from underrepresented backgrounds, this insight can lead to broader organizational changes, such as revising job descriptions or altering how certain qualifications are weighted in the hiring process. By using XAI not just as a diagnostic tool but as a catalyst for change, organizations can make more informed decisions that promote fairness and equity across all aspects of their operations.

In the context of legal compliance, integrating XAI into workflows also provides a safeguard against regulatory scrutiny. For example, in sectors like finance or insurance, where regulators may require explanations for automated decisions, XAI can provide the transparency needed to demonstrate that models are not only effective but also compliant with anti-discrimination laws and ethical standards. Organizations can create audit trails that document how XAI tools were used to evaluate and adjust models, providing evidence that due diligence was performed in ensuring that the models operate fairly.

Leveraging XAI techniques is crucial for a model-centric approach to bias mitigation. By systematically integrating tools like LIME and SHAP into the ML lifecycle—from development and validation to deployment and monitoring—organizations can ensure that their models are transparent, fair, and aligned with ethical standards. This not only helps in identifying and correcting biases but also promotes a culture of accountability and continuous improvement, where fairness and equity are at the forefront of AI development and deployment.

Exploring Algorithmic Debiasing Techniques

Algorithmic debiasing techniques are specifically designed to address biases that arise during the model training process, ensuring that the resulting model operates fairly across various demographic groups. By embedding fairness directly into the algorithms themselves, organizations can create models that are not only effective but also equitable.

One powerful method for algorithmic debiasing is the adversarial debiasing approach, which involves the use of a secondary model, known as the adversary, to identify and exploit biases in the primary model. The adversarial model is trained to recognize and highlight any biases in the predictions made by the primary model. The primary model is then retrained to be robust against the adversary's attempts to exploit these biases, effectively reducing the impact of bias in its predictions. This method is particularly effective in scenarios where biases are subtle and difficult to detect through traditional methods.

Consider a practical example in the context of a facial recognition system. Facial recognition technology has been widely criticized for its performance disparities across different racial groups, particularly its lower accuracy in recognizing individuals with darker skin tones. By employing adversarial debiasing, a secondary model can be trained to detect and exploit these racial biases within the facial recognition system. The primary model, upon retraining, learns to correct these biases, improving its accuracy across all racial groups. The adversarial model acts as a kind of "bias detector," forcing the primary model to become more equitable in its predictions. This process not only enhances the model's fairness but also builds trust in the technology, especially in applications where fair and unbiased recognition is critical, such as in law enforcement or border security.

Another advanced algorithmic debiasing technique is the use of fairness-aware training algorithms, which incorporate fairness constraints directly into the model's objective function. These constraints are designed to promote specific fairness goals, such as statistical parity or equalized odds, by modifying the training process to minimize disparities across different demographic groups. For instance, in a loan approval model, fairness-aware training could involve adjusting the objective function to minimize the difference in loan approval rates between different racial or socioeconomic groups. By incorporating fairness constraints, the model is guided to consider not just accuracy but also equity in its decision-making process.

To illustrate this, imagine a financial institution developing an ML model to predict loan approvals. Traditionally, the model might be optimized solely for accuracy, leading to a situation where certain demographic groups—perhaps those from lower-income backgrounds—are disproportionately denied loans. By integrating fairness-aware constraints into the training process, the institution can ensure that the model also considers the need for equitable treatment across all applicants. The objective function might be adjusted to penalize disparities in approval rates between different groups, thereby encouraging the model to find a balance between accuracy and fairness. This approach not only helps in reducing systemic biases but also aligns with the ethical and legal obligations of the institution, which may be subject to anti-discrimination laws and regulations.

Moreover, fairness-aware training algorithms can be tailored to specific fairness metrics, depending on the context in which the model is deployed. For example, in healthcare, where the stakes are high, and decisions can have life-or-death consequences, equalized odds might be a critical fairness metric. Equalized odds require that a model's true positive and false positive rates are consistent across different demographic groups. In a predictive model designed to identify patients at risk of a particular disease, fairness-aware training can be used to ensure that the model is equally likely to correctly identify at-risk patients across all racial or ethnic groups, thereby preventing disparities in medical care.

Beyond adversarial debiasing and fairness-aware training, another sophisticated algorithmic debiasing technique is the use of reweighting during the training process. Reweighting involves assigning different weights to data points based on their representation or the importance of their demographic group in the training data. This technique is particularly useful in addressing class imbalance, where certain groups are underrepresented in the data. By assigning higher weights to underrepresented groups, the model is encouraged to learn more from these examples, leading to fairer and more balanced predictions.

Consider an example in the context of employment. An ML model used for screening job candidates might be trained on historical hiring data, which could reflect biases such as a preference for candidates from certain prestigious universities or particular demographic backgrounds. If these groups are overrepresented in the training data, the model might learn to favor them, perpetuating existing inequalities. By applying reweighting, the model can be adjusted to give more importance to candidates from underrepresented backgrounds, such as those from less prestigious universities or minority groups. This helps ensure that the model evaluates all candidates more equitably, leading to a more diverse and inclusive hiring process.

Another related technique is the use of data augmentation to enhance the representation of underrepresented groups in the training data. While data augmentation is typically thought of in the context of image processing—where techniques like flipping, rotating, or cropping images are used to create more training examples—it can also be applied to textual or tabular data. For instance, in a natural language processing (NLP) model used to analyze customer feedback, data augmentation might involve generating synthetic text examples that include perspectives from underrepresented demographics. This enriched dataset helps the model learn to better recognize and respond to the diverse range of customer experiences, reducing bias in its predictions.

One of the more cutting-edge developments in algorithmic debiasing is the use of transfer learning with fairness constraints. Transfer learning involves pretraining a model on a large, general dataset and then fine-tuning it on a smaller, specific dataset. When combined with fairness constraints, this technique allows the model to leverage the general knowledge it has acquired while also ensuring that it applies this knowledge equitably across all demographic groups in the specific dataset. For example, a model pretrained on a large dataset of general medical images might be fine-tuned on a smaller dataset of images from a particular demographic group. By incorporating fairness constraints during the fine-tuning process, the model can be guided to perform well across all demographic groups, even those that might be underrepresented in the fine-tuning dataset.

Transfer learning with fairness constraints is particularly valuable in situations where it is challenging to collect large amounts of balanced data. In fields like healthcare or finance, where data is often sensitive and difficult to obtain, transfer learning can provide a way to build robust models while still addressing bias. By pretraining on broad datasets and then applying fairness-aware fine-tuning, organizations can develop models that are both accurate and fair, even in the face of data limitations.

Despite the effectiveness of these algorithmic debiasing techniques, it is essential to recognize that no single method is a silver bullet. Bias in ML models is a complex issue that often requires a combination of techniques and ongoing vigilance. For instance, adversarial debiasing might be highly effective in addressing certain types of bias, such as those related to specific demographic features, but it may not fully resolve issues related to intersectionality—where multiple forms of bias interact and compound each other. Similarly, fairness-aware training can help achieve specific fairness goals, but it may require careful tuning to balance these goals with the model's overall accuracy.

Organizations must therefore take a holistic approach to bias mitigation, incorporating multiple algorithmic debiasing techniques into their model development and deployment processes. This involves not only applying techniques like adversarial debiasing, fairness-aware training, and reweighting but also continuously monitoring the model's performance to detect and address any emerging biases. For instance, after deploying a model, an organization might use fairness metrics to regularly evaluate its performance across different demographic groups, making adjustments as needed to ensure that the model remains fair over time.

Moreover, the integration of algorithmic debiasing techniques into the existing workflows of an organization requires collaboration across various teams, including data scientists, ethicists, legal advisors, and domain experts. Data scientists are responsible for implementing and testing the debiasing techniques, while ethicists and legal advisors ensure that the model's outcomes align with ethical standards and regulatory requirements. Domain experts, such as clinicians in healthcare or financial analysts in banking, provide the contextual knowledge needed to interpret the model's predictions and make informed decisions about bias mitigation.

This collaborative approach is particularly important in high-stakes environments, such as criminal justice, where biased predictions can have serious consequences for individuals and communities. For instance, a predictive policing model that unfairly targets certain neighborhoods or demographic groups can lead to over-policing and exacerbate existing social inequalities. By integrating algorithmic debiasing techniques into the model development process and involving stakeholders from various disciplines, organizations can work to prevent these negative outcomes and build models that contribute to a more just and equitable society.

Deployment and Monitoring Approach

Implementing a robust deployment and monitoring approach for bias mitigation is essential to ensure that machine learning (ML) models remain fair and effective over time. This involves establishing continuous fairness monitoring systems that track model performance across various demographic groups, enabling organizations to detect and address biases that may emerge after deployment. As models interact with dynamic real-world environments, data distributions can shift, leading to model drift—a phenomenon where the model's accuracy and fairness degrade over time. Managing these challenges requires a combination of advanced monitoring techniques, automated alert systems, and a proactive approach to model maintenance.

When an ML model is deployed, it enters a dynamic environment where the data it encounters may differ significantly from the data it was trained on. This is particularly true in industries where user behavior, market conditions, or societal trends evolve rapidly. For example, a recommendation algorithm in an e-commerce platform might initially perform well, offering relevant product suggestions to users across various demographic groups. However, as user preferences shift or new products enter the market, the model may begin to favor certain demographics over others, leading to biased recommendations. Continuous fairness monitoring allows organizations to detect such shifts in performance and address them before they result in significant disparities.

Implementing Continuous Fairness Monitoring

One of the key challenges in continuous fairness monitoring is the identification of appropriate fairness metrics that can be tracked over time. These metrics should be selected based on the specific context of the model and the potential risks of bias in that context. For instance, in a hiring model, relevant metrics might include the selection rates for different gender, racial, or age groups. By continuously tracking these metrics, organizations can identify when the model's performance begins to diverge across these groups, indicating a potential bias. For example, if the percentage of female candidates selected for interviews begins to decline over time, this could signal that the model is becoming biased against female applicants. Such a decline could be due to several factors, including changes in the applicant pool, shifts in the model's feature importance, or even unintended consequences of model updates.

To effectively manage these risks, organizations should establish alert thresholds that trigger investigations when fairness metrics deviate from acceptable levels. These thresholds can be determined based on historical performance data, industry standards, or ethical guidelines. For instance, if the selection rate for a particular demographic group falls below a predetermined threshold, an automated alert can be generated, prompting data scientists and HR professionals to investigate the cause. This might involve analyzing the feature contributions to the model's predictions, checking for data quality issues, or considering whether the model needs to be retrained with more recent or balanced data.

Model drift is another significant challenge that continuous fairness monitoring aims to address. As data distributions shift over time, a model that was once fair and accurate may begin to exhibit biased behavior. For example, consider a credit scoring model used by a bank to assess loan applications. Initially, the model might perform well, accurately predicting default risks across various demographic groups. However, as economic conditions change—such as during a financial crisis or in response to new regulations—the characteristics of loan applicants may also change. The model, trained on historical data, may not adequately account for these new conditions, leading to biased predictions that disproportionately disadvantage certain groups, such as low-income borrowers or minority communities.

To mitigate the risks associated with model drift, organizations can implement advanced monitoring techniques that go beyond simple threshold-based alerts. One such technique is the use of population stability index (PSI) to monitor shifts in the distribution of input features over time. PSI measures the stability of a model's input data distribution, comparing it to the distribution of the training data. A significant change in PSI indicates that the model is encountering data that is different from what it was trained on, which could lead to biased or inaccurate predictions. For instance, if a hiring model shows a high PSI for features related to candidate experience or education, it might suggest that the model is being exposed to a different type of applicant pool than it was trained on. This could result in the model unfairly favoring or penalizing certain candidates, prompting a need for retraining or recalibration.

Another technique to manage model drift and ensure continuous fairness is the use of adversarial validation. Adversarial validation involves training a secondary model to distinguish between the training data and the new data the model encounters in production. If the adversarial model can easily differentiate between the two datasets, it suggests that the production data is significantly different from the training data, indicating potential model drift. This technique is particularly useful in detecting subtle shifts in data distribution that might not be immediately apparent through traditional monitoring methods. For example, in a retail pricing model, adversarial validation could reveal that the model is encountering new types of customer behaviors or purchasing patterns, which might lead to biased pricing recommendations if not addressed.

Automated retraining pipelines can also be integrated into the continuous fairness monitoring system to address model drift proactively. These pipelines automatically trigger the retraining of a model when certain conditions are met, such as when fairness metrics fall below a specified threshold or when significant data shifts are detected. By incorporating fairness constraints into the retraining process, organizations can ensure that the model not only adapts to the new data but also maintains its commitment to fairness across all demographic groups. For instance, in a mortgage approval model, if the monitoring system detects that the model's approval rates for minority applicants have dropped significantly due to changes in economic conditions, the automated retraining pipeline can be triggered to update the model with new data while enforcing fairness constraints to prevent biased outcomes.

One of the advanced topics in continuous fairness monitoring is the integration of explainable AI (XAI) techniques into the monitoring process. XAI tools like SHAP (SHapley Additive exPlanations) and LIME (Local Interpretable Model-agnostic Explanations) can be used to provide real-time explanations for the model's predictions, helping to identify and diagnose biases as they emerge. For instance, if a hiring model begins to show a decline in the selection rate for female candidates, XAI techniques can be used to analyze the feature importance for those candidates, revealing whether certain features are disproportionately influencing the model's decisions. This granular level of insight allows organizations to take targeted actions, such as adjusting the feature weighting or collecting additional data to address the bias.

Organizations must also consider the operational challenges of implementing continuous fairness monitoring in dynamic environments. One challenge is the need for cross-functional collaboration between data scientists, domain experts, and business leaders. Continuous fairness monitoring is not just a technical task but requires input from those who understand the context in which the model is deployed. For example, in a healthcare setting, data scientists might work closely with clinicians to monitor the fairness of a diagnostic model, ensuring that it provides accurate and equitable recommendations for all patients, regardless of their demographic background. This collaboration is crucial for interpreting the results of fairness monitoring and making informed decisions about when and how to intervene.

Another challenge is managing the volume and complexity of data that continuous monitoring generates. As organizations deploy more models across different areas of their operations, the need to monitor multiple fairness metrics across various demographic groups can become overwhelming. To address this, organizations can implement centralized monitoring platforms that aggregate data from different models and provide a unified view of fairness performance. These platforms can include dashboards that visualize fairness metrics in real time, allowing stakeholders to quickly identify trends and take action when needed. For example, a financial institution might use a centralized platform to monitor the fairness of its credit scoring, loan approval, and fraud detection models, ensuring that all are operating fairly across different customer segments.

The deployment of continuous fairness monitoring also raises important ethical considerations. As organizations track fairness metrics and make adjustments to their models, they must be transparent about their processes and decisions. This transparency is essential for building trust with users and stakeholders, who need assurance that the organization is committed to fair and equitable AI practices. For instance, a company deploying an AI-powered recruitment tool might publicly share its fairness monitoring practices, including how it tracks selection rates across demographic groups and the steps it takes to address any biases that emerge. By being open about their fairness monitoring efforts, organizations can demonstrate their commitment to ethical AI and foster greater confidence in their models.

Ultimately, the goal of continuous fairness monitoring is not just to detect and correct biases but to create a feedback loop that enables continuous improvement of ML models. By regularly reviewing fairness metrics, analyzing the causes of bias, and iterating on the model, organizations can ensure that their models remain fair and effective over time. This continuous improvement process is particularly important in industries where the consequences of biased decisions can be severe, such as in healthcare, finance, or criminal justice. For example, in criminal justice, where predictive models are used to assess the risk of recidivism, continuous fairness monitoring can help prevent models from disproportionately impacting certain demographic groups, thereby promoting fairness in sentencing and parole decisions.

Implementing a deployment and monitoring approach for bias mitigation involves more than just tracking fairness metrics; it requires a comprehensive strategy that includes advanced monitoring techniques, automated retraining pipelines, and cross-functional collaboration. By addressing the challenges of monitoring models in dynamic environments and proactively managing model drift, organizations can ensure that their ML models remain fair, transparent, and aligned with ethical standards over time.

Integrating Human-in-the-Loop (HIL) Systems

Integrating human-in-the-loop (HIL) systems into the deployment and monitoring of machine learning (ML) models is a crucial strategy for bias mitigation, particularly in high-stakes applications where fairness, accountability, and human judgment are essential. HIL systems allow for the combination of the efficiency and scalability of ML with the nuanced understanding and ethical considerations that human reviewers bring to the table. By incorporating human oversight into the decision-making process, organizations can mitigate biases that may be present in the model's output and ensure more equitable outcomes.

In critical applications such as criminal justice, where ML models are increasingly used to inform decisions like bail setting or parole recommendations, the stakes are particularly high. A model might be designed to predict the likelihood that a defendant will re-offend or fail to appear for their court date, and based on this prediction, it might recommend a specific bail amount. However, these models are not infallible; they are trained on historical data that may reflect existing biases in the criminal justice system, such as racial or socioeconomic disparities. For instance, if the training data reflects a history of disproportionately higher bail amounts for defendants from certain racial backgrounds, the model may inadvertently perpetuate this bias in its recommendations.

By integrating a human-in-the-loop system, the model's recommendations are not the final decision. Instead, a human judge reviews the recommendation and can adjust it based on additional contextual factors that the model may not have considered. These factors might include the defendant's personal circumstances, such as family responsibilities, employment status, or community ties, which are not easily quantifiable but are crucial for a fair assessment. The human reviewer can also draw on their knowledge of the local community and the nuances of the legal system, ensuring that the decision is not only data-driven but also ethically sound and contextually appropriate.

The integration of HIL systems is particularly valuable in dynamic environments where data distributions and societal conditions change over time, leading to potential model drift. For instance, in the context of public health, an ML model might be used to allocate medical resources during an outbreak of a new disease. The model might recommend distributing resources based on certain demographic factors, such as age or preexisting health conditions, which it deems most at risk based on historical data. However, as the outbreak evolves and new data becomes available, the factors influencing risk might change. A human-in-the-loop system allows healthcare professionals to review and adjust the model's recommendations in real time, taking into account emerging trends, new medical research, and on-the-ground realities that the model might not yet fully incorporate.

The use of HIL systems also plays a critical role in maintaining public trust in AI-driven decision-making processes. In areas like financial services, where ML models are used to approve loans, set interest rates, or assess creditworthiness, decisions can have significant impacts on individuals' lives. For example, a model might recommend denying a loan application based on factors such as the applicant's credit score or employment history. However, the applicant might have recently changed jobs to a higher-paying position or might have a strong record of on-time payments that the model does not adequately weigh. By involving a human reviewer in the decision process, the financial institution can ensure that such nuanced factors are considered, providing a more comprehensive and fair evaluation of the applicant's situation.

Another advanced application of HIL systems is in the area of predictive policing, where ML models are used to forecast where crimes are likely to occur and to allocate law enforcement resources accordingly. While these models can be useful for optimizing patrol routes and resource deployment, they also carry the risk of reinforcing existing biases in policing. For example, if the model's predictions are based on historical crime data that reflects over-policing in certain neighborhoods, it may recommend disproportionately high levels of policing in those areas, perpetuating a cycle of bias and mistrust between law enforcement and the community. By integrating human oversight into the system, law enforcement officers can review the model's predictions and make adjustments based on their knowledge of the community, recent events, or changes in crime patterns that the model might not fully capture. This helps ensure that policing strategies are more balanced and fair, reducing the risk of over-policing certain communities based on biased data.

Implementing HIL systems also provides a mechanism for continuous learning and improvement of the model. Human reviewers can provide feedback on the model's recommendations, identifying areas where the model's predictions were accurate and where they fell short. This feedback can be used to refine the model, retraining it with new data that reflects the human reviewers' insights and decisions. For example, in a hiring process where an ML model is used to screen resumes, human recruiters can review the model's selections and provide feedback on whether the chosen candidates truly align with the company's needs and values. Over time, this feedback loop helps to improve the model's accuracy and fairness, aligning its predictions more closely with human judgment and reducing the likelihood of biased outcomes.

The challenge of implementing HIL systems lies in ensuring that the human reviewers are not merely rubber-stamping the model's decisions but are actively engaged in evaluating and, when necessary, overriding the model's recommendations. This requires training and equipping human reviewers with the tools and knowledge they need to critically assess the model's outputs. For instance, in the context of algorithmic hiring, recruiters should be trained to recognize potential biases in the model's recommendations and to consider a broader set of criteria than those captured by the model. This might involve considering soft skills, cultural fit, or potential for growth—factors that are difficult for an algorithm to quantify but are crucial for making holistic and fair hiring decisions.

Moreover, the implementation of HIL systems must be carefully designed to ensure that the human reviewers' input is not biased itself. This involves selecting a diverse group of reviewers who can bring a range of perspectives to the decision-making process and who are aware of their own potential biases. For example, in a judicial system using HIL for bail decisions, it is important that judges are trained in implicit bias and are encouraged to reflect on how their own experiences and backgrounds might influence their decisions. This helps to ensure that the HIL system does not simply replace one form of bias with another but rather works to mitigate bias overall, leading to more equitable outcomes.

HIL systems also need to be scalable to handle large volumes of decisions without overwhelming the human reviewers. This can be achieved by setting up tiered review systems, where only the most critical or borderline cases are flagged for human review, while the model handles more routine decisions autonomously. For instance, in an e-commerce platform using ML for product recommendations, most recommendations might be made automatically, but a human reviewer could be involved in approving new product categories or handling cases where the model's confidence level is low. This ensures that human resources are focused where they are most needed, allowing the system to operate efficiently while still maintaining a high level of oversight.

The integration of HIL systems into an organization's decision-making processes should be guided by clear policies and ethical frameworks that define the roles and responsibilities of both the ML model and the human reviewers. These policies should outline when and how human intervention is required, how decisions are to be documented, and how accountability is to be maintained. For example, in a financial institution using HIL for loan approvals, there should be clear guidelines on when a loan officer can override the model's recommendation, how the decision is to be justified, and how these decisions are to be audited to ensure consistency and fairness.

Developing Explainable User Interfaces (XUIs)

As machine learning models become more deeply integrated into decision-making processes across various sectors, the need for clear, understandable explanations of how these models work becomes paramount. XUIs bridge the gap between the complexity of ML models and the end-users who rely on their outputs, providing a user-friendly platform that explains model decisions and highlights fairness considerations. This transparency is essential for fostering trust in AI systems and for addressing concerns about potential biases.

Consider a scenario where a financial institution uses an ML model to evaluate loan applications. The decisions made by this model can have significant impacts on applicants' lives, determining whether they can buy a home, start a business, or manage financial emergencies. For applicants, understanding why their loan application was approved or rejected is crucial, not only for accepting the decision but also for knowing how to improve their chances in the future. An Explainable User Interface (XUI) designed for this purpose could present the decision in a clear and concise manner, breaking down the factors that influenced the outcome.

For example, if a loan application is rejected, the XUI might display a detailed explanation, such as: "Your loan application was not approved because your current debt-to-income ratio exceeds our threshold. Additionally, your credit history shows recent missed payments, which impacted your credit score." The XUI would further clarify that the model did not consider sensitive attributes like race, gender, or ethnicity, which are irrelevant to the evaluation. By offering this level of detail, the XUI not only helps the applicant understand the decision but also demonstrates the fairness of the model, reassuring them that the decision was based on objective financial criteria rather than discriminatory factors.

The design of XUIs must balance clarity with the complexity of the information being presented. Users need to understand the key factors influencing decisions without being overwhelmed by technical jargon or unnecessary details. For instance, in a healthcare setting where an ML model is used to recommend treatment plans, an XUI might explain the model's recommendation in terms that a patient can easily understand. If the model suggests a particular medication, the XUI could explain: "This medication is recommended because your recent test results show elevated levels of cholesterol, and your medical history indicates a family risk of heart disease. The model prioritizes treatments with a strong evidence base for patients with similar profiles." This explanation provides patients with a clear rationale for the recommendation, helping them to feel more informed and involved in their healthcare decisions.

XUIs also play a crucial role in addressing and mitigating bias by making the decision-making process more transparent. For instance, in a hiring platform where an ML model screens resume, an XUI could be used to explain why certain candidates were shortlisted for interviews while others were not. The interface could highlight relevant qualifications, experience, and skills that influenced the decision, while also explicitly stating that the model does not consider attributes like gender, age, or ethnicity. If a candidate's application is not selected, the XUI might provide feedback such as: "Your application was not advanced to the interview stage because the role requires five years of experience in a specific programming language, which was not evident in your resume. Consider highlighting relevant experience in your next application." This not only helps the candidate understand the reasoning behind the decision but also promotes fairness by ensuring that the process is perceived as transparent and non-discriminatory.

Advanced XUIs can go a step further by incorporating real-time feedback mechanisms that allow users to interact with the AI system and potentially correct or challenge its outputs. For example, in an educational platform that uses ML to recommend personalized learning paths, an XUI could allow students to provide feedback if they feel the recommendations do not align with their goals or interests. If a student is recommended a set of courses that they feel does not match their career aspirations, they could use the XUI to indicate their preferences, which the system could then consider in future recommendations. This feedback loop not only improves the personalization of the AI but also empowers users to have a say in how the model's outputs are applied to their individual cases.

Developing XUIs also involves significant technical and design challenges, particularly in ensuring that the explanations are both accurate and accessible. For instance, when explaining the outputs of a complex ML model, it is important that the XUI presents information that is not only technically correct but also meaningful to the user. This requires careful consideration of how to translate model outputs into everyday language and how to visualize the information in a way that enhances understanding. In a credit scoring system, for example, instead of simply stating that an applicant's credit score is below the threshold, the XUI might use visual aids such as graphs or charts to show how the applicant's credit score compares to the required range and what specific factors contributed to the score. This approach helps users better understand their financial standing and what steps they might take to improve it.

XUIs must be designed with inclusivity in mind, ensuring that they are accessible to all users, including those with disabilities. This might involve providing explanations in multiple formats, such as text, audio, or video, to cater to different preferences and needs. For example, a visually impaired user might benefit from an audio explanation of the model's decision, while another user might prefer a video that walks them through the decision-making process step by step. By offering these options, XUIs can make AI systems more inclusive and ensure that all users have equal access to the information they need to understand and trust the decisions being made.

The implementation of XUIs also requires a strong focus on privacy and data security, particularly when dealing with sensitive information. In scenarios such as healthcare or finance, where the explanations involve personal data, it is crucial that the XUI is designed to protect user privacy and comply with relevant regulations, such as GDPR or HIPAA. This might involve ensuring that the explanations are delivered in a secure environment, where users can review their information without fear of unauthorized access or data breaches. For instance, a healthcare XUI might require multi-factor authentication to access sensitive explanations about treatment recommendations, ensuring that only the patient and authorized medical professionals can view the details.

Another critical aspect of XUI development is the integration of ethical considerations into the design process. This involves not only ensuring that the explanations are clear and accurate but also that they reflect the ethical principles guiding the AI's deployment. For example, in a hiring platform, the XUI should not only explain why certain candidates were selected but also reflect the organization's commitment to diversity and inclusion by ensuring that the explanations demonstrate how the model aligns with these values. This might include showing how the model actively avoids bias by not considering certain attributes or by ensuring a diverse pool of candidates is represented in the final selection. By embedding these ethical considerations into the XUI, organizations can reinforce their commitment to fairness and build greater trust in their AI systems.

Tools and Communication Strategies

To address bias in machine learning (ML) models effectively, it is essential to utilize advanced tools and strategies tailored to various contexts. This section explores key tools for detecting and mitigating bias, as well as communication strategies to bridge the gap between technical and non-technical stakeholders.

Utilizing Bias Detection Tools and Frameworks

Bias detection tools and frameworks provide powerful capabilities for identifying, explaining, and mitigating biases that may exist within data and models. Among the most notable tools in this space are IBM's AI Fairness 360 Toolkit and Google's What-If Tool for TensorBoard. Each of these tools offers unique strengths and is suited to different scenarios depending on the specific needs of the organization and the characteristics of the data and models being used.

IBM's AI Fairness 360 Toolkit is an open-source framework designed to provide a comprehensive suite of tools for detecting, explaining, and mitigating bias in machine learning models. One of its key strengths is its ability to analyze datasets for bias before model training even begins. This proactive approach allows organizations to identify potential issues in their data that could lead to biased outcomes later in the model lifecycle. For example, in a hiring model, AI Fairness 360 could analyze the distribution of demographic groups in the training data and highlight any imbalances that might lead to a model favoring certain groups over others. This capability is particularly valuable in situations where the training data may reflect historical biases, such as gender or racial disparities in hiring practices.

The AI Fairness 360 Toolkit also includes a variety of fairness metrics that can be applied to evaluate the performance of trained models. These metrics go beyond simple accuracy measures, providing insights into how the model performs across different demographic groups. For instance, in a loan approval model, the toolkit might calculate demographic parity, equalized odds, or disparate impact, revealing whether the model's predictions are consistent and fair across all groups. Additionally, AI Fairness 360 offers explainability tools that help users understand which features in the data contribute most to any detected biases. This is crucial for making informed decisions about how to address and mitigate bias, as it allows data scientists to focus on specific features or subsets of data that are driving unfair outcomes.

While AI Fairness 360 is powerful and comprehensive, it can also be complex to implement, particularly for organizations that do not have a deep bench of data science expertise. The toolkit offers a wide array of features, but navigating them effectively requires a solid understanding of both the underlying data and the specific fairness challenges the organization is facing. For smaller organizations or those new to bias detection, the learning curve associated with AI Fairness 360 might be steep. Moreover, while the toolkit is highly customizable, this flexibility can also be a drawback for users who need more straightforward, out-of-the-box solutions.

Google's What-If Tool is designed with a strong emphasis on usability and integration with TensorFlow models. This tool is particularly well-suited for developers and data scientists who are already working within the TensorFlow ecosystem, as it integrates seamlessly with TensorBoard, TensorFlow's visualization toolkit. The What-If Tool allows users to interactively explore the behavior of their models under different hypothetical scenarios. For example, in a predictive policing model, a user could simulate how the model's predictions change if certain demographic attributes, such as race or socioeconomic status, are altered. This feature is invaluable for identifying potential biases before the model is deployed, as it provides a clear and intuitive way to visualize how the model responds to changes in the input data.

One of the strengths of the What-If Tool is its ability to simulate the effects of counterfactuals—what would happen if certain characteristics of the data were different. This capability is especially useful for testing the robustness of a model's fairness under various scenarios. For instance, if a healthcare model predicts the likelihood of a patient developing a certain condition, the What-If Tool can help explore whether these predictions hold consistently across different demographic groups. By altering patient attributes, such as age or income level, users can assess whether the model is biased toward or against certain groups. This interactive exploration helps data scientists understand the model's fairness in a practical, hands-on way, which can be more accessible and engaging than traditional statistical analysis.

While the What-If Tool is highly user-friendly and accessible, it does have some limitations. It is primarily designed to work with TensorFlow models, which means that organizations using other machine-learning frameworks might find it less useful. Additionally, the tool is more focused on model behavior exploration rather than on comprehensive bias detection and mitigation strategies. While it is excellent for identifying potential biases in specific scenarios, it may not offer the same depth of analysis or range of mitigation techniques as IBM's AI Fairness 360. This makes it a better choice for organizations that need a quick, visual tool for exploring fairness but might not require the extensive functionality provided by AI Fairness 360.

The choice between these tools depends largely on the specific context in which they are being used. For example, a large financial institution concerned with ensuring that its loan approval model is fair across all demographic groups might benefit more from the comprehensive analysis and mitigation capabilities of AI Fairness 360. The toolkit's ability to analyze data and model outputs at a granular level would be invaluable for ensuring compliance with regulations such as the Equal Credit Opportunity Act. On the other hand, a tech startup developing a new product recommendation system using TensorFlow might find the What-If Tool more aligned with its needs. The tool's ease of use and seamless integration with TensorFlow would allow the startup to quickly explore and address potential biases in its model, helping it to deliver a fair and trustworthy product without the need for extensive data science resources.

In advanced scenarios, organizations might even consider using these tools in tandem. For instance, an organization could start with the What-If Tool to quickly identify potential areas of bias during the model development phase. Once these areas are identified, the organization could then use AI Fairness 360 to perform a deeper analysis and implement specific bias mitigation strategies. This combined approach would leverage the strengths of both tools, ensuring both ease of use and comprehensive bias management.

The selection of bias detection tools should be guided by a clear understanding of the organization's goals, the nature of the data, and the specific fairness challenges at hand. Both AI Fairness 360 and the What-If Tool offer valuable capabilities, but their effectiveness will depend on how well they align with the organization's technical environment, expertise, and the complexity of the fairness issues they are trying to address. By carefully evaluating these factors, organizations can choose the right tools to help them build fair, transparent, and trustworthy AI systems that align with their ethical standards and business objectives.

Communicating Explainability for Non-technical Stakeholders

Machine learning (ML) development teams are often composed of highly skilled technical experts who understand the intricacies of algorithms, data processing, and model optimization. However, to effectively address bias and ensure the responsible use of AI, it is crucial to communicate these complex ideas to non-technical stakeholders, such as business leaders, policymakers, and other decision-makers. These individuals play a key role in determining how AI is integrated into business practices and regulatory frameworks, but they may not have the technical background necessary to grasp the nuances of ML systems. Therefore, communication strategies must be tailored to bridge this gap and make the concepts of bias and fairness in AI accessible and relevant to them.

One effective strategy is to avoid the use of technical jargon and complex statistical concepts when discussing bias in ML models with non-technical stakeholders. Instead of delving into the mathematical underpinnings of fairness metrics like equalized odds or demographic parity, it is more productive to focus on the real-world consequences of bias and the steps being taken to mitigate it. For instance, when explaining the potential for bias in a hiring algorithm, rather than discussing the specifics of algorithmic fairness, the conversation could center on the potential outcomes, such as the risk of excluding qualified candidates from underrepresented groups or the legal and reputational risks associated with discriminatory practices. This approach makes the issue more tangible and highlights the importance of bias mitigation in a context that is directly relevant to the stakeholders' roles and responsibilities.

Visual aids, such as charts, graphs, and real-world examples, can be particularly effective in illustrating the concepts of bias and fairness in ML. For example, consider an organization using an ML model to predict employee performance and determine promotions. A visual representation showing the disparity in promotion rates between different demographic groups can immediately convey the impact of bias in the model. By showing how these disparities might arise from biased data or model design, and how fairness mitigation techniques can reduce these disparities, stakeholders can better understand the importance of addressing bias. This visual evidence, coupled with examples from similar organizations that have successfully implemented bias mitigation strategies, can provide compelling justification for investing in fairness initiatives.

Real-world examples also help in grounding the discussion in practical terms. For instance, discussing the case of a financial institution that faced regulatory scrutiny due to biased lending practices highlighted by an ML model can drive home the importance of bias detection tools and fairness audits. Such examples can help non-technical stakeholders appreciate the urgency of bias mitigation, as they can relate the potential consequences to situations that could affect their own organization. This method not only simplifies the complexity of the concepts but also aligns the discussion with the stakeholders' priorities, such as regulatory compliance, customer trust, and ethical governance.

Interactive and engaging communication techniques, like storytelling, can further enhance understanding. For example, telling a story about how a biased algorithm might negatively impact a specific group of people—such as minority communities being unfairly targeted for higher insurance premiums—can humanize the issue and make it more relatable. This approach can be particularly powerful in policymaking contexts, where the ethical implications of AI decisions are paramount. By highlighting the human side of algorithmic bias, communicators can evoke empathy and a deeper sense of responsibility among non-technical stakeholders, motivating them to support fairness initiatives.

Workshops and Training Sessions

Workshops and training sessions provide a structured and interactive environment for educating non-technical stakeholders about the importance of fairness in ML and the techniques used to achieve it. These sessions are crucial for building a shared understanding between technical and non-technical teams, ensuring that everyone involved in AI projects is aligned on the goals of bias mitigation and fairness.

Workshops can be designed to be highly interactive, with participants engaging in hands-on activities that simulate the challenges of bias detection and mitigation. For instance, a workshop might involve a case study where participants are given a dataset and asked to identify potential biases in the data. They could then explore different strategies for mitigating these biases, such as rebalancing the dataset or applying fairness constraints during model training. By working through these challenges in a controlled environment, non-technical stakeholders can gain a practical understanding of the complexities involved in ensuring fairness in ML systems. This hands-on experience helps demystify the technical processes and allows participants to see firsthand how biases can emerge and how they can be addressed.

These workshops can also facilitate deeper discussions between technical and non-technical teams. For example, during a workshop focused on a customer service chatbot, non-technical stakeholders might raise concerns about how the chatbot interacts with users from different cultural backgrounds. This feedback can then be incorporated into the technical team's considerations when designing or refining the model. Such interactions foster collaboration and ensure that the perspectives of non-technical stakeholders—who often have a closer connection to the end users and understand the broader context in which the AI system operates—are integrated into the model development process.

Training sessions can also be tailored to address specific roles within the organization. For instance, business leaders might participate in sessions that focus on the strategic implications of bias in AI, such as how it affects brand reputation, customer loyalty, and compliance with anti-discrimination laws. These sessions could include discussions on how to set organizational policies that prioritize fairness in AI and how to allocate resources effectively to support bias mitigation efforts. On the other hand, sessions for legal teams might delve into the regulatory landscape surrounding AI, helping them understand how to navigate legal risks related to biased decision-making and how to advise the organization on compliance with emerging AI regulations.

To enhance the effectiveness of these sessions, organizations can bring in external experts who specialize in AI ethics, bias mitigation, and fairness auditing. These experts can provide insights into the latest research and best practices, offering a fresh perspective that might not be present within the internal team. For example, an external expert might introduce new fairness metrics or tools that the organization has not yet considered, or they might share case studies of other organizations that have successfully implemented bias mitigation strategies. This external input can be invaluable for expanding the knowledge base of the participants and for encouraging them to think more broadly about the ethical implications of their AI systems.

Explainable AI and Model Interpretability

Explainable AI (XAI) and model interpretability have become increasingly important as machine learning models, particularly deep neural networks, have grown in complexity and power. These sophisticated models, while achieving unprecedented levels of accuracy and performance, often operate as opaque systems—so-called "black boxes"—where the decision-making process is not easily understood. This opacity poses significant challenges for trust, accountability, and fairness in AI systems. Addressing these issues requires a careful balance between leveraging the power of complex models and ensuring that their decisions can be explained and understood by humans.

One of the primary goals of XAI is to provide human-understandable explanations for the predictions made by machine learning models. This is not only crucial for fostering trust among users but also for ensuring that AI systems are used responsibly. When users understand why a model has made a particular decision, they are more likely to trust and rely on the system's outputs. For example, in a healthcare setting, a doctor might be more willing to trust an AI's recommendation for a treatment plan if the system can explain its reasoning in terms of the patient's medical history, lab results, and other relevant factors. Without such explanations, the doctor might be hesitant to follow the AI's advice, especially in critical situations where the stakes are high.

However, achieving explainability in complex models, especially deep learning models, is fraught with challenges. Deep neural networks, for example, consist of multiple layers of interconnected nodes, each performing a series of transformations on the input data. The sheer complexity of these transformations makes it difficult to trace how specific input features influence the final output. While techniques like LIME (Local Interpretable Model-agnostic Explanations) and SHAP (SHapley Additive exPlanations) have been developed to approximate the model's behavior in a way that is more interpretable, these methods have limitations. LIME, for instance, generates local explanations by approximating the complex model with a simpler, interpretable one around the data point of interest. While this can provide valuable insights, it is an approximation and may not fully capture the nuances of the original model's decision-making process.

Similarly, SHAP values offer a way to understand the contribution of each feature to the model's prediction, but interpreting these contributions can be challenging, especially when dealing with highly correlated features or when the model's decision boundary is complex. In such cases, the explanations provided by SHAP might be difficult to translate into actionable insights for non-technical stakeholders. For example, in a financial institution using a deep learning model for credit scoring, SHAP might indicate that a combination of income level, employment history, and loan history significantly influenced the model's decision to reject a loan application. However, the interdependencies between these features and their collective impact on the decision might not be easily understandable to a loan officer or the applicant, even with SHAP's explanations.

The trade-off between model complexity and interpretability is a central concern in the deployment of AI systems, particularly in high-stakes environments where decisions have significant consequences. Complex models like deep neural networks are often favored for their superior performance on tasks such as image recognition, natural language processing, and predictive analytics. However, their lack of transparency can lead to mistrust and potential misuse. For instance, in the criminal justice system, where predictive models are used to assess the likelihood of recidivism, a model's recommendation to deny parole based on opaque reasoning could be seen as unjust or biased, especially if the affected individuals or their advocates cannot understand or challenge the model's decision.

This issue is further compounded by the fact that there is no universally accepted definition of interpretability. Interpretability can mean different things depending on the context: to a data scientist, it might mean understanding the weights and activations within a neural network; to a business leader, it might mean knowing which features are most influential in driving a model's predictions; and to an end-user, it might mean receiving a simple and clear explanation of why a certain decision was made. This diversity in expectations leads to a wide range of techniques and approaches for achieving interpretability, each with its own strengths and limitations. For instance, while decision trees offer inherent interpretability by showing a clear path from input features to the final decision, they often lack the predictive power of more complex models like deep neural networks.

In the domain of data governance, XAI and model interpretability are not just technical considerations but also ethical imperatives. Understanding how a model uses data features is essential for detecting and mitigating biases. For example, if a hiring algorithm disproportionately rejects candidates from certain demographic groups, interpretability tools can help uncover whether the model is unfairly weighting factors like zip code or educational background, which might correlate with race or socioeconomic status. This level of transparency is crucial for ensuring that AI systems treat all individuals equitably and do not perpetuate existing societal biases.

The ethical implications of XAI go beyond just providing transparency. There is a risk that explainability tools could be used to create a false sense of security or to justify decisions that are still fundamentally biased. For instance, a company might use XAI to generate explanations that appear reasonable on the surface but fail to reveal underlying biases in the model's logic. This could lead to a situation where users or regulators are misled into believing that the model is fair and unbiased, when in reality, it is not. Therefore, it is crucial to develop ethical guidelines for the use of XAI, ensuring that explanations are not only accurate but also genuinely reflective of the model's decision-making process.

The limitations of current XAI techniques, particularly in the context of deep learning, underscore the need for ongoing research and development in this area. While existing methods like LIME and SHAP represent significant advances, they are not without their challenges, especially when applied to models that operate in highly complex and dynamic environments. For example, in autonomous driving systems, where deep learning models are used to make real-time decisions, the demand for both high performance and interpretability is exceptionally high. Explaining why a model decided to brake in a particular situation or how it recognized an obstacle on the road is critical for building trust in these systems, yet providing such explanations in a way that is both accurate and understandable remains a significant challenge.

As machine learning continues to advance, the tension between model complexity and interpretability will likely intensify. Striking the right balance will require not only technical innovations in XAI but also a commitment to ethical AI practices that prioritize transparency, fairness, and accountability. This includes developing new tools and frameworks that can provide deeper insights into complex models, as well as fostering a culture of ethical awareness among those who design, deploy, and oversee AI systems. By addressing these challenges head-on, the AI community can work toward creating models that are not only powerful and efficient but also trustworthy and just.

Best Practices for Ethical Implications and Bias Mitigation

To ensure that machine learning (ML) models are fair and unbiased, a few practices are worth noting:

Ethical Data Collection and Preprocessing

Ethical data collection and preprocessing are fundamental to ensuring that machine learning (ML) models are fair and unbiased. The process begins with obtaining informed consent from individuals whose data will be used. Informed consent goes beyond simply getting a checkbox on a form; it involves clearly communicating how data will be collected, stored, used, and potentially shared. For example, in a healthcare context, patients must be informed about how their medical data will contribute to predictive models for treatment outcomes. This transparency builds trust and ensures that individuals are fully aware of how their information will be utilized.

Inclusivity in data collection is another critical factor. Ensuring that data is gathered from diverse sources helps prevent representation bias, where certain groups are underrepresented, leading to models that perform poorly for those groups. For instance, an ML model developed to detect skin cancer might perform well on lighter skin tones but fail on darker tones if the training data primarily consists of images of light-skinned individuals. To mitigate this, the data collection process must consciously include samples that represent all demographics that the model will encounter in real-world scenarios. This inclusivity helps create more robust and equitable models.

Effective data preprocessing is also crucial in bias mitigation. Techniques like normalization and standardization ensure that data is brought onto a consistent scale, preventing models from favoring certain features simply because they are on a larger scale. Handling missing data with techniques such as imputation or excluding incomplete records is another important step. For example, in financial services, missing income data might lead to incorrect predictions if not handled properly. Careful preprocessing ensures that the model's predictions are based on complete and accurate information.

Balancing datasets is essential to address class imbalances that might lead to biased outcomes. For instance, in a criminal justice application, if the dataset includes far more non-recidivist cases than recidivist ones, the model might learn to under-predict recidivism, leading to unfair outcomes. Techniques like oversampling underrepresented classes or using synthetic data generation methods such as SMOTE (Synthetic Minority Over-sampling Technique) can help create a balanced dataset. This ensures that the model has equal exposure to all classes during training, leading to fairer predictions.

Model Development and Evaluation

During the model development phase, incorporating fairness from the outset is crucial. Bias-aware training algorithms, which are designed to minimize biases during the learning process, play a significant role. For example, adversarial debiasing involves training a model alongside an adversary that tries to exploit the model's biases. The main model is then refined to become more resistant to these biases. This technique is particularly useful in applications like hiring, where biases based on gender or race must be minimized to ensure equitable outcomes.

Regularization techniques, which prevent models from overfitting to training data, are another best practice. Overfitting not only reduces a model's generalizability but can also amplify biases present in the training data. For instance, a model that overfits to a specific demographic might perform well for that group but poorly for others, leading to biased outcomes. Cross-validation, where the model's performance is tested on different subsets of the data, helps ensure that the model is robust across various demographics. For example, in a healthcare application predicting patient outcomes, cross-validation can reveal whether the model performs consistently across different age groups, genders, or ethnicities.

Evaluating model fairness involves using metrics specifically designed to detect and measure bias. Demographic parity, for example, ensures that different demographic groups receive similar outcomes. This is crucial in contexts like loan approvals, where demographic groups should have equal access to loans based on their creditworthiness rather than on biased factors. Equalized odds, another metric, ensures that the true positive and false positive rates are similar across groups. This is particularly important in criminal justice, where predictive policing models must not unfairly target certain communities. Disparate impact measures the adverse effects on different groups and is often used in employment contexts to ensure that hiring algorithms do not disproportionately disadvantage certain demographics.

Model interpretability is vital for ensuring transparency and building trust in AI systems. Techniques like SHAP (SHapley Additive exPlanations) and LIME (Local Interpretable Model-agnostic Explanations) help users understand how specific features contribute to a model's predictions. For instance, in a mortgage approval model, SHAP could be used to explain why a particular application was denied, revealing that the decision was primarily influenced by the applicant's credit score and debt-to-income ratio, rather than by irrelevant factors like the applicant's zip code. This transparency allows for greater scrutiny and accountability in model-driven decisions.

Comprehensive documentation of the model's design, assumptions, and limitations is another crucial practice. Documentation ensures that anyone reviewing the model— whether they are data scientists, auditors, or regulators—can understand its workings and identify potential areas of bias. For instance, documenting the sources of training data, the rationale behind feature selection, and the methods used for bias mitigation can provide critical insights during an audit, ensuring that the model complies with ethical standards and regulations.

Deployment and Monitoring

The deployment phase of ML models requires careful planning to ensure that ethical considerations are maintained post-deployment. Establishing ethical review boards that assess ML projects before deployment is a critical step. These boards, which should include ethicists, data scientists, and legal experts, can evaluate whether the model meets ethical standards and whether its deployment might lead to unintended consequences. For example, before deploying a predictive policing model, an ethical review board might examine whether the model could exacerbate existing biases in law enforcement and recommend adjustments to mitigate these risks.

Engaging stakeholders during deployment is also essential. For instance, in deploying an AI-driven recruitment tool, it is important to involve HR professionals, diversity officers, and legal teams to ensure that the tool aligns with the organization's values and complies with employment laws. Continuous monitoring of deployed models is crucial to detect and address biases that may emerge over time. For example, a loan approval model might perform fairly at the outset but could develop biases as new data flows into the system. Regular performance audits, where the model's outputs are compared against fairness metrics, help in identifying any drift in performance.

Incorporating user feedback into the monitoring process is another effective strategy. Users, whether they are customers, employees, or other stakeholders, can provide valuable insights into the model's performance. For instance, if a credit scoring model is perceived as unfair by customers, their feedback can highlight areas where the model might need to be adjusted. This feedback loop is essential for maintaining trust and ensuring that the model remains fair and effective.

Anomaly detection techniques can be used to identify unexpected or biased outcomes in real time. For instance, if a hiring model suddenly starts rejecting a disproportionate number of candidates from a particular demographic group, anomaly detection systems can flag this behavior for further investigation. This proactive approach helps prevent biases from becoming entrenched in the model's decision-making process.

Ethical AI Frameworks

Developing comprehensive ethical AI frameworks is crucial for guiding organizations in the responsible deployment of ML systems. Such frameworks should encompass principles like fairness, transparency, accountability, and privacy. For instance, a fairness principle might require that all models undergo bias testing before deployment, with specific metrics like demographic parity being mandatory for certain applications, such as hiring or lending.

Transparency within an ethical AI framework involves clear communication about how models function, including their limitations. For example, a healthcare AI system that predicts patient outcomes should be transparent about the data it was trained on, the confidence levels of its predictions, and any potential biases. This transparency can be achieved through detailed documentation, explainable AI techniques, and user-friendly interfaces that make complex model decisions understandable to non-experts.

Accountability requires that organizations establish mechanisms to ensure that ethical guidelines are followed and that any negative impacts of ML models are promptly addressed. For example, an organization might set up an internal audit team dedicated to regularly reviewing all AI systems for compliance with ethical standards. If a bias is detected, the team should have the authority to recommend corrective actions, such as retraining the model or adjusting the data used.

Privacy is a critical component of any ethical AI framework. Organizations must adhere to data protection regulations and best practices, such as GDPR or HIPAA, when handling personal data. This includes implementing robust data anonymization techniques and obtaining informed consent from data subjects. For instance, in a healthcare setting, patients should be fully aware of how their data will be used in predictive models, and the data should be anonymized to protect patient confidentiality.

Collaboration and Standards

Collaboration among researchers, policymakers, and industry stakeholders is essential for establishing standards and best practices for ethical ML. Researchers can contribute by developing new techniques for bias detection and mitigation, which can then be integrated into industry practices. For example, advances in adversarial debiasing techniques can be shared with industry practitioners through conferences, publications, and collaborative projects.

Policymakers play a crucial role in creating regulatory frameworks that enforce ethical standards. For instance, laws could be enacted that require companies to conduct bias audits on their AI systems before they are deployed in sensitive areas like finance, healthcare, or law enforcement. These regulations ensure that organizations adhere to ethical principles and protect individuals from the potential harms of biased AI systems.

Industry stakeholders, including businesses and technology providers, bring practical insights and real-world experience to the development of ethical standards. Their involvement ensures that the standards are not only theoretically robust but also feasible for implementation in various industrial contexts. For example, a technology company might work with regulators to develop standards for explainable AI that balance the need for transparency with the technical limitations of current XAI methods.

Organizations can establish internal standards and best practices for ethical ML, which can be tailored to their specific needs and contexts. For example, a financial institution might develop a standard protocol for bias detection and mitigation in its credit scoring models, including regular audits and user feedback mechanisms. By adopting and adapting these standards, organizations can ensure consistency and reliability in the ethical deployment of ML systems.

Measuring Success of Bias Mitigation Practices

Evaluating the effectiveness of bias mitigation practices is crucial to ensure that the implemented strategies are making a meaningful impact. What does "success" look like? Success in this context is not only about meeting regulatory requirements but also about genuinely improving fairness and equity in AI systems. The following sections outline key methods for assessing the success of these practicess.

Continuous Monitoring of Fairness Metrics

One of the most direct ways to measure the success of bias mitigation practices is through the continuous monitoring of fairness metrics. These metrics provide quantitative insights into how well a model is performing across different demographic groups. Common fairness metrics include demographic parity, which ensures that different demographic groups receive similar outcomes, and equalized odds, which balances the true positive and false positive rates across groups. For example, in a hiring model, demographic parity can be monitored by comparing the selection rates of different gender or racial groups. If bias mitigation practices are effective, these metrics should show an improvement over time, indicating that the model is becoming fairer and more equitable.

Moreover, fairness metrics should be tracked over the entire lifecycle of the model, from development through deployment and beyond. This continuous monitoring is critical because biases can emerge or re-emerge as the model interacts with new data or as societal conditions change. For instance, a loan approval model might initially perform fairly, but as economic conditions shift, the model might develop biases against certain demographic groups. Regular monitoring can help detect these shifts early, allowing organizations to adjust their bias mitigation strategies accordingly.

Gathering User Feedback

User feedback is another crucial component of measuring the success of bias mitigation practices. Users—whether they are customers, employees, or other stakeholders—often have valuable insights into how a model's decisions are perceived in the real world. Collecting this feedback can provide qualitative data that complements the quantitative fairness metrics. For example, if customers begin to report fewer instances of perceived unfairness in a loan approval process, it may indicate that the bias mitigation efforts are having a positive impact.

This feedback can be gathered through various channels, such as surveys, focus groups, or customer service interactions. For instance, an organization might conduct regular surveys to assess customer satisfaction with an AI-driven service, asking specific questions about the fairness of the decisions made by the model. Focus groups can also be an effective way to gather more in-depth feedback, as they allow for open-ended discussions where participants can share their experiences and perceptions of the model's fairness.

User feedback is particularly valuable because it can reveal biases or issues that might not be immediately apparent from the fairness metrics alone. For example, a model might perform well according to the metrics but still be perceived as unfair by certain user groups. This discrepancy could arise from factors that are not captured by the metrics, such as the way the model's decisions are communicated to users or how the users experience the outcomes of those decisions. By incorporating user feedback into the evaluation process, organizations can gain a more comprehensive understanding of the effectiveness of their bias mitigation practices.

Conducting Regular Audits

Regular audits are essential for ensuring that bias mitigation practices remain effective over time. These audits should involve a thorough review of the model's performance, as well as the processes used to develop and deploy the model. For example, an audit might include a detailed analysis of the model's output to identify any patterns of bias that have emerged since the last audit. The audit should also examine the data collection and preprocessing methods, the fairness metrics used, and the steps taken to address any biases that were previously identified.

Audits are particularly important for high-stakes applications, such as those used in criminal justice, healthcare, or finance, where biased decisions can have severe consequences. In these contexts, audits can help ensure that the models are not only compliant with ethical standards but also that they continue to perform fairly as they interact with new data and as conditions change. For example, an audit of a predictive policing model might involve a review of arrest rates by demographic group to ensure that the model is not disproportionately targeting certain communities. If biases are detected, the audit should include recommendations for corrective actions, such as retraining the model or adjusting the data used.

Institutionalizing Bias Mitigation Practices

The success of bias mitigation practices can also be measured by the degree to which they become institutionalized within the organization. This involves embedding bias mitigation into the organization's culture, processes, and decision-making frameworks. For example, establishing ethical review boards that regularly assess AI projects for fairness and bias is one way to institutionalize these practices. These boards can provide ongoing oversight, ensuring that bias mitigation remains a priority throughout the lifecycle of the model.

Another indicator of success is the extent to which bias mitigation practices are integrated into the ML development lifecycle. For instance, organizations might adopt a standard protocol that requires bias testing and mitigation at each stage of model development, from data collection and preprocessing to model training and deployment. The regular use of fairness metrics in model evaluation, the incorporation of user feedback, and the commitment to conducting regular audits are all signs that bias mitigation is becoming a standard practice within the organization.

Training and education are also critical components of institutionalization. By providing ongoing training for data scientists, developers, and other stakeholders, organizations can ensure that everyone involved in AI projects understands the importance of bias mitigation and knows how to implement best practices. This training might include workshops on fairness metrics, tutorials on bias detection tools, or seminars on the ethical implications of AI. The more these practices become ingrained in the organization's daily operations, the more likely they are to lead to sustained and effective bias mitigation.

Evaluating Long-Term Impact on Stakeholders

Finally, the success of bias mitigation practices can be measured by evaluating their long-term impact on stakeholders. This involves assessing how the practices affect not only the immediate users of the AI systems but also the broader community. For example, in a healthcare setting, the long-term impact of bias mitigation might be evaluated by tracking health outcomes for different demographic groups over time. If the bias mitigation practices are effective, there should be a reduction in disparities in health outcomes between these groups.

In the context of hiring, long-term success might be measured by examining changes in the diversity of the workforce. If a hiring model is adjusted to mitigate bias, the organization should see an increase in the representation of historically underrepresented groups in its workforce. Similarly, in financial services, the success of bias mitigation might be reflected in more equitable access to credit and financial products across different demographic groups.

Evaluating the long-term impact also involves considering the broader social and ethical implications of AI systems. For example, organizations should assess whether their bias mitigation efforts contribute to greater public trust in AI or whether they help reduce societal inequalities. These broader outcomes are more difficult to measure than fairness metrics or user feedback, but they are essential for understanding the full impact of bias mitigation practices.

Policies and Procedures for Ethical Implications and Bias Mitigation

At this point, it's worth noting some overarching policy values that can help ensure that machine learning (ML) models are fair and unbiased. These polices include evaluation and monitoring policies.

Data Collection Policies

Ethical data collection is foundational for mitigating bias in ML systems. Policies governing data collection must emphasize privacy, inclusivity, and transparency.

Privacy and Consent

Policies must ensure that data is collected in compliance with privacy regulations such as the General Data Protection Regulation (GDPR) and the California Consumer Privacy Act (CCPA). Data subjects should be informed about the purpose of data collection, how their data will be used, and their rights regarding their data. Informed consent is a critical component, requiring clear communication to data subjects about the implications of data sharing.

Inclusivity and Representation

To avoid representation bias, policies should mandate the collection of data from diverse sources. This ensures that the data reflects a broad range of perspectives and experiences, capturing the diversity of the population. Special attention should be paid to underrepresented groups to ensure they are adequately represented in the dataset.

Transparency in Data Collection

Transparency policies require organizations to document and disclose their data collection methods. This includes detailing the sources of data, the criteria for inclusion and exclusion, and any preprocessing steps undertaken before data is used for training ML models. Such transparency helps in auditing and verifying that the data collection processes are ethical and unbiased.

Data Preprocessing Procedures

Once data is collected, preprocessing is crucial to prepare it for model training while addressing potential biases.

Data Cleaning

Data cleaning involves addressing inaccuracies, outliers, and missing values. Procedures should be in place to systematically identify and rectify these issues to improve data quality. Techniques such as imputation for missing values or normalization for scaling data should be documented and consistently applied.

Anonymization and De-identification

To protect privacy, data should be anonymized or de-identified wherever possible. This involves removing personally identifiable information (PII) or using techniques like pseudonymization. Policies should outline the steps for anonymization and the measures to ensure that data cannot be easily re-identified.

Balancing Datasets

Balancing datasets is essential to address class imbalances that can lead to biased model outcomes. Procedures such as oversampling underrepresented classes, undersampling overrepresented classes, or generating synthetic data can be employed. Documentation of these techniques ensures reproducibility and transparency.

Model Development Policies

The development of ML models must adhere to ethical guidelines that prioritize fairness and bias mitigation.

Bias-Aware Algorithms

Policies should promote the adoption of algorithms that are conscious of bias and include fairness constraints during training. These algorithms are specifically crafted to reduce bias and provide fair results across various demographic groups. By incorporating fairness constraints, these algorithms aim to prevent biased outcomes. Their design ensures that all demographic groups are treated equitably. The use of such bias-aware algorithms is crucial for promoting fairness in machine learning models. Implementing these algorithms helps to achieve equitable treatment of different demographic groups. They play a key role in minimizing biased results in AI systems. Encouraging their use is vital for ethical machine learning practices.

Regularization Techniques

To prevent overfitting, it is essential to employ regularization techniques. These techniques help ensure that models can generalize effectively to unseen data. Policies should explicitly mandate the use of regularization methods like L1 and L2 regularization. Incorporating these techniques into model training enhances robustness. Additionally, cross-validation should be a required practice to further strengthen model reliability. By specifying these approaches, policies ensure consistent and reliable performance of machine learning models. Regularization techniques help control model complexity and prevent it from fitting noise in the training data. Cross-validation offers a thorough evaluation of model performance across different data subsets. Together, these practices ensure that models remain robust and generalizable. Adopting these policies is crucial for developing reliable and trustworthy machine learning systems.

Fairness Metrics

Evaluating models using fairness metrics is critical. Policies should mandate the use of metrics like demographic parity, equalized odds, and disparate impact to assess the fairness of model outcomes. These metrics help in identifying and quantifying biases, guiding further refinement of the model.

Model Evaluation and Validation Procedures

Rigorous evaluation and validation procedures are necessary to ensure that models perform fairly and effectively.

Cross-Validation

Cross-validation procedures entail splitting the dataset into several subsets. The model is then trained on different combinations of these subsets. This approach helps evaluate the model's performance on diverse segments of the data. By doing so, it ensures that the model does not favor any specific group. Cross-validation provides a comprehensive assessment of how well the model generalizes to unseen data. It allows for a more robust evaluation by using different parts of the data for training and testing. This method helps identify any potential biases that might exist. It ensures that the model performs consistently across various groups. Consequently, cross-validation enhances the reliability and fairness of the model. Implementing these procedures is crucial for developing unbiased and accurate machine learning models.

Adversarial Testing

Adversarial testing entails designing scenarios that put the model's fairness and robustness to the test. This approach includes evaluating the model on edge cases, which are unusual or extreme instances that it might encounter in real-world applications. Additionally, it involves intentionally introducing biased data to observe the model's reaction. By doing so, this method exposes the model to challenging situations that it might not typically face during standard testing procedures. The primary objective is to uncover any vulnerabilities and understand how well the model maintains fairness under diverse and potentially adverse conditions.

Through adversarial testing, potential weaknesses in the model are identified, providing valuable insights into areas that require improvement. This type of testing is crucial for ensuring that the model performs equitably across all segments, regardless of the biases that might be present in the data. It allows developers to pinpoint specific aspects of the model that need refinement to enhance its robustness and fairness. Ultimately, adversarial testing is a proactive measure to ensure that machine learning models are resilient, reliable, and fair, even when confronted with the most challenging and biased scenarios.

Explainability and Interpretability

Policies should mandate that machine learning models be both interpretable and explainable. This means that the inner workings and decision-making processes of the models should be transparent and understandable to stakeholders. Techniques such as SHAP (SHapley Additive exPlanations) and LIME (Local Interpretable Model-agnostic Explanations) are valuable tools for achieving this transparency. These methods offer detailed insights into how models arrive at their decisions by breaking down the contributions of individual features.

Using SHAP and LIME, developers and users can gain a clearer understanding of the factors influencing the model's predictions. SHAP values, for example, provide a consistent approach to attributing the impact of each feature on the model's output, offering a unified explanation framework. Similarly, LIME approximates the model locally to make its behavior more interpretable in the vicinity of a specific prediction. This level of explanation is crucial for identifying and addressing any biases that may be embedded within the model.

The transparency afforded by these interpretability techniques plays a significant role in ensuring ethical AI practices. By making the decision-making process clear, stakeholders can scrutinize the model's behavior and identify potential areas where bias may exist. This understanding allows for more informed adjustments and improvements to the model, promoting fairness and reliability. Ultimately, requiring models to be interpretable and explainable helps build trust and accountability in AI systems, ensuring that they operate in a manner that is fair, transparent, and justifiable.

Deployment and Monitoring Policies

The deployment of ML models must be governed by policies that ensure continuous monitoring and ethical oversight.

Ethical Review Boards

Establishing ethical review boards to evaluate and authorize ML projects is essential. These boards should include a variety of stakeholders such as ethicists, data scientists, and community representatives. Their main responsibility is to assess the ethical implications of ML model deployments. By ensuring adherence to all relevant policies, these boards play a crucial role in maintaining ethical standards. Including diverse perspectives helps in thoroughly evaluating the potential impacts. This approach guarantees that the ethical considerations are comprehensive and well-rounded. Overall, these review boards are vital for fostering trust and accountability in the development and use of ML technologies.

Stakeholder Engagement

Involving stakeholders throughout the deployment process is crucial for addressing concerns and maintaining transparency. Policies should require ongoing consultations with these stakeholders to obtain their feedback. This feedback is essential for identifying potential issues and making necessary adjustments to the model. Regular engagement with stakeholders ensures that their perspectives and concerns are considered. This approach promotes openness and helps in refining the model to better meet user needs. Ensuring continuous dialogue with stakeholders enhances the model's effectiveness and acceptance. Overall, it fosters a more accountable and responsive deployment process.

Continuous Monitoring

Continuous monitoring policies necessitate regular audits of deployed models to evaluate their performance and fairness. This process involves tracking fairness metrics and performing periodic bias audits. Anomaly detection techniques are also employed to identify any unexpected or biased outcomes. Policies must detail the procedures for retraining and updating models in response to the findings from these monitoring activities. This ensures that models remain accurate and equitable over time. Regular reviews and updates based on monitoring results are essential for maintaining model integrity. By outlining clear procedures for these actions, policies help in addressing potential issues proactively. Continuous oversight is crucial for ensuring that deployed models continue to perform fairly and effectively.

Regulatory Compliance

Organizations are required to adhere to applicable laws and regulations related to data protection, privacy, and AI ethics. Compliance with standards such as the General Data Protection Regulation (GDPR) and the California Consumer Privacy Act (CCPA) is mandatory to ensure the protection of personal data. Additionally, organizations must align their practices with AI ethics guidelines established by influential bodies, including the IEEE and the European Commission. These regulations and guidelines are designed to safeguard individual rights and promote responsible AI use.

Adherence to Legal Standards

Policies should clearly outline the organization's commitment to these standards, specifying how they will be integrated into daily operations. This includes implementing measures to protect personal data, ensure privacy, and uphold ethical practices in AI development and deployment. Adherence to such standards not only ensures legal compliance but also fosters trust and accountability in AI systems. By following these established guidelines, organizations can navigate the complexities of data protection and AI ethics effectively, ensuring that their practices are both lawful and ethically sound.

Documentation and Reporting

Policies should mandate thorough documentation throughout every phase of the machine learning (ML) lifecycle. This encompasses detailed records of data collection methods, preprocessing procedures, model development processes, and evaluation results. By documenting each stage comprehensively, organizations ensure that every aspect of the ML workflow is transparent and traceable.

In addition to internal documentation, regular reporting to regulatory bodies is crucial for maintaining transparency and accountability. This reporting should provide a clear account of how data was collected, processed, and used, as well as how models were developed and assessed. Such transparency helps in verifying compliance with relevant laws and regulations and fosters trust among stakeholders.

Ensuring detailed documentation and timely reporting enables organizations to demonstrate their commitment to ethical and responsible ML practices. It provides a framework for auditing and reviewing the ML processes, helping to identify and address any issues promptly. Overall, these policies contribute to building a robust and accountable ML environment, where practices are consistently monitored and evaluated.

Summary

This chapter delves into the vital role machine learning (ML) systems play across various sectors and the ethical challenges that come with them, especially in data governance and bias. It highlights the need for fairness, transparency, accountability, and privacy when developing and deploying ML models. One of the key discussions is about ensuring fairness by addressing biases that can arise during data collection, training, and deployment. Transparency is also crucial, and explainable AI (XAI) techniques are emphasized as a way to help make AI decisions understandable.

Real-world examples, like the bias issues found in the COMPAS algorithm used in the U.S. criminal justice system, are explored to show how biases in training data can lead to unfair outcomes. The importance of human oversight is also brought into focus, especially in critical areas like autonomous driving and healthcare, where having a human in control is essential to step in when needed.

The chapter extends to different industries such as finance, healthcare, and law enforcement, where the ethical implications of ML models can be profound. Ethical data collection and preprocessing, bias-aware model development, and the continuous monitoring of deployed models are all highlighted as key practices to ensure these systems remain fair and unbiased. Transparency and accountability are central themes, with a strong emphasis on creating clear governance frameworks, using tools to interpret model decisions, and setting up mechanisms to correct any errors or biases that emerge in ML systems.

It provides a well-rounded look at the ethical challenges in ML data governance, offering practical strategies to mitigate bias, promote fairness, and ensure that AI systems are developed and used responsibly.

CHAPTER 6

Model Transparency and Interpretability

Model transparency and interpretability are central themes in the governance of machine learning (ML) systems. These concepts refer to the ability to understand and explain how machine learning models make decisions and are crucial for ensuring that ML systems operate in an ethical and accountable manner. Transparency involves providing insight into the inner workings and processes of ML models, while interpretability refers to the degree to which a human can comprehend the rationale behind a model's predictions or decisions.

Transparency in ML models can be categorized into several levels. At a basic level, it involves providing clear documentation about the model's architecture, including its parameters, algorithms, and the data it uses. More advanced levels of transparency involve elucidating how various inputs influence model outputs and offering insights into the decision-making process. Interpretability, on the other hand, is concerned with the clarity of these explanations. It assesses how well stakeholders, including developers, users, and affected individuals, can understand and trust the model's predictions.

Ensuring Model Interpretability

Interpretability, first and foremost, is concerned with understanding the following. In this subsection, we'll delve deeper into what are interpretability techniques and associated challenges.

A. Nandan Prasad, *Introduction to Data Governance for Machine Learning Systems*,
https://doi.org/10.1007/979-8-8688-1023-7_6

Explainable AI (XAI) Techniques

Explainable AI (XAI) techniques have become a cornerstone in the realm of machine learning (ML) data governance, especially as organizations strive to build transparent and accountable models. With the growing adoption of ML across various industries, the demand for understanding how these models make decisions has surged. Techniques like Local Interpretable Model-agnostic Explanations (LIME) and SHapley Additive exPlanations (SHAP) have emerged as powerful tools in this endeavor. These methods are designed to demystify the decision-making processes of complex models, enabling stakeholders to comprehend, trust, and effectively act upon the outcomes produced by ML systems.

LIME is particularly valuable in providing local interpretability by approximating a complex model with simpler, more interpretable models for individual predictions. This allows users to gain insights into why a model made a specific decision for a particular instance. For example, consider a scenario where a financial institution uses a neural network to assess loan applications. By applying LIME to a rejected application, the institution can generate a simplified model that highlights which features—such as credit score, income level, or loan amount—most heavily influenced the rejection. This level of transparency is crucial for both the applicant and the institution, as it fosters trust and helps ensure compliance with regulatory requirements.

SHAP, on the other hand, offers a more global perspective by assigning importance scores to features for a given prediction based on game theory concepts. It does so by calculating the contribution of each feature to the prediction, treating the prediction as a game where features collaborate to achieve the final outcome. This approach is particularly useful in complex models where interactions between features are not straightforward. For instance, in healthcare, a model predicting the likelihood of a patient developing a certain condition might consider variables like age, medical history, lifestyle factors, and genetic markers. SHAP values can illustrate how each of these factors contributes to the prediction, offering clinicians a deeper understanding of the model's behavior and enabling them to make more informed decisions.

Despite their advantages, integrating XAI techniques like LIME and SHAP into ML data governance frameworks presents significant challenges. One major issue is the computational overhead associated with these methods. Both LIME and SHAP often require multiple model evaluations to generate explanations, which can be resource-intensive. For example, in a real-time fraud detection system where decisions need to be made within milliseconds, the additional processing time required for generating explanations using SHAP could hinder the system's performance. This is especially problematic when dealing with large-scale ML models that are deployed in environments where speed and efficiency are paramount.

Another challenge is the interpretability of the explanations themselves. While XAI techniques are designed to make models more understandable, the explanations they provide can sometimes be complex, particularly for non-expert stakeholders. SHAP values, for example, are rooted in game theory and require an understanding of concepts like Shapley values, which can be a significant barrier for those without a technical background. This creates a paradox where the very techniques intended to enhance transparency may still leave some users confused or unable to fully grasp the model's decisions. In practice, this could mean that a business executive reviewing a model's decision might struggle to understand why a certain feature was assigned a particular SHAP value, undermining the trust that XAI is supposed to build.

XAI techniques often rely on assumptions that may not always hold true in real-world scenarios. LIME, for instance, assumes linearity in the local regions of the model's decision boundary. This assumption can lead to oversimplified or even misleading explanations, especially when applied to complex, non-linear models like deep neural networks. In a model predicting customer churn, for example, LIME might incorrectly suggest that a slight change in one feature, such as customer engagement, has a linear impact on the likelihood of churn, when in reality, the relationship is far more complex. Such oversimplifications can result in decision-makers acting on incorrect insights, potentially leading to adverse outcomes.

The reliability of XAI explanations must be carefully evaluated to ensure that they genuinely contribute to a model's transparency rather than creating a false sense of understanding. This is particularly crucial in high-stakes environments like healthcare or criminal justice, where decisions based on ML models can have significant consequences. For example, in a judicial setting, an XAI tool might explain why a model predicts a higher risk of recidivism for a defendant. However, if the explanation is based on flawed assumptions or overly simplified models, it could lead to unjust outcomes. Regulatory bodies are increasingly recognizing this issue and are demanding that organizations not only provide explanations for ML decisions but also ensure that these explanations are accessible, accurate, and meaningful. This adds another layer of complexity to the deployment of XAI techniques, as organizations must balance the need for transparency with the ethical implications of their models' decisions.

Scalability is another significant challenge when implementing XAI techniques across an organization. As businesses deploy more models in various applications, the demand for scalable explainability solutions grows. However, many XAI techniques, including LIME and SHAP, are not inherently scalable, particularly when applied to ensemble models or deep learning architectures. For instance, in a large retail company using multiple ML models to predict customer preferences, sales trends, and inventory needs, applying SHAP to each model could become computationally prohibitive. This limitation can hinder the widespread adoption of XAI in large organizations where the need for consistent and efficient explainability across various models is critical. Developing scalable XAI solutions is therefore essential for organizations aiming to maintain transparency while managing a diverse range of models.

Integrating XAI techniques into existing ML governance frameworks can also be cumbersome. Organizations often face challenges in standardizing XAI methods across different teams and projects, leading to inconsistencies in how transparency and interpretability are addressed. For example, one team might use LIME to explain their model's decisions, while another relies on SHAP, resulting in different interpretations and understandings of model behavior across the organization. This lack of standardization can complicate efforts to ensure that all models meet the same transparency standards, particularly when regulatory compliance is at stake. Establishing best practices and developing unified guidelines for XAI implementation are therefore critical tasks that require careful coordination and continuous updates as the field evolves.

Trade-offs Between Interpretability and Accuracy

The trade-off between interpretability and accuracy is a fundamental challenge in the field of machine learning (ML). Simpler models, such as linear regression or decision trees, offer a high degree of interpretability, making it easier for stakeholders to understand how decisions are made. These models often align well with the need for transparency, especially in domains where decisions have significant consequences, such as healthcare or finance. However, the simplicity that affords these models their interpretability also limits their ability to capture complex patterns in data, often resulting in lower accuracy compared to more sophisticated models like deep neural networks or gradient boosting machines.

A classic example of this trade-off can be observed in the healthcare industry, where the use of interpretable models is often crucial due to the high stakes involved. Consider a case where a hospital must decide between using a logistic regression model and a deep neural network for predicting patient readmission rates. Logistic regression is straightforward, with clear coefficients indicating the influence of each predictor variable on the outcome. This transparency is essential for doctors and medical staff who need to understand and trust the model's predictions. However, logistic regression might fail to capture non-linear relationships or interactions between variables, potentially leading to less accurate predictions. In contrast, a deep neural network could significantly improve predictive accuracy by modeling these complex relationships, but at the cost of interpretability, as it becomes challenging to explain how specific predictions are derived.

In practice, this trade-off was exemplified by a study conducted by Obermeyer et al. (2019), which investigated the use of an ML model to predict which patients would benefit from extra medical resources. The study revealed that the model, which was designed to prioritize accuracy, inadvertently exhibited racial bias. The black patients who needed additional care were less likely to be identified than their white counterparts. The complexity and opacity of the model made it difficult for the developers to detect this bias initially. This case highlights the danger of prioritizing accuracy over interpretability in contexts where fairness and transparency are critical. The lesson here is that while sophisticated models can offer higher accuracy, they must be approached with caution, especially in sensitive applications where the consequences of errors or biases can be severe.

On the other hand, there are scenarios where the trade-off tilts in favor of accuracy, with interpretability taking a backseat. For example, in algorithmic trading, where speed and precision are paramount, the use of complex models like deep learning or ensemble methods is often justified. These models can capture subtle patterns and make predictions in real time, offering a competitive edge in the market. The opacity of these models is less of a concern because the primary goal is to maximize returns, and the decisions do not directly impact individuals in a personal or ethical sense. However, even in such contexts, organizations must be mindful of the potential risks associated with model opacity, such as overfitting or the propagation of undetected biases that could lead to financial losses or regulatory scrutiny.

A notable example of balancing this trade-off can be seen in the finance industry's approach to credit scoring. Traditionally, credit scores were determined using interpretable models like logistic regression, where the contribution of each factor (e.g., income, credit history) to the final score could be clearly understood by both lenders and borrowers. However, as financial institutions sought to improve the accuracy of their predictions, more complex models like gradient boosting machines were introduced. These models, while offering better predictive performance, are harder to interpret. To address this, some institutions have employed XAI techniques like SHAP to provide explanations for individual credit scores generated by these complex models. This approach allows them to retain the accuracy benefits of advanced models while still offering some level of interpretability, which is crucial for meeting regulatory requirements and maintaining customer trust.

Navigating the trade-off between interpretability and accuracy is not merely a technical challenge; it also involves ethical considerations and a deep understanding of the specific context in which the model is deployed. For instance, in the development of autonomous vehicles, engineers face the dilemma of choosing between highly accurate, complex models and simpler, more interpretable ones. A highly accurate model might be better at navigating complex environments but could be difficult to interpret in the event of an accident, making it challenging to determine the cause of the failure. On the other hand, a simpler model might be easier to troubleshoot but could lack the precision needed for safe operation. This trade-off becomes even more pronounced in scenarios where autonomous vehicles must make real-time decisions that could impact human lives, such as in the case of collision avoidance systems.

The successful navigation of these trade-offs often requires a hybrid approach, where interpretability and accuracy are balanced according to the specific needs of the application. One such example can be seen in the use of ensemble learning techniques in medical diagnostics. By combining multiple models—some highly interpretable, others more complex—healthcare providers can achieve a balance between accuracy and transparency. For instance, a diagnostic system might use a decision tree model to provide an initial, interpretable assessment of a patient's condition, followed by a more complex neural network to refine the diagnosis. The decision tree offers a clear explanation that can be communicated to the patient, while the neural network ensures that the diagnosis is as accurate as possible. This layered approach allows for the strengths of both interpretability and accuracy to be leveraged, mitigating the downsides of relying solely on one type of model.

The decision of how to balance interpretability and accuracy should be guided by the specific goals and constraints of the application, as well as the ethical considerations involved. Organizations must be willing to engage in a continuous dialogue with stakeholders, including end users, regulators, and the broader community, to ensure that their models are not only effective but also fair and transparent. This involves not only choosing the right models but also implementing robust governance frameworks that include regular audits, bias detection, and the use of XAI techniques to explain and justify model decisions. By carefully considering these factors, organizations can successfully navigate the trade-offs between interpretability and accuracy, ensuring that their ML models are both effective and trustworthy.

Implementing and Monitoring Transparency and Interpretability

Implementing and monitoring transparency and interpretability within machine learning (ML) frameworks is a multifaceted process that demands a robust and systematic approach. This process should align seamlessly with the broader goals of data governance, ensuring that the models deployed are not only effective but also accountable and understandable. To achieve this, organizations must embed transparency and interpretability considerations into the earliest stages of model development. This foundational step influences key decisions around model architecture, data preprocessing, and feature selection, ensuring that these principles are not an afterthought but an integral part of the ML pipeline.

A specific example of how this early integration can be achieved is through the adoption of tools like Model Card Toolkit (MCT) by Google. MCT provides a framework for documenting key details about a model, including its intended use, performance metrics, ethical considerations, and interpretability features. By using such tools from the outset, organizations can standardize the documentation process, making it easier to monitor and adjust models as they evolve. This approach not only enhances transparency but also ensures that models are easier to audit throughout their lifecycle. Embedding transparency and interpretability at the design phase, therefore, sets the stage for the development of ML models that are inherently more explainable and easier to manage.

Once transparency and interpretability are integrated into the design phase, establishing clear and comprehensive documentation practices becomes the next critical step. Documentation serves as the cornerstone of transparency, providing a detailed record of the decisions made during model development and the rationale behind them. For instance, by utilizing frameworks like the Open Neural Network Exchange (ONNX), organizations can document and share models across different platforms, ensuring consistency and transparency in how models are trained, tested, and deployed. ONNX also allows for the tracking of changes and updates to models, making it easier to understand how a model has evolved over time and how those changes impact its interpretability.

This documentation should extend to all aspects of the model, including data sources, preprocessing steps, model selection criteria, and the specific interpretability techniques employed. Thorough documentation enables stakeholders to trace the model's development process, providing clarity on the factors that influenced its design and behavior. It also plays a crucial role in regulatory compliance, as it offers the necessary evidence to demonstrate that the model adheres to transparency and fairness standards. For example, in the financial sector, documentation may include details on how fairness was evaluated and ensured, particularly in models used for credit scoring or loan approvals. This level of detail helps build trust with both regulators and users by providing a clear and traceable path of the model's development.

As the model progresses to the training phase, organizations must employ explainable AI (XAI) techniques to enhance interpretability. These techniques, such as SHAP (SHapley Additive exPlanations) and LIME (Local Interpretable Model-agnostic Explanations), are essential tools in demystifying complex models. By offering insights into how individual features contribute to predictions, these techniques make it possible to understand and explain the behavior of even the most sophisticated models. For instance, SHAP can be used in models predicting patient outcomes in healthcare settings. By identifying which factors—such as age, medical history, or treatment type—contribute most significantly to a prediction, SHAP helps clinicians make informed decisions based on model outputs.

Implementing these techniques during training allows organizations to identify and address potential biases or transparency issues before the model is deployed. This proactive approach ensures that the model remains interpretable and that any deviations from expected behavior are promptly detected and rectified. For example, if SHAP reveals that a model is disproportionately weighting certain features in a way that could lead to biased outcomes, developers can adjust the model or its inputs to mitigate this risk. Continuous evaluation using these techniques is critical for maintaining a high standard of interpretability throughout the model's lifecycle.

Continuous monitoring is another vital aspect of maintaining transparency and interpretability. This involves establishing policies that require regular audits of deployed models to assess their performance and adherence to transparency standards. A specific tool that can be utilized for this purpose is Microsoft Azure's ML Ops suite, which provides a comprehensive framework for monitoring ML models in production. Azure ML Ops allows organizations to track key metrics related to model performance and interpretability, such as feature importance scores, prediction confidence levels, and the consistency of explanations across different data segments. By systematically monitoring these metrics, organizations can ensure that the model continues to operate in a transparent and understandable manner, even as it encounters new data or undergoes updates.

Handling discrepancies that arise during monitoring is another critical component of maintaining transparency. When anomalies or unexpected outcomes are detected, organizations must have a clear protocol for investigating and addressing these issues. For instance, if monitoring reveals that a model's predictions are becoming less accurate or are showing signs of bias, organizations can use tools like IBM's AI Fairness 360 toolkit to diagnose and correct these issues. This toolkit offers a range of algorithms and metrics for assessing fairness in AI models, allowing organizations to identify the root causes of discrepancies and implement corrective actions. This not only helps in maintaining the model's transparency but also ensures that it remains fair and unbiased over time.

Fostering a culture of transparency within the organization is also crucial for the successful implementation of these practices. This involves encouraging open communication and feedback from all stakeholders, including users, data scientists, ethicists, and regulatory bodies. Engaging with these groups helps to gather insights into how well the model's transparency measures are functioning in practice. For example, regular focus groups or user testing sessions can provide valuable feedback on the interpretability of model outputs, revealing areas where explanations may be unclear or insufficient. By incorporating this feedback into the model's development and monitoring processes, organizations can continually refine and improve the model's transparency and interpretability.

Anomaly detection plays a key role in effective monitoring by identifying unexpected or biased outcomes that may indicate a loss of transparency. Techniques such as statistical process control (SPC) or more advanced methods like unsupervised machine learning can be employed to detect shifts in model behavior. For instance, SPC can be used to monitor the output distribution of a model over time, flagging any significant deviations that warrant further investigation. In a real-world scenario, a financial institution might use SPC to track the distribution of credit scores assigned by a model, ensuring that any sudden shifts—such as a disproportionate increase in scores for a particular demographic group—are promptly addressed.

To support ongoing transparency efforts, organizations must establish clear procedures for retraining and updating models based on monitoring results. This is where tools like Google's TensorFlow Extended (TFX) can be invaluable. TFX provides a framework for managing the entire lifecycle of an ML model, including monitoring, updating, and retraining. With TFX, organizations can automate the retraining process based on predefined triggers, such as when a model's performance drops below a certain threshold or when new data becomes available. By doing so, they ensure that models are continuously optimized for both accuracy and interpretability, with a clear audit trail documenting all changes.

Transparency and interpretability are dynamic goals that require continuous adaptation to changing circumstances. As new data becomes available or as models are exposed to new contexts, the transparency measures initially implemented may need to be revisited and adjusted. This dynamic approach can be supported by implementing a flexible monitoring framework that allows for the periodic reassessment of transparency practices. For example, in the context of evolving regulatory requirements, organizations may need to update their transparency measures to remain compliant. Tools like Fiddler AI's Explainable Monitoring platform can help in this regard, providing real-time insights into model behavior and alerting organizations to potential compliance risks.

Effective monitoring of transparency also involves regular reporting to both internal stakeholders and external regulatory bodies. These reports should offer a comprehensive overview of the model's transparency metrics, any identified issues, and the actions taken to address them. For instance, in the healthcare industry, regular transparency reports might be required to demonstrate compliance with patient data protection regulations. These reports not only provide accountability but also help to build trust with external stakeholders by showing a commitment to transparency and ethical AI practices.

The successful implementation and monitoring of transparency and interpretability also require a commitment to ongoing education and training. As the field of AI and ML continues to evolve, new techniques and best practices for transparency will emerge. Organizations must ensure that their teams are equipped with the latest knowledge and skills to implement these practices effectively. Regular training sessions, workshops, and access to resources on XAI techniques and transparency best practices are essential for maintaining high standards of interpretability in ML models. For example, companies might invest in training programs on emerging tools like H2O.ai's Driverless AI, which offers advanced interpretability features. By doing so, they ensure that their teams are prepared to implement cutting-edge transparency techniques, keeping their models aligned with both organizational goals and societal expectations.

Regulatory and Ethical Considerations

Regulatory and ethical considerations in the context of machine learning (ML) data governance and model transparency and interpretability are becoming increasingly critical as organizations deploy more sophisticated AI systems. One of the primary regulatory concerns is compliance with data protection laws, such as the General Data Protection Regulation (GDPR) in Europe or the California Consumer Privacy Act (CCPA) in the United States. These regulations require that individuals have the right to understand and contest automated decisions that significantly impact them. Consequently, organizations must ensure that their ML models are transparent enough to provide clear explanations for their outputs. Failure to comply with these regulations can result in significant legal and financial penalties, as well as damage to the organization's reputation.

In addition to compliance with data protection laws, there are specific regulations emerging that directly address the use of AI and ML. For example, the European Union's proposed Artificial Intelligence Act aims to regulate the use of AI systems based on their risk level. High-risk AI systems, which include those used in critical areas such as healthcare, law enforcement, and employment, will be subject to stringent requirements, including the need for transparency and explainability. These regulatory frameworks emphasize the importance of model interpretability, as they require that organizations provide clear documentation and evidence that their models are fair, unbiased, and understandable to both regulators and affected individuals.

Ethical considerations extend beyond regulatory compliance and encompass broader societal impacts. One of the key ethical principles in AI is fairness, which requires that ML models do not discriminate against individuals or groups based on protected characteristics such as race, gender, or socioeconomic status. Ensuring fairness is closely linked to model transparency, as opaque models can obscure biases and prevent stakeholders from identifying and addressing discriminatory practices. To uphold ethical standards, organizations must prioritize the development and deployment of interpretable models that allow for thorough scrutiny of their decision-making processes. This commitment to transparency helps to prevent and mitigate the potential harms caused by biased or unfair AI systems.

Another ethical consideration is the accountability of ML models. As AI systems are increasingly used to make or inform decisions with significant consequences, it is essential that organizations can hold the models—and, by extension, their developers—accountable for the outcomes. This requires a level of transparency that allows stakeholders to understand how a model arrived at a particular decision and to trace the steps taken during the model's development and deployment. Accountability is not only an ethical obligation but also a practical necessity in maintaining public trust in AI systems. Transparent and interpretable models make it possible to identify who is responsible for decisions, enabling appropriate actions to be taken in cases of error or harm.

Transparency and interpretability also play a crucial role in ensuring informed consent in the use of AI systems. Individuals affected by ML decisions should have the right to know how these decisions are made and what factors influence the outcomes. Ethical AI practice demands that organizations provide clear and accessible explanations that allow individuals to make informed choices about their interactions with AI systems. This is particularly important in contexts such as healthcare, where patients must understand the basis of AI-driven recommendations or diagnoses to give truly informed consent to treatment options. Providing interpretable models and transparent explanations is thus a fundamental aspect of respecting individuals' autonomy and rights.

The concept of transparency also intersects with the ethical principle of beneficence, which requires that AI systems are designed and used to benefit individuals and society as a whole. Transparent models enable stakeholders to assess the potential risks and benefits of deploying AI systems in various contexts. For instance, in the financial sector, transparent credit scoring models can help identify and address potential risks of exclusion or discrimination, ensuring that the system benefits all users equitably. By making the inner workings of AI models accessible and understandable, organizations can better ensure that their AI systems contribute positively to society and do not inadvertently cause harm.

The ethical principle of non-maleficence, which obligates organizations to avoid causing harm, is closely tied to the use of transparent and interpretable models. Opaque models can lead to unintended consequences, such as biased or harmful decisions, which may not be immediately apparent to users or developers. By prioritizing transparency and interpretability, organizations can more effectively identify and mitigate these risks before they result in harm. This proactive approach is essential for maintaining ethical standards in AI deployment and for protecting the welfare of individuals and communities affected by AI decisions.

Regulatory bodies are increasingly recognizing the importance of transparency and interpretability in maintaining the ethical integrity of AI systems. As a result, there is a growing trend toward requiring organizations to provide clear explanations and justifications for the decisions made by their AI models. This regulatory shift reflects a broader societal demand for accountability and transparency in the use of AI, particularly in high-stakes areas such as criminal justice, healthcare, and finance. Organizations must stay abreast of these evolving regulatory expectations and ensure that their ML governance frameworks are equipped to meet the demands for transparency and interpretability.

Organizations must also navigate the ethical implications of trade-offs between model accuracy and interpretability. While complex models like deep neural networks may offer higher accuracy, they are often less interpretable than simpler models such as decision trees or linear regression. Ethical AI practice requires organizations to carefully consider these trade-offs and to prioritize interpretability in contexts where understanding the model's decision-making process is crucial for ethical compliance. This may involve opting for slightly less accurate but more interpretable models in situations where transparency is essential, such as in medical diagnosis or legal decision-making.

The ethical and regulatory landscape surrounding AI is dynamic and continuously evolving. Organizations must remain vigilant in their efforts to stay informed about new developments in AI ethics and regulations and to update their practices accordingly. This involves ongoing education and training for all stakeholders involved in the development and deployment of ML models, as well as continuous monitoring and evaluation of the models themselves. By embedding transparency and interpretability into the core of their ML governance frameworks, organizations can better navigate the complex regulatory and ethical challenges of AI and ensure that their AI systems are both responsible and trustworthy.

Best Practices on Model Transparency and Interpretability

This section explores key practices that organizations can adopt to enhance the transparency and interpretability of their models, from comprehensive documentation and data handling to the implementation of explainable AI techniques. Each of these practices can help ensure that stakeholders understand, evaluate, and trust the decisions made by these models.

Comprehensive Documentation Practices

A solid foundation for ensuring model transparency lies in maintaining comprehensive documentation. It is essential for organizations to implement robust policies that mandate thorough documentation at each phase of the machine learning (ML) lifecycle. This documentation should begin with the initial design of the model and continue through all subsequent stages, including any updates or modifications made over time. Every aspect of the model's development should be meticulously recorded, covering the methods used for data collection, the steps taken in data preprocessing, the specific architecture chosen for the model, and the procedures followed during training. Additionally, the results of all evaluations should be thoroughly documented to provide a clear understanding of the model's performance and decision-making processes.

Such documentation plays a critical role in making the model's operations transparent to both internal and external stakeholders. By capturing the details of how data is collected and processed, the rationale behind the choice of model architecture, and the methodologies used during training, organizations can ensure that their models are open to scrutiny and can be effectively audited. This not only aids in regulatory compliance but also builds trust among users and other stakeholders who rely on the ML model's outputs. Furthermore, maintaining detailed records of all updates and modifications ensures that any changes to the model over time are transparent, enabling stakeholders to understand how and why the model has evolved. In essence, comprehensive documentation serves as a vital tool for achieving and sustaining model transparency throughout the entire ML lifecycle.

Data Documentation

Documenting data sources is an essential aspect of achieving transparency in machine learning (ML) processes. The importance of this practice cannot be overstated, as it forms the backbone of understanding how a model operates and ensuring that its decisions can be traced back to their origins. This documentation begins with a thorough record of where and how the data was collected. For every dataset used in training or testing the model, it is imperative to note the precise sources, whether they are internal data repositories, publicly available datasets, or information acquired from third-party vendors. This level of detail ensures that anyone reviewing the model's development can understand the foundational data that drives its outputs.

Equally important is the documentation of the quality and completeness of the data. Transparency requires that any gaps, inconsistencies, or potential biases in the data are clearly identified and recorded. This documentation should include an assessment of the data's overall quality, highlighting any missing values, outliers, or errors that may have been present before preprocessing. It should also address the completeness of the data, noting whether the dataset fully represents the population or scenario it is intended to model. By keeping detailed records of these aspects, organizations can provide stakeholders with a clear picture of the strengths and limitations of the data used in the model, which is crucial for evaluating the model's reliability and fairness.

It is critical to record any preprocessing steps that were applied to the data before it was used for model training. This includes procedures such as data cleaning, where errors and inconsistencies are corrected, and normalization, where data is scaled or transformed to fit a specific range or distribution. Other preprocessing techniques, like data augmentation, where synthetic data points are generated to enhance the dataset, should also be meticulously documented. By detailing these processes, organizations can offer insights into how the raw data was prepared and transformed, which is vital for understanding the model's behavior and ensuring that the preprocessing steps do not introduce unintended biases or distortions.

The rationale behind the selection of specific datasets should be carefully documented. This includes explaining why certain datasets were chosen over others, what criteria were used to determine their relevance, and how they align with the model's objectives. When third-party data sources are involved, it is important to document the agreements or permissions that govern the use of this data, as well as any limitations or constraints associated with it. By providing comprehensive records of the data sources and preprocessing procedures, organizations can ensure that their ML models are transparent, making it possible for stakeholders to critically evaluate the entire data pipeline. This level of transparency not only fosters trust but also supports ongoing monitoring and refinement of the model, ensuring that it remains fair, reliable, and aligned with its intended purpose.

Model Architecture and Training

Documentation of the model architecture is a crucial component in achieving transparency within machine learning (ML) projects. This documentation should provide a comprehensive outline of the model's structure, detailing the arrangement and configuration of layers, the number of nodes within each layer, and the activation functions utilized. Such documentation is not merely a technical necessity but a vital tool for enabling stakeholders to grasp the underlying complexity of the model. By clearly laying out these structural elements, organizations offer insight into the design choices that shape the model's behavior, allowing for a deeper understanding of how the model processes input data and generates predictions.

Understanding the model's architecture extends beyond just knowing the components; it also involves comprehending the rationale behind specific design decisions. For instance, the choice of particular activation functions, such as ReLU or Sigmoid, can significantly influence the model's performance, and these decisions should be transparently documented. Special configurations, such as the use of dropout layers to prevent overfitting or convolutional layers in image processing tasks, should also be clearly described. By meticulously recording these architectural elements, stakeholders can better evaluate the model's suitability for its intended application and assess whether the design aligns with best practices in the field.

Training documentation is equally important in ensuring transparency and reproducibility of ML models. This documentation should include a detailed account of the hyperparameter settings used during training, such as learning rates, batch sizes, and the number of epochs. These settings are critical as they directly impact the model's ability to learn from data and converge on an optimal solution. In addition to hyperparameters, the duration of the training process, including the time taken for each training session and the total time required to train the model, should be documented.

This information provides context for the model's development, allowing stakeholders to understand the resources and computational effort involved in bringing the model to its final form.

Any methods employed for tuning and optimization during the training phase should be thoroughly recorded. Techniques such as grid search or random search for hyperparameter tuning and optimization methods like stochastic gradient descent or Adam play a pivotal role in refining the model's performance. By documenting these methods, organizations ensure that the model's development process is not only transparent but also reproducible. Reproducibility is key in ML, as it allows other researchers or stakeholders to replicate the results and verify the model's efficacy. In summary, comprehensive documentation of both the model's architecture and training process is essential for fostering transparency, enabling informed evaluation, and supporting the continuous improvement of ML models.

Evaluation and Performance Metrics

Policies should require meticulous documentation of the procedures used to evaluate machine learning models, with a particular emphasis on the metrics employed to assess their performance. It is essential that organizations keep detailed records that outline the specific metrics utilized, such as accuracy, precision, recall, and other pertinent measures. This documentation should clearly describe the methodology behind these calculations, providing a transparent view of how each metric was derived. By doing so, stakeholders are given a comprehensive understanding of the criteria used to judge the model's effectiveness, which is crucial for ensuring the integrity and reliability of the evaluation process.

Understanding the model's performance metrics is fundamental to grasping its strengths and weaknesses. Accuracy, for example, gives a general idea of how well the model performs across all classes, but it may not fully capture performance nuances in cases of class imbalance. Thus, documenting precision and recall offers additional insights, particularly in understanding how well the model handles false positives and false negatives, respectively. This nuanced approach to documenting performance metrics allows stakeholders to see beyond just a single measure, providing a fuller picture of how the model is expected to perform in various scenarios.

In addition to the basic performance metrics, it is crucial to document any advanced evaluation techniques used to gauge the model's generalization capabilities. Techniques like cross-validation, where the model is trained and tested on different subsets of the data, should be carefully recorded. Cross-validation provides a robust method for assessing how well the model is likely to perform on unseen data, and its documentation should include the specific cross-validation method used, such as k-fold or leave-one-out, along with the rationale behind the choice. Similarly, if split-testing techniques were employed, where data is divided into distinct training and testing sets, the details of this process should be clearly outlined.

The documentation should also address how these evaluation techniques contribute to a better understanding of the model's performance across different data distributions and contexts. For instance, the results from cross-validation might reveal potential overfitting, where the model performs exceptionally well on training data but poorly on validation data. By thoroughly documenting these findings, organizations can identify and address potential issues before the model is deployed in real-world settings. This level of detail in recording evaluation procedures not only enhances transparency but also serves as a valuable resource for ongoing model refinement and improvement.

Maintaining detailed records of the entire model evaluation process ensures that the performance assessments are both transparent and verifiable. Stakeholders, including regulatory bodies, internal auditors, and external partners, can refer to this documentation to understand precisely how performance was measured and validated. This transparency is crucial for building trust in the model's capabilities and ensuring that it meets the necessary standards for deployment. In essence, comprehensive documentation of model evaluation procedures safeguards the credibility of the machine learning process, providing a clear, reproducible path for assessing and verifying the model's performance.

Implementing Explainable AI Techniques

Implementing explainable AI (XAI) techniques as part of best practices for model transparency and interpretability in machine learning (ML) data governance is essential for ensuring that AI systems are understandable, trustworthy, and aligned with ethical standards. In the complex landscape of AI and ML, models often operate as black boxes, making decisions based on patterns that are not easily discernible by humans. This opacity can lead to challenges in understanding, validating, and trusting these models, particularly in high-stakes domains like healthcare, finance, and criminal justice. By adopting XAI techniques, organizations can bring transparency to these processes, enabling stakeholders to comprehend how decisions are made and to ensure that these decisions are fair, accurate, and reliable.

One of the critical aspects of XAI is local interpretability, which focuses on explaining individual predictions made by a model. Local interpretability is crucial when stakeholders need to understand why a particular decision was made for a specific instance. Techniques such as LIME (Local Interpretable Model-agnostic Explanations) and SHAP (SHapley Additive exPlanations) are commonly used to provide local explanations. LIME works by approximating the model locally around a prediction, creating a simpler, interpretable model that can be analyzed. SHAP, on the other hand, assigns an importance value to each feature for a given prediction, indicating how much each feature contributed to the outcome. These local interpretability methods allow users to understand and trust individual predictions, which is especially important in cases where decisions have significant consequences for individuals.

In addition to local interpretability, global interpretability is another key component of XAI. Global interpretability involves understanding the overall behavior of the model rather than just individual predictions. It aims to provide insights into how the model works across the entire dataset, highlighting the general rules and patterns that the model has learned. Techniques for global interpretability include feature importance ranking, partial dependence plots, and surrogate models. Feature importance ranking helps identify which features have the most significant impact on the model's decisions across all data points, while partial dependence plots show how changes in a particular feature affect the model's predictions. Surrogate models, which are simpler models that approximate the behavior of more complex models, can also be used to provide a global understanding of the model's decision-making process. By offering a broader view of how the model operates, global interpretability ensures that stakeholders can understand and validate the model's overall logic and assumptions.

Visualization tools play a crucial role in both local and global interpretability by making the results of XAI techniques more accessible and understandable to a broader audience. Visualization techniques, such as heatmaps, decision trees, and feature importance charts, help to translate complex model behaviors into more intuitive and user-friendly formats. For example, SHAP values can be visualized using summary plots that show the impact of each feature across multiple predictions, providing a clear picture of which features are driving the model's decisions. Similarly, LIME explanations can be visualized through bar charts that represent the contribution of each feature to a specific prediction. These visual tools are invaluable for communicating model behavior to stakeholders who may not have a deep technical background but still need to understand and trust the AI system.

Another important aspect of implementing XAI is ensuring that these techniques are integrated into the broader ML lifecycle, from model development to deployment and monitoring. This requires establishing policies and procedures that mandate the use of XAI techniques at each stage of the model lifecycle. During model development, for instance, developers should be required to generate and review local and global explanations to ensure that the model is making decisions in a transparent and fair manner. During deployment, these explanations should be made available to end users, allowing them to understand and evaluate the model's decisions in real time. Ongoing monitoring should also include regular audits of the model's explanations to detect any changes in behavior that could indicate potential biases or other issues.

Implementing XAI as part of ML data governance also involves addressing the challenges associated with interpretability and transparency. One of the primary challenges is the trade-off between model complexity and interpretability. Complex models, such as deep neural networks, often provide higher accuracy but at the cost of being less interpretable. This trade-off necessitates careful consideration of the balance between accuracy and transparency, depending on the specific use case and the level of trust required by stakeholders. In some cases, it may be necessary to opt for slightly less accurate but more interpretable models, particularly in situations where transparency is crucial for regulatory compliance or ethical reasons. Organizations must establish guidelines that help navigate these trade-offs, ensuring that models are both effective and understandable.

The implementation of XAI techniques must consider the potential for over-reliance on explanations that may not fully capture the model's behavior. While tools like LIME and SHAP provide valuable insights, they are approximations and may not always reflect the true underlying decision-making process of the model. This limitation highlights the importance of using multiple XAI techniques in conjunction, as well as combining them with domain expertise to accurately interpret and validate model behavior. Organizations should train their data scientists and decision-makers to understand the strengths and limitations of different XAI methods, ensuring that explanations are used appropriately and not taken at face value without further analysis.

The integration of XAI techniques into ML data governance practices also necessitates ongoing education and training for all stakeholders involved in the ML lifecycle. Data scientists, developers, and decision-makers need to be proficient in the use of XAI tools and techniques, understanding how to generate and interpret explanations effectively. Regular training sessions and workshops can help build this expertise, ensuring that the organization has the necessary skills to implement and maintain explainable AI systems. Additionally, end users of AI systems should be educated on how to interpret the explanations provided by XAI tools, enabling them to make informed decisions based on the model's outputs.

The successful implementation of XAI techniques requires a strong commitment to transparency and accountability at the organizational level. This means that XAI should not be seen as an optional add-on but as an integral part of the organization's ML data governance strategy. Leadership must prioritize transparency and interpretability as core values, embedding them into the culture and operations of the organization. This includes establishing clear policies that mandate the use of XAI techniques, setting up governance structures that oversee their implementation, and creating channels for stakeholders to provide feedback and raise concerns about model transparency. By fostering a culture of transparency, organizations can ensure that their AI systems are not only effective but also trustworthy and aligned with ethical standards.

Implementing explainable AI techniques as part of best practices for model transparency and interpretability is essential for responsible and effective ML data governance. By focusing on both local and global interpretability, leveraging visualization tools, integrating XAI throughout the ML lifecycle, and addressing the challenges of interpretability, organizations can build AI systems that are transparent, understandable, and aligned with the needs of stakeholders. This commitment to transparency and accountability not only enhances trust in AI systems but also ensures that they are used in a manner that is ethical, fair, and compliant with regulatory standards.

Addressing the Trade-offs Between Interpretability and Accuracy

Addressing the trade-offs between interpretability and accuracy is central to ensuring that machine learning (ML) models align with both technical requirements and ethical standards. The inherent tension between these two attributes often presents a complex challenge. Highly accurate models, such as deep neural networks, can deliver exceptional performance in handling intricate tasks, but often at the expense of interpretability. In contrast, simpler models like decision trees or linear regressions offer greater transparency but may lack the accuracy required for complex data patterns. Balancing these trade-offs requires a nuanced approach that takes into account the specific needs of the application and the context in which the model will be deployed.

Organizations must begin by carefully selecting models that align with their primary objectives. If accuracy is the primary goal, as it might be in predictive maintenance or financial forecasting, complex models like deep learning architectures might be the preferred choice. These models are capable of capturing complex, non-linear relationships within the data, leading to high predictive accuracy. However, the complexity that gives these models their power also makes them difficult to interpret, raising concerns about transparency and trust, particularly when the model's decisions have significant impacts on individuals or society.

For example, consider a financial institution using a deep learning model to predict loan defaults. While the model's accuracy is critical for the institution's financial health, it also needs to ensure that the decisions are explainable to customers and regulators. The institution might find itself in a situation where the high accuracy of the model is undercut by its lack of interpretability, leading to potential challenges in regulatory compliance and customer trust. This scenario highlights the importance of carefully considering whether the accuracy benefits of a complex model justify the potential drawbacks in transparency.

On the other hand, when interpretability is paramount, such as in healthcare or legal decisions, simpler models may be more appropriate. A linear regression model used to predict patient outcomes, for example, might not capture every nuance in the data but offers clear and understandable relationships between inputs and predictions. This transparency is crucial in scenarios where decisions must be explained to non-technical stakeholders, such as doctors, patients, or legal authorities. Here, the trade-off is a conscious decision to prioritize interpretability over potential gains in accuracy, ensuring that the model's decisions are transparent and understandable, which is essential for trust and accountability.

However, in many cases, neither extreme—pure accuracy nor pure interpretability—adequately addresses the needs of the organization. This is where hybrid approaches come into play, offering a flexible solution that balances the strengths of both complex and simple models. A hybrid approach might involve using a complex model, such as a gradient-boosting machine, for making predictions, while employing a simpler, interpretable model to provide explanations. This strategy allows the organization to benefit from the predictive power of the complex model while ensuring that the decision-making process remains transparent. For instance, in an online recommendation system, a deep learning model could be used to generate personalized recommendations based on a vast array of data points, while a decision tree could be used to explain the most influential factors in generating a specific recommendation to the user.

An alternative hybrid approach involves the use of surrogate models. These are simpler models that approximate the behavior of more complex models, providing a more interpretable view of the decision-making process without sacrificing overall performance. For example, a random forest model might be used to predict credit risk, with a decision tree serving as a surrogate to explain the model's predictions in a more understandable way. This approach allows stakeholders to gain insights into the model's behavior while still leveraging the accuracy benefits of the more complex model.

Integrating interpretability into the model development process from the outset is crucial for achieving a balance between accuracy and transparency. Setting clear goals for both accuracy and interpretability during the design phase helps guide decisions about model architecture, feature selection, and evaluation criteria. For instance, feature importance analysis and partial dependence plots can be integrated into the development process to ensure that the final model not only performs well but also provides meaningful insights into its decision-making process. Regularly revisiting these goals throughout the model lifecycle ensures that the balance between interpretability and accuracy is maintained as the model evolves.

In practice, this means continuously evaluating the model's performance and interpretability against the organization's objectives and making adjustments as needed. For example, if a model initially designed for high accuracy starts to exhibit signs of opacity or bias, it may be necessary to revisit the design and incorporate more interpretability-focused techniques. This could involve re-weighting features, adjusting the model's architecture, or even switching to a different modeling approach that better aligns with the need for transparency.

The decision-making framework for choosing the right approach to balancing interpretability and accuracy involves several key considerations. Organizations should start by defining the primary objective of the model—is the goal to achieve the highest possible accuracy, or is transparency more critical due to the need for regulatory compliance or ethical considerations? If accuracy is the priority, the next step is to assess the level of interpretability required and whether the model's complexity can be justified in the context of the application's risk profile. For models where transparency is essential, organizations might consider simpler models or hybrid approaches that combine the strengths of both complex and simple models.

The flowchart provided illustrates this decision-making process, guiding organizations through the key steps in choosing the right approach based on their specific needs. Starting with the primary objective, the flowchart branches into paths that explore the trade-offs between high accuracy and high interpretability. It also considers the possibility of hybrid approaches, suggesting combinations of complex and simple models or the use of interpretable surrogates to achieve a balance between performance and transparency.

This structured approach ensures that the trade-offs between interpretability and accuracy are addressed in a way that aligns with both the technical requirements and ethical standards of the organization. By carefully evaluating these trade-offs, leveraging hybrid approaches, and integrating interpretability into the development process, organizations can build AI systems that are not only high-performing but also transparent and trustworthy.

Figure 6-1 features a flowchart that guides organizations through the process.

Decision-Making Framework for Balancing Interpretability and Accuracy

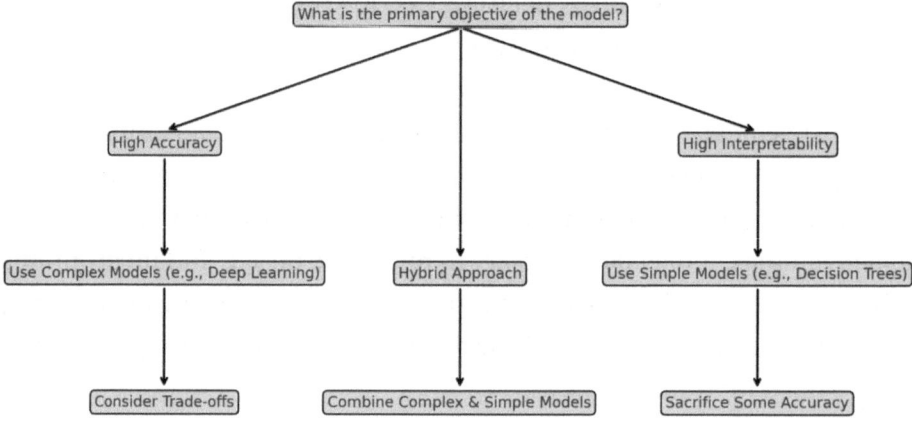

Figure 6-1. *Flowchart illustrating the navigating complexities of machine learning models*

This decision-making framework provides a structured approach to navigating the complex trade-offs between interpretability and accuracy in machine learning models. The framework begins with the critical question: "What is the primary objective of the model?" This question directs the organization to consider whether the primary focus should be on achieving the highest possible accuracy, which may be necessary in scenarios like financial forecasting or autonomous systems, or whether interpretability is paramount, as is often the case in healthcare, legal, or regulatory contexts.

For organizations that prioritize accuracy, complex models, such as deep neural networks or ensemble methods, are suggested which are capable of capturing intricate patterns and relationships within large datasets. These models are often indispensable in high-stakes environments where predictive performance can directly impact outcomes, such as fraud detection systems or precision medicine applications. However, organizations may have to assess whether the level of accuracy achieved by these models justifies the potential loss of transparency. This is particularly important in scenarios where decisions must be explained to non-technical stakeholders, such as patients, customers, or regulatory bodies.

When interpretability is identified as the primary objective, organizations may go toward simpler models, such as decision trees, linear regression, or logistic regression, which offer a higher degree of transparency. These models allow stakeholders to understand and trust the decision-making process, which is crucial in environments where transparency is legally mandated or ethically required. For example, in a legal setting, a simple model used to predict recidivism rates might be preferred over a complex model, as the decisions made by the model must be easily interpretable and defensible in court. The framework encourages organizations to consider whether the potential reduction in accuracy is an acceptable trade-off in exchange for the benefits of greater transparency.

Organizations may also explore hybrid approaches for situations where neither pure accuracy nor pure interpretability adequately addresses the organization's needs. Hybrid models combine the strengths of both complex and simple models, enabling organizations to leverage the accuracy of sophisticated algorithms while maintaining an interpretable decision-making process. This approach is particularly useful in scenarios where the model's decisions must be both accurate and explainable, such as in financial lending or personalized medicine. For instance, a hybrid approach might involve using a complex model to generate predictions and then applying a simpler, interpretable model to explain the most influential factors behind those predictions. This allows the organization to retain the predictive power of the complex model while ensuring that the decision-making process is transparent and understandable to stakeholders.

Integration and Monitoring of Transparency Practices

Integration and monitoring of transparency practices are crucial for maintaining model transparency and interpretability within machine learning (ML) data governance. As machine learning models are deployed and used in real-world applications, ensuring that these models remain transparent and interpretable throughout their lifecycle is essential for fostering trust, ensuring fairness, and meeting regulatory requirements. Implementing best practices for integrating transparency practices and continuously monitoring their effectiveness involves several key components, including continuous monitoring and updates, as well as active stakeholder engagement.

Continuous monitoring of ML models is essential for maintaining transparency and interpretability over time. Once a model is deployed, it is important to regularly assess its performance and behavior to ensure that it continues to operate as intended and adheres to the established transparency standards. This includes tracking key performance metrics and interpretability indicators to identify any deviations or issues that may arise. Monitoring should be conducted using automated tools and processes that can detect changes in model performance or decision-making patterns. Automated monitoring tools such as IBM's Watson OpenScale can be invaluable in this process. Watson OpenScale offers real-time monitoring of model fairness, bias detection, and transparency, providing alerts when a model's behavior deviates from expected norms. For instance, in a case study involving a financial institution, Watson OpenScale was used to monitor a credit scoring model. The tool detected a drift in the model's predictions, where it began disproportionately favoring certain demographic groups. This early detection allowed the institution to intervene, retraining the model with adjusted data inputs to restore fairness and transparency.

Monitoring tools like Microsoft's Fairlearn dashboard also play a critical role in identifying transparency issues. Fairlearn is designed to assess and mitigate fairness concerns in machine learning models by evaluating model outputs across different demographic groups. A real-world application of Fairlearn involved a healthcare provider using a model to predict patient readmission rates. The tool revealed that the model's predictions were biased against older patients, leading to a higher likelihood of readmission for this group. By leveraging Fairlearn, the healthcare provider was able to identify and address this bias, recalibrating the model to ensure more equitable treatment across all patient demographics.

Updates to ML models are another critical aspect of maintaining transparency and interpretability. As new data becomes available or as requirements evolve, models may need to be updated or retrained to reflect these changes. It is important to document and communicate any modifications to the model, including changes in its architecture, features, or training data. This documentation should be comprehensive and include explanations of why changes were made and how they impact the model's transparency and interpretability. By providing clear and detailed records of model updates, organizations can ensure that stakeholders are informed and can understand how the model's behavior has been affected. Additionally, organizations should establish procedures for validating and verifying the updated model to ensure that it continues to meet transparency and interpretability standards.

Active stakeholder engagement is a vital component of integrating and monitoring transparency practices. Engaging with stakeholders, including end users, regulatory bodies, and data scientists, helps ensure that transparency and interpretability practices are aligned with the needs and expectations of all parties involved. Regular communication with stakeholders can provide valuable feedback on the effectiveness of transparency practices and identify areas for improvement. For example, user feedback on the clarity of model explanations can help guide enhancements to interpretability techniques. Additionally, involving stakeholders in the monitoring process can help ensure that any issues related to transparency are promptly addressed and that the model remains accountable to its users.

Stakeholder engagement also involves educating and training users on the transparency practices and interpretability techniques used in the model. Providing training sessions and resources can help users understand how to interpret model outputs and explanations effectively. This education is crucial for ensuring that users can make informed decisions based on the model's predictions and insights. Additionally, engaging stakeholders in discussions about the ethical implications and potential biases of the model can help foster a culture of transparency and accountability. By actively involving users in the transparency process, organizations can build trust and ensure that the model's decisions are understood and accepted.

Implementing transparency practices and monitoring their effectiveness also requires a strong governance framework. Organizations should establish clear policies and procedures for integrating transparency practices into the ML lifecycle, including guidelines for documentation, monitoring, and stakeholder engagement. This governance framework should outline the roles and responsibilities of various stakeholders, including data scientists, model developers, and compliance officers. By creating a structured approach to transparency and interpretability, organizations can ensure that best practices are consistently applied and that any issues are addressed in a timely manner.

Organizations should leverage advanced tools and technologies to support the integration and monitoring of transparency practices. For example, machine learning operations (MLOps) platforms can provide automated monitoring and reporting features that help track model performance and interpretability in real time. These tools can also facilitate the documentation of model changes and updates, making it easier to maintain transparency and accountability. For instance, DataRobot's MLOps platform offers comprehensive monitoring and governance features that help organizations track model performance, detect anomalies, and maintain transparency. In a case study involving a telecommunications company, DataRobot's platform was used to monitor an ML model designed to predict customer churn. The platform's real-time monitoring capabilities enabled the company to quickly identify and address a drop in model accuracy, which was traced back to changes in customer behavior due to a new marketing campaign. By swiftly updating the model and documenting the changes, the company was able to maintain transparency and continue delivering accurate predictions.

Regular audits and reviews are also important for assessing the effectiveness of transparency practices. Conducting periodic audits can help identify any gaps or shortcomings in the current practices and provide recommendations for improvements. These audits should include a review of documentation, monitoring processes, and stakeholder engagement efforts to ensure that they are meeting the established transparency standards. By regularly evaluating transparency practices, organizations can ensure that they are continuously improving and adapting to changes in the ML landscape.

The integration and monitoring of transparency practices are essential for maintaining model transparency and interpretability in ML data governance. By implementing continuous monitoring and updates, actively engaging with stakeholders, and establishing a robust governance framework, organizations can ensure that their AI systems remain transparent, understandable, and accountable. These practices not only help build trust and confidence among users but also ensure that models are used responsibly and ethically. Through ongoing evaluation and enhancement of transparency practices, organizations can navigate the complexities of AI and uphold the highest standards of transparency and interpretability.

Regulatory and Ethical Considerations

Regulatory and ethical considerations are integral to the best practices for model transparency and interpretability in machine learning (ML) data governance. Ensuring that AI systems are both compliant with legal standards and aligned with ethical principles is crucial for fostering trust, safeguarding rights, and maintaining accountability. As machine learning models become increasingly prevalent in decision-making processes across various sectors, adhering to regulatory requirements and addressing ethical implications become essential for responsible and transparent AI governance.

Compliance with legal standards is a fundamental aspect of implementing best practices for model transparency and interpretability. Regulations such as the General Data Protection Regulation (GDPR) in the European Union, the California Consumer Privacy Act (CCPA) in the United States, and various other regional and national data protection laws impose strict requirements on how personal data is collected, processed, and used. These regulations often include mandates for data transparency, such as the right to explanation, which grants individuals the ability to understand and challenge automated decisions made by AI systems. To comply with these standards, organizations must ensure that their models are designed and documented in a way that allows for clear explanations of how decisions are made. This includes maintaining detailed records of model development processes, data sources, and decision criteria, as well as providing accessible explanations of model outputs to users.

Another crucial aspect of regulatory compliance is adherence to industry-specific regulations that govern the use of AI and ML technologies. For example, in sectors such as finance, healthcare, and criminal justice, there are additional regulatory requirements that address the use of AI in high-stakes decisions. Financial institutions must comply with regulations that mandate transparency in credit scoring models, while healthcare providers must follow guidelines that ensure the interpretability of diagnostic tools. Organizations operating in these regulated industries must not only meet general data protection and privacy requirements but also ensure that their AI systems comply with sector-specific standards. This involves understanding and integrating relevant regulations into the model development and deployment processes, as well as engaging with regulatory bodies to stay informed about evolving requirements.

Ethical implications are equally important when considering model transparency and interpretability. Ethical considerations in AI involve addressing issues such as fairness, bias, and accountability. Ensuring that AI systems operate transparently and interpretably helps to mitigate the risks of bias and discrimination. For example, by making model decisions more understandable, organizations can identify and address potential biases in the data or decision-making process. This is particularly important in areas such as hiring, lending, and law enforcement, where biased decisions can have significant social and legal consequences. Organizations must implement practices that promote fairness and equity, including regular audits of model performance, transparency in data handling, and mechanisms for addressing ethical concerns.

The ethical use of AI also requires that organizations consider the broader societal impact of their models. Transparency and interpretability are not just about compliance but also about fostering trust and ensuring that AI systems are used in a way that aligns with societal values. This includes engaging with diverse stakeholders, including affected communities, to understand their concerns and expectations regarding AI systems. Organizations should strive to develop AI models that are not only effective but also respectful of human rights and ethical principles. This involves incorporating ethical considerations into every stage of the model lifecycle, from design and development to deployment and monitoring.

Addressing ethical implications involves developing and adhering to ethical guidelines and frameworks for AI. Many organizations and professional bodies have established ethical principles for AI, such as fairness, accountability, and transparency. Adopting these guidelines and incorporating them into organizational policies can help ensure that AI systems are developed and used in a responsible manner. Organizations should also provide training and resources to their teams to promote ethical practices and raise awareness about the potential ethical challenges associated with AI.

The interplay between regulatory compliance and ethical considerations often necessitates a proactive approach to model transparency and interpretability. Organizations must not only meet existing legal requirements but also anticipate and address emerging ethical issues. This proactive approach includes staying informed about regulatory changes, participating in industry discussions, and adapting practices to address new ethical challenges. By taking a forward-thinking approach, organizations can better navigate the complexities of AI governance and ensure that their models remain transparent, fair, and accountable.

Transparency and interpretability can enhance organizational accountability by providing mechanisms for auditing and reviewing AI systems. Regular audits and evaluations can help identify and rectify any issues related to compliance or ethical standards. By maintaining clear documentation of model development processes, decision criteria, and performance metrics, organizations can facilitate external reviews and demonstrate their commitment to responsible AI practices. This transparency in auditing processes not only helps to ensure regulatory compliance but also reinforces trust among stakeholders and the public.

Policies and Procedures for Model Transparency and Interpretability

This section explores the policy values that can anchor an organization's efforts to enhance the transparency and interpretability of their models.

Comprehensive Documentation as a Foundation

Comprehensive documentation is the cornerstone of transparency in machine learning. Effective policies should mandate detailed documentation at every stage of the ML lifecycle, from data collection to model deployment and monitoring. This documentation should provide a clear and accessible record of all processes, decisions, and assumptions made during the development and operation of an ML model.

The first step in this documentation process involves recording data sources and collection methods. This includes documenting where and how data was obtained, the quality and completeness of the data, and any preprocessing steps applied before the data was used for training. For instance, organizations should keep detailed records of the origin of the data, including any third-party sources, and provide justifications for selecting specific datasets. Additionally, preprocessing procedures, such as data cleaning, normalization, and augmentation, should be meticulously documented to offer insight into how the data was prepared for training.

Next, model architecture and design choices should be thoroughly documented. This includes outlining the structure of the model, such as the layers, nodes, activation functions, and any special configurations used. Such information helps stakeholders understand the model's complexity and the rationale behind design decisions. Training documentation should also cover hyperparameter settings, training duration, and methods used for tuning and optimization. By recording these details, organizations ensure that the model's development process is transparent and reproducible.

Finally, the documentation should include detailed records of model evaluation procedures. This involves documenting the metrics used to assess model performance, such as accuracy, precision, recall, and other relevant indicators. It is also important to record any cross-validation or split-testing techniques employed to evaluate the model's generalization capabilities. This practice ensures that stakeholders can understand how performance was measured and verified, which is critical for maintaining trust and accountability.

413

Continuous Monitoring and Model Updates

While comprehensive documentation lays the groundwork for transparency in machine learning (ML) models, the real challenge lies in maintaining this transparency throughout the model's lifecycle. Continuous monitoring and regular updates are crucial components of this ongoing process. However, these practices are not without their challenges. Organizations often face several pitfalls when trying to maintain transparency through continuous monitoring and updates, and overcoming these obstacles requires a strategic and thoughtful approach.

One common challenge in continuous monitoring is the risk of model drift, which occurs when the statistical properties of the input data change over time, leading to a deterioration in model performance. For example, a retail company might deploy an ML model to predict customer demand based on historical sales data. Over time, changes in consumer behavior, market trends, or external factors such as economic shifts can cause the model's predictions to become less accurate. This drift can compromise both the accuracy and transparency of the model, as stakeholders may struggle to understand why the model's predictions are deviating from expectations. To mitigate this risk, organizations need to implement robust monitoring tools that can detect early signs of drift. Tools like Amazon SageMaker Model Monitor provide real-time insights into data distribution and model predictions, enabling organizations to identify and address drift before it significantly impacts model performance.

Another challenge is the introduction of bias during model updates. When a model is retrained with new data, there is a risk that the updated model may inadvertently introduce or amplify biases that were not present in the original version. For instance, an HR department using an ML model for hiring might update the model with recent data that reflects current hiring trends. If this new data includes implicit biases—such as a preference for candidates from certain backgrounds or demographics—the updated model might begin to favor these groups, leading to biased hiring decisions. To prevent this, organizations must rigorously evaluate the fairness and bias metrics of the updated model before deployment. Techniques such as adversarial debiasing or the use of fairness constraints during training can help ensure that the model remains fair and transparent even after updates.

Transparency challenges also arise when models are updated frequently, especially in dynamic environments where rapid changes are necessary. For example, in financial markets, ML models used for trading or risk assessment might require frequent updates to reflect the latest market conditions. However, with each update, the complexity of maintaining transparency increases, as stakeholders need to understand not only how the model operates but also how each update affects its behavior. This can lead to confusion and a loss of trust if updates are not communicated clearly and effectively. To address this, organizations should establish clear and detailed documentation processes that record every change made to the model, including the rationale behind updates and their expected impact on transparency. This documentation should be easily accessible to all stakeholders, ensuring that they can track the evolution of the model and maintain confidence in its decisions.

The validation and verification of models before and after updates pose significant challenges. Ensuring that an updated model meets transparency and interpretability standards requires thorough testing in real-world scenarios, which can be both time-consuming and resource-intensive. For instance, in the healthcare sector, an ML model used for diagnosing diseases might be updated with new patient data or diagnostic techniques. Before deploying the updated model, it must undergo extensive testing to ensure that its predictions are accurate, fair, and interpretable. This process might involve running the model on a test dataset that simulates real-world conditions, comparing its outputs with those of the previous version, and conducting user testing with healthcare professionals to assess the interpretability of the new predictions. Organizations must allocate sufficient resources and time for these validation processes, as shortcuts can lead to significant transparency issues down the line.

Another pitfall in maintaining transparency during continuous updates is the potential loss of institutional knowledge. Over time, as models are updated and new versions are deployed, the individuals who originally developed and understood the model may leave the organization or move to different roles. This can create a knowledge gap, where the current team lacks a deep understanding of how the model was originally designed and how it has evolved. This gap can make it difficult to maintain transparency, as the rationale behind past decisions and updates may be lost. To overcome this, organizations should prioritize comprehensive knowledge transfer and documentation practices. For instance, by using version control systems like Git for model code and configurations, teams can ensure that all changes are recorded and can be easily reviewed by new team members. Additionally, regular training sessions and workshops can help keep the entire team up-to-date on the model's history and current state, ensuring that transparency is maintained across different iterations.

Organizations may struggle with the challenge of balancing the need for continuous updates with the requirement for regulatory compliance. In sectors such as finance or healthcare, models are often subject to strict regulations that require them to be transparent, explainable, and fair. However, frequent updates can complicate compliance, as each change must be thoroughly documented and justified to meet regulatory standards. For example, a financial institution updating its credit scoring model must ensure that the updated version complies with regulations such as the Fair Credit Reporting Act (FCRA), which mandates that consumers have the right to understand how their credit scores are determined. To manage this, organizations should establish a governance framework that integrates regulatory compliance into the update process. This might involve creating a cross-functional team that includes compliance officers, data scientists, and legal experts to review and approve all model updates. By embedding compliance considerations into the update process, organizations can maintain transparency while ensuring that they meet all regulatory requirements.

Continuous monitoring and model updates are essential for maintaining transparency and interpretability in ML models, but they are fraught with challenges. Organizations must be proactive in identifying and addressing potential pitfalls, such as model drift, bias introduction, and knowledge loss, to ensure that their models remain transparent, fair, and trustworthy over time. By implementing robust monitoring tools, establishing clear documentation processes, and prioritizing validation and verification, organizations can navigate these challenges and uphold the highest standards of transparency throughout the model's lifecycle.

Stakeholder Engagement and Communication

Engaging with stakeholders is a cornerstone of effective policies and procedures for model transparency and interpretability. The involvement of stakeholders, including users, regulators, and data scientists, is crucial in ensuring that transparency practices meet the needs and expectations of those affected by a model's decisions. To achieve meaningful engagement, organizations must not only maintain regular communication with stakeholders but also implement strategies that effectively convey complex transparency issues, particularly to non-technical stakeholders.

One of the most significant challenges in stakeholder engagement is bridging the gap between the technical complexity of machine learning models and the understanding of non-technical stakeholders. These stakeholders, such as business leaders, customers, or regulatory officials, may not have the technical expertise to grasp the intricacies of model algorithms, data processing, or interpretability techniques. Therefore, organizations need to develop tailored communication strategies that demystify these complexities without oversimplifying the critical details.

A practical strategy for communicating complex transparency issues involves the use of visual aids and analogies that relate to familiar concepts. For instance, when explaining how a model like a neural network makes decisions, an analogy comparing the model to a human decision-making process can be helpful. Just as a person might weigh different factors before making a decision, a neural network evaluates various inputs to arrive at an output. Visual tools such as flowcharts, decision trees, or even simple diagrams can illustrate how different features influence a model's decision, making the process more tangible for non-technical audiences. For example, using a decision tree visualization to show how a model arrives at a loan approval decision can help stakeholders understand the factors that play a crucial role in the outcome, such as income level, credit score, and employment history.

Another effective communication strategy is to translate technical terms and concepts into plain language. For instance, instead of discussing "feature importance scores" in a technical context, it can be more impactful to describe how the model "considers certain factors more heavily than others" when making predictions. This approach helps stakeholders relate to the model's functionality in a way that aligns with their everyday experiences. For example, in a healthcare setting, rather than explaining that a model uses a "random forest algorithm" to predict patient outcomes, it might be more effective to describe the model as "a system that compares multiple different scenarios to determine the most likely outcome for a patient, similar to how a doctor might consider multiple symptoms before diagnosing a condition."

Organizations can also leverage interactive tools that allow stakeholders to engage directly with the model and explore its decision-making process. For example, an interactive dashboard that visualizes how changes in input variables affect the model's predictions can empower stakeholders to experiment with the model and see firsthand how it works. This type of engagement not only demystifies the model's inner workings but also builds trust by allowing stakeholders to verify that the model behaves in a transparent and consistent manner. A case in point could be a financial institution using an interactive dashboard to demonstrate to regulators how changes in a customer's financial profile, such as a sudden increase in income or a decrease in debt, would affect their credit score prediction. This hands-on experience can make the model's decision-making process more relatable and understandable to those who might otherwise find the model's complexity intimidating.

Engaging stakeholders effectively also involves creating opportunities for two-way communication. Instead of merely presenting information about the model, organizations should actively solicit feedback and involve stakeholders in discussions about transparency practices. One way to facilitate this is through workshops or focus groups where stakeholders can express their concerns, ask questions, and suggest improvements. For example, an organization deploying an ML model in a public sector context might hold a series of workshops with community leaders and public officials to gather their input on the model's transparency. These sessions can help identify areas where the model's outputs might be unclear or where its decision-making process could be better explained. By incorporating this feedback, the organization not only improves the model's transparency but also strengthens its relationship with the community.

Incorporating case studies or real-world examples into communication efforts can also enhance stakeholder understanding. Describing how similar models have been used in other organizations or industries can provide context and demonstrate the practical implications of transparency practices. For example, explaining how an ML model in healthcare was successfully updated to address biases that were initially overlooked can help stakeholders appreciate the importance of ongoing monitoring and transparency in their own context. Such examples make the abstract concept of transparency more concrete, showing stakeholders how it directly impacts outcomes and decision-making in real-world scenarios.

Another key aspect of stakeholder engagement is the use of storytelling to convey the implications of transparency practices. Storytelling can humanize the impact of machine learning models, making the consequences of decisions more relatable. For instance, rather than simply stating that a model's decision-making process is transparent, an organization might tell the story of how a transparent model helped prevent a discriminatory outcome in a hiring process, leading to a fairer and more diverse workplace. This approach can be particularly powerful in communicating with non-technical stakeholders, as it frames transparency not just as a technical requirement but as a value that aligns with broader ethical and social goals.

Effective stakeholder engagement also requires clear communication channels that are accessible to all parties involved. Organizations should establish dedicated platforms for ongoing dialogue between model developers, users, and regulatory bodies. This might include regular updates on the model's performance, changes to its architecture, or explanations of how it handles new types of data. For example, a quarterly newsletter or an online portal could provide updates on the model's transparency practices, recent changes, and upcoming enhancements. These updates should be written in a way that is accessible to non-technical readers, avoiding jargon and focusing on the practical implications of the information being shared. By keeping stakeholders informed through clear and consistent communication, organizations can ensure that any concerns are addressed promptly and that trust in the model remains strong.

Compliance with Legal Standards

Compliance with legal standards is a critical consideration in developing policies and procedures for model transparency and interpretability. Various data protection and privacy laws, such as the General Data Protection Regulation (GDPR) in the European Union and the California Consumer Privacy Act (CCPA) in the United States, impose strict requirements on how personal data is collected, processed, and used in machine learning models. These regulations often include mandates for data transparency and the right to explanation, which grant individuals the ability to understand and challenge automated decisions made by AI systems.

To comply with these legal standards, organizations must ensure that their models are designed and documented in a way that allows for clear explanations of how decisions are made. This includes maintaining detailed records of data sources, preprocessing steps, model architecture, and evaluation metrics, as well as providing accessible explanations of model outputs to users. Additionally, organizations should establish procedures for responding to data subject requests related to transparency, such as requests for information about how personal data was used in the model's decision-making process.

In addition to data protection and privacy laws, organizations must also comply with industry-specific regulations that govern the use of AI and ML technologies. For example, in the financial sector, regulations may require transparency in credit scoring models, while in healthcare, guidelines may mandate the interpretability of diagnostic tools. Policies should outline procedures for ensuring compliance with these sector-specific standards, including conducting regular audits and engaging with regulatory bodies to stay informed about evolving requirements.

Ethical Considerations and Bias Mitigation

Ethical considerations are central to the development of policies and procedures for model transparency and interpretability. Ensuring that AI systems operate transparently and interpretably helps mitigate the risks of bias, discrimination, and other ethical concerns. Organizations must implement practices that promote fairness and equity, including regular audits of model performance, transparency in data handling, and mechanisms for addressing ethical concerns.

One of the key ethical considerations in ML is the potential for bias in model decisions. Bias can arise from various sources, including biased data, model design choices, and unintended interactions between features. To address this issue, organizations should implement procedures for detecting and mitigating bias throughout the ML lifecycle. This includes conducting regular bias audits, using fairness metrics to assess model performance, and applying techniques such as adversarial testing to identify and address potential sources of bias.

Ethical considerations also involve ensuring that AI systems are used in a manner that aligns with societal values and respects human rights. This includes engaging with diverse stakeholders, including affected communities, to understand their concerns and expectations regarding AI systems. Organizations should strive to develop AI models that are not only effective but also respectful of human rights and ethical principles. This involves incorporating ethical considerations into every stage of the model lifecycle, from design and development to deployment and monitoring.

Regular Audits and Reviews

Regular audits and reviews are indispensable for ensuring that transparency and interpretability practices within machine learning (ML) models are maintained and continually improved. These audits serve as a critical checkpoint, assessing whether an organization's practices align with established transparency standards and identifying areas that require enhancement. A thorough audit process not only ensures compliance with current regulations but also prepares the organization to address emerging challenges in the evolving field of AI governance.

The frequency of audits is a key factor in maintaining effective transparency practices. Organizations must determine an audit schedule that balances the need for thorough oversight with the practicalities of resource allocation. For models deployed in high-stakes environments, such as healthcare or finance, where decisions can have significant consequences, audits may be necessary on a quarterly basis to ensure that the model's behavior remains consistent with transparency and fairness standards. In less critical applications, biannual or annual audits might suffice. The frequency of these audits should be explicitly defined in organizational policies, taking into account the potential risks associated with the model's use, the pace of change in the operational environment, and the availability of new data or explainability techniques.

A comprehensive audit must include several key components to ensure that all aspects of transparency and interpretability are adequately covered. First, the audit should evaluate the documentation associated with the model. This includes reviewing how well the development process, data sources, feature selection, and model updates have been documented. The documentation should clearly outline the rationale behind key decisions and demonstrate how these decisions align with transparency goals. For example, if a model was updated to improve accuracy, the documentation should explain how this update affects interpretability and what measures were taken to maintain or enhance transparency.

The audit should also assess the effectiveness of the ongoing monitoring processes. This involves examining whether the tools and methods used to monitor the model's performance and fairness are functioning as intended. For instance, if the organization uses tools like SHAP or LIME to provide model explanations, the audit should verify that these tools are being applied correctly and consistently and that their outputs are being appropriately communicated to stakeholders. The audit might also review whether the monitoring tools have successfully detected any instances of model drift or bias and whether corrective actions were promptly implemented.

Stakeholder engagement is another critical area that should be evaluated during the audit. The audit should review how well the organization has communicated with stakeholders about the model's transparency practices. This includes assessing whether stakeholders have been provided with adequate training and resources to understand and interpret the model's outputs. The audit should also consider the feedback mechanisms in place, determining whether stakeholder concerns have been effectively addressed and whether this feedback has led to meaningful improvements in transparency practices.

To ensure a comprehensive audit, organizations should develop a checklist or set of guidelines that auditors can follow. This checklist should cover all aspects of transparency and interpretability, including documentation, monitoring, stakeholder engagement, and compliance with relevant regulations. For example, the checklist might include items such as "Review model update logs for completeness and clarity," "Verify the accuracy and relevance of feature importance explanations," and "Assess stakeholder feedback and the organization's response to it." By using a standardized checklist, organizations can ensure that audits are thorough and consistent, reducing the risk of overlooking critical transparency issues.

The audit process should also be forward-looking, considering not only current compliance but also the organization's preparedness for future challenges. As new explainability techniques or regulatory requirements emerge, the audit should assess whether the organization's practices are flexible enough to adapt. For example, if a new regulation requires that all AI-driven decisions be explainable to end users in plain language, the audit should evaluate whether the organization's current tools and processes can meet this requirement and what steps may be necessary to achieve compliance.

The outcomes of the audit should be documented and communicated to all relevant stakeholders. This includes not only reporting the findings but also outlining the corrective actions that will be taken to address any identified gaps or weaknesses. Regular audits should be seen as an opportunity for continuous improvement rather than merely a compliance exercise. By fostering a culture of transparency and accountability, organizations can ensure that their ML models remain not only effective but also fair, interpretable, and aligned with both current and future standards.

Training and Education for Transparency

Training and education are foundational to ensuring that transparency and interpretability practices are effectively implemented and maintained within machine learning (ML) models. These efforts are essential not only for data scientists and model developers but also for end users and other non-technical stakeholders who interact with ML outputs. By fostering a deep understanding of transparency practices across all levels of the organization, companies can enhance trust in their models and ensure that these models are used responsibly and effectively.

For data scientists and model developers, training programs should delve into the technical aspects of transparency and interpretability. One highly regarded resource is the "Explainable AI" course offered by Coursera in partnership with the University of California, Berkeley. This course provides in-depth knowledge of various explainability techniques, such as SHAP (SHapley Additive exPlanations), LIME (Local Interpretable Model-agnostic Explanations), and model-agnostic methods, equipping data scientists with the tools they need to make their models more transparent. Participants learn not only how to implement these techniques but also how to evaluate their effectiveness in different contexts. This technical foundation is critical for ensuring that models are not only accurate but also understandable to a broad audience.

To complement these technical skills, organizations should also emphasize the ethical and legal implications of transparency in their training programs. The AI Ethics course offered by Harvard University is a valuable resource in this regard. This program explores the intersection of AI, ethics, and law, helping data scientists and model developers understand the broader impact of their work. By engaging with case studies and real-world examples, participants learn how to navigate complex ethical dilemmas and ensure that their models comply with legal standards for transparency and fairness. This holistic approach to training helps data scientists appreciate the significance of their work beyond the technical realm, fostering a commitment to ethical AI practices.

For end users and other non-technical stakeholders, training should focus on demystifying ML models and providing practical guidance on how to interpret model outputs. The AI for Everyone course by Andrew Ng, available on Coursera, is an excellent resource that introduces non-technical audiences to the fundamentals of AI and ML. This course helps stakeholders understand the basic principles of how models work, what to expect from them, and how to use their outputs responsibly. By using real-world examples and avoiding technical jargon, the course makes complex concepts accessible to a broader audience, empowering non-technical stakeholders to engage confidently with ML models.

Workshops and hands-on training sessions are also crucial for non-technical stakeholders. Organizations might consider hosting interactive sessions where participants can explore how changes in input data affect model outputs. For instance, a healthcare organization could conduct workshops for clinicians, demonstrating how an ML model predicts patient outcomes based on various health metrics. By allowing clinicians to manipulate input variables and observe the resulting changes in predictions, these workshops can help them understand the model's decision-making process. This hands-on experience is invaluable for building trust in the model and ensuring that its outputs are used effectively in decision-making.

Another effective strategy for educating stakeholders is the use of tailored user guides and documentation. For example, Google's Model Card Toolkit provides a standardized way to document and communicate key information about ML models. These model cards include details on the model's intended use, performance metrics, and limitations, presented in a format that is accessible to non-technical users. By providing clear and concise documentation, organizations can help stakeholders understand the capabilities and constraints of the models they interact with, reducing the risk of misinterpretation and misuse.

Organizations should also consider offering ongoing education and training to keep stakeholders up to date with the latest developments in transparency and interpretability. The fast-paced nature of AI and ML means that new techniques, tools, and regulatory requirements are constantly emerging. Regularly scheduled training sessions, webinars, and access to online resources can help ensure that all stakeholders remain informed about the best practices in the field. For instance, subscribing to platforms like O'Reilly, which offers a wide range of technical books, courses, and live training sessions, can be an effective way to keep teams current on the latest trends and advancements in AI transparency.

Moreover, organizations can benefit from creating internal learning communities or forums where stakeholders can share knowledge, ask questions, and discuss challenges related to transparency and interpretability. These communities can be supported by tools like Slack or Microsoft Teams, where dedicated channels can be set up to facilitate ongoing dialogue about transparency issues. By fostering a culture of continuous learning and open communication, organizations can ensure that transparency remains a priority throughout the ML lifecycle.

Incorporating feedback mechanisms into training programs is another important aspect of effective education for transparency. After training sessions or workshops, organizations should gather feedback from participants to assess the effectiveness of the training and identify areas for improvement. This feedback can help tailor future training sessions to better meet the needs of stakeholders. For example, if non-technical users express difficulty in understanding certain aspects of model interpretability, the organization might develop additional resources or simplify the language used in future sessions to enhance comprehension.

Training and education are not one-time efforts but ongoing processes that must evolve alongside the advancements in AI and ML. By investing in comprehensive and continuous training programs, organizations can build a knowledgeable and engaged stakeholder base that understands the importance of transparency and is equipped to uphold these standards in practice. Through a combination of technical courses, ethical training, practical workshops, and accessible documentation, organizations can ensure that all stakeholders—from data scientists to end users—are empowered to contribute to a transparent and interpretable ML ecosystem.

Challenges and Trade-offs in Transparency

Transparency and interpretability are fundamental to responsible AI governance, yet achieving these goals presents significant challenges and trade-offs, particularly in the context of complex machine learning models. As organizations strive to balance the demands of accuracy and transparency, they encounter numerous obstacles that require thoughtful strategies and robust frameworks to navigate successfully.

One of the primary challenges lies in reconciling the complexity of modern machine learning models with the need for transparency. Models such as deep neural networks are renowned for their ability to capture intricate patterns and relationships within data, often leading to superior predictive performance. However, this complexity comes at the cost of interpretability, making it difficult for stakeholders to understand how these models arrive at their decisions. For instance, a deep learning model used to predict financial risks might be highly accurate, but the opacity of its decision-making process can hinder trust and raise concerns among regulators and users. To address this, organizations must develop strategies that allow them to leverage the power of complex models while still maintaining a level of transparency that meets both ethical and regulatory standards.

One effective strategy is the use of hybrid models that combine the strengths of both complex and simple models. For example, an organization might deploy a deep learning model for its predictive capabilities but use a simpler, more interpretable model, such as a decision tree or logistic regression, to provide explanations for the model's outputs. This approach enables the organization to maintain high accuracy in predictions while offering transparency in decision-making. For instance, a healthcare provider using a complex model to predict patient outcomes could employ a decision tree to explain the factors influencing a particular prediction, such as age, medical history, and treatment options. This explanation can be communicated to clinicians and patients in a way that is understandable and actionable, thereby enhancing trust and facilitating informed decision-making.

Organizations can also adopt a framework that prioritizes transparency in high-stakes decisions while allowing for greater complexity in less critical areas. This framework involves categorizing decisions based on their impact and the need for transparency. For high-stakes decisions, such as those involving legal, financial, or healthcare outcomes, the framework would prioritize the use of models that are inherently more interpretable, even if this means sacrificing some degree of accuracy. In contrast, for decisions that are less impactful or where transparency is not as critical, the framework might permit the use of more complex models with less interpretability. By tailoring the model selection to the specific context of the decision, organizations can ensure that they maintain appropriate levels of transparency where it matters most.

Another significant challenge is ensuring that transparency practices are scalable across a diverse range of models and use cases. As organizations deploy various models in different domains, maintaining consistent transparency standards can become increasingly difficult. A key strategy to overcome this challenge is the implementation of standardized guidelines that can be applied across different models and use cases. These guidelines should outline the minimum requirements for transparency, such as the use of model documentation, the application of explainability techniques, and the regular monitoring of model behavior. For instance, an organization might mandate that all models, regardless of complexity, must be accompanied by a model card—a structured documentation format that provides information about the model's purpose, performance, and limitations. This ensures that transparency practices are consistently applied, even as models vary in complexity and application.

To further enhance scalability, organizations can employ automated tools that facilitate the application of transparency practices across different models. Tools like IBM's AI Fairness 360 and Microsoft's Fairlearn provide frameworks for evaluating and mitigating bias in machine learning models, which is a critical aspect of transparency. These tools allow organizations to assess the fairness and interpretability of their models in a systematic and repeatable manner, ensuring that transparency standards are met across all deployed models. For example, a financial institution using multiple models for credit scoring might use these tools to ensure that each model is fair and transparent, regardless of the differences in their underlying algorithms.

Another challenge in maintaining transparency is the dynamic nature of machine learning models, which often require regular updates to reflect new data or changing conditions. Each update has the potential to introduce new complexities or obscure previously transparent decision-making processes. To navigate this challenge, organizations should establish a robust governance framework that includes procedures for documenting and communicating model updates. This framework should ensure that any changes made to a model are thoroughly documented, including the reasons for the update, the changes in the model's architecture or inputs, and the expected impact on transparency and interpretability. For example, if an organization updates its customer segmentation model to incorporate new behavioral data, the documentation should explain how this data was integrated and how it affects the model's decision-making process.

A well-defined governance framework should also include regular audits and reviews of transparency practices. These audits should assess whether the model continues to meet transparency standards following updates and whether any new risks or challenges have emerged. For instance, an audit might reveal that a model update has inadvertently reduced the interpretability of the model's outputs, prompting a review and potential revision of the update. By embedding transparency checks into the update process, organizations can ensure that their models remain transparent and trustworthy over time.

The trade-offs between transparency and accuracy also present significant challenges, particularly when it comes to stakeholder trust. Stakeholders, including regulators, customers, and internal users, may have varying expectations regarding the transparency of machine learning models. While some stakeholders may prioritize accuracy and performance, others may be more concerned with understanding how decisions are made, especially in contexts where those decisions have significant personal or financial implications. Navigating these divergent expectations requires a careful balancing act where organizations must communicate the reasons behind their model choices and the trade-offs involved.

One effective way to manage stakeholder expectations is through transparent communication and education. Organizations should proactively engage with stakeholders to explain the rationale behind the use of certain models and the trade-offs between transparency and accuracy. For example, a company using a complex model for loan approvals might hold informational sessions with customers and regulators to explain how the model works, why it was chosen, and what measures are in place to ensure fairness and transparency. By involving stakeholders in these discussions, organizations can build trust and foster a greater understanding of the complexities involved in machine learning.

Organizations should consider adopting transparency frameworks that are adaptable to different regulatory environments and industry standards. As regulations surrounding AI and machine learning continue to evolve, organizations must be prepared to adjust their transparency practices to comply with new requirements. A flexible transparency framework should allow organizations to quickly adapt to changes in the regulatory landscape while maintaining high standards of transparency and interpretability. For example, an organization operating in multiple jurisdictions might develop a framework that aligns with the most stringent regulatory standards, ensuring compliance across all regions while upholding a consistent level of transparency.

Navigating the challenges and trade-offs in transparency requires a strategic approach that balances the demands of accuracy, interpretability, and stakeholder trust. By adopting hybrid models, implementing scalable transparency practices, establishing robust governance frameworks, and engaging in proactive communication with stakeholders, organizations can overcome these challenges and maintain high standards of transparency in their machine learning models.

Summary

This chapter explores the intricacies of achieving transparency and interpretability in machine learning (ML) models, focusing on the delicate balance between these principles and model accuracy. It begins by emphasizing the importance of transparency and interpretability in responsible AI governance, noting the challenge that complex models like deep neural networks, while highly accurate, often lack transparency, making it difficult for stakeholders to fully grasp their decision-making processes.

To navigate these challenges, strategies for balancing accuracy with interpretability are discussed, including the use of hybrid models that combine the strengths of both simple and complex approaches. There's a strong emphasis on the importance of embedding transparency and interpretability into the ML development process from the start, ensuring that these considerations are integral to the design rather than added as an afterthought.

Continuous monitoring and regular updates are highlighted as essential practices for maintaining transparency over time. The discussion covers the risks of model drift and bias during updates and offers strategies to mitigate these issues effectively. Additionally, the significance of stakeholder engagement and clear communication is underscored, particularly when addressing complex transparency issues with non-technical audiences.

Regular audits and reviews are presented as critical tools for ensuring that transparency practices remain robust and evolve in response to new challenges. Guidelines for conducting thorough audits are provided, focusing on the evaluation of documentation, monitoring processes, and stakeholder engagement. The section wraps up by highlighting the importance of ongoing training and education to foster a culture of transparency within organizations, ensuring that all stakeholders are well-prepared to engage with ML models responsibly and effectively.

CHAPTER 7

Monitoring and Maintaining Machine Learning Systems

Effective monitoring and maintenance of machine learning (ML) systems is a crucial component of ensuring their reliability, robustness, and relevance over time. Given the dynamic nature of ML models and the environments in which they operate, ongoing oversight is necessary to detect and address issues such as model drift, bias, and performance degradation. This chapter explores the multifaceted processes involved in monitoring and maintaining ML systems from a data governance perspective. We will discuss the policies, best practices, challenges, and tools required to sustain ML models in production, ensuring that they continue to function effectively and ethically over time.

Establishing a Monitoring Framework

Establishing a monitoring framework for machine learning (ML) systems is crucial for ensuring that these systems continue to perform as intended and provide reliable, actionable insights over time. The complexity of ML models, coupled with the dynamic nature of the data they process, necessitates a comprehensive approach to monitoring that goes beyond basic oversight. Such a framework is not simply a collection of metrics and tools but a strategic, integrated process that aligns closely with the specific goals and operational environment of the ML system. This holistic approach ensures that every element of the ML lifecycle, from data ingestion to model inference and output, is continuously scrutinized for performance, stability, and alignment with business objectives.

© Aditya Nandan Prasad 2024
A. Nandan Prasad, *Introduction to Data Governance for Machine Learning Systems*,
https://doi.org/10.1007/979-8-8688-1023-7_7

The foundation of an effective monitoring framework lies in the careful identification and definition of key performance indicators (KPIs) that serve as benchmarks for evaluating the ML system's success. These KPIs must be meticulously chosen to reflect the specific tasks and objectives that the model is designed to achieve. For instance, in classification tasks, common KPIs include metrics like accuracy, precision, recall, and the F1 score. These metrics are critical in understanding how well the model distinguishes between different classes and how it balances false positives and false negatives. In contrast, regression models might rely on KPIs such as mean squared error or R-squared, which offer insights into the model's ability to predict continuous outcomes accurately.

The selection of KPIs should not be limited to traditional performance metrics. In applications where ML systems impact critical decisions—such as those related to human resources, finance, or customer relations—fairness and bias metrics become equally important. These metrics are essential for ensuring that the model's decisions do not disproportionately affect certain demographic groups, thereby maintaining ethical standards and regulatory compliance. For example, in a hiring algorithm, it would be important to monitor whether the model's predictions are biased against candidates of a certain gender or ethnicity. Including fairness metrics in the monitoring framework helps identify and mitigate such biases, ensuring that the ML system operates equitably across all user groups.

Once KPIs have been clearly defined, the next step involves implementing automated monitoring tools that are capable of real-time tracking. These tools play a crucial role in maintaining the continuous oversight of the ML system's performance. By continuously monitoring the defined metrics, these tools provide a dynamic and up-to-date view of the system's health. Real-time monitoring is particularly valuable in detecting anomalies—sudden deviations from expected behavior that could indicate underlying issues. For instance, if a model that typically maintains an accuracy rate of 90% suddenly drops to 70%, this could signal a problem that requires immediate investigation. Automated monitoring tools should be equipped with sophisticated anomaly detection algorithms that can identify such irregularities in real time, allowing for swift corrective action.

The ability to respond quickly to anomalies is further enhanced by the integration of alert systems within the monitoring framework. These alerts, when properly configured, can notify relevant stakeholders—such as data scientists, engineers, and business leaders—of significant deviations from expected performance metrics. For example, an alert could be triggered if there is a significant drop in model accuracy, a sudden change in data distribution, or a spike in prediction errors. The timely notification of such events allows for rapid diagnosis and resolution of issues, preventing them from escalating into more severe problems. This proactive approach to monitoring ensures that the ML system remains reliable and continues to deliver accurate and trustworthy results.

A robust monitoring framework must also encompass the data pipelines that feed into the ML system. Data quality is a critical determinant of model performance, and any issues in the data pipeline—such as missing values, outliers, or data drift—can have a direct impact on the accuracy and reliability of the model's outputs. Therefore, the monitoring framework should include tools and processes that continuously assess the quality of incoming data. For example, automated checks can be implemented to detect missing or inconsistent data, while statistical tests can identify shifts in data distributions that could affect model performance. By ensuring that the data entering the system is accurate, complete, and consistent, the monitoring framework helps maintain the integrity of the entire ML pipeline.

System stability is another key consideration in the design of a monitoring framework. ML systems often operate in dynamic environments where the underlying infrastructure—such as servers, databases, and networks—must be constantly monitored to ensure stability and responsiveness. Any instability in the system's infrastructure can lead to performance degradation, increased latency, or even downtime, all of which can have serious consequences, particularly in mission-critical applications. For example, in an e-commerce platform that uses ML to personalize customer experiences, a failure in the underlying infrastructure could lead to delays in product recommendations or incorrect pricing, negatively impacting customer satisfaction and sales. To prevent such scenarios, the monitoring framework should include tools that continuously track the performance of the system's infrastructure, allowing for the early detection and resolution of potential issues.

One of the challenges in maintaining a monitoring framework is ensuring that it remains effective as the ML system evolves. As models are retrained, new data is introduced, and the operational environment changes, the monitoring framework must be adaptable enough to accommodate these shifts. This requires regular reviews and updates to the monitoring processes and tools, ensuring that they remain aligned with the current state of the ML system. For example, as new KPIs are introduced or as the model's performance goals are adjusted, the monitoring framework must be updated to reflect these changes. Continuous improvement is therefore a key principle in maintaining a monitoring framework that can keep pace with the evolving needs of the ML system.

Stakeholder engagement is another critical element of a successful monitoring framework. Different stakeholders—ranging from data scientists and engineers to business leaders and compliance officers—have different needs and perspectives when it comes to monitoring ML systems. Effective communication and reporting mechanisms are therefore essential for ensuring that all stakeholders are kept informed of the system's performance and any issues that arise. For example, regular reports that summarize the monitoring results, highlight potential risks, and outline corrective actions can help build trust and ensure that everyone is aligned in maintaining the system's integrity. In cases where the ML system affects critical business operations, such as customer service or financial forecasting, timely and transparent communication of monitoring results is essential for maintaining stakeholder confidence.

Advanced technologies such as machine learning-based anomaly detection and predictive analytics can further enhance the effectiveness of the monitoring framework. These technologies leverage the power of machine learning to identify patterns and trends that might not be immediately apparent through conventional monitoring techniques. For example, predictive analytics can be used to forecast potential issues before they occur, allowing for preemptive action to be taken. Machine learning-based anomaly detection can identify subtle deviations from normal behavior that might indicate the early stages of a problem, even before it becomes apparent in the performance metrics. By incorporating these advanced tools into the monitoring framework, organizations can gain deeper insights into their ML systems and stay ahead of potential issues, ensuring that their models continue to deliver value without unexpected disruptions.

Integrating these tools into an existing ML pipeline requires careful planning and consideration. The first step is to evaluate the specific needs of the ML system and select the tools that best align with these needs. For example, if the primary concern is data quality, a tool that specializes in data validation and preprocessing might be the best fit. If the focus is on model performance, an automated monitoring platform that tracks key metrics and provides real-time alerts would be more appropriate. Once the tools have been selected, they must be integrated into the existing ML pipeline in a way that minimizes disruption and maximizes efficiency. This might involve setting up data feeds to the monitoring tools, configuring alert systems, and establishing protocols for responding to identified issues. Throughout this process, it is important to engage with stakeholders to ensure that the monitoring framework aligns with their expectations and provides the insights they need to make informed decisions.

As the ML system continues to evolve, the monitoring framework must also be regularly reviewed and updated to ensure that it remains effective. This might involve revisiting the selected KPIs to ensure that they still reflect the system's performance goals, updating the monitoring tools to accommodate new data sources or model architectures, and refining the alert systems to better address emerging issues. By adopting a continuous improvement approach, organizations can ensure that their monitoring framework remains robust and capable of supporting the ongoing success of their ML systems. This commitment to continuous improvement is essential for maintaining the long-term reliability and effectiveness of the ML system, allowing organizations to derive maximum value from their investments in machine learning technology.

Data Pipeline Monitoring

Data serves as the lifeblood of any machine learning (ML) system, providing the critical inputs necessary for generating reliable and actionable outputs. The process of monitoring the data pipeline, which encompasses the entire journey of data from its initial collection to its final utilization within ML models, is not merely a technical requirement; it is a strategic imperative. Ensuring data integrity, accuracy, and timeliness throughout this journey is essential for maintaining the overall effectiveness of ML models. Each stage in the data pipeline—whether it involves data collection, transformation, cleaning, or storage—presents potential points of failure. Issues such as data drift, quality degradation, or bottlenecks can significantly impact the performance of the ML models that rely on this data. Therefore, a comprehensive and proactive approach to monitoring the data pipeline is crucial for sustaining the efficacy and reliability of ML systems in production environments.

Addressing the Challenge of Data Drift

One of the most significant challenges in monitoring the data pipeline is addressing the phenomenon of data drift. Data drift occurs when the statistical properties of the input data change over time, leading to a potential mismatch between the data on which the model was trained and the data it encounters during inference. This drift can result from various factors, including shifts in user behavior, changes in market conditions, or the introduction of new data sources. For example, a model trained on consumer behavior data before the COVID-19 pandemic may not perform as expected when applied to data collected during or after the pandemic due to drastic changes in purchasing patterns and economic conditions.

To effectively combat data drift, organizations must implement advanced monitoring tools that continuously track data distributions and compare them against the baseline distributions used during model training. One widely used method for detecting data drift is the Kolmogorov–Smirnov test, a statistical test that measures the distance between two probability distributions. This test can identify shifts in the distribution of new data compared to the training data, signaling a potential drift that could affect model performance. For instance, a financial institution monitoring transaction data for fraud detection might use such tests to identify changes in transaction patterns, which could indicate a shift in consumer behavior or the emergence of new types of fraudulent activities. Upon detecting a significant drift, the monitoring system could trigger alerts to notify data scientists and engineers, prompting them to investigate and, if necessary, retrain the model with updated data. This real-time detection and response capability is critical for maintaining the accuracy and reliability of ML models in dynamic environments.

Ensuring Data Quality Throughout the Pipeline

Detecting data drift is just one aspect of a robust data pipeline monitoring strategy; ensuring the quality of the data is equally important. High-quality data is the foundation of any successful ML model, as the accuracy of the model's predictions is directly tied to the integrity of the data it processes. Data quality issues can arise from various sources, including incomplete data, outliers, and inconsistencies introduced during data collection or preprocessing. For example, an e-commerce company relying on ML models for personalized product recommendations might encounter data quality issues if customer profiles are missing key attributes or if there are discrepancies in transaction records due to system integration problems.

To mitigate these risks, organizations must implement automated data validation checks at multiple stages of the data pipeline. Automated data validation serves as a first line of defense against data quality issues. These checks can be tailored to the specific needs of the ML system, ensuring that only data meeting the necessary standards are allowed to enter the model. For instance, a healthcare provider using ML for predictive analytics might implement validation checks to ensure that all patient records include critical fields such as age, diagnosis, and treatment history. Similarly, the system might validate that numerical values such as blood pressure or glucose levels fall within medically acceptable ranges. By automating these checks, the organization can prevent poor-quality data from contaminating the model, thereby safeguarding the reliability of its predictions.

Data quality is not a static attribute, and even data sources that initially provide high-quality inputs can degrade over time. Continuous monitoring of data quality is therefore essential for identifying issues that may arise after the initial validation stage. For example, a telecommunications company might monitor call data records for inconsistencies or missing information that could affect the accuracy of its ML models for network optimization. Over time, data quality issues might emerge due to changes in the data collection process, such as the introduction of new sensors or software updates. Continuous monitoring tools, such as those that track the percentage of missing values or the frequency of outliers, can help detect these issues early, allowing the organization to take corrective action before the data quality degradation impacts the model's performance.

Advanced Techniques for Data Pipeline Monitoring

While basic monitoring techniques are essential for ensuring data integrity, more advanced methods can provide deeper insights and more robust protection against data-related issues. One such technique is the use of machine learning itself to monitor data pipelines. By applying ML algorithms to analyze patterns in data flow and quality metrics, organizations can predict potential failures or anomalies before they occur.

For instance, predictive analytics can be employed to forecast data pipeline performance based on historical data and usage patterns. In a cloud-based ML system, for example, predictive models might analyze past pipeline failures to identify trends or conditions that typically precede a failure, such as increased latency or resource contention. By recognizing these patterns early, the system can take preemptive actions, such as reallocating resources or rerouting data flows, to prevent a potential failure from disrupting the pipeline.

Another advanced approach involves the use of graph-based anomaly detection. In this method, the data pipeline is represented as a graph, with nodes representing data sources, transformation processes, and storage points, and edges representing data flows between these components. Anomalies can be detected by analyzing changes in the structure or behavior of the graph over time. For example, a sudden change in the volume of data flowing between two nodes—such as a significant increase in data transferred from a data source to a processing unit—might indicate a potential issue, such as a data surge caused by a system misconfiguration or an external event. By identifying these anomalies early, organizations can take corrective actions before they impact the ML models that rely on the data pipeline.

Real-time analytics platforms, such as **Apache Kafka** combined with **Apache Flink**, provide the infrastructure necessary for implementing such advanced monitoring techniques. Kafka's distributed streaming architecture allows for the continuous collection and processing of large volumes of data in real time, while Flink's powerful analytics capabilities enable the detection of patterns, anomalies, and trends in the data flow. By leveraging these tools, organizations can build sophisticated monitoring systems that not only detect and respond to issues as they arise but also predict and prevent potential problems before they affect the data pipeline.

Collaborative Efforts in Data Pipeline Monitoring

Effective data pipeline monitoring requires collaboration across different teams within the organization. Data engineers, data scientists, and IT professionals must work together to establish and maintain the monitoring framework, ensuring that it covers all aspects of the data pipeline, from ingestion and processing to model training and deployment. This collaborative approach is essential for creating a comprehensive monitoring system that addresses the specific needs and challenges of each stage in the data pipeline.

For example, in a large e-commerce company, cross-functional teams might be established to oversee the monitoring of data pipelines that feed into various ML models, such as recommendation engines, pricing algorithms, and fraud detection systems. Data engineers would focus on ensuring the smooth operation and scalability of the data pipelines, while data scientists would concentrate on the quality and relevance of the data being processed. IT professionals, on the other hand, would ensure that the necessary infrastructure and tools are in place to support real-time monitoring and anomaly detection. By fostering a culture of collaboration and shared responsibility, organizations can create a robust monitoring system that supports the long-term success of their ML initiatives.

Building a Sustainable Monitoring Framework

Building a sustainable monitoring framework for data pipelines involves not only the implementation of advanced tools and techniques but also the establishment of processes and protocols that ensure continuous improvement and adaptation. As ML systems evolve and new data sources are introduced, the monitoring framework must be regularly reviewed and updated to reflect these changes.

For instance, as an organization expands its ML capabilities and begins to incorporate new types of data—such as unstructured text or image data—into its models, the monitoring framework must be adapted to accommodate the unique challenges associated with these data types. This might involve the development of new validation checks, the integration of specialized anomaly detection algorithms, or the implementation of additional monitoring tools that can handle the increased complexity of the data pipeline.

Continuous improvement should be a core principle of the monitoring framework, with regular reviews and audits conducted to assess the effectiveness of the monitoring processes and identify areas for enhancement. These reviews should involve all stakeholders, including data engineers, data scientists, IT professionals, and business leaders, to ensure that the monitoring framework remains aligned with the organization's strategic goals and objectives.

Case Studies

The importance of data pipeline monitoring is underscored by the success stories of companies that have implemented comprehensive monitoring frameworks to ensure the reliability and performance of their ML models. Netflix and Airbnb are prime examples of organizations that have built sophisticated monitoring systems to maintain the integrity and timeliness of the data feeding into their ML models.

Netflix has developed an advanced data pipeline monitoring system that continuously tracks the quality and timeliness of the data feeding into its recommendation algorithms. The system is designed to detect anomalies in the data, such as unexpected spikes in viewing patterns or inconsistencies in user profiles, and trigger alerts for further investigation. For example, if a sudden surge in viewing activity is detected for a specific genre or show, the monitoring system would flag this as a potential anomaly. Data scientists at Netflix would then investigate to determine whether this spike is due to a legitimate trend—such as the release of a highly anticipated show—or if it might be caused by a data processing error. This proactive approach allows Netflix to maintain the accuracy of its recommendations, even as user behavior and content offerings evolve.

Similarly, Airbnb has implemented a comprehensive data pipeline monitoring system that includes real-time validation checks and anomaly detection tools. The system is designed to ensure that the data used to power its pricing algorithms and search ranking models is accurate, complete, and up-to-date. For instance, Airbnb's monitoring tools continuously track the quality of listing data, such as availability, pricing, and guest reviews. If any inconsistencies or errors are detected—such as a sudden drop in the availability of listings in a popular location—the system triggers an alert for further investigation. By continuously monitoring its data pipelines, Airbnb can quickly identify and resolve data quality issues, ensuring that its ML models deliver reliable and accurate predictions.

Another illustrative case is that of LinkedIn, which relies heavily on ML models to drive various features on its platform, including job recommendations, connection suggestions, and content personalization. To ensure that its models operate effectively, LinkedIn has developed a robust data pipeline monitoring framework that focuses on data freshness and consistency. For example, LinkedIn monitors the timeliness of its data pipelines to ensure that real-time data, such as user interactions and profile updates, is processed and incorporated into its models without delay. Any bottlenecks or delays in the data pipeline are immediately flagged and addressed, preventing outdated data from influencing model predictions. This approach has allowed LinkedIn to maintain high levels of engagement and relevance across its platform.

Model Performance Monitoring

Ensuring the continuous effectiveness of machine learning (ML) models is a multifaceted task that demands vigilant monitoring of their performance over time. This process involves a systematic approach to tracking how well a model meets the established key performance indicators (KPIs) while remaining alert to any signs of performance decline. Various factors can contribute to such degradation, including changes in the underlying data, shifts in the target population, or the introduction of new variables. By closely monitoring these aspects, organizations can maintain the relevance and accuracy of their ML models, ensuring they continue to provide value in dynamic environments.

Continuous Performance Monitoring

At the core of effective model performance monitoring is the implementation of tools that can continuously assess the model's predictions against real-world outcomes. These tools are designed to track various performance metrics, such as accuracy, precision, recall, F1 score, and other relevant KPIs, depending on the specific use case. Continuous monitoring allows organizations to detect deviations from expected performance metrics promptly, providing an early warning system for potential issues.

For instance, consider a credit scoring model used by a financial institution to assess loan applications. The model's performance is crucial for minimizing risk while approving as many valid loans as possible. If the model begins to approve an increasing number of high-risk applicants or reject low-risk ones, continuous monitoring tools can detect these shifts quickly. By comparing the model's predictions with actual repayment behaviors, the institution can identify when the model's accuracy is declining, signaling the need for further investigation or retraining.

Continuous monitoring is particularly effective when coupled with real-time feedback mechanisms. For example, in online retail, recommendation systems are constantly fed new data as users interact with the platform. By continuously comparing predicted user behaviors—such as clicking on or purchasing recommended products— with actual behaviors, these systems can adjust recommendations in real time, ensuring they remain relevant and personalized. This dynamic adaptation helps maintain high user engagement and satisfaction.

The Role of Retraining in Performance Monitoring

One of the critical aspects of continuous monitoring is the ability to identify when a model's performance has degraded to the point where retraining is necessary. Retraining involves updating the model with new data, reflecting the most recent trends and patterns. This process helps the model stay aligned with the current data environment, ensuring its predictions remain accurate and reliable.

Take the example of a sentiment analysis model used by a company to gauge public opinion on its products. Over time, language evolves, new slang terms emerge, and the context in which words are used can shift. If the model is not retrained periodically, it may start to misclassify sentiments, leading to inaccurate insights. Continuous monitoring tools can track the model's performance on a rolling basis, identifying when its accuracy in classifying sentiments begins to wane. By retraining the model with up-to-date data, the company can ensure that its analysis remains relevant and accurate.

Retraining can be triggered by specific events identified through monitoring, such as a significant drop in accuracy or the detection of data drift. Data drift occurs when the statistical properties of the input data change over time, which can negatively impact the model's performance. For example, a predictive maintenance model in a manufacturing plant may rely on sensor data to predict equipment failures. If the operating conditions of the equipment change—perhaps due to new maintenance practices or upgrades—the model may no longer perform as expected. Continuous monitoring would detect this data drift, prompting a retrain of the model with the new data to restore its predictive power.

Periodic Audits

While continuous monitoring provides an ongoing assessment of a model's performance, periodic audits serve as a more comprehensive review. These audits involve a thorough examination of the model's predictions, the errors it generates, and any biases that may have emerged over time. Audits are typically conducted at regular intervals, such as quarterly or biannually, depending on the criticality of the model and the dynamics of the environment in which it operates.

A key benefit of periodic audits is their ability to uncover long-term trends in model performance that might not be immediately apparent through day-to-day monitoring. For example, a predictive policing model used by law enforcement agencies might initially perform well but start to exhibit biased behavior as it encounters new data distributions over time. A periodic audit could reveal that the model has begun disproportionately targeting certain communities, prompting an investigation into the underlying causes and the implementation of corrective measures.

Audits also provide an opportunity to assess the model's interaction with the data at a deeper level. By analyzing the model's errors and predictions, organizations can gain insights into how the model processes different types of data and where it might be struggling. For instance, an audit of a healthcare diagnostic model might reveal that it performs well for the general population but has lower accuracy for specific demographic groups. This insight could lead to targeted improvements in the model, such as incorporating more diverse training data or adjusting the model's architecture to better handle these cases.

The introduction of new variables into the model's data can significantly impact performance, making the role of audits even more critical. When new features are added to the model, they can interact with existing variables in complex and sometimes unpredictable ways. Regular monitoring allows organizations to quickly assess the impact of these new variables, while audits provide a more in-depth analysis to ensure that the model continues to operate as intended. For example, an e-commerce company might add a new variable to its recommendation engine, such as social media activity. An audit might reveal that while this variable improves recommendations for some user segments, it introduces biases for others, leading to a more refined feature selection process in future iterations.

Integrating Continuous Monitoring with Periodic Audits

To create a comprehensive framework for maintaining ML models over time, it is essential to integrate continuous monitoring with periodic audits. This dual approach ensures that models remain effective, fair, and aligned with business objectives while allowing for the detection and resolution of both short-term and long-term issues.

For example, a financial institution deploying an ML model for credit risk assessment might implement a continuous monitoring system that tracks model performance metrics such as default rates and approval rates in real time. This system would alert stakeholders to any immediate performance issues, such as a sudden spike in defaults among approved applicants. At the same time, the institution might conduct quarterly audits to review the model's performance in more detail, examining trends in bias, changes in the demographic makeup of the applicants, and the impact of any new data sources or variables introduced into the model.

This integrated approach not only allows for the timely detection of performance issues but also provides a strategic overview that informs future model development and deployment strategies. By combining the strengths of continuous monitoring and periodic audits, organizations can ensure that their ML models remain robust and trustworthy, even as the data environment and business needs evolve.

Tools and Platforms for Model Performance Monitoring

Implementing an effective model performance monitoring framework requires the right tools and platforms. These tools should be capable of handling the complexities of modern ML systems, including real-time data processing, advanced analytics, and integration with various data sources.

441

One such tool is **Azure Machine Learning**, a cloud-based platform that offers robust model monitoring capabilities. Azure Machine Learning allows organizations to track key performance metrics, detect data drift, and automate the retraining process. The platform provides detailed dashboards that offer insights into model performance across different dimensions, enabling data scientists and engineers to quickly identify and address issues.

Another example is **Google Cloud AI Platform**, which offers end-to-end ML lifecycle management, including model monitoring. The platform provides tools for tracking model predictions, comparing them against actual outcomes, and identifying performance degradation over time. It also supports custom metrics, allowing organizations to tailor monitoring to their specific needs. For instance, a retail company using the Google Cloud AI Platform might monitor its recommendation model's precision and recall, adjusting its algorithms as new products and customer preferences emerge.

DataRobot is another platform that provides extensive model monitoring and management capabilities. It allows for continuous tracking of model performance, offering alerts and detailed reports that help organizations stay on top of potential issues. DataRobot's monitoring tools are designed to work seamlessly with its automated ML capabilities, enabling organizations to quickly retrain models and deploy updates as needed. This integration is particularly valuable for industries with rapidly changing data environments, such as finance and healthcare, where maintaining model accuracy is critical.

For organizations that require a more customizable solution, **MLflow** offers an open-source platform for managing the ML lifecycle, including monitoring. MLflow allows data scientists to log and track experiments, monitor model performance, and manage deployment workflows. Its flexibility makes it an ideal choice for organizations that need to integrate monitoring with existing infrastructure or develop custom monitoring solutions tailored to their unique requirements.

Case Studies

To illustrate the practical application of model performance monitoring, consider the case of a global logistics company that implemented a predictive model to optimize delivery routes. The model was designed to predict traffic patterns and suggest the most efficient routes for delivery trucks, aiming to reduce fuel consumption and improve delivery times.

Continuous monitoring of the model's performance revealed that its accuracy began to decline during the winter months, a period characterized by unpredictable weather conditions and road closures. By integrating real-time weather data into the monitoring framework, the company was able to identify the root cause of the decline and retrain the model with this new data. The retrained model improved delivery accuracy during winter, demonstrating the value of continuous monitoring in adapting to changing conditions.

Periodic audits of the model also uncovered long-term trends that continuous monitoring did not detect. For example, an audit revealed that the model's predictions were less accurate in rural areas compared to urban centers. This discrepancy was traced back to differences in the quality and granularity of the data available for these regions. By addressing this issue—through the incorporation of higher-resolution data for rural areas—the company was able to improve the model's overall performance and ensure more consistent delivery times across different geographies.

Another example comes from the financial sector, where a bank used ML models to detect fraudulent transactions. Continuous monitoring of the model's performance allowed the bank to quickly identify changes in fraud patterns, such as the rise of new tactics by fraudsters. By regularly retraining the model with data reflecting the latest fraud trends, the bank was able to maintain high detection rates and minimize losses.

Periodic audits of the fraud detection model provided additional insights, such as identifying biases in the model's predictions. For instance, the model was found to be more likely to flag transactions from certain geographic regions as fraudulent, even when the actual incidence of fraud in those regions was no higher than elsewhere. This bias was addressed by rebalancing the training data and adjusting the model's features, leading to more equitable and accurate fraud detection.

Bias and Fairness Monitoring

Ensuring fairness and eliminating bias in machine learning (ML) models is a fundamental responsibility for organizations that seek to maintain effective and ethical ML systems. Bias in ML can manifest in various ways, such as through biased training data, the selection of features that inherently favor certain groups, or the model's predictions, which may be skewed against particular demographics. Addressing these biases requires a proactive approach, where the model's performance is continually evaluated across different demographic segments to ensure that it does not unfairly disadvantage any group.

Implementing Fairness Metrics

One of the primary strategies for monitoring bias involves the implementation of fairness metrics designed to assess the model's performance across various subgroups. These metrics are critical in detecting subtle forms of bias that might not be evident through standard performance measures. For example, demographic parity ensures that the model's predictions are equally distributed across different demographic groups, while the equal opportunity metric assesses whether the model is providing the same chance of a positive outcome to all groups. Another useful measure is the disparate impact ratio, which compares the rate of favorable outcomes between different groups. By regularly tracking these metrics, organizations can identify and address biases before they become entrenched in the model's operations.

Consider the case of a hiring algorithm used by a large corporation. The algorithm was initially trained on historical hiring data, which inadvertently included biases favoring candidates from certain educational backgrounds. By implementing fairness metrics such as demographic parity, the company was able to identify that candidates from less prestigious institutions were systematically being rated lower. This discovery prompted the company to adjust its model by including a more diverse set of features that better represented the candidates' potential, such as relevant work experience and skills rather than just educational pedigree. As a result, the revised model produced more equitable outcomes, increasing the hiring rate for underrepresented groups.

Another example is in the financial sector, where a bank utilized an ML model to assess creditworthiness. The bank implemented the equal opportunity metric to ensure that applicants from different ethnic backgrounds had the same likelihood of being approved for a loan if they met the relevant financial criteria. Continuous monitoring of this metric revealed that the model was less likely to approve loans for applicants from certain minority groups, even when they had similar financial profiles to other applicants. By retraining the model with a more balanced dataset and revising the feature selection process to eliminate variables that indirectly reflected racial disparities, the bank was able to reduce this bias, leading to fairer lending practices.

Continuous Monitoring and Real-Time Bias Detection

Continuous monitoring of fairness metrics is essential for the timely detection of bias in ML models. This approach allows organizations to observe the model's behavior in real time and make necessary adjustments to mitigate bias as soon as it is detected.

For instance, if a model shows a significant drop in equal opportunity for a particular subgroup, this could indicate the need for immediate intervention, such as retraining the model with more representative data or adjusting the feature selection process. The goal is to ensure that the model remains equitable and does not inadvertently harm certain groups.

One advanced technique for continuous bias monitoring is the use of real-time dashboards that track fairness metrics alongside standard performance metrics like accuracy and precision. These dashboards provide data scientists and decision-makers with a holistic view of the model's performance, allowing them to quickly identify and respond to any emerging biases. For example, a predictive policing model used by a city's law enforcement agency might be monitored through such a dashboard. If the model begins to disproportionately target certain neighborhoods or demographic groups for increased police presence, the dashboard would immediately highlight this disparity, prompting a review and adjustment of the model.

Another approach involves integrating bias detection algorithms directly into the ML pipeline. These algorithms can be designed to flag potential biases as the model processes new data. For example, in an automated content moderation system used by a social media platform, bias detection algorithms could be employed to ensure that content from minority groups is not unfairly targeted for removal. If the system detects that posts from a particular demographic are being flagged more frequently despite adhering to the platform's guidelines, the model could be adjusted in real time to correct this bias, ensuring a more equitable moderation process.

Conducting Fairness Audits

Conducting fairness audits is a crucial component of bias management. These audits provide a comprehensive review of the model's performance across different demographic groups, focusing on uncovering and mitigating any sources of bias. Unlike continuous monitoring, which provides real-time insights, fairness audits offer a deeper, more thorough analysis of how the model interacts with different segments of the population. They involve a detailed examination of the data, the model's internal mechanics, and the outcomes it produces.

Fairness audits can take various forms, depending on the complexity and impact of the ML model being evaluated. One effective approach is the use of counterfactual analysis, where the model is tested on hypothetical scenarios to assess how it might behave under different conditions. For example, a fairness audit of a mortgage approval model might involve creating synthetic data that represents applicants from a variety of demographic backgrounds, income levels, and credit histories. By running this synthetic data through the model, auditors can identify any patterns of bias that might not be evident in the model's real-world performance.

Another technique used in fairness audits is the exploration of proxy variables— those that are not directly related to protected attributes like race or gender but are closely correlated with them. For instance, in an employment screening model, zip codes might be used as a proxy for race due to the socioeconomic and demographic patterns associated with different regions. During the fairness audit, the model's reliance on such proxies would be scrutinized to ensure that it does not lead to discriminatory outcomes. If the audit reveals that the model is unfairly penalizing candidates from certain zip codes, adjustments can be made to either remove or de-emphasize these variables in the decision-making process.

The Role of Diverse Stakeholder Engagement

The involvement of diverse stakeholders is another key element in conducting effective fairness audits. It is important to include input from ethicists, who can provide guidance on the ethical implications of the model's behavior, and legal experts, who can ensure that the model complies with relevant laws and regulations. Additionally, representatives from the communities impacted by the model's decisions should be involved in the audit process. Their perspectives are invaluable in identifying biases that might not be apparent to technical teams but could have significant real-world consequences.

For example, consider a healthcare provider implementing an ML model to allocate resources for patient care. A fairness audit of this model might involve consultations with medical ethicists to ensure that the model's decisions align with ethical standards, such as prioritizing care for the most vulnerable populations. Legal experts would review the model to ensure compliance with healthcare regulations and anti-discrimination laws. Meanwhile, patient advocacy groups would be engaged to provide insights into how the model's decisions affect different patient demographics, particularly those who are underserved or marginalized. This inclusive approach ensures that the model not only functions effectively but also aligns with broader social and ethical goals.

Another example can be seen in the education sector, where an ML model is used to predict student performance and identify those at risk of falling behind. A fairness audit of this model might include input from educators, parents, and student advocacy groups. Educators can provide context on the diverse learning needs of students, while parents and advocacy groups can highlight potential biases that might affect students from disadvantaged backgrounds. For instance, the audit might reveal that the model is disproportionately flagging students from low-income families as being at risk, possibly due to biases in the training data. By involving these stakeholders in the audit process, the education institution can refine the model to ensure it supports all students equitably.

Structuring and Conducting Effective Fairness Audits

Fairness audits provide a structured approach to bias mitigation, systematically evaluating the model's performance across different demographic groups. These audits should be conducted regularly to ensure that the model remains unbiased over time. The frequency of audits can depend on the model's application and the sensitivity of its outcomes. For high-stakes scenarios, such as those involving human resources, finance, or healthcare, more frequent audits may be necessary to safeguard against bias.

For example, in the context of HR, a fairness audit of an ML-based hiring tool might be conducted quarterly, particularly during peak hiring seasons. This audit would involve a comprehensive review of the tool's performance metrics, disaggregated by demographic groups such as gender, race, and age. The audit team would examine the model's hiring recommendations, looking for any patterns of bias or discrimination. If the audit reveals that the model is consistently favoring one demographic group over others, the hiring process would be adjusted to ensure that all candidates are evaluated fairly.

In the financial sector, fairness audits of credit scoring models might be conducted on a biannual basis. These audits would assess the model's performance across different demographic groups, focusing on key metrics such as loan approval rates and interest rates offered to different groups. The audit might also explore the impact of new regulations or economic conditions on the model's fairness. If the audit identifies any biases, such as a lower approval rate for minority applicants, the financial institution would take corrective action, such as retraining the model with more diverse data or revising the feature selection process.

Building Trust Through Fairness Audits

Regular fairness audits help build trust in ML systems, both within the organization and among external stakeholders. When stakeholders know that a model has been thoroughly vetted for bias and fairness, they are more likely to trust its decisions and outcomes. This trust is crucial, especially in contexts where the model's decisions have significant consequences, such as in hiring, lending, or criminal justice. By demonstrating a commitment to fairness through regular audits, organizations can strengthen their reputation and foster greater confidence in their ML systems.

For instance, a tech company that uses an ML model for content moderation might face public scrutiny if the model is perceived to unfairly target certain groups or viewpoints. By conducting regular fairness audits and publicly sharing the results, the company can demonstrate its commitment to unbiased content moderation. This transparency helps build trust among users, who can see that the company is actively working to ensure that its platform is fair and inclusive.

In the criminal justice system, where ML models are increasingly used to assess the risk of reoffending, fairness audits are essential for maintaining public trust. If a fairness audit reveals that a risk assessment model is biased against certain demographic groups—such as assigning higher risk scores to individuals from minority backgrounds—the justice system must take immediate action to rectify this bias. By doing so, the system not only ensures fairer outcomes for individuals but also upholds the integrity and credibility of the judicial process.

Addressing Systemic Bias Through Fairness Audits

Fairness audits can reveal systemic biases that go beyond the model itself. For instance, if a model consistently underperforms for a particular demographic group, this could indicate broader issues in the data collection process, the feature selection criteria, or even the underlying business practices. By uncovering these deeper issues, organizations can take corrective action not only to improve the model but also to address the root causes of bias within their operations.

Consider a case where an ML model used by a healthcare provider consistently underestimates the risk of complications for patients from certain ethnic backgrounds. A fairness audit might reveal that this bias stems from a lack of representative data in the training set, as well as potential biases in the clinical guidelines used to develop the model. To address this systemic issue, the healthcare provider might partner with community organizations to collect more representative data, revise its clinical guidelines to better account for diverse patient populations, and retrain the model accordingly. This comprehensive approach not only improves the model's fairness but also enhances the overall quality of care provided to patients.

In the retail sector, a fairness audit of a customer segmentation model might reveal that the model is biased against older customers, leading to less personalized marketing efforts for this demographic. This bias might be traced back to the feature selection process, which may have overemphasized digital engagement metrics that are less relevant for older customers. By revising the feature selection process to include a broader range of customer behaviors, the retailer can create a more inclusive model that better serves all customer segments.

System Stability and Scalability

Ensuring the stability and scalability of machine learning (ML) systems is as crucial as monitoring data quality, model performance, and bias. Stability and scalability are fundamental aspects that ensure an ML system can effectively manage increasing workloads, adapt to changing demands, and maintain consistent availability over time. Without a robust approach to these elements, even the most advanced ML models can fail in production environments, leading to system downtimes, performance bottlenecks, and ultimately, a loss of trust from users and stakeholders.

System stability is the foundation upon which reliable ML operations are built. It involves ensuring that the infrastructure supporting the ML system, including servers, databases, and network connections, operates consistently and without interruptions. A stable system is critical because ML models often require significant computational resources, particularly during tasks such as training, real-time inference, and data processing. Any instability in the underlying infrastructure can lead to slow responses, inaccurate results, or complete system failures.

Consider a real-time fraud detection system used by a financial institution. This system must analyze transactions as they occur, flagging potentially fraudulent activity within milliseconds to prevent financial losses. If the servers supporting this system experience high latency or overload, the model might fail to analyze transactions quickly enough, allowing fraudulent activities to slip through. To prevent such scenarios, stability monitoring tools continuously track performance metrics like server response times, CPU utilization, and network bandwidth. These tools can alert administrators when performance deviates from acceptable levels, enabling them to take corrective actions, such as balancing server loads or scaling resources to maintain system integrity.

Tools and Best Practices for Stability Monitoring

Various tools are available to help organizations monitor the stability of their ML systems. One popular tool is **Prometheus**, an open-source monitoring and alerting toolkit designed for reliability and scalability. Prometheus is particularly well-suited for monitoring time-series data, which is essential for tracking the performance of ML systems over time. It can collect and store metrics, execute queries, and trigger alerts when predefined thresholds are exceeded.

Grafana is often used in conjunction with Prometheus to visualize metrics in real time. Grafana dashboards provide an at-a-glance view of system health, making it easier to spot trends and anomalies. For instance, in an e-commerce platform utilizing ML for product recommendations, a Grafana dashboard might display real-time data on server load, response times, and error rates. If the system starts to lag during a high-traffic event, the dashboard would highlight these issues, allowing the technical team to respond before the problem escalates.

Datadog is another comprehensive monitoring service that integrates with cloud environments and offers features like anomaly detection, root cause analysis, and infrastructure monitoring. For ML systems deployed on cloud platforms such as AWS or Azure, Datadog can provide detailed insights into the performance of virtual machines, databases, and network components. This is crucial for maintaining the stability of cloud-based ML systems, where resources are often scaled dynamically based on demand.

Kubernetes, the popular container orchestration platform, plays a significant role in ensuring system stability by automating the deployment, scaling, and management of containerized applications. For ML systems, Kubernetes can monitor the health of containers running ML models, ensuring that they are automatically restarted or rescheduled if they fail. This self-healing capability is vital for maintaining stability in complex, distributed ML environments.

Scalability in Machine Learning Systems

Scalability is the ability of an ML system to handle increased workloads, whether in the form of more data, a higher number of concurrent users, or more complex models. As organizations grow and their ML applications become more integral to operations, scalability becomes a critical factor in sustaining performance and efficiency. Scalability in ML systems involves several dimensions:

Data Scalability: As the volume of data increases, the system must be capable of processing and analyzing larger datasets without compromising speed or accuracy. This often requires distributed data processing frameworks like **Apache Spark**, which can efficiently handle large-scale data processing tasks across multiple nodes.

Model Scalability: As ML models become more complex, they require more computational power. This might involve scaling up the infrastructure to include more powerful GPUs or TPUs, which are specialized processors designed for ML workloads. Cloud platforms like **Google Cloud AI** and **AWS SageMaker** provide on-demand access to these resources, allowing organizations to scale their models as needed without significant upfront investments in hardware.

User Scalability: As the number of users interacting with an ML system increases, the system must be able to handle concurrent requests efficiently. Load balancers and auto-scaling groups are commonly used to distribute traffic across multiple servers, ensuring that no single server becomes a bottleneck. For instance, on a popular social media platform, the recommendation engine must scale to serve millions of users simultaneously, providing personalized content in real time. Technologies like **Nginx** for load balancing and **AWS Elastic Load Balancing** ensure that the system can handle this traffic without degradation in performance.

Load Testing and Capacity Planning

Regular load testing is a best practice for assessing the scalability of ML systems. Load testing involves subjecting the system to simulated high-traffic scenarios to evaluate its response under pressure. Tools like **Apache JMeter** and **Gatling** are commonly used for this purpose. These tools can simulate thousands of users interacting with the system, allowing organizations to observe how it handles increased data loads and user requests.

During load testing, metrics such as response time, throughput, and error rate are closely monitored. For example, an online retail company might use JMeter to simulate a Black Friday event, where traffic to the site surges dramatically. By observing how the ML models powering the recommendation engine and inventory management system perform under these conditions, the company can identify potential bottlenecks and optimize its infrastructure accordingly.

Capacity planning is another essential aspect of scalability management. This involves predicting future infrastructure needs based on current usage trends and business growth projections. For instance, if an organization expects a 50% increase in user traffic over the next year, capacity planning would involve determining the additional resources required to support this growth without sacrificing performance. Cloud platforms offer elasticity, allowing organizations to scale resources up or down as needed, which is a significant advantage for managing costs while ensuring scalability.

Managing Complexity as ML Systems Scale

As ML systems scale, they often become more complex, with multiple models, data pipelines, and microservices interacting within a distributed environment. Managing this complexity requires robust monitoring and orchestration tools to ensure that all components work together seamlessly.

Apache Kafka is a distributed streaming platform that plays a crucial role in managing the data flows within large-scale ML systems. Kafka enables real-time data ingestion and processing, ensuring that ML models receive up-to-date information. For instance, in a large financial institution using ML for fraud detection, Kafka might be used to stream transaction data from various sources to the ML models in real time. This ensures that the models are always working with the latest data, allowing for immediate detection of fraudulent activities.

Kubernetes also plays a vital role in managing the complexity of scaling ML systems. By automating the deployment and scaling of containerized applications, Kubernetes ensures that each component of the ML system operates efficiently, even as the system grows. For example, in a machine translation service used by a global enterprise, Kubernetes might be used to scale the number of language models dynamically based on user demand. This ensures that translation services remain fast and accurate, even during peak usage periods.

Cost Efficiency in Scaling

While scaling up infrastructure is often necessary to handle increased workloads, it can also be expensive. Therefore, organizations must find a balance between performance and cost efficiency. This involves not only optimizing existing resources but also exploring more efficient algorithms and architectures that require less computational power.

For instance, **model distillation** is a technique where a large, complex model (the teacher model) is used to train a smaller, more efficient model (the student model) that approximates the performance of the original model. This approach allows organizations to deploy models that are less resource-intensive, thereby reducing the cost of scaling while maintaining high accuracy.

Cloud-based solutions offer another avenue for cost-efficient scaling. Cloud platforms like **AWS**, **Google Cloud**, and **Microsoft Azure** provide a range of services that allow organizations to scale their ML infrastructure on-demand. By leveraging cloud services, organizations can avoid the capital expenditure associated with purchasing and maintaining physical hardware. Instead, they can pay for only the resources they use, scaling up during periods of high demand and scaling down when demand decreases.

Auto-scaling is a feature offered by most cloud platforms that automatically adjusts the number of active servers based on current demand. This ensures that the system always has enough capacity to handle the workload without the need to maintain excess resources during low-traffic periods. For example, a streaming service might use auto-scaling to handle spikes in traffic during the release of a popular new show, ensuring a smooth user experience without incurring unnecessary costs.

High Availability and Disaster Recovery

Ensuring system stability and scalability contributes to maintaining high availability, which is crucial for ML systems that support critical operations. High availability means that the system remains operational and accessible at all times, even in the face of hardware failures, network issues, or other disruptions. Achieving high availability requires a combination of redundancy, failover mechanisms, and robust monitoring.

Redundancy involves having multiple instances of critical system components, such as servers and databases, so that if one instance fails, others can take over without interruption. For example, in a healthcare application that uses ML to assist in diagnosis, high availability is vital. The system might be deployed across multiple data centers, with each data center running redundant instances of the ML models and databases. If one data center experiences a failure, the others can continue to operate, ensuring that the application remains available to healthcare providers.

Failover mechanisms are automated processes that detect failures and switch operations to a backup system. For example, in a financial trading platform, failover mechanisms might automatically redirect transactions to a secondary server if the primary server goes down, ensuring that trades can continue without disruption. These mechanisms are often integrated with monitoring tools that continuously check the health of the system and trigger failover processes when necessary.

Disaster recovery planning is another critical aspect of high availability. This involves creating a strategy for restoring system operations in the event of a catastrophic failure, such as a data center outage or a major cyberattack. For ML systems, disaster recovery planning might include maintaining regular backups of training data, models, and system configurations, as well as setting up secondary data centers in geographically dispersed locations to ensure that the system can be quickly restored in the event of a disaster.

Predictive Monitoring and Trend Analysis

Regular monitoring of system stability and scalability enables organizations to identify trends and patterns in system usage. This information can be used to predict future needs and plan for infrastructure upgrades accordingly. For instance, if monitoring data reveals a consistent increase in user activity during certain times of the year, the organization can prepare by scaling up resources in advance.

Predictive analytics can be applied to monitoring data to forecast future demands on the system. For example, a retail company might use predictive models to analyze historical traffic patterns and predict the expected load on its e-commerce platform during the holiday season. By understanding these trends in advance, the company can ensure that its ML models and underlying infrastructure are prepared to handle the anticipated increase in traffic.

Capacity planning based on predictive monitoring helps organizations avoid the pitfalls of both under-provisioning and over-provisioning resources. Under-provisioning can lead to system failures and poor performance, while over-provisioning can result in unnecessary costs. By using predictive analytics to guide capacity planning, organizations can strike the right balance, ensuring that their ML systems are both scalable and cost-effective.

Anomaly detection in monitoring data is another important tool for maintaining system stability. Machine learning algorithms can be used to detect unusual patterns in system performance that may indicate emerging issues. For instance, a sudden increase in server load might be an early warning sign of a denial-of-service attack. By detecting such anomalies early, organizations can take preventive measures to protect their ML systems from potential threats.

Continuous Model Improvement

One of the cornerstones of continuous model improvement is the implementation of a robust model versioning system. Model versioning allows organizations to systematically track, evaluate, and compare different iterations of a model. This practice is critical in environments where models are frequently updated or where multiple models are deployed simultaneously.

Consider a recommendation engine used by an e-commerce platform. As consumer behavior evolves, the platform's ML models must adapt to these changes to provide relevant product suggestions. By employing a model versioning system, the platform can maintain multiple versions of the recommendation model, each trained on different datasets or using various algorithms. For example, one version might prioritize recent user interactions, while another might focus on long-term preferences. By comparing the performance of these versions through A/B testing or offline evaluation, the platform can identify which model delivers the most effective recommendations under current conditions.

In another example, a financial institution may use model versioning to manage credit risk models. Different versions of the model might be tailored to various economic scenarios, such as a recession or a booming economy. By systematically evaluating these versions, the institution can deploy the most appropriate model based on current market conditions, thereby optimizing risk management.

Automated Model Retraining

Automated model retraining is another critical strategy for continuous improvement. In rapidly changing environments, where data patterns evolve quickly, manual model updates can be inefficient and prone to delays. Automated retraining addresses this challenge by enabling models to learn from new data regularly without requiring human intervention.

For instance, in the context of a real-time fraud detection system used by a bank, the model must continuously adapt to new fraud tactics that evolve over time. Automated retraining allows the model to integrate the latest transaction data, identifying emerging patterns of fraudulent behavior. By setting up a retraining schedule, such as daily or weekly, the model can remain effective in detecting new types of fraud, even as fraudsters change their strategies.

A practical example of automated retraining can be seen in the deployment of chatbots. As users interact with the chatbot, new conversational patterns emerge, and the language model must adapt to these changes to provide accurate responses. By automatically retraining the model on the latest interaction data, the chatbot improves its ability to understand and respond to user queries, thereby enhancing the overall user experience.

Evaluating Model Performance

Regular evaluation of model performance is essential to ensure that models continue to meet their intended objectives. This ongoing assessment involves comparing the model's predictions against actual outcomes and adjusting the model as needed to address any discrepancies.

One approach to performance evaluation is the use of rolling windows for model validation. In this method, a model is continually tested on the most recent data, ensuring that its performance is aligned with current trends. For example, a demand forecasting model used by a retailer might be evaluated weekly using the latest sales data. If the model's predictions consistently deviate from actual sales, this could indicate a need for retraining or feature refinement.

Another critical aspect of performance evaluation is the identification of concept drift, where the statistical properties of the target variable change over time. For example, a customer churn model may become less effective if customer behavior changes due to external factors like economic downturns or new competitors. Continuous monitoring for concept drift allows organizations to detect these shifts early and take corrective actions, such as updating the model or incorporating new features.

Integrating Feedback Loops

Incorporating feedback loops into the continuous improvement process is crucial for aligning models with real-world outcomes. Feedback loops involve collecting data from the model's predictions and actual outcomes, then using this information to refine the model.

For instance, in a personalized marketing campaign, the effectiveness of targeted ads can be evaluated by tracking user engagement metrics, such as click-through rates and conversion rates. This feedback can reveal which segments of the audience are responding positively to the ads and which are not. By feeding this information back into the model, marketers can adjust the targeting strategy, optimize ad content, and improve the overall effectiveness of the campaign.

Another example is in predictive maintenance systems used in manufacturing. Sensors collect data on machine performance, and the model predicts when a machine is likely to fail. By comparing these predictions with actual machine failures, the system can improve its accuracy over time. If the model consistently overestimates the time until failure, the feedback loop can be used to adjust the model's parameters, leading to more precise maintenance schedules and reduced downtime.

Experimentation and Feature Engineering

Continuous improvement often involves experimenting with different algorithms and feature sets to uncover new insights and enhance model capabilities. Experimentation allows data scientists to test various approaches and identify the most effective methods for optimizing model performance.

In the healthcare industry, for example, ML models are used to predict patient outcomes based on electronic health records (EHRs). By experimenting with different feature engineering techniques, such as incorporating genetic data or social determinants of health, data scientists can improve the model's predictive power. This iterative process of experimentation and refinement is essential for developing models that provide accurate and actionable insights for healthcare providers.

Experimentation can also involve the use of ensemble methods, where multiple models are combined to produce a more accurate prediction. For instance, in weather forecasting, ensemble models that aggregate predictions from different meteorological models can provide more reliable forecasts. By continuously refining the ensemble components and adjusting the weighting of each model based on real-time performance, the overall predictive accuracy can be enhanced.

Integrating New Data Sources

As data evolves, integrating new data sources is crucial for continuous model improvement. New data sources can provide additional insights and improve model accuracy, but they also present challenges, such as data quality issues, compatibility with existing models, and increased computational complexity.

For example, in the retail industry, integrating social media data into demand forecasting models can provide valuable insights into consumer sentiment and emerging trends. However, social media data is often noisy, unstructured, and subject to rapid changes. Addressing these challenges requires sophisticated data preprocessing techniques, such as natural language processing (NLP) for text analysis and sentiment scoring.

In financial services, incorporating alternative data sources, such as satellite imagery or weather data, into credit risk models can enhance their predictive power. However, these data sources often require specialized processing pipelines and feature extraction techniques to be effectively integrated with traditional financial data. For instance, satellite imagery might be used to assess agricultural yields, which in turn can inform credit decisions for farmers. The challenge lies in transforming this imagery into actionable features that can be used by the model.

Data integration also raises concerns about data governance and compliance, particularly when dealing with sensitive or personally identifiable information (PII). Organizations must ensure that new data sources comply with regulations such as GDPR or CCPA. This often involves implementing data anonymization techniques and obtaining explicit user consent before using new data sources in the model.

Monitoring the Impact of Model Changes

After making improvements or integrating new data sources, it is essential to monitor the impact of these changes on the model's performance. This involves conducting rigorous A/B testing or holdout validation to compare the new version of the model against the previous version.

For example, a ride-sharing platform might update its pricing model to incorporate new data on fuel prices and traffic conditions. Before fully deploying the updated model, the platform could run an A/B test where the old and new models are used simultaneously in different regions. By comparing the outcomes—such as ride prices, driver earnings, and customer satisfaction—the platform can assess whether the changes have led to the desired improvements.

Another approach to monitoring the impact of changes is through post-deployment performance tracking. This involves continuously evaluating the model's predictions against real-world outcomes after the updated model has been deployed. For instance, in a recommendation system for streaming services, the updated model's recommendations can be tracked to see if they lead to increased user engagement and longer viewing times. If the new model performs worse than the previous version, it might be necessary to roll back the changes or conduct further refinements.

Documentation and Systematic Improvement Processes

Ensuring that the continuous improvement process is systematic and well-documented is essential for maintaining transparency, accountability, and reproducibility. Detailed documentation of model versions, retraining schedules, performance evaluations, and feedback received helps in maintaining a clear overview of the model's evolution.

For instance, in a pharmaceutical company using ML models for drug discovery, documenting the evolution of the model—including changes in algorithms, data sources, and feature sets—is crucial for regulatory compliance and intellectual property protection. This documentation provides a comprehensive history of how the model has been developed, tested, and improved over time, which is essential for audits, patent filings, and knowledge transfer within the organization.

Systematic improvement processes also involve the use of automated pipelines that integrate all aspects of continuous improvement, from data ingestion to model deployment. **MLOps** (machine learning operations) frameworks, such as **Kubeflow** or **MLflow**, provide tools for automating the lifecycle of ML models, ensuring that all steps in the continuous improvement process are consistent, repeatable, and scalable. For example, an MLOps pipeline might automatically retrain models, perform validation, and deploy the best-performing version, all while logging the entire process for future reference.

Fostering a Culture of Continuous Improvement

A culture of continuous improvement within an organization is crucial for the ongoing success of ML systems. This culture encourages teams to regularly review and refine models, experiment with new approaches, and stay informed about the latest advancements in ML technology.

For example, a technology company that fosters a culture of continuous improvement might organize regular hackathons or workshops where data scientists and engineers collaborate on improving existing models or developing new ones. These events not only drive innovation but also promote knowledge sharing and cross-functional collaboration.

Leadership support is also critical in fostering this culture. When leaders prioritize continuous improvement and allocate resources—such as time, tools, and training—teams are more likely to engage in ongoing model enhancement efforts. For instance, providing data scientists with access to the latest ML tools, cloud computing resources, and professional development opportunities can empower them to drive continuous improvement initiatives effectively.

Governance and Compliance Monitoring

Governance and compliance monitoring in machine learning (ML) systems is a critical aspect of ensuring that these technologies operate within the bounds of legal, ethical, and industry-specific standards. As ML models increasingly influence decision-making across sectors, maintaining rigorous governance and compliance protocols is essential to prevent legal breaches, protect data privacy, and uphold ethical standards. This involves not only implementing automated tools for compliance checks but also regularly conducting audits to assess the system's alignment with regulatory requirements.

Automated Compliance Checks

Automated compliance checks are foundational to maintaining adherence to regulatory and governance standards in ML systems. These systems are designed to continuously monitor various aspects of the ML pipeline, from data processing to model deployment, ensuring that all operations comply with relevant laws and guidelines.

One of the primary advantages of automated compliance checks is their ability to provide real-time monitoring and immediate feedback. For example, in a healthcare application where patient data is processed by ML models, automated compliance tools can ensure that data handling practices align with regulations such as the Health Insurance Portability and Accountability Act (HIPAA). These tools can automatically verify that sensitive patient information is encrypted during storage and transmission, access to data is appropriately restricted, and audit logs are maintained to track data access and modifications.

Software solutions like **BigID** and **Collibra** are specifically designed to automate compliance monitoring. BigID, for instance, helps organizations discover and classify sensitive data, ensuring that data handling practices comply with privacy regulations like the General Data Protection Regulation (GDPR) and the California Consumer Privacy Act (CCPA). By integrating such tools into the ML pipeline, organizations can continuously assess compliance with data protection laws, automatically flagging any violations for immediate remediation.

Another critical aspect of automated compliance is ensuring that ML models themselves do not perpetuate bias or discrimination, which is particularly important in sectors like finance or criminal justice. Tools like **Fairness Indicators** from Google provide automated checks for bias, enabling continuous monitoring of model predictions to ensure fairness across different demographic groups. For instance, in a lending model, Fairness Indicators can help ensure that loan approval rates do not disproportionately favor or disadvantage specific racial or gender groups, aligning with fair lending laws and ethical guidelines.

Regular Compliance Audits

While automated compliance checks provide continuous monitoring, regular compliance audits offer a more comprehensive and in-depth evaluation of an ML system's adherence to governance standards. Audits are typically conducted at set intervals, such as quarterly or annually, and involve a thorough review of the system's processes, data management practices, and model outputs.

Compliance audits are especially important for identifying gaps that automated systems might overlook. For example, an audit might reveal that while a model complies with current data protection laws, it is not fully aligned with the organization's ethical guidelines on transparency or explainability. In such cases, auditors may recommend adjustments, such as enhancing the interpretability of model decisions or improving documentation practices to better explain how decisions are made.

An illustrative example of the importance of compliance audits can be seen in the financial sector, where institutions are required to meet stringent regulations around anti-money laundering (AML). While automated tools can monitor transactions for suspicious patterns, a compliance audit might uncover deeper issues, such as inadequate documentation of model updates or insufficient training for staff on the latest regulatory changes. By addressing these issues, the organization can not only ensure regulatory compliance but also strengthen its overall governance framework.

Handling Compliance in Different Regulatory Environments

Managing compliance across different regulatory environments presents significant challenges, particularly for global organizations that operate in multiple jurisdictions. Each region may have its own set of regulations governing data privacy, model transparency, and ethical considerations, requiring a flexible and adaptable compliance strategy.

Another challenge in managing compliance across different regions is the variation in requirements for model transparency and accountability. In Europe, the GDPR emphasizes the "right to explanation," where individuals affected by automated decisions have the right to understand the logic behind those decisions. This necessitates the use of explainable AI (XAI) techniques, which can be integrated into the ML system to provide clear and understandable explanations of model outputs. Tools like LIME (Local Interpretable Model-agnostic Explanations) and SHAP (SHapley Additive exPlanations) can be employed to meet these requirements by offering insights into how models make decisions, which is critical for compliance in jurisdictions with strict transparency regulations.

Responding to Compliance Issues

Establishing clear protocols for responding to compliance issues is crucial for mitigating risks and ensuring swift remediation. When a compliance breach is detected, whether through automated checks or audits, the organization must act promptly to investigate the issue, implement corrective actions, and document the resolution process.

A well-defined response plan begins with incident detection and assessment, where the nature and scope of the compliance breach are evaluated. For example, if an ML model inadvertently violates data privacy regulations by exposing sensitive information, the organization must quickly assess the extent of the breach and determine the affected parties. Tools like **Splunk** and **LogRhythm** can assist in this phase by providing real-time monitoring and detailed logs that help trace the source of the issue.

Once the breach is assessed, the next step is to implement corrective actions. This might involve retraining the model with anonymized data, revising access controls, or updating the compliance monitoring tools to prevent future violations. In the case of a financial institution that detects biased outcomes in its credit scoring model, corrective actions could include recalibrating the model to remove biased features, conducting additional fairness audits, and enhancing training for the data science team on ethical AI practices.

Documentation of the entire process is essential for transparency and accountability. Detailed records of the incident, the actions taken, and the final resolution should be maintained to demonstrate compliance with regulatory requirements. This documentation is not only crucial for internal governance but also for external audits and regulatory inquiries. Platforms like **ServiceNow** offer incident management tools that help organizations document and track compliance issues from detection to resolution, ensuring a structured and transparent process.

Training and Awareness

Ensuring compliance in ML systems requires more than just tools and audits; it also necessitates a culture of compliance across the organization. Training and awareness programs are critical for educating staff involved in managing ML systems about governance standards, data protection laws, and ethical guidelines.

Regular training sessions can help ensure that team members are up-to-date on the latest regulatory developments and understand the importance of compliance in their daily work. For example, data scientists and engineers should be trained on the ethical implications of their work, including how to recognize and mitigate biases in ML models. Legal and compliance teams should also receive training on the technical aspects of ML systems to better understand how to evaluate and enforce compliance.

Organizations like **Ethena** provide compliance training platforms that offer customizable courses on topics such as data privacy, ethical AI, and regulatory compliance. By integrating these training programs into the organization's broader governance strategy, companies can foster a culture of compliance that permeates all levels of the organization.

Integrating Compliance Monitoring into ML System Management

Integrating compliance monitoring into the broader management strategy of ML systems is essential for creating a cohesive and effective oversight mechanism. Compliance should not be viewed as a standalone function but rather as an integral part of the system's lifecycle, from development to deployment and beyond.

One approach to integration is the adoption of MLOps practices, which combine ML development and operations with robust governance and compliance protocols. MLOps platforms like Kubeflow and Azure Machine Learning offer tools for automating compliance checks, managing model versioning, and conducting regular audits, all within a unified framework. By embedding compliance monitoring into the MLOps pipeline, organizations can ensure that governance considerations are addressed at every stage of the ML lifecycle.

Another important aspect of integration is the alignment of compliance monitoring with business objectives. Compliance should support the organization's goals, not hinder them. By aligning compliance strategies with business objectives, organizations can ensure that their ML systems not only meet regulatory requirements but also drive value and innovation.

For instance, a healthcare provider using ML for patient diagnosis must ensure compliance with health data regulations while also striving to improve patient outcomes. By integrating compliance monitoring into the ML system management strategy, the provider can achieve both objectives: protecting patient privacy and enhancing the quality of care.

Stakeholder Engagement and Communication

A central aspect of stakeholder engagement is the implementation of a regular reporting process. These reports serve as a critical touchpoint for communicating the system's status, performance metrics, and any emerging issues. To be effective, reports must be tailored to the specific needs and expectations of different stakeholder groups, ensuring that the information provided is relevant, comprehensible, and actionable.

For data scientists and engineers, reports should include detailed technical information such as key performance indicators (KPIs), error rates, data drift observations, and model performance metrics. These stakeholders need access to granular data that enables them to diagnose issues, optimize models, and make informed decisions about system adjustments. For example, a weekly report might highlight a drop in model accuracy, prompting data scientists to investigate potential causes such as changes in input data distributions or newly introduced biases.

Business leaders, on the other hand, require a high-level summary of the system's performance that emphasizes business impact. These reports should focus on how the ML system is contributing to organizational goals, such as increasing revenue, improving customer satisfaction, or reducing operational costs. For instance, a quarterly report for executives might showcase how the implementation of an ML-based recommendation engine has led to a 15% increase in sales conversions, providing clear evidence of the system's value to the business.

Regulators and compliance officers need reports that demonstrate adherence to relevant laws and ethical guidelines. These reports should include information on data privacy practices, bias mitigation efforts, and compliance with industry-specific regulations. For example, a compliance report might detail the steps taken to ensure that an ML model used in credit scoring does not discriminate against protected demographic groups, thereby ensuring that the organization remains in good standing with regulatory bodies.

Creating Effective Communication Pathways

Beyond regular reporting, it is crucial to establish clear and effective communication pathways for stakeholders. These channels ensure that stakeholders can quickly report issues, provide feedback, and stay informed about the system's ongoing status. Several strategies can be employed to create these communication pathways, each tailored to the specific needs and preferences of the stakeholder group.

One strategy is the creation of a dedicated support team that serves as the first point of contact for any issues or queries related to the ML system. This team should be equipped to handle a wide range of concerns, from technical glitches reported by engineers to questions about data usage raised by compliance officers. By centralizing support, organizations can ensure that all stakeholder concerns are addressed efficiently and consistently.

Another effective communication pathway is the development of an online portal or dashboard that provides stakeholders with real-time access to key metrics and system updates. For example, a custom dashboard for business leaders might display real-time KPIs, alerting them to any significant changes in system performance. This proactive approach allows stakeholders to stay informed without needing to wait for the next scheduled report, empowering them to make timely decisions based on the most current information.

Regular meetings, whether virtual or in-person, also play a vital role in maintaining open lines of communication. These meetings can be scheduled weekly, monthly, or quarterly, depending on the needs of the project and the stakeholders involved. For example, a monthly meeting with data scientists and engineers might focus on technical deep dives, where team members discuss recent challenges, share insights, and collaborate on potential solutions. On the other hand, a quarterly meeting with business leaders and executives might be more strategic, focusing on long-term goals, emerging opportunities, and the overall impact of the ML system on the organization.

Tailoring Communication Strategies to Different Audiences

Tailoring communication strategies to different audiences is essential for ensuring that the information provided is both relevant and accessible. This customization not only facilitates better understanding but also encourages active participation from all stakeholders, fostering a collaborative environment where everyone feels engaged and valued.

For technical stakeholders, such as data scientists and engineers, communication should be data-driven and detail-oriented. These stakeholders need access to raw data, in-depth analyses, and technical explanations that allow them to fully understand the system's behavior and performance. For example, a technical report might include data visualizations that show the distribution of errors across different model inputs, helping engineers identify patterns that could indicate underlying issues.

In contrast, communication with business leaders should focus on the strategic implications of the ML system's performance. Rather than delving into the technical details, reports and updates for this audience should highlight how the system's outcomes align with business objectives. For instance, a presentation to the executive team might include a case study demonstrating how the ML system has optimized supply chain operations, resulting in significant cost savings and faster delivery times.

When communicating with regulators and compliance officers, it is important to emphasize transparency, accountability, and adherence to legal and ethical standards. Reports for this audience should include clear documentation of compliance efforts, such as how data is anonymized, how bias is monitored and mitigated, and how the organization ensures that its ML models operate within the bounds of relevant regulations. Providing this level of detail not only helps build trust with regulatory bodies but also protects the organization from potential legal and ethical risks.

Templates for Stakeholder Reports

To streamline communication and ensure consistency, organizations can develop templates for stakeholder reports. These templates should be designed to meet the specific needs of each stakeholder group, ensuring that all relevant information is included in a format that is easy to understand and act upon.

Template for Technical Stakeholders (Data Scientists and Engineers)

For data scientists and engineers, a template might include sections for model performance metrics, error analysis, data drift reports, and recommendations for improvement. This template would ensure that every report provides a comprehensive overview of the technical aspects of the ML system, allowing these stakeholders to quickly identify and address any issues.

Title: ML System Performance Report – [Date]

1. **Overview:**

   - **System Name:** [ML System Name]

   - **Report Date:** [Date]

   - **Prepared By:** [Name/Team]

   - **Purpose:** To provide a detailed technical overview of the ML system's performance, including key metrics, error analysis, and recommendations.

2. **Model Performance Metrics:**

   - **Model Accuracy:** [Accuracy %]

   - **Precision:** [Precision %]

   - **Recall:** [Recall %]

   - **F1 Score:** [F1 Score %]

- **AUC-ROC:** [AUC-ROC Value]

- **Model Drift Detection:** [Description of detected drifts and their impact]

3. **Error Analysis:**

- **Error Rate:** [Error Rate %]

- **Confusion Matrix:** [Insert Confusion Matrix Image/Table]

- **Top Sources of Error:** [List of features or scenarios causing the most errors]

- **Root Cause Analysis:** [Detailed analysis of why these errors are occurring]

4. **Data Drift Reports:**

- **Drift Detection Method:** [Method Used]

- **Detected Drifts:** [Details of data drift – when, where, how much]

- **Impact on Model:** [Explanation of how drift affects model performance]

5. **Recommendations for Improvement:**

- **Model Tuning:** [Suggested tuning of hyperparameters or retraining schedules]

- **Data Collection Adjustments:** [Suggestions for collecting more representative or diverse data]

- **Feature Engineering:** [Recommendations for new or revised feature sets]

6. **Action Items:**

- **Immediate Actions:** [Actions that need to be taken immediately]

- **Future Considerations:** [Long-term improvements or changes to consider]

Attachments:

- [Additional data files, logs, charts, etc.]

Template for Business Leaders

A template for business leaders, on the other hand, might focus on business impact metrics, such as ROI, customer engagement, and operational efficiency. This template could include a section for key takeaways, where the most critical insights are summarized, and a section for action items, where recommendations for future initiatives are outlined. By using this template, business leaders can quickly grasp the strategic value of the ML system and make informed decisions about resource allocation and project prioritization.

Title: Quarterly ML System Business Impact Report – [Date]

1. **Executive Summary:**

   - **System Name:** [ML System Name]

   - **Report Date:** [Date]

   - **Prepared By:** [Name/Team]

   - **Overview:** Summary of the ML system's performance and impact on business goals.

2. **Key Business Metrics:**

   - **Revenue Impact:** [Amount or % increase in revenue attributable to the ML system]

   - **Customer Satisfaction:** [Customer feedback scores or engagement metrics]

   - **Operational Efficiency:** [Metrics such as time saved, cost reductions, or process optimizations]

   - **Market Performance:** [Any market share gains or competitive advantages realized]

3. **Strategic Insights:**

   - **Growth Opportunities:** [Identify areas where the ML system could be further leveraged]

   - **Challenges and Risks:** [Highlight any risks or obstacles and proposed mitigation strategies]

   - **Impact on KPIs:** [Discuss how the ML system has influenced key business KPIs]

4. **Case Study/Success Story:**

    – **Example:** [Provide a specific instance where the ML system drove significant value]

    – **Outcome:** [Results achieved and their impact on the business]

5. **Recommendations for Leadership:**

    – **Investment Needs:** [Suggestions for where further investment could yield greater returns]

    – **Strategic Adjustments:** [Any changes to business strategy that could enhance the effectiveness of the ML system]

    – **Resource Allocation:** [Recommendations for shifting resources to support the ML system]

6. **Action Items:**

    – **Executive Actions:** [Decisions needed from leadership]

    – **Operational Actions:** [Tasks for implementation teams based on the report findings]

    Attachments:

    – [Charts, graphs, and additional documentation]

Template for Regulators and Compliance Officers

For regulators and compliance officers, a template might include sections for legal compliance, ethical considerations, data privacy practices, and audit results. This template would ensure that all reports provide a thorough account of the organization's efforts to maintain compliance with relevant regulations, offering transparency and accountability that can help build trust with external stakeholders.

Title: ML System Compliance and Governance Report – [Date]

1. **Overview:**

    – **System Name:** [ML System Name]

    – **Report Date:** [Date]

- **Prepared By:** [Name/Team]

- **Purpose:** To provide a comprehensive report on the compliance and governance aspects of the ML system.

2. **Legal Compliance:**

- **Data Privacy:** [Details of how data privacy laws (GDPR, CCPA, etc.) are being complied with]

- **Data Protection Measures:** [Encryption, anonymization techniques, access controls, etc.]

- **Audit Logs:** [Summary of audit logs tracking data access and model changes]

- **Model Transparency:** [Details on how the system adheres to transparency regulations]

3. **Ethical Considerations:**

- **Bias and Fairness:** [Summary of bias detection and mitigation efforts]

- **Explainability:** [Overview of how the system ensures model explainability for end users]

- **Ethical Audits:** [Results of any recent ethical audits or reviews]

4. **Compliance Audits:**

- **Recent Audits:** [Details of recent compliance audits – who conducted them, what was found]

- **Audit Findings:** [Key findings from the audits]

- **Corrective Actions:** [Actions taken to address any issues identified in the audits]

5. **Data Governance:**

- **Data Management Practices:** [Overview of data governance practices]

- **Data Lifecycle Management:** [Details of how data is collected, stored, processed, and deleted]

- **Compliance with Data Regulations:** [How the system ensures compliance with specific data regulations]

6. **Recommendations for Compliance Improvement:**

   - **Gaps Identified:** [Any gaps in compliance that need to be addressed]

   - **Proposed Solutions:** [Recommendations for addressing these gaps]

   - **Training Needs:** [Suggestions for additional training on compliance for the team]

   Attachments:

   - [Audit reports, legal documents, and additional compliance-related materials]

These templates can be adapted and expanded to fit specific organizational needs, ensuring that all stakeholders receive the information they require in a format that best suits their roles and responsibilities.

Building a Collaborative Environment

Effective stakeholder engagement is not just about communication; it is also about building a collaborative environment where stakeholders feel empowered to contribute their insights, feedback, and expertise. By fostering a culture of collaboration, organizations can ensure that all voices are heard, leading to more innovative solutions and a more robust ML system.

One way to build this collaborative environment is by involving stakeholders in the decision-making process. For example, when planning a major update to the ML system, organizations can hold workshops or brainstorming sessions where stakeholders from different departments come together to discuss potential changes, share concerns, and offer suggestions. This collaborative approach ensures that the final decision reflects a broad range of perspectives, increasing the likelihood of successful implementation.

Another strategy for fostering collaboration is by recognizing and rewarding stakeholder contributions. For instance, an organization might implement a recognition program where stakeholders who provide valuable feedback or identify critical issues are publicly acknowledged and rewarded. This not only encourages active participation but also reinforces the importance of stakeholder engagement in the success of the ML system.

Adapting Communication Strategies over Time

As the ML system evolves and stakeholder needs change, it is important to regularly review and adapt communication strategies. This ensures that the engagement process remains effective and continues to meet the needs of all stakeholders.

For example, as the ML system matures, technical stakeholders may require more advanced metrics and deeper analyses, while business leaders might shift their focus to new areas of impact, such as sustainability or customer retention. By regularly reviewing stakeholder feedback and staying attuned to these changing needs, organizations can adjust their communication strategies accordingly, ensuring that all stakeholders remain informed and engaged.

Regularly updating communication templates and channels is also essential for maintaining relevance. For instance, as new communication technologies emerge, organizations might adopt more interactive reporting formats, such as dashboards that allow stakeholders to explore data in real time or virtual reality meetings that facilitate more immersive collaboration. By embracing these innovations, organizations can enhance the effectiveness of their stakeholder engagement efforts and ensure that their communication strategies remain cutting edge.

Tools and Technologies for Monitoring

Monitoring machine learning (ML) systems is a critical task that ensures the smooth operation, reliability, and effectiveness of these complex technologies. Given the multifaceted nature of ML systems, monitoring requires a comprehensive approach that covers model performance, data pipeline integrity, and overall system health. A wide array of tools and technologies have been developed to address these needs, each offering unique features and functionalities tailored to different aspects of the monitoring process. The integration of these tools into a cohesive monitoring framework is essential for maintaining robust ML systems that consistently deliver accurate and reliable results.

Model Monitoring Platforms

Model monitoring platforms are specialized tools designed to track and assess the performance of machine learning models in real time. These platforms are crucial for organizations that rely on ML models to drive decision-making processes, as they provide continuous insights into how well models are performing and whether they remain aligned with their intended goals. The key features of these platforms typically include real-time alerts, interactive dashboards, and automated reporting, which together enable proactive management of ML models.

One prominent example of a model monitoring platform is **Fiddler AI**, which offers a comprehensive suite of tools for monitoring model performance, ensuring fairness, and maintaining transparency. Fiddler AI's platform allows organizations to track key performance indicators (KPIs) such as accuracy, precision, and recall while also monitoring for potential biases that could affect model outputs. For instance, in the context of a credit scoring model, Fiddler AI can continuously assess whether the model is providing equitable credit decisions across different demographic groups, thereby ensuring compliance with fairness regulations.

Another powerful platform is **WhyLabs**, which specializes in monitoring for model drift—a situation where the performance of an ML model degrades over time due to changes in the underlying data. WhyLabs offers automated drift detection and diagnostics, alerting data scientists when a model's predictions begin to diverge from expected patterns. For example, an e-commerce platform using ML to recommend products might experience a shift in consumer behavior during a holiday season, leading to changes in purchasing patterns. WhyLabs would detect this drift, allowing the organization to retrain or adjust the model to maintain its effectiveness.

Data Pipeline Monitoring Solutions

Data pipeline monitoring solutions are designed to oversee the flow of data through an ML system, ensuring that data remains accurate, consistent, and relevant as it moves from source to model. These tools are particularly important for maintaining the quality of data that feeds into ML models, as any issues with data integrity can directly impact the accuracy and reliability of model predictions.

Great Expectations is an example of a data pipeline monitoring tool that focuses on data validation. It allows organizations to define, test, and enforce expectations about their data, ensuring that it meets predefined quality standards before being processed by ML models. For instance, a healthcare organization using ML to predict patient outcomes might use Great Expectations to validate that incoming data is complete, correctly formatted, and within expected ranges. This prevents scenarios where missing or erroneous data could lead to inaccurate predictions, ultimately safeguarding patient care.

Monte Carlo is another tool that specializes in data observability, providing end-to-end visibility into data pipelines. Monte Carlo's platform can automatically detect and alert teams to data anomalies, such as unexpected spikes in missing values or sudden changes in data distributions. For a financial institution using ML to detect fraudulent transactions, Monte Carlo could identify unusual patterns in transaction data that might indicate a breach in data quality, allowing the organization to investigate and address the issue before it affects the model's performance.

Data lineage tracking is another critical feature offered by data pipeline monitoring tools, enabling organizations to trace the origin and transformation of data throughout the pipeline. **OpenLineage** is an example of a tool that provides this capability, helping organizations understand how data flows through various stages of processing and how it influences model outputs. This is particularly important for regulatory compliance, as it allows organizations to demonstrate transparency and accountability in their data practices.

General-Purpose Monitoring and Logging Tools

While model monitoring platforms and data pipeline solutions focus on specific aspects of ML systems, general-purpose monitoring and logging tools provide a broader view of the overall system health and infrastructure performance. These tools are essential for ensuring that the underlying infrastructure supporting ML systems—such as servers, databases, and networks—remains stable and efficient.

Prometheus is a widely used open-source monitoring tool that specializes in collecting and querying time-series data, making it ideal for tracking infrastructure metrics such as CPU usage, memory consumption, and network latency. In the context of an ML system deployed in a cloud environment, Prometheus can monitor the resource utilization of virtual machines running ML models, helping organizations ensure that their infrastructure is scalable and responsive to changing workloads.

Grafana is often used in conjunction with Prometheus to create interactive dashboards that visualize system performance metrics in real time. For example, an organization using ML for real-time fraud detection might set up a Grafana dashboard to display key metrics such as transaction processing times, alert generation rates, and model inference latencies. This visual representation allows stakeholders to quickly assess the health of the ML system and respond to any emerging issues.

Elasticsearch is another powerful tool that supports the logging and analysis of large volumes of data generated by ML systems. By indexing and searching logs, Elasticsearch helps organizations troubleshoot issues, monitor for security breaches, and optimize system performance. For an e-commerce platform using ML to personalize user experiences, Elasticsearch could be used to analyze logs from user interactions, identifying patterns that indicate performance bottlenecks or areas for improvement.

Comparative Analysis of Monitoring Tools

When selecting monitoring tools for ML systems, organizations must consider their specific needs, the complexity of their ML models, and the regulatory environment in which they operate. A comparative analysis of the tools mentioned can help organizations determine which combination of platforms will provide the most effective monitoring framework.

For instance, an organization in the healthcare sector might prioritize data validation and compliance features, making tools like Great Expectations and OpenLineage essential components of their monitoring strategy. These tools ensure that data feeding into ML models meets stringent quality standards and that the organization can trace data lineage for regulatory purposes. In contrast, a technology company focused on real-time applications might prioritize tools like Prometheus and Grafana to monitor infrastructure performance and ensure low-latency responses from their ML systems.

Integrating model monitoring platforms like Fiddler AI or WhyLabs with general-purpose tools such as Elasticsearch can also provide a comprehensive monitoring solution that covers both the performance of ML models and the health of the supporting infrastructure. This combination allows organizations to detect issues at multiple levels, from data quality problems that could affect model predictions to infrastructure failures that might disrupt system availability.

Case Study: Integrated Use of Monitoring Tools

Consider a financial institution that has implemented a comprehensive ML system to detect fraudulent transactions in real time. The system leverages multiple data sources, including transaction history, user behavior, and geolocation data, to identify potentially fraudulent activity. To ensure the effectiveness and reliability of this system, the institution integrates various monitoring tools into a cohesive framework.

First, the institution uses WhyLabs to monitor the performance of its fraud detection models. WhyLabs provides real-time alerts for model drift, ensuring that the models remain accurate as transaction patterns evolve. For example, during a major shopping holiday, the institution notices an increase in false positives due to changes in customer behavior. WhyLabs detects this drift and alerts the data science team, who quickly retrain the models using updated data.

Simultaneously, the institution employs Great Expectations to validate incoming transaction data before it is processed by the ML models. This tool ensures that data is complete, correctly formatted, and within expected ranges, preventing issues that could compromise the accuracy of fraud predictions.

Prometheus and Grafana are used to monitor the underlying infrastructure supporting the ML system. The institution sets up Grafana dashboards to visualize server performance metrics, such as CPU usage and memory consumption, ensuring that the system remains responsive even during peak transaction periods.

Elasticsearch is used to log and analyze system activity, including user interactions and model inference results. By indexing these logs, the institution can quickly investigate any anomalies, such as sudden increases in false positives or delayed transaction processing times, and take corrective actions.

This integrated use of monitoring tools allows the financial institution to maintain a robust, reliable, and compliant ML system that effectively detects and prevents fraudulent transactions. By covering all aspects of the ML lifecycle—from data validation to model performance and infrastructure health—the institution ensures that its system operates smoothly and delivers consistent results, even in the face of evolving threats and changing data patterns.

Challenges and Solutions in Monitoring and Maintenance

Monitoring and maintaining machine learning (ML) systems is a complex and ongoing process that presents various challenges. These challenges include managing data drift, dealing with model complexity, and ensuring scalability. Each of these areas requires targeted solutions to maintain the robustness and reliability of ML systems. Successfully navigating these challenges demands both technical expertise and strategic planning, as organizations must be prepared to adapt to evolving circumstances while ensuring that their ML models continue to perform at high standards.

Managing Data Drift

One of the most significant challenges in ML system maintenance is managing data drift. Data drift occurs when the statistical properties of the input data change over time, which can lead to performance degradation in the model. This issue is particularly problematic because it can result in biased predictions and reduced accuracy, undermining the effectiveness of the model. Detecting data drift is crucial to maintaining the model's reliability.

Organizations can address data drift by implementing advanced data pipeline monitoring tools capable of detecting changes in data distributions. These tools provide real-time alerts when substantial drift is detected, enabling data scientists to respond quickly. Regular model retraining is another effective strategy to counteract data drift. By frequently updating the model with new data, organizations can ensure that the model adapts to evolving patterns, maintaining its accuracy and relevance.

For example, a retail company might use ML models to predict sales trends. If consumer behavior shifts due to seasonal changes or economic factors, the input data will also change. Without monitoring for data drift, the model's predictions could become increasingly inaccurate. By using a tool that monitors for shifts in the distribution of purchasing patterns and retrains the model accordingly, the company can keep its sales forecasts reliable.

Tackling Model Complexity

Model complexity is another challenge in the monitoring and maintenance of ML systems, especially with models that involve deep learning or large-scale algorithms. These models can be difficult to monitor and maintain due to their intricate structures and the vast number of parameters they encompass. The opaque nature of such models often makes it challenging to understand their decision-making processes, which can be a barrier to ensuring transparency and trust.

To address this challenge, organizations should employ explainability techniques such as SHAP (SHapley Additive exPlanations) and LIME (Local Interpretable Model-agnostic Explanations). These techniques provide insights into how the model arrives at its predictions, enhancing transparency and making it easier to identify and mitigate biases. Simplifying model architectures, where feasible, or adopting hybrid approaches that combine interpretable models with more complex ones can also help strike a balance between performance and transparency.

Consider a healthcare organization that uses a deep learning model to predict patient outcomes based on medical records. The complexity of the model might make it difficult for healthcare providers to understand how specific predictions are generated, potentially leading to trust issues. By integrating SHAP or LIME into the model's monitoring framework, the organization can provide explanations for individual predictions, making the model's decisions more transparent and increasing trust among healthcare professionals.

Ensuring Scalability

Scalability is a critical challenge as organizations deploy more ML models and manage larger datasets. As the demand for ML applications grows, the need for scalable monitoring solutions becomes increasingly important. These solutions must handle larger workloads and provide real-time insights to prevent bottlenecks and system failures.

Implementing scalable monitoring tools is essential for maintaining system performance as the volume of data and the number of models increase. Organizations must also optimize both the model and the underlying infrastructure to ensure stability and responsiveness under heavy loads. Regular load testing and capacity planning are crucial components of this strategy, helping organizations anticipate scalability issues and address them before they impact performance.

For instance, a financial services firm using ML to detect fraudulent transactions might experience a significant increase in transaction volume during peak shopping periods. To ensure the system can handle this increased load, the firm would need to implement scalable monitoring tools that can provide real-time insights into system performance. By optimizing the model and infrastructure, the firm can maintain system stability and continue to detect fraud effectively, even under heavy traffic conditions.

Addressing Data Quality

Data quality is fundamental to the success of any ML system, and maintaining it is an ongoing challenge. Ensuring data integrity across the entire pipeline is crucial, as any errors or inconsistencies in the data can directly affect the accuracy of the model's predictions. Effective data pipeline monitoring involves more than just tracking the flow of data; it requires validation checks at various stages to catch issues such as missing values or inconsistencies before they impact the model.

Organizations should employ monitoring tools that can detect anomalies in data distributions and alert stakeholders to potential issues. Regular audits of the data pipeline can also help identify and resolve data quality problems. By maintaining high standards of data integrity, organizations can support the reliability and accuracy of their ML models.

For example, a telecommunications company might use ML models to predict network failures based on sensor data. If the data feeding into the model contains errors or inconsistencies, the predictions could be inaccurate, leading to costly downtime. By implementing a robust data pipeline monitoring system that validates data at each stage and conducts regular audits, the company can ensure that the data used by the model is accurate and consistent, thereby improving the reliability of the predictions.

Navigating Compliance

Compliance with regulations and ethical standards is a critical aspect of ML system monitoring and maintenance. Organizations must ensure that their ML systems adhere to data privacy requirements, such as those outlined in GDPR or CCPA, and maintain audit trails and access controls to manage compliance effectively. Automated compliance checks integrated into the monitoring framework can continuously assess the system's adherence to these standards, providing timely alerts if any issues are detected.

Regular compliance audits complement these automated checks by offering a thorough review of the system's adherence to governance standards. These audits should involve detailed assessments by legal and regulatory experts to identify areas where improvements are needed. By integrating audit findings into the monitoring framework, organizations can address compliance issues proactively and adjust practices as necessary to stay aligned with evolving regulations.

For instance, a multinational corporation using ML for customer analytics across different regions must navigate varying data privacy regulations. Automated compliance checks can ensure that the system adheres to each region's specific requirements, while regular audits conducted by a compliance team can provide a deeper review of data handling practices. This combined approach helps the organization manage compliance effectively, reducing the risk of regulatory breaches.

Fostering Stakeholder Engagement

Effective communication with stakeholders is essential for the successful monitoring and maintenance of ML systems. Stakeholders, including data scientists, engineers, business leaders, and end users, need to be informed about the system's performance and any issues that arise. Establishing clear communication channels and regular reporting processes helps keep stakeholders updated on key metrics and developments.

Tailoring reports to the specific needs of different stakeholders enhances the relevance of the information shared. For example, engineers may require detailed technical reports, while business leaders might prefer high-level summaries that focus on the impact of the ML system on business objectives. Regular meetings and feedback sessions also facilitate ongoing engagement and collaboration, helping to address concerns and improve system performance.

Consider a technology company that uses ML to personalize customer experiences on its platform. Engaging stakeholders through regular updates ensures that data scientists understand the technical performance of the models, while business leaders are aware of how these models are driving customer engagement and revenue. This collaborative approach helps the organization align technical and business goals, leading to more effective ML system management.

Continuous Improvement

Continuous improvement is key to maintaining the effectiveness of ML systems over time. Implementing a model versioning system allows organizations to track and compare different versions of the model, providing insights into which version performs best. This iterative approach helps refine the model and ensures that improvements are systematically incorporated.

Automated model retraining is another crucial strategy for continuous improvement. By setting up processes for automatic retraining, organizations can ensure that the model is regularly updated with new data, allowing it to adapt to changing patterns and trends. This proactive approach helps mitigate the impact of data drift and other changes that might affect model performance.

For example, an e-commerce platform might use a recommendation model that needs to be regularly updated as new products are added and customer preferences evolve. By implementing a model versioning system and automated retraining, the platform can continuously refine the model, ensuring that it remains effective in driving sales and improving customer satisfaction.

Integrating Diverse Monitoring Tools

Utilizing general-purpose monitoring tools such as Prometheus, Grafana, and Elasticsearch enhances the overall monitoring framework. Prometheus offers powerful capabilities for collecting and storing metrics, enabling detailed analysis and visualization of system performance. Grafana complements Prometheus by providing advanced visualization options, allowing users to create customizable dashboards that reflect various performance indicators. Elasticsearch plays a crucial role in log management, enabling efficient indexing, searching, and analysis of logs generated by the system. Together, these tools provide a comprehensive view of system health and performance, supporting effective monitoring and issue detection.

Integrating diverse monitoring tools into a cohesive framework ensures comprehensive coverage of all critical aspects of ML system maintenance. Combining specialized model monitoring platforms with data pipeline solutions and general-purpose tools creates a robust system capable of addressing a wide range of monitoring needs. This holistic approach enables organizations to maintain visibility into model performance, data quality, and infrastructure health. By leveraging the strengths of different tools and technologies, organizations can enhance their monitoring capabilities and ensure that their ML systems remain effective and reliable. This integration supports ongoing improvements and adjustments, helping organizations meet the demands of evolving machine learning operations.

Summary

This chapter focuses on the essential practices for monitoring and maintaining machine learning (ML) systems to ensure they remain reliable, robust, and aligned with their intended goals. ML systems and their environments are constantly evolving, which makes continuous oversight critical to address issues such as model drift, bias, and performance degradation. A comprehensive monitoring framework is necessary, incorporating real-time tracking, automated alerts, and well-defined key performance indicators (KPIs) that help organizations keep their models effective and responsive to changes.

It also highlights the importance of data pipeline monitoring, system stability, and scalability. Ensuring the integrity and accuracy of the data feeding into ML models is vital, as any disruption in the data pipeline can severely impact model performance. Advanced tools and techniques, such as machine learning-based anomaly detection and predictive analytics, are essential for maintaining data quality and consistency. Furthermore, the stability and scalability of the underlying infrastructure must be managed using tools like Prometheus, Grafana, and Kubernetes, which help organizations handle increasing workloads while maintaining system responsiveness.

Lastly, the chapter emphasizes the need for continuous improvement and effective stakeholder communication. Regular model retraining and updates are crucial to adapt to evolving data and operational environments, ensuring the models remain accurate and relevant. Clear communication with stakeholders—ranging from data scientists to business leaders—is necessary to keep everyone informed about the system's performance and any issues that arise. This collaborative approach, combined with advanced monitoring techniques and a focus on continuous improvement, enables organizations to sustain their ML systems over time, maintaining their effectiveness and ethical integrity.

CHAPTER 8

Regulatory Compliance and Risk Management

Introduction to Regulatory Compliance for Machine Learning

Regulatory compliance in the context of machine learning (ML) encompasses adhering to laws, standards, and guidelines designed to ensure that ML systems operate within legal and ethical boundaries. This compliance is critical due to the complex nature of ML systems, which involve processing vast amounts of data and making automated decisions that can significantly impact individuals and organizations. As ML technology becomes increasingly pervasive, understanding and implementing regulatory requirements is essential for maintaining trust and mitigating risks.

One of the primary regulations impacting ML systems is the General Data Protection Regulation (GDPR), which governs data protection and privacy for individuals within the European Union (EU). GDPR sets stringent requirements for how personal data is collected, stored, and used, emphasizing transparency, data subject rights, and accountability. For ML systems, compliance with GDPR involves ensuring that data used for training models is collected with explicit consent, providing individuals with access to their data, and implementing measures to protect data from breaches. Additionally, GDPR mandates that organizations conduct impact assessments to evaluate how their ML models affect privacy.

The California Consumer Privacy Act (CCPA) is another significant regulation that affects ML systems, particularly for organizations operating in California. CCPA provides consumers with rights regarding their personal data, including the right to know what data is collected, the right to request deletion of their data, and the right to opt out of

485

A. Nandan Prasad, *Introduction to Data Governance for Machine Learning Systems*, https://doi.org/10.1007/979-8-8688-1023-7_8

data sales. For ML systems, compliance with CCPA involves implementing mechanisms for data access and deletion requests, ensuring transparency in data practices, and providing options for consumers to manage their data preferences.

The Health Insurance Portability and Accountability Act (HIPAA) is a critical regulation for ML systems that handle health information in the United States. HIPAA sets standards for the protection of patient health information, emphasizing confidentiality, integrity, and security. For ML systems used in healthcare, compliance with HIPAA requires implementing stringent security measures to protect health data, conducting risk assessments, and ensuring that ML models adhere to privacy and security requirements. This includes safeguarding data during collection, processing, and storage and ensuring that models do not expose sensitive health information.

In addition to these data protection regulations, the Arvind, Justin, and Laura (AJL) Model Bias Framework is an important consideration for ensuring fairness and accountability in ML systems. This framework focuses on identifying and mitigating bias in ML models, which can lead to discriminatory outcomes and undermine the effectiveness of the systems. The AJL Framework provides guidelines for assessing model fairness, conducting bias audits, and implementing strategies to address and reduce bias. Compliance with this framework involves regularly evaluating ML models for potential biases, adjusting algorithms and data practices to promote fairness, and ensuring that models do not disproportionately impact specific groups.

Other considerations for regulatory compliance in ML include adhering to industry-specific regulations and standards. For instance, financial institutions may need to comply with regulations such as the Payment Card Industry Data Security Standard (PCI DSS) to protect payment information, while organizations in the education sector must consider regulations related to student data privacy. Each industry may have unique requirements that impact how ML systems should be designed, implemented, and monitored to ensure compliance.

Ethical considerations also play a crucial role in regulatory compliance for ML systems. Ensuring that ML models are transparent, explainable, and accountable is essential for maintaining public trust and addressing ethical concerns. Organizations must implement practices that allow for transparency in how models make decisions, provide explanations for automated outcomes, and establish accountability mechanisms for addressing errors or issues. This involves adopting best practices for model documentation, implementing explainable AI techniques, and engaging with stakeholders to address ethical concerns.

Organizations must stay informed about evolving regulatory requirements and industry standards. As regulations continue to develop and new guidelines emerge, organizations must adapt their compliance practices to meet changing expectations. This involves monitoring updates from regulatory bodies, participating in industry forums, and engaging with legal and compliance experts to ensure that ML systems remain aligned with current regulations and best practices. Regularly reviewing and updating compliance policies and procedures is essential for maintaining adherence to regulatory requirements and managing potential risks.

Importance of Compliance for Mitigating Risks and Building Trust

One of the primary risks addressed by compliance is data privacy violations. Data privacy laws, such as the General Data Protection Regulation (GDPR) and the California Consumer Privacy Act (CCPA), are designed to protect individuals' personal information. Non-compliance with these regulations can lead to severe consequences, including legal actions, substantial fines, and reputational damage. By implementing robust compliance measures, organizations can safeguard personal data, uphold privacy rights, and mitigate the risk of data breaches that could compromise sensitive information.

Algorithmic bias is another significant concern that compliance can help address. ML models can inadvertently perpetuate or amplify biases present in the training data, leading to unfair or discriminatory outcomes. Compliance with frameworks like the AJL Model Bias Framework ensures that organizations actively monitor and address bias in their algorithms. By adhering to best practices for bias detection and mitigation, organizations can promote fairness, avoid discriminatory practices, and enhance the ethical integrity of their ML systems.

Model explainability and transparency are critical aspects of compliance that contribute to building trust with users and stakeholders. Transparent ML systems that provide clear explanations for their predictions and decisions help users understand how outcomes are generated. Compliance with transparency standards requires organizations to implement explainable AI techniques, document model decisions, and offer insights into the model's inner workings. This transparency fosters trust by enabling users to validate and challenge automated decisions, ensuring that ML systems operate in an accountable and responsible manner.

Security vulnerabilities are another area where compliance plays a vital role in risk mitigation. ML systems are susceptible to various security threats, including data breaches, adversarial attacks, and unauthorized access. Adhering to security regulations and implementing best practices for data protection can help organizations safeguard their systems against these threats. Compliance with standards such as the Health Insurance Portability and Accountability Act (HIPAA) and the Payment Card Industry Data Security Standard (PCI DSS) ensures that sensitive data is protected through robust security measures, reducing the risk of breaches and enhancing overall system integrity.

Compliance fosters a culture of accountability and ethical responsibility within organizations. By adhering to regulatory requirements and ethical guidelines, organizations demonstrate their commitment to responsible practices and stakeholder interests. This commitment builds trust with customers, partners, and regulators, as it reflects a dedication to operating within legal and ethical boundaries. Establishing a culture of compliance also encourages ongoing vigilance and continuous improvement, further strengthening the organization's reputation and operational resilience.

Compliance can enhance operational efficiency by providing clear guidelines and frameworks for managing ML systems. Regulatory requirements often include specific procedures for data handling, model development, and performance monitoring. By following these guidelines, organizations can streamline their processes, reduce the likelihood of errors, and ensure that their ML systems operate effectively and ethically. This structured approach to compliance not only mitigates risks but also contributes to the overall efficiency and effectiveness of ML initiatives.

Another important consideration is the impact of compliance on organizational reputation. Organizations that prioritize compliance and demonstrate a commitment to ethical practices are more likely to earn the trust and confidence of their stakeholders. Positive public perception and trust can lead to competitive advantages, increased customer loyalty, and stronger business relationships. Conversely, non-compliance and associated risks can result in reputational damage that may be difficult to recover from, underscoring the importance of maintaining high standards of compliance.

Compliance with regulatory and ethical standards is essential for adapting to evolving legal and societal expectations. As technology advances and new regulations emerge, organizations must stay informed and agile in their compliance efforts. This involves monitoring regulatory changes, engaging with industry experts, and updating compliance practices to reflect new requirements. By proactively addressing these changes, organizations can navigate the evolving landscape of ML regulation, manage emerging risks, and continue to build trust with their stakeholders.

Compliance is a cornerstone of effective risk management and trust-building in ML systems. Addressing risks such as data privacy violations, algorithmic bias, and security vulnerabilities and ensuring model explainability are crucial for maintaining the integrity and effectiveness of ML initiatives. By adhering to regulatory requirements and ethical guidelines, organizations can protect sensitive data, promote fairness, enhance transparency, and safeguard against security threats. Ultimately, a strong commitment to compliance fosters a culture of accountability, operational efficiency, and trust, supporting the long-term success and sustainability of ML systems.

Case Studies
Adapting to GDPR in Machine Learning Practices

The General Data Protection Regulation (GDPR) represents one of the most comprehensive and stringent data protection laws globally. Its primary focus is to safeguard the personal data of individuals within the European Union (EU) and to provide them with greater control over how their data is used. For companies employing ML, GDPR compliance requires a thorough understanding of data processing activities and the ability to demonstrate that these activities are conducted in accordance with the law.

One leading company that has successfully navigated GDPR compliance is Google. Given its vast data-driven operations, Google has had to implement numerous measures to ensure that its ML systems adhere to GDPR principles. For instance, Google developed robust data anonymization techniques to minimize the risk of identifying individuals from processed data. The company also implemented transparency mechanisms that allow users to understand how their data is used in ML models, along with offering users the ability to exercise their rights to access, rectify, or delete their personal data. Google's approach to GDPR compliance in ML underscores the importance of integrating privacy by design, ensuring that data protection principles are embedded into the development and deployment of ML models from the outset.

Another example is the financial services company HSBC, which has leveraged ML for fraud detection and customer risk profiling. To comply with GDPR, HSBC had to ensure that their ML models did not infringe on the privacy rights of individuals. This was particularly challenging given the nature of the data used in these models, which often included sensitive financial information. HSBC responded by implementing strict data access controls, ensuring that only authorized personnel could access certain types of data. They also employed techniques such as data minimization, where only the necessary data required for the ML models was processed. HSBC's experience highlights the importance of a comprehensive data governance strategy to achieve GDPR compliance, especially when dealing with sensitive financial data.

Navigating CCPA Regulations in Machine Learning

The California Consumer Privacy Act (CCPA) imposes similar yet distinct requirements compared to GDPR, with a particular focus on giving California residents greater control over their personal information. CCPA compliance in ML requires companies to be transparent about how they collect, use, and share personal data, and to provide consumers with the ability to opt out of data sales.

Netflix, a major player in the entertainment industry, has adapted its ML practices to comply with CCPA requirements. Netflix uses ML algorithms extensively to personalize content recommendations and enhance user experiences. To align with CCPA, Netflix had to ensure that its data processing activities were transparent and that users had the ability to control their data. The company provided clear privacy notices explaining how data is used in its recommendation algorithms and offered users the option to delete their data if they so choose. Moreover, Netflix implemented strict controls to prevent the unauthorized sharing of user data with third parties. The company's approach to CCPA compliance emphasizes the need for transparency and user control in ML practices, particularly when personalizing services based on user data.

Retail giant Walmart also faced challenges in aligning its ML practices with CCPA regulations. Walmart uses ML models for various purposes, including inventory management, customer insights, and targeted marketing. To comply with CCPA, Walmart had to re-evaluate its data-sharing practices, particularly those involving third-party vendors and partners. The company introduced mechanisms that allowed consumers to opt out of data sharing and ensured that all third-party agreements were reviewed and updated to reflect CCPA requirements. Walmart's experience showcases the complexities of managing data sharing and third-party relationships in the context of CCPA compliance, highlighting the need for rigorous oversight and continuous monitoring.

Ensuring HIPAA Compliance in Healthcare Machine Learning

The Health Insurance Portability and Accountability Act (HIPAA) is a critical regulation for companies operating in the healthcare sector, particularly those utilizing ML for medical diagnostics, patient care, and health data analytics. HIPAA mandates stringent safeguards to protect the privacy and security of health information, and non-compliance can result in severe penalties.

IBM Watson Health, a leader in healthcare technology, has successfully integrated HIPAA compliance into its ML operations. IBM Watson uses ML models to analyze vast amounts of healthcare data, providing insights for personalized medicine and clinical decision support. To comply with HIPAA, IBM Watson implemented encryption and access controls to protect health information. The company also adopted data de-identification techniques to ensure that health data used in ML models could not be traced back to individual patients. IBM Watson's approach illustrates the importance of robust security measures and data protection techniques in achieving HIPAA compliance, particularly when dealing with sensitive health data.

Another example is the Mayo Clinic, which employs ML for predictive analytics and patient outcome forecasting. To ensure HIPAA compliance, the Mayo Clinic established a rigorous framework for data governance and security. This framework included the use of encrypted data storage, regular audits of data processing activities, and the implementation of access controls to limit who could view or manipulate sensitive health information. The Mayo Clinic's experience underscores the need for continuous oversight and the importance of aligning ML practices with regulatory requirements to protect patient data effectively.

Fairness and Non-discrimination Regulations

Within the framework of machine learning (ML) governance, fairness and non-discrimination are essential pillars that ensure ML models are developed and deployed responsibly. Governance frameworks establish the policies, processes, and guidelines that organizations must follow to prevent bias and ensure equitable outcomes in automated decision-making systems. As ML applications become more pervasive in critical sectors such as finance, healthcare, and employment, it is imperative that these governance structures robustly address the risks of discrimination and bias. However, implementing fairness in ML presents significant technical challenges, including issues with biased datasets and the limitations of current debiasing methods.

The Challenge of Biased Datasets

One of the most significant technical challenges in achieving fairness in ML is the presence of bias in the datasets used to train models. Bias in data can arise from various sources, including historical discrimination, societal inequalities, and sampling errors. When biased data is used to train ML models, the models are likely to perpetuate or even exacerbate these biases, leading to unfair or discriminatory outcomes.

For example, in the context of employment, historical data might reflect past discriminatory practices that favored certain demographic groups over others. If an ML model is trained on such data, it may learn to replicate these biases, leading to decisions that unfairly disadvantage underrepresented groups. This issue is particularly concerning in high-stakes applications such as hiring, lending, and law enforcement, where biased outcomes can have severe consequences for individuals and communities.

Addressing the challenge of biased datasets requires a multifaceted approach. One strategy is to curate and preprocess the data to remove or mitigate biases before training the model. This may involve techniques such as data resampling, where the dataset is adjusted to ensure a more balanced representation of different demographic groups. However, this approach is not without its limitations. Resampling can lead to a loss of valuable information and may not fully eliminate the underlying biases in the data. Moreover, identifying and correcting for all potential biases in a dataset is a complex and often subjective task, as biases can be subtle and deeply embedded in the data.

Another approach is to incorporate fairness constraints directly into the model training process. This involves modifying the model's objective function to penalize biased outcomes, encouraging the model to produce fairer predictions. While this method can be effective in reducing bias, it often comes with trade-offs in terms of model accuracy. Achieving a balance between fairness and accuracy is a significant challenge, as models that are too heavily constrained may fail to capture important patterns in the data, leading to reduced performance.

Technical Limitations of Current Debiasing Methods

Despite ongoing research and development, current debiasing methods in ML have notable limitations that make achieving fairness a complex endeavor. One of the primary challenges is that most debiasing techniques are designed for specific types of bias and may not generalize well to different contexts or datasets. For example, methods that are effective in reducing gender bias in a particular dataset may not be applicable to addressing racial bias in another dataset. This lack of generalizability complicates the implementation of fairness across diverse ML applications.

One common debiasing technique is adversarial debiasing, where an additional model (the adversary) is trained to predict the sensitive attribute (such as race or gender) from the model's predictions. The main model is then trained to minimize this prediction, effectively removing the information about the sensitive attribute from its outputs. While adversarial debiasing can be effective in reducing bias, it often introduces instability in the model training process, leading to challenges in model convergence and performance. Additionally, adversarial debiasing requires careful tuning of hyperparameters, and its effectiveness can vary significantly depending on the specific dataset and application.

Another limitation of current debiasing methods is their reliance on well-defined fairness metrics. In practice, fairness is a multidimensional concept that can be difficult to quantify. Different fairness metrics, such as demographic parity, equalized odds, and predictive parity, capture different aspects of fairness, and optimizing for one metric may lead to trade-offs with others. For instance, achieving demographic parity, where the model's predictions are independent of the sensitive attribute, might result in a loss of accuracy for certain groups, leading to unfair outcomes. This complexity makes it challenging to implement a one-size-fits-all solution for fairness in ML, requiring organizations to carefully consider the specific context and goals of their applications when selecting and applying fairness metrics.

Addressing Bias in Model Development and Deployment

Effective ML governance frameworks must include processes for addressing bias not only during model training but also throughout the entire ML lifecycle, from data collection to model deployment. This requires a holistic approach that integrates technical solutions with organizational policies and practices.

During the data collection phase, organizations should prioritize the use of diverse and representative datasets that capture a wide range of perspectives and experiences. This involves actively seeking out and including data from underrepresented groups, as well as continuously monitoring the data for potential biases. In some cases, it may be necessary to supplement existing datasets with additional data from diverse sources to ensure that the model is trained on a more balanced and representative dataset.

During the model development phase, organizations should implement fairness audits that evaluate the model's performance across different demographic groups. These audits can help identify potential biases in the model's predictions and inform the development of strategies to mitigate these biases. For example, if a fairness audit reveals that the model is disproportionately favoring one demographic group over others, the organization may choose to retrain the model with adjusted data or apply fairness constraints to the model's objective function.

During the deployment phase, organizations must continuously monitor the model's performance and fairness in real-world applications. This involves setting up automated monitoring tools that can detect potential biases in the model's predictions and alert stakeholders to any issues. Regular audits should also be conducted to review the model's fairness and ensure that it continues to operate in compliance with relevant regulations and ethical guidelines. By integrating fairness considerations into every stage of the ML lifecycle, organizations can better manage the risks of bias and discrimination in their models.

Ethical and Regulatory Considerations

Beyond technical challenges, implementing fairness in ML also involves navigating a complex landscape of ethical and regulatory considerations. Different industries are governed by specific regulations that address fairness and non-discrimination in the use of ML models. For instance, the financial sector must adhere to the Equal Credit Opportunity Act (ECOA), while the healthcare sector must comply with the Health Insurance Portability and Accountability Act (HIPAA) and the Affordable Care Act (ACA). An effective ML governance framework must integrate these sector-specific regulations into its policies and procedures, ensuring that models are developed and deployed in a manner that aligns with legal and ethical standards.

Regulatory compliance is essential for mitigating the risks of discrimination and bias in ML models, but it is not sufficient on its own. Organizations must also consider the broader ethical implications of their models and take proactive steps to ensure that their ML systems are used in a way that promotes equity and justice. This includes engaging with external stakeholders, such as advocacy groups and impacted communities, to understand the potential impacts of the model and to incorporate diverse perspectives into the model development process.

Ethical governance also involves establishing transparency measures that allow stakeholders to understand how decisions are made by the model and to challenge these decisions if they believe they are unfair. This is particularly important in high-stakes applications, such as lending, hiring, and law enforcement, where the consequences of biased or discriminatory outcomes can be severe. By prioritizing transparency and accountability, organizations can build trust with their stakeholders and ensure that their ML models are used in a responsible and ethical manner.

Continuous Monitoring and Auditing for Fairness

Continuous monitoring and auditing are critical components of ML governance that ensure ongoing compliance with fairness and non-discrimination regulations. Governance frameworks should require that organizations implement automated monitoring tools that can detect bias in real time and alert stakeholders to potential issues. Regular audits should also be conducted to thoroughly review the fairness of ML models, particularly in high-stakes applications such as hiring, lending, and law enforcement. These audits should be documented and reported to governance bodies to ensure transparency and accountability in the organization's use of ML.

Continuous monitoring is particularly important in the context of ML, as models can degrade over time due to changes in the underlying data or shifts in the distribution of inputs. This phenomenon, known as model drift, can lead to unintended biases and unfair outcomes if not properly managed. By implementing automated monitoring tools that can detect and correct for model drift, organizations can ensure that their models continue to operate fairly and effectively over time.

Regular audits are also essential for maintaining the fairness of ML models. These audits should be conducted by independent teams that are not involved in the development or deployment of the model, ensuring an unbiased assessment of the model's performance. Audits should include a thorough review of the model's predictions across different demographic groups, as well as an evaluation of the model's compliance with relevant regulations and ethical guidelines. By conducting regular audits, organizations can identify and address potential biases before they result in discriminatory outcomes, ensuring that their models continue to operate in a fair and responsible manner.

Education and Training for Fairness in Machine Learning

Effective ML governance requires that all stakeholders, including data scientists, engineers, and decision-makers, are educated on the importance of fairness and non-discrimination. Governance frameworks should mandate regular training programs that cover the legal and ethical aspects of ML, as well as best practices for ensuring fairness in model development and deployment. These programs should be designed to raise awareness of the potential risks of bias and discrimination and to equip stakeholders with the knowledge and tools needed to address these issues proactively. By fostering a culture of fairness and compliance, organizations can better manage the complexities of ML governance.

Training programs should include practical exercises that allow stakeholders to apply fairness principles in real-world scenarios. This may involve analyzing case studies of biased ML models and developing strategies to mitigate these biases as well as working with fairness-enhancing tools and techniques. By providing stakeholders with hands-on experience in addressing bias and discrimination in ML, organizations can ensure that their teams are equipped to develop and deploy models that align with legal and ethical standards.

Education and training should also emphasize the importance of interdisciplinary collaboration in addressing fairness issues in ML. This involves bringing together experts from different fields, including data science, ethics, law, and social sciences, to work together on developing fair and equitable ML models. By fostering a collaborative and interdisciplinary approach to fairness, organizations can better navigate the complexities of implementing fairness in ML and ensure that their models are used for the benefit of all stakeholders.

The implementation of fairness and non-discrimination in ML is a complex and challenging task that requires a comprehensive and multifaceted approach. Technical challenges, such as biased datasets and the limitations of current debiasing methods, must be addressed alongside ethical and regulatory considerations to ensure that ML models are developed and deployed in a fair and responsible manner. By integrating fairness considerations into every stage of the ML lifecycle, from data collection to model deployment, organizations can better manage the risks of bias and discrimination and build trust with their stakeholders.

Continuous monitoring and auditing are essential for maintaining the fairness of ML models over time, while education and training programs are critical for ensuring that all stakeholders are equipped to address fairness issues proactively. By prioritizing fairness and non-discrimination in their ML governance frameworks, organizations can ensure that their models are used in a way that aligns with legal and ethical standards, promoting equity and justice in automated decision-making systems. As ML continues to play an increasingly significant role in decision-making processes, organizations must prioritize fairness and non-discrimination to build trust, avoid legal risks, and ensure that their systems are used for the benefit of all stakeholders.

Techniques for Ensuring Compliance

One of the foundational techniques for ensuring fairness and non-discrimination is **rigorous data auditing**. Within an ML Data Governance framework, data audits should be conducted regularly to assess the quality and representativeness of the data used to train ML models. Auditing involves evaluating datasets for biases that could lead to discriminatory outcomes. This includes checking for skewed distributions, underrepresented groups, and historical biases that may influence model predictions. Automated bias detection tools can be integrated into the governance framework to continuously monitor data inputs and flag potential issues before they affect model outputs.

Implementing fairness metrics is another essential technique within ML Data Governance. These metrics, such as demographic parity, equalized odds, and disparate impact ratio, are used to measure how well ML models perform across different demographic groups. By continuously monitoring these metrics, organizations can detect instances where a model's predictions disproportionately affect a particular group. Governance frameworks should mandate the use of these metrics in both the development and deployment phases of ML models, ensuring that fairness is consistently evaluated and maintained over time.

To ensure compliance with fairness and non-discrimination regulations, ML Data Governance frameworks must prioritize **model explainability and interpretability**. Techniques such as SHAP (SHapley Additive exPlanations) and LIME (Local Interpretable Model-agnostic Explanations) can be used to provide insights into how models make decisions. By understanding the factors that influence model predictions, organizations can identify and correct biases that may lead to unfair outcomes. Governance policies should require that all deployed models have a level of transparency that allows for the detection of bias and discrimination, thereby supporting compliance with regulatory standards.

Within an ML Data Governance framework, **algorithmic debiasing techniques** are employed to mitigate biases that may exist in training data or that arise during the model training process. Techniques such as re-weighting, adversarial debiasing, and bias correction algorithms can be implemented to adjust model outcomes in a way that promotes fairness. These techniques are particularly useful in scenarios where data is inherently biased due to historical or systemic inequalities. Governance frameworks should include guidelines for when and how to apply debiasing techniques, ensuring that models are fair and do not perpetuate discrimination.

A critical aspect of ensuring fairness and non-discrimination is the collection of diverse and representative data. ML Data Governance frameworks must establish policies that prioritize the **inclusion of diverse data sources** that reflect the characteristics of all relevant populations. This involves going beyond traditional data collection methods to actively seek out data from underrepresented groups. By ensuring that training data is diverse, organizations can reduce the risk of bias and create models that are more equitable in their predictions. Governance frameworks should also include protocols for continuously updating datasets to reflect changing demographics.

Fairness-aware model training is a technique that integrates fairness considerations directly into the model training process. This involves adjusting the training algorithm to minimize disparities in model performance across different groups. Techniques such as fairness constraints and regularization can be applied to ensure that the model's objectives align with fairness goals. Within an ML Data Governance framework, organizations should implement fairness-aware training protocols that require models to be evaluated not just on accuracy but also on their fairness metrics. This ensures that models are optimized for both performance and equity.

Regular **bias audits** are a crucial technique for ensuring ongoing compliance with fairness and non-discrimination regulations. Within the ML Data Governance framework, these audits should be scheduled at regular intervals to assess the fairness of ML models in production. Compliance checks involve reviewing model outputs and comparing them against established fairness metrics and regulatory requirements. Governance policies should mandate that bias audits are documented and reviewed by a cross-functional team, including legal, ethical, and technical experts, to ensure that any identified issues are promptly addressed.

Ensuring compliance with fairness and non-discrimination regulations also requires the **involvement of diverse stakeholders** in the governance process. This includes engaging ethicists, legal experts, and representatives from affected communities in the development and deployment of ML models. Governance frameworks should establish mechanisms for stakeholder input, ensuring that the perspectives of those who may be impacted by ML decisions are considered. Ethical oversight bodies should be empowered to review and approve models before deployment, with a focus on preventing discrimination and promoting fairness.

Transparency is a key principle of ML Data Governance that supports compliance with fairness and non-discrimination regulations. Techniques for ensuring transparency include comprehensive documentation of the data, models, and algorithms used in ML systems. Governance frameworks should require that all decisions made by ML models are traceable and that detailed records are kept of the data sources, model parameters, and fairness metrics used. This transparency not only facilitates internal audits and external reviews but also builds trust with stakeholders by demonstrating a commitment to ethical practices.

Protecting user rights is a critical aspect of compliance within the ML Data Governance framework. Techniques such as enabling **user feedback mechanisms** and providing avenues for individuals to challenge model decisions are essential for ensuring fairness. Governance policies should include provisions for model accountability, requiring that organizations have processes in place to respond to user complaints and rectify any unfair or discriminatory outcomes. Additionally, governance frameworks should mandate that users are informed about how their data is being used and the potential impact of ML models on their lives.

Effective ML Data Governance requires **collaboration across various departments** within an organization, including data science, legal, compliance, and ethics teams. Techniques for fostering this collaboration include the formation of cross-functional governance committees that oversee fairness and non-discrimination efforts. These committees should be responsible for coordinating audits, reviewing model fairness, and ensuring that all stakeholders are aligned with the organization's ethical and regulatory commitments. By promoting cross-departmental collaboration, governance frameworks can ensure that fairness and non-discrimination are prioritized throughout the ML lifecycle.

Managing the entire lifecycle of ML models is another technique that supports compliance with fairness and non-discrimination regulations. This involves **monitoring models from development through deployment and beyond** with a focus on maintaining fairness at each stage. ML Data Governance frameworks should include policies for model versioning, regular retraining, and decommissioning of models that no longer meet fairness standards. Lifecycle management ensures that models remain compliant with regulatory requirements and that any changes in data or societal context are reflected in model performance.

Continuous training and education for all stakeholders involved in ML development and deployment are essential for ensuring compliance with fairness and non-discrimination regulations. Governance frameworks should require regular training sessions on the legal, ethical, and technical aspects of fairness in ML. These programs should be designed to raise awareness of the risks of bias and discrimination and to equip teams with the knowledge and tools needed to develop fair and transparent models. By fostering a culture of fairness through education, organizations can ensure that their ML governance practices are aligned with regulatory expectations.

Finally, **continuous improvement and adaptation** are key techniques within an ML Data Governance framework for ensuring long-term compliance with fairness and non-discrimination regulations. Governance frameworks should include processes for regularly reviewing and updating fairness policies, tools, and techniques to reflect new developments in technology, law, and society. This involves staying informed about emerging best practices, regulatory changes, and advances in fairness-aware ML techniques. By committing to continuous improvement, organizations can ensure that their ML systems remain compliant and fair in a rapidly evolving landscape.

Resource Implications of Compliance Techniques in Fairness and Non-discrimination Regulations

Automated bias detection tools can streamline the auditing process, but these tools require significant upfront investment in software, hardware, and ongoing maintenance. Human oversight is crucial, as automated tools alone cannot fully address the nuances of bias detection. Skilled data scientists and domain experts are needed to interpret audit results and make informed decisions about data quality, adding to personnel costs. Moreover, thorough data audits are time-consuming, potentially slowing down the ML development cycle and delaying model deployment. Organizations may need to allocate additional resources to expedite the audit process, including hiring more personnel or investing in faster computing infrastructure, to ensure both speed and thoroughness.

Fairness-aware training protocols integrate fairness considerations directly into the ML model development process, often requiring more complex and computationally intensive training methods. This increases the demand for advanced cloud computing services or on-premises hardware, driving up costs. Developing and fine-tuning fairness-aware algorithms also necessitates specialized expertise in both ML techniques and fairness metrics, leading to additional financial outlays for training or hiring skilled personnel. The iterative nature of fairness-aware training, which involves multiple cycles of training and evaluation to balance fairness and model performance, extends the time required to develop and deploy models. This can be particularly challenging in industries where speed to market is crucial, necessitating careful allocation of time and resources.

Stakeholder involvement requires engaging ethicists, legal experts, and community representatives, which requires time and resources. Organizations must allocate funds for convening oversight bodies, compensating participants, and facilitating regular meetings. This stakeholder engagement is vital for building trust and ensuring that ML models are developed and deployed fairly, but it comes with significant costs.

Continuous improvement and adaptation are ongoing requirements for maintaining compliance with fairness and non-discrimination regulations. Staying informed about emerging best practices and regulatory changes necessitates continuous investment in research and professional development. Organizations must allocate resources for attending conferences, participating in industry groups, and engaging with academic research to stay at the forefront of fairness-aware ML practices. Adapting ML models and governance practices in response to new information or changing societal expectations also requires significant resources. This may involve retraining models, revising governance policies, or implementing new tools and techniques, all of which contribute to the cumulative costs of compliance.

Ensuring compliance with fairness and non-discrimination regulations in ML systems requires substantial financial and human resources. The costs associated with rigorous data auditing, fairness-aware training protocols, regular bias audits, and stakeholder involvement are significant but necessary for maintaining the fairness, transparency, and regulatory compliance of ML models. Organizations must carefully consider these resource implications and strategically allocate their time and funds to achieve long-term success in their ML governance efforts.

Interpretability and Explainability Requirements

Interpretability and explainability are essential components for ensuring that machine learning models are not only effective but also aligned with regulatory and ethical standards. These requirements are increasingly becoming a focal point as governments and regulatory bodies introduce frameworks like the Right to Explanation (RtX) and the EU General Data Protection Regulation (GDPR), which mandate transparency in automated decision-making processes. These regulations, along with other industry-specific guidelines, underscore the need for organizations to develop machine learning models that are not only accurate but also interpretable and explainable to various stakeholders, including end users, regulators, and internal auditors.

The Right to Explanation (RtX) is a pivotal concept within the GDPR, emphasizing that individuals have the right to obtain meaningful information about the logic, significance, and consequences of automated decision-making processes that affect them. This regulation is particularly relevant for ML systems that make decisions with significant impacts, such as credit scoring, hiring, or law enforcement. Ensuring compliance with RtX requires organizations to implement models that can provide clear, understandable explanations of how decisions are made. This often necessitates the use of interpretable models or the application of post hoc explanation techniques, such as LIME or SHAP, which can help unpack the decision-making process of more complex models like deep learning networks.

The Right to Explanation (RtX) within the General Data Protection Regulation (GDPR) mandates that individuals impacted by automated decision-making processes have the right to receive meaningful information about how these decisions are made. This requirement is particularly crucial for machine learning (ML) systems that influence critical aspects of people's lives, such as credit scoring, hiring decisions, or law enforcement actions. Ensuring compliance with RtX requires organizations to adopt models that can provide clear and understandable explanations of their decision-making processes. To achieve this, many organizations turn to explainability tools and frameworks like SHAP (SHapley Additive exPlanations) and LIME (Local Interpretable Model-agnostic Explanations). These tools are designed to make complex ML models more interpretable, but they come with their own sets of advantages and limitations.

SHAP is a widely used framework based on cooperative game theory, specifically the Shapley value concept. It provides a unified measure of feature importance by attributing the change in the model's output to each feature, considering all possible combinations of features. SHAP values offer consistency, meaning that if a model changes such that a feature contributes more to the model output, the SHAP value for that feature increases accordingly. This consistency makes SHAP particularly valuable for explaining complex models like deep learning networks or ensemble models.

SHAP provides consistent feature importance values, ensuring that the contribution of each feature is fairly represented. It's values can be used to understand both the overall model behavior (global interpretability) and individual predictions (local interpretability). It can be applied to any model type, making it a versatile tool for various ML applications. However, calculating SHAP values can be computationally expensive, particularly for large datasets or complex models with many features. While SHAP provides clear quantitative insights, interpreting these insights in the context of specific decisions can still be challenging, especially for non-technical stakeholders.

LIME is another popular tool for explaining ML models. Unlike SHAP, which provides a unified global explanation, LIME focuses on local interpretability by approximating the model with an interpretable surrogate model around a specific prediction. LIME creates perturbations around the data point of interest, observes how the model predictions change, and fits a simple, interpretable model (like linear regression) to explain the model's decision in that local context.

LIME excels at providing explanations for individual predictions, making it useful for understanding specific outcomes in detail. Like SHAP, LIME can be used with any type of ML model, offering flexibility across different applications. It's approach of using simple models to explain complex predictions can make it easier for stakeholders to grasp the reasoning behind individual decisions. However, the explanations generated by LIME can vary significantly with different perturbations, leading to potential inconsistencies in the explanations. LIME relies on a surrogate model to approximate the decision boundary of the complex model, which may not always capture the nuances of the original model accurately. It may not scale well with very large datasets or highly complex models, as generating local explanations for every prediction can be resource-intensive.

Table 8-1 summarizes the key differences between SHAP, LIME, and other commonly used explainability techniques.

Table 8-1. *Differences between explainability techniques*

Technique	Type of Interpretability	Model Agnostic	Computational Complexity	Strengths	Weaknesses
SHAP	Global and local	Yes	High	Consistency, comprehensive insights	Computationally expensive, complex interpretation
LIME	Local	Yes	Moderate	Local explanations, simplicity	Stability issues, approximation accuracy
Integrated gradients	Local and global	No	Moderate	Handles deep networks well, provides clear attribution	requires differentiable models, may be less intuitive
Partial dependence plots (PDPs)	Global	No	Low	Easy to implement, clear global understanding	Can be misleading for correlated features, not suitable for local interpretability
Feature importance (e.g., Gini importance)	Global	No	Low	Provides quick insights into feature relevance	May not capture interactions, model dependent

When choosing an explainability technique, organizations must consider the specific needs of their application, including the type of model, the importance of global versus local interpretability, and the resources available for implementation. Each technique offers a unique balance of strengths and limitations, and the choice of technique should align with the overall goals of transparency, fairness, and accountability in ML systems.

The GDPR, as a broader framework, places several obligations on organizations that use machine learning models, particularly in relation to data processing, consent, and the rights of data subjects. One of the key aspects of GDPR compliance is ensuring that data subjects are informed about how their data is used, which includes understanding how decisions are made using their data. This requirement pushes organizations to adopt models that are not only effective but also transparent. Organizations may need to document their models thoroughly, ensure that they can provide explanations upon request, and conduct regular audits to verify that the models remain compliant with GDPR standards.

Interpretability and explainability are not only regulatory requirements but also crucial for building trust with stakeholders. When stakeholders, including customers, employees, and regulators, understand how a model works and can trust that it operates fairly and accurately, they are more likely to accept and support its use. This is particularly important in sensitive areas like healthcare, finance, and criminal justice, where the consequences of decisions made by ML models can be significant. Ensuring that models are interpretable and explainable helps mitigate the risk of bias and discrimination, which are key concerns in the application of machine learning in these fields.

To meet interpretability and explainability requirements, organizations often need to balance the use of complex models with simpler, more interpretable alternatives. While complex models like deep neural networks may offer higher accuracy, they can be difficult to interpret. In contrast, simpler models like decision trees or linear regressions are easier to explain but may not achieve the same level of performance. Organizations must carefully evaluate the trade-offs between accuracy and interpretability, particularly in contexts where regulatory compliance and ethical considerations are paramount.

One approach to addressing these challenges is the use of hybrid models that combine interpretable components with more complex algorithms. For example, an organization might use a decision tree to provide an initial, interpretable decision, which is then refined by a more complex model. Alternatively, organizations can use interpretable models for critical decision points and reserve complex models for less sensitive tasks. This approach can help balance the need for accuracy with the requirement for transparency.

Organizations should invest in tools and techniques that enhance the interpretability of complex models. Techniques such as feature importance analysis, partial dependence plots, and surrogate models can provide insights into how a model makes decisions, even when the model itself is complex. These tools can be integrated into the ML development process to ensure that models remain interpretable and that explanations are available when needed.

Another key consideration is the documentation of models and decision-making processes. Thorough documentation is essential for ensuring that models are transparent and that their decisions can be explained to stakeholders. This documentation should include details about the model's architecture, the data used for training, the feature selection process, and any assumptions or limitations. Regular updates to this documentation are necessary to ensure that it remains accurate and reflective of the model's current state.

Organizations must also consider the role of human oversight in the deployment of machine learning models. Even with the most interpretable models, human judgment is necessary to ensure that the decisions made by these models align with ethical standards and regulatory requirements. Implementing processes that allow for human review of automated decisions, particularly in cases where those decisions have significant impacts, is crucial for maintaining compliance and ensuring fairness.

Ongoing monitoring and evaluation are critical components of ensuring that machine learning models continue to meet interpretability and explainability requirements. As models are updated and as new data becomes available, organizations must continuously assess whether the models remain compliant with regulatory standards and whether they continue to operate transparently. This may involve regular audits, performance evaluations, and updates to the models or the tools used to explain their decisions.

Sector-Specific Regulations and Standards

Different sectors face unique regulatory landscapes that shape how ML models are developed, deployed, and maintained. These sector-specific regulations and standards are designed to address the distinct challenges and risks associated with the use of ML in diverse industries such as financial services, healthcare, autonomous vehicles, and more. From data privacy laws to safety standards, understanding and complying with these regulations is vital for organizations to mitigate risks, avoid legal penalties, and build trust with stakeholders.

Financial Services

The financial services sector is one of the most heavily regulated industries globally, and the introduction of ML technologies has added new layers of complexity to regulatory compliance. Financial institutions are increasingly using ML models for a variety of purposes, including credit scoring, fraud detection, algorithmic trading, and customer service automation. However, these applications must comply with stringent regulations that govern data usage, model transparency, and decision-making fairness. Key regulations in this sector include the General Data Protection Regulation (GDPR) in Europe, the Fair Credit Reporting Act (FCRA) in the United States, and the Basel Committee on Banking Supervision (BCBS) guidelines.

Compliance with GDPR in Financial Services

The GDPR, which came into effect in 2018, has significant implications for financial services companies that utilize ML. GDPR mandates strict requirements for data processing, including the necessity for obtaining explicit consent from data subjects, ensuring data minimization, and allowing individuals the right to access and correct their data. For ML models, this means that financial institutions must ensure that their models are transparent and that decisions made by these models can be explained to customers. Furthermore, the GDPR's "Right to Explanation" necessitates that financial institutions be able to provide meaningful information about how decisions affecting individuals are made, particularly in cases where automated decision-making is involved.

Fair Credit Reporting Act (FCRA) and Its Impact on ML

In the United States, the Fair Credit Reporting Act (FCRA) is a crucial regulation that impacts the use of ML in credit scoring and reporting. The FCRA requires that credit scoring models be fair, transparent, and non-discriminatory. It mandates that consumers have the right to know the information that was used to make credit decisions about them and the ability to dispute incorrect or incomplete information. For ML models, compliance with FCRA means ensuring that the models do not introduce bias or unfairly disadvantage certain groups of consumers. Financial institutions must also provide clear and understandable explanations for adverse actions, such as loan denials, based on ML-driven credit scores.

Basel Committee on Banking Supervision (BCBS) Guidelines

The Basel Committee on Banking Supervision (BCBS) has issued a series of guidelines aimed at ensuring the soundness and stability of the global financial system. These guidelines include principles for the use of ML models in risk management, such as ensuring that models are robust, reliable, and subject to regular validation. The BCBS also emphasizes the importance of governance frameworks that oversee the development, implementation, and monitoring of ML models. Financial institutions must ensure that their ML models are not only compliant with regulatory standards but also resilient to changes in data, market conditions, and other external factors.

Healthcare

In the healthcare sector, the use of ML presents significant opportunities for improving patient outcomes, optimizing clinical workflows, and advancing medical research. However, these benefits come with strict regulatory requirements designed to protect patient privacy, ensure the safety and efficacy of ML-driven medical devices, and maintain the integrity of clinical data. Key regulations in this sector include the Health Insurance Portability and Accountability Act (HIPAA) in the United States, the Medical Device Regulation (MDR) in Europe, and various standards from the International Organization for Standardization (ISO).

Compliance with HIPAA in Healthcare

The Health Insurance Portability and Accountability Act (HIPAA) is the primary regulation governing the use of patient data in the United States. HIPAA sets stringent standards for the protection of health information, requiring healthcare organizations to implement robust data security measures, ensure patient consent for data usage, and provide patients with access to their health information. For ML applications, compliance with HIPAA involves ensuring that ML models do not compromise patient privacy, that data used in model training is de-identified or anonymized, and that the models are secure against unauthorized access. Additionally, healthcare organizations must be able to demonstrate that their ML models are compliant with HIPAA requirements through thorough documentation and regular audits.

Medical Device Regulation (MDR) in Europe

In Europe, the Medical Device Regulation (MDR) governs the use of ML in medical devices, including diagnostic tools, monitoring systems, and therapeutic applications. The MDR requires that medical devices undergo rigorous testing and validation to ensure their safety and efficacy before they can be marketed. For ML-driven medical devices, this means that manufacturers must demonstrate that their models are reliable, accurate, and capable of providing consistent results. Additionally, the MDR mandates that manufacturers provide detailed documentation of the model's development, validation, and risk management processes. This documentation is critical for obtaining regulatory approval and ensuring that the device can be safely used in clinical settings.

ISO Standards for ML in Healthcare

The International Organization for Standardization (ISO) has developed several standards that are relevant to the use of ML in healthcare. These include ISO 13485, which outlines requirements for quality management systems in medical device manufacturing, and ISO 14971, which provides guidelines for risk management in the development of medical devices. Compliance with these standards is essential for ensuring that ML models used in healthcare are safe, effective, and reliable. Healthcare organizations and medical device manufacturers must implement rigorous quality management and risk management processes to ensure that their ML models meet ISO standards and are fit for use in clinical practice.

Autonomous Vehicles

The development and deployment of autonomous vehicles (AVs) represent a significant area of innovation in the automotive industry. However, the use of ML in AVs introduces complex regulatory challenges related to safety, liability, and data privacy. Regulatory bodies around the world are developing frameworks to govern the testing, deployment, and operation of AVs, with a focus on ensuring that these vehicles are safe, reliable, and compliant with existing road safety laws. Key regulations in this sector include the United Nations Economic Commission for Europe (UNECE) regulations, the National Highway Traffic Safety Administration (NHTSA) guidelines in the United States, and various national and regional standards.

UNECE Regulations for Autonomous Vehicles

The United Nations Economic Commission for Europe (UNECE) has established a series of regulations that set the standards for the safety and performance of autonomous vehicles. These regulations include requirements for vehicle testing, certification, and ongoing compliance monitoring. For ML systems in AVs, compliance with UNECE regulations involves ensuring that the models are capable of making safe and reliable driving decisions in a wide range of scenarios. This includes demonstrating that the models can handle complex driving environments, detect and respond to potential hazards, and operate in a manner that is consistent with human drivers. Additionally, manufacturers must provide detailed documentation of the model's development, testing, and validation processes to obtain regulatory approval.

NHTSA Guidelines for Autonomous Vehicles

In the United States, the National Highway Traffic Safety Administration (NHTSA) has issued guidelines for the development and deployment of autonomous vehicles. These guidelines emphasize the importance of safety, transparency, and accountability in the use of ML for autonomous driving. Compliance with NHTSA guidelines involves implementing rigorous testing and validation processes to ensure that the ML models used in AVs are safe and reliable. This includes conducting extensive simulations, real-world testing, and continuous monitoring of vehicle performance. Additionally, manufacturers must provide clear and transparent information about how their ML models make driving decisions, particularly in situations where the vehicle may encounter unexpected or unusual conditions.

Data Privacy Considerations in Autonomous Vehicles

Data privacy is a significant concern in the development and deployment of autonomous vehicles, particularly as these vehicles collect and process large amounts of data about drivers, passengers, and their surroundings. Compliance with data privacy regulations, such as the GDPR and the California Consumer Privacy Act (CCPA), is essential for ensuring that personal data is handled responsibly and securely. For ML systems in AVs, this involves implementing robust data governance practices to ensure that data is collected, processed, and stored in compliance with privacy laws. Manufacturers must also provide mechanisms for individuals to access, correct, and delete their data and ensure that data is protected against unauthorized access or breaches.

Telecommunications

The telecommunications industry is another sector where ML is being increasingly utilized, particularly in areas such as network optimization, customer service automation, and fraud detection. However, the use of ML in telecommunications is subject to a complex regulatory landscape that includes data privacy laws, cybersecurity regulations, and industry-specific standards. Key regulations in this sector include the Federal Communications Commission (FCC) rules in the United States, the European Electronic Communications Code (EECC), and the ISO/IEC standards for information security management.

FCC Regulations for Telecommunications

The Federal Communications Commission (FCC) in the United States regulates the telecommunications industry, with a focus on ensuring that communications networks are reliable, secure, and accessible to all. For ML systems used in telecommunications, compliance with FCC regulations involves ensuring that models are designed and deployed in a manner that supports network reliability and security. This includes implementing robust testing and validation processes to ensure that ML models do not introduce vulnerabilities or degrade network performance. Additionally, telecommunications companies must ensure that their ML models are transparent and that they can provide clear explanations of how decisions, such as network optimizations or service prioritizations, are made.

European Electronic Communications Code (EECC)

In Europe, the European Electronic Communications Code (EECC) governs the telecommunications sector, setting out rules for network security, data privacy, and consumer protection. Compliance with the EECC is essential for telecommunications companies that utilize ML models for network management, customer service, and other applications. The EECC requires that companies implement strong data governance practices, including ensuring the security and privacy of customer data, providing transparency about how data is used, and allowing customers to exercise their rights under data protection laws. Additionally, the EECC mandates that telecommunications companies provide clear and accessible information to consumers about their services, including any decisions made by ML models that impact service delivery.

ISO/IEC Standards for Information Security Management

The ISO/IEC 27000 series of standards provides guidelines for information security management, which are particularly relevant for the telecommunications industry. Compliance with these standards is essential for ensuring that ML models used in telecommunications are secure and that they do not introduce risks to the confidentiality, integrity, or availability of information. Telecommunications companies must implement comprehensive information security management systems (ISMS) that address the unique challenges posed by ML, including the need for secure data storage, access controls, and monitoring of model performance. Regular audits and assessments are also necessary to ensure ongoing compliance with ISO/IEC standards and to identify and address any potential security vulnerabilities.

Retail Industry

The retail sector is increasingly adopting ML technologies for applications such as personalized marketing, demand forecasting, and inventory management. However, the use of ML in retail is subject to a range of regulatory requirements, particularly in relation to data privacy, consumer protection, and competition law. Key regulations in this sector include the GDPR, the CCPA, and various national and regional consumer protection laws.

Consumer Protection and ML in Retail

Consumer protection laws are a critical consideration for retailers that use ML models, particularly in areas such as personalized marketing and pricing strategies. These laws are designed to ensure that consumers are treated fairly and that their rights are protected. For ML models, compliance with consumer protection laws involves ensuring that decisions, such as personalized pricing or targeted advertising, are transparent and non-discriminatory. Retailers must also provide consumers with clear information about how their data is used and ensure that they have the ability to opt out of certain data processing activities. Additionally, retailers must be vigilant in avoiding practices that could be seen as unfair or deceptive, such as price discrimination based on ML-driven segmentation.

Competition Law and ML in Retail

Competition law is another important consideration for retailers that use ML, particularly in relation to pricing algorithms and market analysis. Competition authorities in many jurisdictions have raised concerns about the potential for ML models to facilitate anti-competitive behavior, such as price-fixing or market manipulation. To comply with competition law, retailers must ensure that their ML models do not engage in practices that could be seen as collusive or anti-competitive. This includes implementing safeguards to prevent models from automatically adjusting prices in a way that mirrors competitors' actions or from using data in a manner that restricts competition. Regular monitoring and auditing of ML models are essential for ensuring compliance with competition law and for identifying any potential risks.

The integration of ML technologies into various sectors brings significant opportunities for innovation and efficiency, but it also introduces complex regulatory challenges. Sector-specific regulations and standards are designed to address the unique risks associated with the use of ML in different industries, from financial services to healthcare, autonomous vehicles, telecommunications, and retail. Compliance with these regulations is essential for ensuring that ML models are safe, reliable, and aligned with legal and ethical standards. Organizations must invest in robust data governance frameworks, implement thorough testing and validation processes, and maintain transparency and accountability in their ML practices. By doing so, they can mitigate risks, avoid legal penalties, and build trust with their stakeholders in an increasingly data-driven world.

Considerations for High-Risk or Safety-Critical Applications

High-risk applications are subject to a complex regulatory landscape designed to mitigate the potential risks associated with the deployment of ML models. Regulatory bodies across various sectors have established guidelines and standards to ensure that ML systems are safe, reliable, and operate within acceptable risk parameters. For instance, the healthcare sector is governed by regulations like the Health Insurance Portability and Accountability Act (HIPAA) and the Medical Device Regulation (MDR), while the automotive sector is regulated by bodies such as the National Highway Traffic Safety Administration (NHTSA) and the United Nations Economic Commission for Europe (UNECE). Understanding these regulations is critical for organizations to design and implement compliant ML systems.

In healthcare, the use of ML in applications such as diagnostic tools, treatment planning, and patient monitoring falls under the category of safety-critical systems. These applications are governed by strict regulations to protect patient safety and privacy. For instance, HIPAA mandates rigorous data protection and privacy standards, ensuring that patient data used in ML models is securely handled and that the models do not introduce biases or errors that could harm patients. Additionally, the MDR requires that ML-driven medical devices undergo extensive testing and validation to demonstrate their safety and efficacy before they can be approved for use in clinical settings.

Autonomous vehicles (AVs) represent another high-risk application of ML, where the consequences of system failures can be life-threatening. Regulatory bodies such as the UNECE and NHTSA have established stringent guidelines to ensure that ML models used in AVs are safe, reliable, and capable of making decisions that prioritize human safety. These regulations require extensive testing and validation of ML models under a wide range of driving conditions, including edge cases that may not occur frequently but have severe consequences if not properly managed. Additionally, AV manufacturers must ensure that their models are transparent and that they can explain the decision-making processes to regulators, especially in the event of an accident.

The aviation sector is another area where ML applications are considered safety-critical, particularly in areas such as autopilot systems, air traffic control, and maintenance scheduling. Regulatory bodies like the Federal Aviation Administration (FAA) and the European Union Aviation Safety Agency (EASA) have established rigorous standards for the use of ML in aviation to ensure the safety and reliability of these systems. Compliance with these standards involves extensive testing, validation, and continuous monitoring of ML models to detect and mitigate potential risks. Additionally, the aviation industry must adhere to strict data governance practices to ensure that data used in ML models is accurate, up-to-date, and securely stored.

ML applications in critical infrastructure, such as power grids, water supply systems, and telecommunications networks, are also considered high-risk due to the potential for widespread disruption in the event of a failure. Regulatory bodies have established guidelines to ensure that ML systems used in these sectors are secure, reliable, and resilient to cyberattacks. For example, the North American Electric Reliability Corporation (NERC) provides standards for the reliability and security of power grids, which must be followed by utilities that use ML for grid management and optimization. Compliance with these standards requires robust data governance practices, including secure data storage, access controls, and continuous monitoring for anomalies.

In high-risk applications, the explainability of ML models is a critical consideration, as it allows stakeholders to understand how decisions are made and to identify potential risks or biases in the system. Regulatory bodies often require that ML models used in safety-critical applications be transparent and that their decision-making processes can be explained in clear and understandable terms. This is particularly important in sectors like healthcare and autonomous vehicles, where the ability to explain decisions can be the difference between life and death. Techniques such as SHAP (SHapley Additive exPlanations) and LIME (Local Interpretable Model-agnostic Explanations) are commonly used to provide insights into ML models' decision-making processes.

Bias and fairness are significant concerns in high-risk ML applications, as biased models can lead to discriminatory outcomes or exacerbate existing inequalities. Regulatory bodies have increasingly focused on ensuring that ML models in safety-critical sectors are fair and do not introduce or perpetuate biases. For example, in the healthcare sector, biased ML models could lead to unequal treatment of patients based on race, gender, or socioeconomic status, which could have severe consequences for patient outcomes. Organizations must implement rigorous bias detection and mitigation strategies as part of their ML data governance frameworks to ensure that their models are fair and comply with relevant regulations.

Security is a paramount concern in high-risk ML applications, particularly in sectors such as critical infrastructure and autonomous vehicles. These systems are often targeted by cyberattacks, which can have catastrophic consequences if successful. Regulatory bodies require that organizations implement robust security measures to protect ML models from unauthorized access, tampering, or manipulation. This includes the use of encryption, secure data storage, and access controls, as well as continuous monitoring for potential security breaches. Additionally, organizations must ensure that their ML models are resilient to adversarial attacks, where malicious actors attempt to manipulate the model's inputs to produce incorrect or harmful outputs.

In high-risk ML applications, continuous monitoring and risk management are essential for maintaining compliance with sector-specific regulations and ensuring the ongoing safety and reliability of ML systems. Regulatory bodies often require that organizations implement monitoring systems that can detect and respond to changes in model performance, data quality, or external conditions that could impact the safety or reliability of the system. This includes the use of automated monitoring tools, regular audits, and the establishment of incident response protocols to address potential risks or failures promptly. By maintaining a proactive approach to monitoring and risk management, organizations can mitigate the risks associated with high-risk ML applications and ensure compliance with regulatory requirements.

Given the complexity and high stakes of safety-critical ML applications, cross-sector collaboration and standardization are crucial for ensuring consistent and effective data governance practices. Regulatory bodies, industry groups, and organizations must work together to develop and implement standards that address the unique challenges of high-risk ML applications. This includes the development of industry-specific guidelines, best practices, and certification programs that can help organizations ensure that their ML systems are compliant with relevant regulations and capable of operating safely and reliably. Additionally, cross-sector collaboration can facilitate the sharing of knowledge and expertise, helping to advance the state of ML data governance across different industries.

Ethics plays a crucial role in the governance of high-risk ML applications, particularly in sectors where decisions made by ML models can have significant impacts on individuals or society as a whole. Organizations must consider the ethical implications of their ML systems and ensure that they are designed and deployed in a manner that respects human rights, promotes fairness, and minimizes harm. This includes conducting ethical impact assessments, engaging with stakeholders to understand their concerns, and implementing governance frameworks that prioritize ethical considerations in the development and deployment of ML models. By integrating ethics into their ML data governance practices, organizations can build trust with stakeholders and ensure that their systems operate in a socially responsible manner.

Transparency and accountability are key principles in the governance of high-risk ML applications, particularly in sectors where the consequences of errors or failures can be severe. Regulatory bodies often require that organizations provide clear and transparent information about their ML models, including how they were developed, how they make decisions, and what measures are in place to mitigate risks. This transparency is essential for building trust with stakeholders, including regulators, customers, and the public. Additionally, organizations must establish clear lines of accountability for their ML systems, ensuring that there are processes in place for identifying and addressing issues and that individuals or teams are responsible for overseeing the safety and compliance of the system.

In high-risk applications, sector-specific data governance frameworks are essential for ensuring that ML systems are designed, developed, and deployed in a manner that complies with relevant regulations and standards. These frameworks should be tailored to the unique needs and challenges of each sector, taking into account factors such as the nature of the data being used, the potential risks associated with the application, and the regulatory landscape. For example, in healthcare, a data governance framework might focus on ensuring patient privacy and data security, while in autonomous vehicles, the framework might prioritize safety and reliability. By developing sector-specific data governance frameworks, organizations can ensure that their ML systems are compliant with relevant regulations and capable of operating safely and effectively.

Regular audits and assessments are critical components of ML data governance in high-risk applications, as they help ensure that ML systems remain compliant with regulatory requirements and continue to operate safely and reliably. Regulatory bodies often require that organizations conduct regular audits of their ML models, including assessments of data quality, model performance, and compliance with relevant standards. These audits can help identify potential risks or issues with the ML system, allowing organizations to take corrective action before problems escalate. Additionally, organizations should conduct regular risk assessments to evaluate the potential impact of their ML systems and ensure that appropriate safeguards are in place to mitigate risks.

Compliance with sector-specific regulations and standards can be particularly challenging in high-risk ML applications due to the complexity and evolving nature of the technology. Organizations must navigate a rapidly changing regulatory landscape, where new guidelines and standards are constantly being developed to address emerging risks and challenges. Additionally, the inherent complexity of ML models, particularly in areas like deep learning, can make it difficult to ensure that these systems are fully compliant with regulatory requirements. To address these challenges, organizations must invest in robust data governance frameworks, continuous monitoring, and ongoing training and education for their teams to ensure that they are up-to-date with the latest regulatory developments and best practices.

As the use of ML in high-risk applications continues to grow, regulatory bodies are likely to develop more stringent and comprehensive regulations to address the unique risks associated with these systems. This could include the introduction of new standards for explainability, fairness, and security, as well as more rigorous testing and validation requirements. Additionally, there may be increased focus on the ethical implications of ML systems, with regulators requiring organizations to demonstrate that their models are designed and deployed in a manner that promotes fairness, transparency, and social responsibility. Organizations must stay ahead of these developments by continuously updating their data governance practices and engaging with regulators and industry groups to ensure that they are prepared for future regulatory changes.

Advances in technology can play a significant role in helping organizations ensure compliance with sector-specific regulations and standards in high-risk ML applications. For example, the development of more sophisticated explainability tools can help organizations demonstrate how their ML models make decisions, while advanced monitoring and auditing tools can provide real-time insights into model performance and compliance. Additionally, technologies like blockchain can be used to create immutable records of data and model decisions, providing a transparent and auditable trail that can be used to demonstrate compliance with regulatory requirements. By leveraging these technologies, organizations can enhance their ML data governance frameworks and ensure that their systems are compliant with relevant regulations.

Ensuring compliance with sector-specific regulations and standards in high-risk ML applications requires collaboration across multiple disciplines, including data science, engineering, legal, and ethics. Data scientists and engineers must work closely with legal and compliance teams to ensure that ML models are developed and deployed in a manner that complies with regulatory requirements. Additionally, ethical considerations must be integrated into the development process, with input from ethicists and other stakeholders to ensure that ML systems are designed to promote fairness and minimize harm. By fostering cross-disciplinary collaboration, organizations can ensure that their ML data governance frameworks are robust, comprehensive, and capable of addressing the unique challenges of high-risk applications.

The governance of ML systems in high-risk or safety-critical applications is a complex and challenging task that requires careful consideration of sector-specific regulations and standards. As ML continues to be integrated into critical sectors like healthcare, autonomous vehicles, aviation, and critical infrastructure, the importance of robust data governance practices will only increase. Organizations must invest in developing and implementing comprehensive data governance frameworks that address the unique risks associated with these applications, while also ensuring compliance with relevant regulations and standards. By doing so, they can mitigate risks, build trust with stakeholders, and ensure that their ML systems are safe, reliable, and aligned with legal and ethical standards.

Case Studies on Successful Compliance Strategies in Different Industries

Healthcare: IBM Watson Health and HIPAA Compliance

IBM Watson Health has been at the forefront of applying artificial intelligence and machine learning in healthcare, aiming to revolutionize clinical decision-making and personalized treatment plans. The system analyzes vast amounts of medical data, including patient records, clinical studies, and genetic information, to provide insights that assist healthcare professionals in diagnosing conditions and recommending treatment options. Given the sensitive nature of the data involved, particularly patient health information, IBM Watson Health operates in a highly regulated environment. The Health Insurance Portability and Accountability Act (HIPAA) is one of the primary regulations governing the handling of such data in the United States.

HIPAA imposes stringent requirements on organizations that manage health information, mandating the protection of patient data through security, privacy, and breach notification rules. For IBM Watson Health, ensuring compliance with HIPAA is not just a legal obligation but also a critical component of maintaining trust with healthcare providers and patients. The stakes are high, as non-compliance could lead to severe penalties, legal action, and damage to the organization's reputation. IBM Watson Health's approach to compliance involves integrating advanced technological solutions with rigorous operational practices to ensure that patient data remains secure and private.

Compliance Strategy

IBM Watson Health's compliance strategy revolves around implementing robust encryption protocols to secure patient data both at rest and in transit. Encryption ensures that even if unauthorized access to the data occurs, the information remains unreadable without the proper decryption keys. This strategy is complemented by strict access controls that limit data access to authorized personnel only. These controls are enforced through role-based access mechanisms, where each user's access to information is restricted based on their role within the organization. This approach minimizes the risk of data breaches by ensuring that sensitive health information is only accessible to those who need it for legitimate purposes.

Another key component of IBM Watson Health's strategy is the establishment of comprehensive auditing mechanisms. These mechanisms continuously monitor access to and usage of patient data, providing a real-time overview of how information is handled within the system. Audits are critical for detecting any unauthorized access attempts or potential breaches, allowing the organization to respond swiftly to security incidents. IBM Watson Health also uses these audits to regularly assess its compliance with HIPAA requirements, making adjustments as needed to address emerging risks or regulatory changes.

Collaboration across departments is essential for IBM Watson Health to maintain compliance with HIPAA. The organization fosters close cooperation between its IT, legal, and healthcare teams to ensure that both technical and regulatory aspects are addressed effectively. This cross-functional collaboration enables the organization to integrate legal requirements into its technological solutions seamlessly, ensuring that compliance is maintained without hindering the system's functionality. By adopting a holistic approach that combines advanced technology with rigorous operational practices, IBM Watson Health successfully navigates the complex regulatory landscape of healthcare data management.

Lessons Learned

IBM Watson Health's experience highlights several key lessons in achieving compliance with healthcare regulations like HIPAA. First, the implementation of strong encryption and access control measures is crucial for protecting sensitive data and maintaining compliance. These measures ensure that patient information remains secure, even in the event of unauthorized access. Second, continuous monitoring and regular audits are essential for detecting potential breaches and maintaining compliance over time. These practices allow organizations to identify and address security risks proactively, reducing the likelihood of significant compliance failures. Finally, cross-functional collaboration between IT, legal, and healthcare teams is vital for integrating regulatory requirements into technological solutions, ensuring that compliance is achieved without compromising system performance.

Finance: HSBC and GDPR Compliance

HSBC, one of the world's largest financial institutions, utilizes machine learning models to enhance its operations in areas such as fraud detection, credit scoring, and customer risk assessment. Given its global presence, HSBC must comply with a variety of data protection regulations, including the General Data Protection Regulation (GDPR) in Europe. The GDPR is one of the most comprehensive data protection laws, imposing strict requirements on how personal data is collected, processed, and stored. For financial institutions like HSBC, where the processing of personal data is central to operations, ensuring compliance with GDPR is paramount.

The GDPR's emphasis on the right to privacy and the Right to Explanation for automated decisions directly impacts HSBC's use of machine learning. The regulation requires that individuals have the right to understand how decisions affecting them are made, especially in contexts like credit scoring and loan approvals. Non-compliance could result in hefty fines and significant reputational damage, making it essential for HSBC to develop a compliance strategy that integrates GDPR requirements into its machine learning operations. HSBC's approach to GDPR compliance involves a combination of data anonymization, interpretability tools, and proactive governance to protect customer data and maintain transparency.

Compliance Strategy

HSBC's primary strategy for GDPR compliance focuses on the use of data anonymization techniques to protect personal data while ensuring that it remains useful for machine learning purposes. By anonymizing data, HSBC can reduce the risk of privacy breaches, as the data can no longer be traced back to specific individuals. This approach is particularly important for large-scale machine learning models that require extensive datasets for training. Anonymization allows HSBC to continue leveraging customer data for insights and decision-making without violating GDPR's strict data protection requirements.

To comply with the GDPR's Right to Explanation, HSBC has also integrated advanced interpretability tools such as SHAP (SHapley Additive exPlanations) into its machine learning models. These tools provide transparent explanations for how automated decisions are made, which is critical for customer-facing applications like credit scoring. By using SHAP, HSBC can offer clear and understandable insights into the factors that influence a model's decision, ensuring that customers understand how their personal data is used and how decisions about them are made. This transparency is a key component of building and maintaining customer trust.

HSBC's compliance strategy also includes proactive data governance practices to ensure ongoing adherence to GDPR requirements. The company regularly reviews its data processing activities, updates its data protection policies, and conducts audits to identify and mitigate potential compliance risks. These governance practices are supported by cross-functional teams that include data scientists, legal experts, and compliance officers, who work together to ensure that GDPR requirements are fully integrated into the company's machine learning operations. By adopting a proactive and collaborative approach to data governance, HSBC successfully maintains GDPR compliance while continuing to innovate with machine learning.

Lessons Learned

HSBC's experience with GDPR compliance underscores the importance of data anonymization and transparency in protecting customer privacy and maintaining regulatory compliance. Anonymization techniques allow organizations to use personal data responsibly, minimizing the risk of breaches while still enabling valuable insights. The use of interpretability tools like SHAP demonstrates the necessity of providing clear explanations for automated decisions, which is crucial for building trust and ensuring that customers understand how their data is used. Finally, proactive data governance and regular reviews are essential for maintaining compliance over time, allowing organizations to adapt to evolving regulations and emerging risks.

Automotive: Tesla and Autonomous Vehicle Regulations

Tesla has emerged as a leader in the development of autonomous vehicles, leveraging advanced machine learning models to enable features like self-driving, navigation, and collision avoidance. The automotive industry, particularly in the context of autonomous driving, is subject to stringent regulatory oversight due to the potential safety implications of automated decision-making systems. Regulations from bodies like the National Highway Traffic Safety Administration (NHTSA) in the United States set the standards for vehicle safety, which Tesla must comply with to ensure that its vehicles are both safe and legally operable on public roads.

Autonomous vehicles rely on real-time processing of vast amounts of data from various sensors, including cameras, radar, and lidar, to make split-second decisions. These decisions must be both reliable and explainable to meet regulatory requirements. Compliance with these regulations is critical for Tesla, not only to avoid legal repercussions but also to ensure public trust in autonomous vehicle technology. Tesla's approach to compliance involves rigorous testing, real-time monitoring, and detailed documentation of its autonomous driving systems to meet the high standards set by regulatory bodies.

Compliance Strategy

Tesla's strategy for ensuring compliance with autonomous vehicle regulations begins with rigorous testing and validation of its machine learning models. The company subjects its vehicles to extensive simulations and real-world driving scenarios to evaluate how the models perform under various conditions. This testing is crucial for identifying potential safety issues and ensuring that the models meet the strict safety standards required by regulators. Tesla also conducts over-the-air updates to continuously improve its models, ensuring that the vehicles remain compliant as new challenges or regulations arise.

Real-time monitoring is another critical component of Tesla's compliance strategy. The company equips its vehicles with a robust data collection system that continuously records driving data, which is then analyzed to monitor the performance of the autonomous systems. This real-time data collection allows Tesla to detect and address issues as they occur, ensuring that the vehicles operate safely and effectively. In the event of an accident or malfunction, the data can be used to understand what happened and to make necessary adjustments to prevent future incidents.

Tesla's compliance strategy also includes maintaining detailed documentation of its autonomous driving systems. This documentation includes records of all model decisions, testing results, and updates, which are essential for regulatory reporting and post-incident analysis. By keeping thorough documentation, Tesla can demonstrate its compliance with safety regulations and provide evidence of its commitment to vehicle safety. This documentation is also critical for building trust with both regulators and the public, as it shows that Tesla is transparent about how its autonomous systems operate and how safety is maintained.

Lessons Learned

Tesla's approach to compliance in the automotive industry highlights the importance of rigorous testing, real-time monitoring, and thorough documentation. Rigorous testing ensures that machine learning models used in autonomous vehicles meet the high safety standards required by regulators. Real-time monitoring allows for the continuous assessment of vehicle performance, enabling Tesla to detect and address issues promptly. Detailed documentation is essential for regulatory compliance and transparency, providing evidence of the company's commitment to safety. Tesla's experience demonstrates that a proactive and comprehensive approach to compliance is critical for success in the highly regulated automotive industry.

Telecommunications: Vodafone and Data Privacy Compliance

Vodafone, a leading global telecommunications provider, uses machine learning to enhance various aspects of its operations, including customer segmentation, predictive maintenance, and network optimization. The telecommunications industry is heavily regulated, with companies required to comply with data privacy laws that vary by region. In Europe, Vodafone must adhere to the General Data Protection Regulation (GDPR), while in other regions, it must comply with local data protection laws. Given the massive amounts of customer data that Vodafone processes daily, ensuring compliance with these regulations is critical to protecting customer privacy and maintaining the company's reputation.

Data privacy compliance is particularly challenging in the telecommunications industry due to the continuous flow of data across networks and the need to process this data in real-time. Vodafone's approach to compliance involves implementing robust data governance frameworks and advanced data protection technologies to ensure that customer data is handled in accordance with applicable regulations. The company's strategy emphasizes transparency, customer control over data, and the use of encryption to protect data throughout its lifecycle.

Compliance Strategy

Vodafone's primary compliance strategy centers on implementing data minimization techniques to reduce the amount of personal data collected and processed. By limiting data collection to what is necessary for specific purposes, Vodafone minimizes the risk of non-compliance with data privacy laws. The company also employs advanced encryption methods to protect customer data both at rest and in transit. Encryption ensures that even if data is intercepted, it remains secure and unreadable without the appropriate decryption keys.

Transparency is another key element of Vodafone's compliance strategy. The company has adopted explainability tools to provide customers with clear insights into how their data is used and how automated decisions are made. This transparency is critical for building trust with customers and ensuring compliance with regulations that require companies to provide clear explanations of data processing activities. Vodafone also provides customers with tools to manage their data preferences, allowing them to view, delete, or opt-out of data sharing, thus ensuring that customers have control over their personal information.

Vodafone's strategy includes regular audits and reviews of its data processing activities to ensure ongoing compliance with data privacy regulations. These audits are conducted by cross-functional teams that include legal, compliance, and technical experts. The teams work together to identify potential compliance risks and implement corrective measures as needed. By conducting regular audits and maintaining a proactive approach to data privacy, Vodafone ensures that its operations remain compliant with regional and international data protection laws.

Lessons Learned

Vodafone's experience in telecommunications compliance underscores the importance of data minimization, encryption, and transparency in protecting customer privacy. Data minimization reduces the risk of non-compliance by limiting the amount of personal data collected and processed. Encryption is critical for safeguarding data throughout its lifecycle, ensuring that it remains secure even if intercepted. Transparency and customer control over data build trust and ensure that Vodafone meets its regulatory obligations. Regular audits and a proactive approach to compliance are essential for maintaining adherence to data privacy regulations over time.

Retail: Walmart and CCPA Compliance

Walmart, as one of the largest retailers globally, leverages machine learning to enhance various aspects of its business, including personalized marketing, inventory management, and customer service. With operations in California, Walmart must comply with the California Consumer Privacy Act (CCPA), one of the most stringent data privacy laws in the United States. The CCPA grants consumers extensive rights over their personal data, including the right to know what data is collected, the right to delete data, and the right to opt out of data sales. For Walmart, ensuring compliance with the CCPA is critical to maintaining customer trust and avoiding legal penalties.

The retail industry faces unique challenges in data privacy compliance due to the vast amounts of personal data collected from customers, both online and in-store. Walmart's approach to CCPA compliance involves implementing comprehensive data governance frameworks and providing customers with tools to manage their data preferences. The company's strategy emphasizes transparency, customer empowerment, and rigorous data protection measures to ensure that personal data is handled responsibly and in accordance with CCPA requirements.

Compliance Strategy

Walmart's compliance strategy for CCPA starts with revising its data collection and processing practices to ensure that they align with the law's requirements. The company has implemented robust data governance frameworks that track and manage data throughout its lifecycle, ensuring that customer data is collected, stored, and processed in compliance with CCPA. These frameworks include policies for data minimization, ensuring that only the necessary data is collected for specific business purposes, thereby reducing the risk of non-compliance.

Walmart has also developed and deployed tools that allow customers to easily manage their data preferences. These tools enable customers to view what data Walmart has collected about them, request the deletion of their data, or opt out of data sharing with third parties. By empowering customers to control their personal information, Walmart not only complies with CCPA but also strengthens customer trust and loyalty.

Regular reviews and audits are integral to Walmart's compliance strategy. The company conducts frequent assessments of its data processing activities to ensure ongoing compliance with CCPA. These audits involve cross-functional teams, including legal, compliance, and technical experts, who work together to identify potential risks and implement corrective actions. By maintaining a proactive approach to compliance, Walmart ensures that its operations remain aligned with CCPA requirements and that customer data is protected at all times.

Lessons Learned

Walmart's experience with CCPA compliance highlights the importance of customer empowerment, data governance, and proactive auditing. Providing customers with tools to manage their data preferences ensures that their rights under the CCPA are respected and builds trust in the brand. Robust data governance frameworks help track and manage data responsibly, reducing the risk of non-compliance. Regular audits and proactive risk management are essential for maintaining compliance with data privacy laws and ensuring that customer data is protected at all times.

Risk Management Framework for ML Systems

A risk management framework (illustrated in Figure 8-1) for machine learning (ML) systems is essential for identifying, assessing, mitigating, and monitoring risks throughout the ML lifecycle.

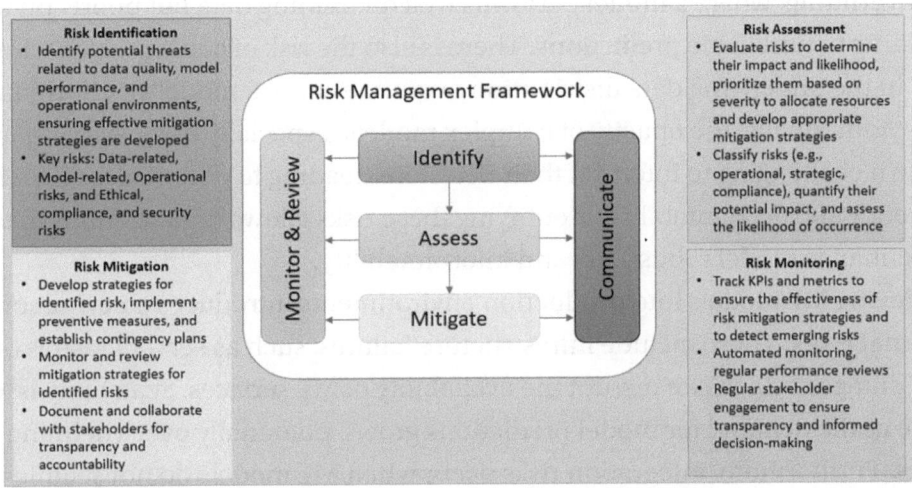

Figure 8-1. *Machine learning risk management framework*

Risk Identification

Risk identification is the first and arguably one of the most critical steps in the risk management process for machine learning (ML) systems. It involves systematically pinpointing potential threats that could adversely impact the performance, reliability, or ethical deployment of ML models. Given the complexity and variability of ML systems, risk identification must be thorough and context-specific, taking into account the unique challenges posed by data, models, infrastructure, and operational environments. Understanding the broader context, including the specific use cases, regulatory environment, and organizational goals, is essential for effective risk identification.

Data is the foundation of any ML system, and consequently, **data-related risks** are among the most significant threats. These risks can stem from poor data quality, such as inaccurate, incomplete, or outdated information. Additionally, biases in training data can lead to unfair or discriminatory model outcomes, particularly if the data reflects historical prejudices. Another critical data-related risk is data privacy breaches, where sensitive or personal information is exposed or misused. Organizations must identify these risks early in the ML lifecycle to implement appropriate safeguards and ensure data integrity, fairness, and compliance with regulations like GDPR and CCPA.

Model-related risks are another crucial area of focus during the risk identification process. These risks are associated with the ML algorithms themselves, including issues such as overfitting, where a model performs well on training data but poorly on unseen data, leading to unreliable predictions. There is also the risk of concept drift, where changes in the underlying data distribution over time cause a model's performance to degrade. Additionally, the opacity of complex models, especially deep learning models, can make it challenging to interpret their decisions, leading to potential issues with transparency and accountability. Identifying these risks allows organizations to take steps to enhance model robustness and interpretability.

Deploying ML models into production environments introduces a new set of **operational risks**. These include infrastructure failures, such as server crashes or network outages, which can disrupt the availability of ML services. Scalability issues can also arise as the demand for model predictions grows, potentially overwhelming system resources. Furthermore, integration risks occur when ML models do not seamlessly integrate with existing IT systems, leading to data flow interruptions or incorrect outputs. Identifying these operational risks is essential to ensuring the smooth and reliable functioning of ML systems in real-world scenarios.

Ethical and compliance risks are increasingly important in the context of ML systems, particularly as these technologies are applied in sensitive areas such as healthcare, finance, and law enforcement. These risks involve the potential for ML models to make biased or unethical decisions, either due to biased training data or flawed algorithms. Compliance risks also arise when ML systems fail to meet legal and regulatory requirements, such as those outlined in GDPR, HIPAA, or industry-specific standards. Identifying these risks requires a deep understanding of both the ethical implications of ML and the regulatory landscape in which the technology operates.

Security risks are a significant concern for ML systems, particularly as they become more integrated into critical business processes. These risks include the potential for adversarial attacks, where malicious actors manipulate input data to deceive or mislead the model. For instance, an attacker could subtly alter data to cause a model to make incorrect predictions, potentially leading to severe consequences in areas like autonomous driving or financial trading. Another security risk is the potential exposure of sensitive model parameters or intellectual property through model inversion attacks. Identifying these security risks is crucial for protecting ML systems from malicious threats.

As ML models become more complex, the challenge of ensuring explainability and transparency has grown. The risk here is that stakeholders, including users, regulators, and even the developers themselves, may not fully understand how a model arrives at its decisions. This lack of transparency can lead to mistrust, especially in high-stakes applications such as credit scoring or criminal justice. Furthermore, regulatory frameworks like the EU's General Data Protection Regulation (GDPR) include provisions like the Right to Explanation, which mandates that organizations provide clear explanations of automated decisions. Identifying the risks associated with poor model explainability and transparency is therefore critical to maintaining trust and compliance.

A lack of proper governance over ML models poses significant risks, particularly as organizations scale their use of AI technologies. Governance risks include the absence of standardized procedures for model development, deployment, and monitoring, which can lead to inconsistencies in model quality and performance. Moreover, without proper governance, there is a risk that models could be deployed without adequate testing or oversight, increasing the likelihood of failures or ethical breaches. Identifying these governance-related risks helps organizations establish robust frameworks for managing the lifecycle of ML models, ensuring they are developed and deployed responsibly.

Human–machine interaction is a critical area where risks can emerge, especially as ML systems increasingly influence decision-making processes. Risks in this domain include the potential for over-reliance on ML models, where users may blindly trust the model's output without applying critical judgment. Conversely, there is also the risk of underutilization, where users might ignore valuable insights provided by the model due to a lack of trust or understanding. Additionally, the risk of cognitive bias can arise if users unknowingly reinforce their biases through interactions with the model. Identifying these risks is essential for designing ML systems that effectively complement human decision-making while minimizing negative impacts.

The rapid pace of technological evolution in the ML field presents its own set of risks. As new algorithms, tools, and techniques emerge, there is a risk that previously deployed models may become outdated or obsolete, potentially leading to performance degradation or security vulnerabilities. Moreover, the adoption of cutting-edge technologies without thorough testing or understanding can introduce new risks, such as unforeseen interactions between different ML components or compatibility issues with existing systems. Identifying these risks requires organizations to stay abreast of technological developments and continuously assess the implications for their ML systems.

Managing the entire lifecycle of an ML model—from development through deployment and eventual retirement—introduces several risks. During the development phase, there is a risk of selecting inappropriate algorithms or hyperparameters, which can result in suboptimal model performance. In the deployment phase, risks include inadequate testing or failure to monitor the model effectively, leading to potential failures in production. Finally, as models age, there is a risk that they may become less relevant due to changes in the underlying data or application context, necessitating their timely retirement or replacement. Identifying these lifecycle-related risks helps organizations manage their models effectively over time.

ML systems, particularly those involving proprietary algorithms or data, are subject to legal and intellectual property (IP) risks. These risks include potential violations of IP rights, such as using copyrighted data without proper authorization or infringing on patents related to specific ML techniques. Additionally, legal risks can arise if ML systems inadvertently generate outputs that violate laws or regulations, such as producing discriminatory results that contravene anti-discrimination laws. Identifying these legal and IP risks is essential for ensuring that ML systems operate within the bounds of the law and respect the rights of others.

As organizations increasingly collaborate with external partners in the development and deployment of ML systems, new risks related to these partnerships must be identified. Risks include potential misalignment of objectives between partners, leading to conflicts or suboptimal outcomes. There is also the risk of dependency on third-party vendors or platforms, which could expose the organization to vulnerabilities if the vendor's systems fail or are compromised. Additionally, risks related to data sharing and privacy must be considered, especially if sensitive data is exchanged between partners. Identifying these risks helps organizations manage collaborations effectively while protecting their interests.

The regulatory landscape for ML and AI is continually evolving, with new laws and guidelines being introduced regularly. This dynamic environment poses risks for organizations that must ensure their ML systems remain compliant with current regulations. There is a risk that regulatory changes could render existing models non-compliant, requiring costly updates or even the suspension of certain applications. Additionally, the introduction of new regulations, such as those governing data privacy or algorithmic transparency, may impose additional burdens on ML systems. Identifying these regulatory risks is crucial for staying ahead of compliance requirements and avoiding legal penalties.

Organizations must identify strategic risks associated with the broader adoption of ML technologies. These risks include the potential for ML initiatives to fail to deliver the expected business value, either due to unrealistic expectations or poor implementation. There is also the risk of negative public perception or backlash if ML systems are perceived as invasive, biased, or unethical. Moreover, the rapid adoption of ML may outpace the organization's ability to manage and govern these technologies effectively, leading to operational challenges. Identifying these strategic risks helps organizations align their ML efforts with their overall business strategy and mitigate potential pitfalls.

By systematically identifying these risks, organizations can develop targeted strategies to manage and mitigate them, ensuring the responsible and successful deployment of ML systems.

Risk Assessment

Risk assessment in machine learning (ML) systems is a pivotal step in ensuring the successful implementation and operation of these systems within an organization. This process involves evaluating the identified risks from the risk identification phase to determine their potential impact and likelihood. The aim is to prioritize risks based on their severity, enabling organizations to allocate resources effectively and develop appropriate mitigation strategies.

The first step in risk assessment is to classify and categorize the risks. This involves determining whether risks are operational, strategic, financial, or compliance-related. For ML systems, risks might include issues related to model performance, data quality, bias, and regulatory compliance. Each category of risk requires a different approach for assessment and management, tailored to its specific characteristics and implications.

Quantifying the potential impact of each risk is a crucial aspect of the assessment process. This involves estimating the potential consequences of a risk event occurring, such as financial losses, reputational damage, or legal implications. For instance, a risk of model bias could lead to unfair treatment of certain groups, potentially resulting in legal action or loss of customer trust. Quantification helps in understanding the gravity of each risk and aids in prioritizing mitigation efforts.

Assessing the likelihood of a risk occurring is equally important. This involves evaluating how probable it is that a particular risk will materialize based on historical data, industry trends, and the current operational environment. For example, if an

ML model is trained on biased data, the likelihood of it producing biased outcomes is high. Understanding the probability helps in estimating the frequency and potential recurrence of risks.

Combining the impact and likelihood assessments provides a risk matrix or risk score for each identified risk. This matrix helps in visualizing which risks are most critical and should be addressed first. Risks with high impact and high likelihood are typically prioritized, as they pose the greatest threat to the organization's objectives and operations.

Risk assessment also involves analyzing the effectiveness of existing controls and mitigation strategies. This entails reviewing current practices and measures in place to manage risks and determining their adequacy. For instance, if an organization has implemented regular model audits to detect bias, assessing the effectiveness of these audits is crucial to ensure they are adequately addressing the risk of bias.

Scenario analysis is another important tool in risk assessment. This involves creating hypothetical scenarios to understand how different risk factors might interact and impact the ML system. Scenario analysis helps in preparing for various possible outcomes and developing contingency plans to address potential issues effectively.

Risk assessment should also consider the evolving nature of ML systems and their operational environments. As technology advances and regulatory requirements change, new risks may emerge, and existing risks may evolve. Regularly updating the risk assessment process ensures that it remains relevant and effective in addressing current and future challenges.

Stakeholder input is vital in the risk assessment process. Engaging with various stakeholders, including data scientists, engineers, business leaders, and regulatory experts, provides diverse perspectives and insights. This collaborative approach ensures a comprehensive understanding of potential risks and enhances the accuracy of the assessment.

Documenting the risk assessment process is essential for transparency and accountability. This includes recording the identified risks, their impact and likelihood assessments, and the rationale behind the prioritization of risks. Documentation serves as a reference for future assessments and helps in demonstrating due diligence to regulators and stakeholders.

Risk Mitigation

Once risks have been assessed, organizations need to create and execute plans to address these risks proactively, ensuring the robustness and reliability of their ML systems. Effective risk mitigation involves several key steps, each aimed at minimizing potential adverse effects and enhancing overall system performance. The first step in risk mitigation is to develop risk mitigation strategies tailored to each identified risk. These strategies should address both the likelihood of the risk occurring and its potential impact. For instance, if a risk involves model bias, strategies might include diversifying training data, implementing fairness constraints, and regularly auditing model predictions for bias. Tailoring mitigation strategies ensures that each risk is managed effectively based on its specific characteristics and severity.

Implementing preventive measures is a crucial aspect of risk mitigation. These measures aim to reduce the likelihood of risks occurring in the first place. For example, to mitigate data quality issues, organizations can implement rigorous data validation procedures and establish protocols for data cleaning and preprocessing. Preventive measures help in avoiding risks before they materialize, reducing the need for corrective actions later.

Organizations should establish contingency plans for managing risks that cannot be entirely eliminated. Contingency plans outline steps to be taken if a risk event occurs, including response actions, communication strategies, and recovery procedures. For instance, if an ML model experiences performance degradation due to data drift, a contingency plan might involve temporarily reverting to a previous model version while addressing the drift.

Monitoring and reviewing risk mitigation strategies are essential to ensuring their effectiveness. Regular monitoring helps in tracking the performance of implemented measures and identifying any gaps or issues. For example, if a risk mitigation strategy involves regular model retraining, monitoring should include evaluating the effectiveness of retraining intervals and adjusting them as needed. Continuous review ensures that mitigation strategies remain relevant and effective over time.

Training and awareness programs are important components of risk mitigation. Educating team members about potential risks and their mitigation strategies enhances their ability to manage and address risks effectively. Training programs can include workshops on best practices for model development, data handling, and compliance requirements. By fostering a culture of risk awareness, organizations can improve their overall risk management capabilities.

Documentation of risk mitigation activities is crucial for transparency and accountability. This includes recording the implemented strategies, preventive measures, contingency plans, and monitoring results. Comprehensive documentation serves as a reference for evaluating the effectiveness of mitigation efforts and provides evidence of due diligence to regulators and stakeholders.

Collaboration with external experts and stakeholders can enhance risk mitigation efforts. Engaging with industry experts, consultants, and regulatory bodies provides valuable insights and feedback on risk management practices. For example, external audits and reviews can identify potential weaknesses in risk mitigation strategies and recommend improvements. Collaboration helps in ensuring that risk management practices align with industry standards and regulatory requirements.

Continuous improvement is a fundamental principle of risk mitigation. Organizations should regularly assess and refine their risk management strategies based on new insights, emerging risks, and changes in the operational environment. For instance, as new ML technologies and methodologies develop, updating risk mitigation strategies to address new risks or opportunities is essential for maintaining effective risk management.

Integrating risk mitigation strategies into the ML system's lifecycle is important for ensuring comprehensive management of risks. This involves incorporating risk management practices into the stages of model development, deployment, and monitoring. For example, risk assessments and mitigation strategies should be part of the model development phase to address potential issues before deployment.

Compliance with regulatory and ethical standards is a key aspect of risk mitigation. Ensuring that risk mitigation strategies align with relevant laws, regulations, and ethical guidelines helps in managing compliance risks. For example, implementing measures to address algorithmic bias and ensure transparency can help organizations comply with regulations such as the GDPR and the CCPA.

Risk Monitoring

Risk monitoring in machine learning (ML) systems involves the continuous oversight of risk mitigation strategies and the detection of emerging risks. By systematically tracking various metrics and indicators, organizations can ensure that risks are effectively managed and that any deviations from expected performance are promptly addressed. A robust risk monitoring process is critical for maintaining the stability and reliability of ML systems over time. Integrating a feedback loop into this process allows for continuous refinement of risk management strategies, ensuring that lessons learned from monitoring are applied to improve ML system development.

The foundation of effective risk monitoring begins with establishing key performance indicators (KPIs) and risk metrics that align with the identified risks and mitigation strategies. These KPIs should encompass various aspects of the ML system, such as model performance, data quality, and system stability. For example, metrics like accuracy, precision, recall, and F1 score are crucial for monitoring model performance, while data quality indicators such as missing values or outliers should be tracked regularly. Clearly defined and relevant KPIs ensure that risk monitoring efforts are focused on critical areas that impact system performance and risk levels. The data collected through these KPIs is vital for creating a feedback loop where findings from risk monitoring are continually integrated back into the ML development process.

Continuous monitoring is facilitated by automated tools and systems that track KPIs and risk metrics in real time. These tools provide instant feedback on system performance, allowing for rapid identification and response to potential issues. For instance, real-time monitoring solutions can detect sudden drops in model accuracy or deviations in data distributions, triggering alerts that prompt immediate investigation and corrective action. The ability to respond quickly to anomalies or risks is a crucial aspect of the feedback loop. By incorporating real-time data into the ML development process, organizations can make timely adjustments to models and strategies, preventing the recurrence of similar issues in the future.

Regular performance reviews and assessments are integral to the risk monitoring process. These reviews involve periodic evaluations of the ML system's performance against established KPIs and risk metrics. For example, quarterly performance reviews might include an analysis of model accuracy trends, data quality issues, and system stability metrics. These reviews help identify long-term trends, potential risks, and areas where risk mitigation strategies may need adjustment. They also provide an opportunity to evaluate the effectiveness of implemented risk management practices and make necessary improvements. The insights gained from these reviews are fed back into the ML development cycle, ensuring that risk management strategies are continuously refined and adapted to evolving conditions.

Effective risk monitoring also requires integration with broader system monitoring practices. Coordinating risk monitoring efforts with other aspects of system performance monitoring, such as infrastructure health, application performance, and user experience, provides a comprehensive view of the ML system's overall health. For example, monitoring tools that track server load, network latency, and database performance should be integrated with risk monitoring systems. This integration helps in identifying correlations between system issues and potential risks, enabling a more holistic approach to risk management. By feeding these broader system metrics back into the ML development process, organizations can refine both the ML models and the governance practices surrounding them.

Historical data analysis plays a crucial role in risk monitoring by revealing patterns and trends that might indicate emerging risks or areas of concern. Analyzing past incidents of model degradation or data quality issues provides insights into recurring problems and helps in developing more effective risk mitigation strategies. Historical data not only supports proactive risk management but also feeds into the feedback loop, where lessons learned from previous incidents are integrated into the ML development process. This ongoing refinement ensures that the risk management strategies evolve based on empirical evidence, enhancing the resilience of ML systems over time.

Stakeholder engagement is an essential aspect of risk monitoring, ensuring that all relevant parties are informed about the ML system's risk status and any potential issues. Regular communication with stakeholders, including data scientists, engineers, business leaders, and regulatory bodies, facilitates transparency and informed decision-making. Regular risk monitoring reports and updates should be shared with stakeholders to provide a clear understanding of the current risk landscape. Engaging stakeholders in the feedback loop allows for the incorporation of diverse perspectives into the ML development process, ensuring that risk management strategies are aligned with organizational goals and regulatory requirements.

Documentation and reporting are critical components of risk monitoring. Comprehensive documentation of monitoring activities, including risk metrics, performance trends, and incident reports, provides a record of the ML system's risk management efforts. Detailed logs of risk monitoring activities and findings support audits and reviews, ensuring transparency and accountability. The documentation also serves as a key element in the feedback loop, where the recorded data and insights are used to refine risk management strategies and improve the ML development process. By maintaining thorough documentation, organizations can track progress over time and demonstrate the effectiveness of their risk management practices.

Evaluating the effectiveness of risk mitigation strategies is a continuous process that ensures these strategies remain relevant and effective in addressing identified risks. Regular assessments help determine whether the implemented strategies are successfully mitigating risks and whether any adjustments are needed. For example, if risk monitoring reveals that certain mitigation measures are not achieving the desired results, new strategies or adjustments may be required. The feedback loop plays a crucial role here, as the results of these evaluations are fed back into the ML development process, leading to the ongoing refinement of both the risk management strategies and the ML models themselves.

Risk monitoring in ML systems must be dynamic and adaptable to changes in both the ML system and its environment. As ML technologies and methodologies evolve and as new risks emerge, the risk monitoring framework should be updated to address these changes. For example, new risks related to advanced algorithms or emerging data privacy regulations may require adjustments to the monitoring approach. The feedback loop ensures that these updates are not merely reactive but are integrated into the ML development process in a way that anticipates and mitigates future risks. This dynamic approach to risk management, supported by a robust feedback loop, ensures that ML systems remain resilient and capable of adapting to a rapidly evolving landscape.

Incident Response

Incident response in machine learning (ML) systems is a critical process designed to address and manage the consequences of unexpected events that impact the performance, security, or functionality of ML systems. This process encompasses a range of activities, from the initial detection of an anomaly to the final recovery and restoration of normal operations. Effective incident response is essential for maintaining the integrity and reliability of ML systems, particularly given the complex and dynamic nature of these systems. As ML systems increasingly become integral to various sectors, including finance, healthcare, and autonomous vehicles, the ability to promptly and effectively respond to incidents is paramount.

An incident in an ML system can range from performance degradation and model drift to data breaches and security vulnerabilities. The consequences of such incidents can be significant, potentially leading to incorrect predictions, loss of sensitive data, or compliance violations. Therefore, a well-structured incident response plan is crucial to minimize the impact of these incidents and ensure a swift resolution. The response process involves several stages, each requiring careful planning and execution to effectively manage the incident and restore normal operations.

The importance of incident response extends beyond merely addressing technical issues; it also involves managing communication, legal implications, and compliance with regulatory requirements. For instance, incidents involving data breaches may require notifications to affected individuals and regulatory bodies, while incidents affecting model performance may need a thorough analysis to prevent future occurrences. Therefore, incident response in ML systems must be comprehensive and encompass all aspects of incident management.

The evolving landscape of threats and challenges in ML systems necessitates continuous improvement in incident response strategies. As new vulnerabilities and attack vectors emerge, organizations must adapt their incident response plans to address these evolving risks. This dynamic environment underscores the need for a proactive and adaptable approach to incident response in ML systems.

Defining the Scope of Incident Response

Defining the scope of incident response is a foundational step that involves outlining the boundaries and parameters of the incident response plan. This includes specifying the types of incidents that fall under the scope of the plan, such as data breaches, model performance issues, and security incidents. By clearly defining the scope, organizations can ensure that the incident response plan is targeted and relevant, addressing the specific risks and challenges associated with their ML systems.

The scope of incident response should also include the identification of stakeholders involved in the response process. This typically includes data scientists, engineers, IT security professionals, and business leaders. Each stakeholder has a role to play in managing the incident, and their responsibilities should be clearly defined to ensure a coordinated and efficient response. Additionally, the scope should outline the communication channels and procedures for escalating incidents to higher levels of management if necessary.

Another critical aspect of defining the scope is determining the thresholds for incident classification and escalation. This involves setting criteria for what constitutes a significant incident that requires immediate attention and distinguishing it from less critical issues that can be handled through routine maintenance. Establishing these thresholds helps prioritize incidents and allocate resources effectively, ensuring that high-impact incidents receive the necessary attention and resources.

The scope of incident response should include considerations for compliance and regulatory requirements. This involves identifying the specific regulations and standards that apply to the organization's ML systems and ensuring that the incident response plan addresses these requirements. Compliance considerations may include data protection laws, industry standards, and contractual obligations, all of which play a crucial role in shaping the incident response strategy.

Establishing Incident Detection Mechanisms

Establishing effective incident detection mechanisms is crucial for identifying and addressing issues in ML systems before they escalate into significant problems. Incident detection involves implementing tools and techniques to monitor system performance, data integrity, and security. This can include the use of monitoring software, anomaly detection algorithms, and real-time alerts to identify deviations from normal system behavior.

Monitoring tools should be configured to track various metrics relevant to ML systems, such as model accuracy, data drift, and system resource utilization. For example, monitoring tools can provide alerts when model performance degrades beyond a certain threshold, indicating a potential issue with the model or its input data. Similarly, anomaly detection algorithms can identify unusual patterns in data that may signal a data breach or other security incident.

Real-time alerts are a critical component of incident detection, providing immediate notifications to relevant stakeholders when an issue is detected. These alerts should be configured to provide actionable information, such as the nature of the anomaly, its potential impact, and recommended actions for addressing the issue. Effective alerting mechanisms help ensure that incidents are promptly identified and managed, reducing the risk of extended downtime or data loss.

Manual monitoring and review processes can complement automated tools. Regular reviews of system logs, performance reports, and security assessments can help identify potential issues that automated tools may not detect. Combining automated and manual detection approaches provides a comprehensive strategy for identifying and addressing incidents in ML systems.

Incident Identification and Classification

Once an incident is detected, it is essential to accurately identify and classify the incident to determine its nature and severity. Incident identification involves analyzing the alerts and anomalies generated by detection mechanisms to confirm that an incident has occurred. This step requires careful evaluation of the evidence to understand the scope and impact of the incident.

Incident classification involves categorizing the incident based on its type, impact, and urgency. Classification helps prioritize incidents and allocate resources effectively. For example, a data breach involving sensitive customer information may be classified as a high-priority incident due to its potential legal and reputational consequences. In contrast, a minor performance issue with a non-critical model may be classified as a lower priority.

The classification process should include criteria for assessing the severity of the incident, such as the extent of data loss, the number of affected users, and the potential impact on system operations. This assessment helps determine the appropriate response actions and resources required to address the incident. Accurate classification is crucial for ensuring that incidents are managed effectively and that high-impact issues receive prompt attention.

Incident identification and classification should involve collaboration with relevant stakeholders, including data scientists, engineers, and security experts. Input from these stakeholders can provide valuable insights into the nature of the incident and help refine the classification process. Collaborative identification and classification ensure a comprehensive understanding of the incident and support effective incident management.

Assessing the Impact and Scope of the Incident

Assessing the impact and scope of an incident is a critical step in the incident response process, as it helps determine the extent of the damage and the resources required for resolution. This assessment involves evaluating how the incident affects various components of the ML system, including the model, data pipeline, and infrastructure.

Impact assessment involves analyzing the consequences of the incident on system performance, data integrity, and security. For example, if an incident involves a model performance degradation, the impact assessment should evaluate how this affects the accuracy of predictions and the potential implications for business operations. Similarly, if the incident involves a data breach, the assessment should consider the sensitivity of the compromised data and the potential impact on affected individuals.

Scope assessment involves determining the extent of the incident, including the affected systems, users, and data. This assessment helps identify which components of the ML system are impacted and the degree of disruption caused. Understanding the scope of the incident is crucial for developing a targeted response strategy and allocating resources effectively.

The assessment process should also involve collaboration with stakeholders to gather information and insights about the incident. Input from data scientists, engineers, and security experts can provide a comprehensive understanding of the incident's impact and scope. This collaborative approach ensures that all aspects of the incident are considered and that the response strategy is well-informed.

Incident Containment Strategies

Containment is a crucial phase in the incident response process, aimed at limiting the spread and impact of the incident to prevent further damage. Effective containment strategies involve isolating affected components, implementing temporary fixes, and stabilizing the system to prevent additional issues.

One common containment strategy involves isolating compromised systems or components to prevent the incident from spreading to other parts of the ML system. For example, if a data breach is detected, isolating the affected data sources or servers can help prevent unauthorized access to additional data. Similarly, if a model is experiencing performance issues, it may be temporarily disabled or replaced with a backup model to maintain system functionality.

Implementing temporary fixes or workarounds can help stabilize the system and mitigate the immediate impact of the incident. For example, applying patches to address security vulnerabilities or modifying system configurations to address performance issues can help contain the incident. These temporary measures should be carefully managed to ensure that they do not introduce new issues or compromise system stability.

Effective containment also involves communication with stakeholders to keep them informed about the incident and the actions being taken. Providing timely updates and instructions helps ensure that stakeholders are aware of the situation and can take appropriate actions to support the response. Containment strategies should be coordinated with all relevant stakeholders to ensure a unified approach to managing the incident.

Eradication of the Root Cause

Eradication involves removing the underlying cause of the incident to prevent recurrence and restore the ML system to a secure and stable state. This phase requires a thorough investigation to identify the root cause of the incident and implement measures to address it.

The eradication process begins with a detailed analysis of the incident to understand its origins and contributing factors. This may involve reviewing system logs, analyzing data patterns, and conducting forensic investigations to identify the root cause. For example, if a model is compromised due to a security vulnerability, the investigation should determine how the vulnerability was exploited and what steps are needed to address it.

Once the root cause is identified, appropriate measures should be implemented to eliminate it. This may include applying patches to fix security vulnerabilities, cleaning up corrupted data, or updating flawed models. Eradication efforts should be carefully planned and executed to ensure that the underlying issue is fully addressed and that the system is restored to a secure state.

Verification of eradication efforts is essential to ensure that the root cause has been effectively removed and that the incident is resolved. Testing and validation should be conducted to confirm that the measures taken have successfully addressed the issue and that no residual problems remain. Effective eradication is critical for preventing the recurrence of the incident and ensuring the long-term stability of the ML system.

Recovery and Restoration of Normal Operations

The recovery phase involves restoring normal operations and functionality to the ML system after an incident. This phase includes restoring affected components, validating system performance, and ensuring that the system is operating as intended.

Restoration of normal operations begins with bringing affected systems and components back online. This may involve reactivating models, reconnecting data pipelines, and reconfiguring system settings to resume normal functionality. The recovery process should be carefully managed to ensure that systems are restored in a controlled and systematic manner.

Validation of system performance is a critical step in the recovery phase. This involves testing the restored systems to ensure that they are functioning correctly and that any issues have been resolved. For example, if a model has been restored, its performance should be evaluated to confirm that it meets the required accuracy and reliability standards.

Communication with stakeholders during the recovery phase is essential to provide updates on the status of the system and any ongoing activities. Stakeholders should be informed about the steps taken to restore normal operations and any additional actions required. Effective communication helps ensure that all parties are aware of the system's status and can plan accordingly.

Communication and Reporting During an Incident

Effective communication and reporting are essential components of incident response, ensuring that stakeholders are informed and that relevant information is shared in a timely manner. Communication during an incident involves coordinating with internal teams, external partners, and regulatory bodies to manage the situation effectively.

Internal communication involves keeping all relevant teams and stakeholders informed about the incident, including its status, impact, and response actions. Regular updates should be provided to ensure that everyone is aware of the situation and can take appropriate actions. Internal communication should be clear, concise, and focused on providing actionable information.

External communication may involve notifying external partners, customers, and regulatory bodies about the incident. For example, if a data breach occurs, affected individuals and regulatory authorities may need to be notified in accordance with legal requirements. External communication should be managed carefully to ensure that accurate information is provided and that any potential reputational damage is minimized.

Reporting during an incident involves documenting all relevant details, including the nature of the incident, response actions, and any outcomes. Detailed incident reports should be prepared to provide a comprehensive record of the incident and the response process. These reports are valuable for post-incident analysis, regulatory compliance, and continuous improvement efforts.

Documentation and Record-Keeping

Documentation and record-keeping are critical for maintaining a comprehensive record of the incident response process and ensuring that all relevant information is captured and retained. Effective documentation supports transparency, accountability, and continuous improvement.

Incident documentation should include detailed records of the incident, including its nature, impact, and response actions. This documentation should capture all relevant information, such as incident timelines, communication logs, and technical analyses. Accurate and detailed records are essential for understanding the incident, evaluating the response, and identifying areas for improvement.

Record-keeping should also include documentation of any decisions made during the incident response process. This includes the rationale for response actions, any deviations from established procedures, and the outcomes of those actions. Documenting decision-making processes helps ensure accountability and provides a basis for evaluating the effectiveness of the response.

Effective documentation and record-keeping also involve maintaining records of any regulatory or compliance-related activities. This includes documentation of notifications to regulatory bodies, compliance with legal requirements, and any required reports. Proper record-keeping supports compliance efforts and helps ensure that all regulatory obligations are met.

Post-incident Analysis and Improvement

Post-incident analysis is a critical phase in the incident response process, focusing on evaluating the incident and the response efforts to identify lessons learned and areas for improvement. This analysis helps organizations refine their incident response strategies and enhance their overall security posture.

The post-incident analysis process begins with a thorough review of the incident, including its causes, impacts, and response actions. This review should involve all relevant stakeholders and may include analyzing incident documentation, conducting interviews, and reviewing system data. The goal is to understand what happened, how it was handled, and what can be improved.

Identifying lessons learned from the incident is a key aspect of post-incident analysis. This involves evaluating the effectiveness of the response, identifying any gaps or weaknesses, and determining what changes are needed to improve future incident management. Lessons learned should be documented and used to update incident response plans, policies, and procedures.

Continuous improvement efforts should focus on implementing changes based on the findings from the post-incident analysis. This may include updating incident response plans, enhancing detection and monitoring mechanisms, and improving communication and coordination. Ongoing improvement helps ensure that the organization is better prepared for future incidents and can respond more effectively.

Compliance and Regulatory Considerations

Compliance with regulatory requirements is a critical aspect of incident response, particularly in industries with stringent data protection and privacy regulations. Regulatory considerations involve ensuring that incident response efforts meet legal and regulatory obligations and that required notifications and reports are submitted in a timely manner.

Regulatory compliance involves understanding and adhering to relevant laws and regulations, such as data protection regulations, industry standards, and contractual obligations. For example, regulations such as the General Data Protection Regulation (GDPR) and the California Consumer Privacy Act (CCPA) impose specific requirements for handling data breaches and notifying affected individuals.

Incident response plans should be designed to address these regulatory requirements by including procedures for notifying regulatory bodies, affected individuals, and other stakeholders. Compliance considerations should be integrated into all phases of the incident response process, from detection and containment to recovery and reporting.

Ensuring compliance also involves staying informed about changes in regulatory requirements and updating incident response plans accordingly. Organizations should monitor regulatory developments and adjust their incident response strategies to address new or evolving requirements. Compliance with regulatory obligations is essential for avoiding legal penalties and maintaining trust with stakeholders.

Ethical and Legal Implications

Ethical and legal implications play a significant role in incident response, influencing how incidents are managed and resolved. Addressing these implications involves considering the ethical and legal aspects of incident management, including privacy concerns, data protection, and accountability.

Ethical considerations in incident response involve ensuring that responses are conducted in a manner that respects individuals' rights and privacy. For example, handling personal data in accordance with data protection laws and avoiding unnecessary exposure of sensitive information are critical ethical considerations. Ethical incident management also involves transparency and honesty in communication with affected parties and stakeholders.

Legal implications of incident response involve adhering to legal requirements and regulations related to data protection, security, and incident reporting. This includes understanding and complying with applicable laws, such as data breach notification requirements and industry-specific regulations. Legal considerations also involve managing potential liabilities and ensuring that incident management practices support legal compliance.

Balancing ethical and legal considerations with the practical aspects of incident response requires careful planning and decision-making. Organizations should develop policies and procedures that address both ethical and legal aspects of incident management, ensuring that responses are conducted in a responsible and compliant manner. Ethical and legal considerations should be integrated into all phases of the incident response process to support effective and responsible incident management.

Training and Preparedness

Training and preparedness are vital for ensuring that incident response teams are equipped to handle incidents effectively. Ongoing training and preparedness activities help build the skills and knowledge needed for effective incident management and ensure that teams are ready to respond to a wide range of scenarios.

Training programs should cover various aspects of incident response, including detection, classification, containment, eradication, and recovery. Training should be tailored to the specific roles and responsibilities of team members and should include both theoretical knowledge and practical exercises. Regular training sessions and simulations help ensure that team members are familiar with incident response procedures and can respond confidently to incidents.

Preparedness activities involve developing and maintaining incident response plans, conducting regular drills and exercises, and ensuring that all necessary resources and tools are available. Preparedness also includes reviewing and updating incident response plans based on lessons learned from previous incidents and changes in the threat landscape. Ongoing preparedness efforts help ensure that teams are well-prepared to handle incidents and minimize the impact on the organization.

Training and preparedness should also involve collaboration with external experts and partners. Engaging with third-party consultants, industry groups, and regulatory bodies can provide additional insights and best practices for incident management. Collaboration with external experts helps enhance the effectiveness of training and preparedness efforts and supports a comprehensive approach to incident response.

Role of Automation in Incident Response

Automation plays a critical role in enhancing the efficiency and effectiveness of incident response in machine learning (ML) environments by streamlining processes and reducing the time required to detect, assess, and address incidents. Automated tools and systems are integral to managing various aspects of incident response, from monitoring and detection to analysis and remediation, thereby enabling quicker and more precise responses to potential threats.

Automated Tools in Detection and Alerting

Automated detection and alerting systems are pivotal in providing real-time notifications of potential incidents, allowing for rapid response and mitigation. Tools like Splunk and IBM QRadar, which are security information and event management (SIEM) systems, are widely used in incident response. These tools are designed to monitor vast amounts of data across an organization's infrastructure, identifying anomalies or unusual patterns that could indicate a security incident. For instance, Splunk can analyze logs and network traffic to detect deviations from normal behavior, automatically triggering alerts when predefined thresholds are exceeded. QRadar, on the other hand, offers advanced correlation capabilities that can detect complex attack patterns across various sources, enhancing the speed and accuracy of threat detection. Automated alerting ensures that incidents are promptly identified and addressed, significantly reducing the risk of extended downtime or data loss, which is crucial in environments where machine learning models are deployed and rely on continuous data processing.

Enhanced Analysis and Investigation with Automation

Automation also supports incident analysis and investigation by providing tools for data aggregation, correlation, and visualization. For example, Splunk's machine learning capabilities can automatically correlate data from multiple sources, helping to identify patterns and trends that might otherwise go unnoticed. This is particularly beneficial in understanding the root cause and impact of an incident, especially in complex ML environments where data flows from numerous sources and systems. The automated analysis provided by these tools supports a more efficient and accurate assessment of incidents, which is essential for developing effective response strategies. IBM QRadar also excels in this area by offering integrated threat intelligence, which can correlate incident data with known threats, providing deeper insights into the nature of the attack and potential vulnerabilities that were exploited. Automation in analysis enables organizations to quickly understand the scope of an incident, reducing the time to resolution and minimizing the potential for damage.

Automated Remediation and Case Study Illustration

Automation in incident remediation is another crucial aspect of enhancing incident response capabilities. Tools like Ansible, integrated with SIEM systems like Splunk and QRadar, allow for automated deployment of security patches or configuration changes across affected systems. For example, if an incident involves a vulnerability that can be mitigated by a software update, an automated patch management system can deploy the necessary updates across all impacted systems without requiring manual intervention. This not only speeds up the remediation process but also ensures that the response is consistent and thorough, minimizing the risk of human error.

A case study that illustrates the benefits of automation in incident response involves a financial services company that experienced a significant security breach due to a misconfigured server. The company's incident response team used IBM QRadar to detect unusual login attempts across multiple servers. QRadar's automated correlation and alerting capabilities quickly identified the breach, allowing the team to contain the incident within minutes. Splunk was then used to analyze the logs and pinpoint the exact misconfiguration that led to the breach. By integrating Ansible with these tools, the company was able to automatically deploy configuration updates across its entire server infrastructure, preventing similar incidents in the future. This case highlights how automation can significantly reduce response times, improve accuracy in incident analysis, and ensure a swift, consistent remediation process.

The integration of automated tools in incident response not only enhances the speed and accuracy of detecting and addressing incidents but also significantly reduces the operational burden on IT and security teams. By automating routine tasks such as monitoring, alerting, and patch management, these teams can focus on more strategic aspects of incident response, such as improving security policies and conducting post-incident reviews. This approach not only strengthens the overall security posture of an organization but also ensures that machine learning systems, which are increasingly critical to business operations, remain secure and resilient against emerging threats.

Developing and Maintaining Incident Response Plans

Developing and maintaining incident response plans is a critical component of an organization's overall strategy for managing and mitigating the impact of security incidents. An incident response plan provides a structured and methodical approach to detecting, responding to, and recovering from various types of incidents. By defining clear procedures and protocols, the plan serves as a roadmap for managing scenarios that could otherwise lead to significant disruptions or losses.

Aligning Incident Response with Organizational Strategies

Incident response plans should not operate in isolation; they must be carefully aligned with broader organizational strategies, such as business continuity and disaster recovery plans. This alignment ensures that incident response efforts are integrated into the overall strategy for maintaining operational resilience and minimizing downtime. Business continuity plans focus on ensuring that essential functions can continue during and after a disruption, while disaster recovery plans outline the steps for restoring systems and data after an incident. By integrating incident response plans with these strategies, organizations can ensure a coordinated response that addresses immediate threats while supporting long-term recovery efforts.

When developing an incident response plan, it is essential to conduct a thorough risk assessment to identify potential incidents that could impact the organization. This assessment should consider the likelihood and potential impact of various threats, such as cyberattacks, data breaches, or system failures. Understanding these risks allows the organization to prioritize its incident response efforts and allocate resources effectively. The plan should clearly outline the roles and responsibilities of team members, ensuring that everyone understands their specific duties during an incident. Communication protocols should also be established to ensure that information flows efficiently between team members, management, and external stakeholders.

Maintaining and Updating Incident Response Plans

Maintaining an incident response plan is an ongoing process that requires regular updates and reviews. The threat landscape is constantly evolving, with new vulnerabilities and attack vectors emerging regularly. Organizations must ensure that their incident response plans remain dynamic and responsive to these changes. This involves incorporating lessons learned from previous incidents, adapting to new technologies, and staying current with regulatory requirements. For example, as new data protection laws are enacted, the incident response plan may need to be updated to include specific procedures for handling breaches of personal information.

Regular reviews of the incident response plan help to identify areas that may need improvement or adjustment. These reviews should involve collaboration with internal stakeholders, such as IT and security teams, as well as external experts who can provide insights into emerging threats and best practices. By keeping the plan up to date, organizations can ensure that they are prepared to respond effectively to incidents, reducing the risk of significant operational disruptions or compliance violations.

Testing and validation are essential components of maintaining an effective incident response plan. Regular drills and exercises should be conducted to simulate various incident scenarios and assess the plan's effectiveness. These exercises allow the organization to evaluate how well team members understand and can execute the plan's procedures and protocols. Testing also provides an opportunity to identify any gaps or weaknesses in the plan, which can then be addressed through revisions and additional training. For example, a drill that simulates a ransomware attack might reveal that the communication protocol needs to be streamlined to ensure quicker decision-making during an actual incident.

Ensuring Responsiveness to Evolving Threats and Regulatory Requirements

One of the key challenges in maintaining an incident response plan is ensuring that it remains responsive to evolving threats and regulatory requirements. The cyber threat landscape is highly dynamic, with attackers continually developing new techniques and tactics. To keep pace with these changes, organizations must adopt a proactive approach to incident response planning. This includes staying informed about the latest threats and vulnerabilities through threat intelligence feeds, industry reports, and collaboration with other organizations. By incorporating this information into the incident response plan, organizations can better anticipate and prepare for emerging threats.

Regulatory compliance is another critical consideration in developing and maintaining incident response plans. Many industries are subject to specific regulations that dictate how organizations must respond to security incidents. For example, the General Data Protection Regulation (GDPR) in the European Union requires organizations to report data breaches within 72 hours of discovery. Failure to comply with these regulations can result in significant fines and reputational damage. Therefore, the incident response plan must include procedures for ensuring that the organization meets all relevant regulatory requirements. This may involve designating a compliance officer to oversee the incident response process and ensure that all necessary reporting and documentation are completed.

To ensure that the incident response plan remains effective, organizations should establish a process for continuous improvement. This process should involve regular feedback from incident response team members, as well as input from external experts and stakeholders. By continually refining and updating the plan, organizations can enhance their ability to respond to incidents quickly and effectively, minimizing the impact on operations and maintaining compliance with regulatory requirements.

Case Study: Incident Response Alignment with Business Continuity

A financial services firm provides a practical example of how aligning incident response plans with broader organizational strategies can enhance overall resilience. The firm faced a significant data breach that threatened to disrupt its operations. Fortunately, the firm had integrated its incident response plan with its business continuity and disaster recovery strategies. As a result, the response team was able to quickly isolate the affected systems and initiate the disaster recovery process, which included restoring critical data from backups and rerouting transactions through alternative channels. The seamless coordination between incident response and business continuity efforts allowed the firm to minimize downtime and maintain customer trust, illustrating the importance of a holistic approach to incident management.

Developing and maintaining an incident response plan is not a one-time effort but a continuous process that requires alignment with broader organizational strategies, such as business continuity and disaster recovery plans. By ensuring that the incident response plan remains dynamic and responsive to evolving threats and regulatory requirements, organizations can better protect their operations, assets, and reputation in the face of increasingly sophisticated cyber threats. Regular testing, validation, and updates, combined with a commitment to continuous improvement, are essential for maintaining an effective incident response plan that supports the organization's long-term resilience and success.

Measuring Incident Response Effectiveness

Measuring the effectiveness of incident response is crucial for evaluating the success of the response process and identifying areas for improvement. Effective measurement involves assessing various metrics and performance indicators to determine how well incidents are managed and resolved.

Key metrics for measuring incident response effectiveness include response time, resolution time, and the impact of the incident. Response time measures how quickly the incident is detected and addressed, while resolution time measures the duration required to resolve the incident and restore normal operations. Assessing the impact of the incident involves evaluating the extent of damage, data loss, or operational disruption.

Other important metrics include the accuracy of incident classification, the effectiveness of containment and remediation efforts, and stakeholder satisfaction. Evaluating the accuracy of classification helps ensure that incidents are appropriately prioritized and managed. Assessing the effectiveness of containment and remediation efforts provides insights into the efficiency of response actions and identifies areas for improvement. Stakeholder satisfaction surveys can provide valuable feedback on the communication and management of the incident.

Continuous improvement efforts should be based on the findings from the effectiveness measurement process. This includes updating incident response plans, refining procedures, and implementing changes based on lessons learned. Measuring and improving incident response effectiveness helps enhance the organization's overall incident management capabilities and supports a proactive approach to handling future incidents.

Continuous Improvement in Incident Response

Continuous improvement in incident response is crucial for organizations to adapt to new challenges, enhance response capabilities, and maintain the effectiveness of incident management practices. As cyber threats evolve, so too must the strategies and tools used to detect, respond to, and recover from incidents. Continuous improvement ensures that incident response remains dynamic, responsive, and aligned with the organization's broader risk management strategies. This process involves a structured framework that includes steps for analyzing incidents, identifying lessons learned, implementing changes, and monitoring outcomes.

Analyzing Incidents and Response Efforts

The first step in the continuous improvement process is a thorough analysis of incidents and the corresponding response efforts. This analysis is fundamental to understanding what worked well and what did not during an incident. It begins with a detailed review of incident documentation, which includes logs, timelines, and any records of decisions made during the incident response. This documentation provides a comprehensive overview of the incident, offering insights into the sequence of events and the effectiveness of the response.

Assessing response metrics is also a critical component of this analysis. Metrics such as the time to detect, respond, and recover from an incident can reveal how efficiently the incident was handled. For example, if the time to detect an incident was significantly longer than expected, it may indicate a need for improved monitoring or alerting systems. Similarly, if recovery took longer than anticipated, there may be gaps in the disaster recovery plan that need to be addressed.

Gathering feedback from stakeholders is another essential part of the analysis process. Stakeholders include not only the incident response team but also other departments affected by the incident, such as IT, legal, and communications. Their feedback can provide valuable perspectives on how the incident was managed and where improvements can be made. For example, IT staff may highlight technical challenges encountered during the response, while legal teams may identify compliance issues that arose.

Through this comprehensive analysis, organizations can identify strengths and weaknesses in their incident response practices. Strengths should be reinforced, and weaknesses should be targeted for improvement. The goal is to build on what is working well while addressing any gaps or issues that could hinder future incident responses.

Identifying Lessons Learned

Once the analysis is complete, the next step is to identify the lessons learned from the incident. Lessons learned are the key takeaways that can inform improvements to the incident response process. These lessons should be documented and shared across the organization to ensure that all relevant parties are aware of the insights gained.

Identifying lessons learned involves more than just recognizing mistakes; it also includes acknowledging what went right. For example, if a particular detection tool was instrumental in identifying the incident early, that tool should be highlighted as a critical asset. On the other hand, if communication between teams was slow or ineffective, this could be identified as an area for improvement.

The process of identifying lessons learned should be systematic and structured. It often involves debriefing sessions or post-incident reviews where team members can discuss their experiences and observations. These sessions should be conducted in an open and constructive manner, encouraging honest feedback and collaboration. The focus should be on learning and improvement rather than assigning blame.

The lessons learned should also be categorized based on their relevance to different aspects of the incident response process. For example, some lessons may relate to technical aspects such as detection and remediation, while others may focus on communication or decision-making processes. By categorizing lessons learned, organizations can more easily identify patterns or recurring issues that need to be addressed.

Implementing Changes

After identifying the lessons learned, the next step is to implement changes to the incident response plan, procedures, and tools. This is where the insights gained from the analysis and lessons learned are translated into actionable improvements. Implementing changes is a critical part of the continuous improvement process, as it ensures that the organization is better prepared for future incidents.

The implementation process begins with updating the incident response plan. The plan should be revised to incorporate the lessons learned and address any identified weaknesses. For example, if the analysis revealed that the organization was slow to detect the incident, the plan might be updated to include more rigorous monitoring and alerting protocols. Similarly, if communication was identified as an issue, the plan might be revised to include clearer communication channels and protocols.

In addition to updating the incident response plan, organizations should also update their incident response procedures. Procedures are the specific steps that team members follow during an incident, and they should be as clear and concise as possible. If any procedures were found to be lacking during the incident, they should be revised or replaced with more effective ones. For example, if the procedure for escalating an incident was unclear or inefficient, it should be streamlined to ensure that incidents are escalated quickly and effectively.

The tools used in incident response should also be reviewed and updated as necessary. This might involve upgrading existing tools, implementing new ones, or retiring tools that are no longer effective. For example, if a particular detection tool was found to be inadequate, it might be replaced with a more advanced solution. Similarly, if a tool used for communication or coordination was found to be lacking, it should be upgraded or replaced.

Implementing changes also involves training and development for the incident response team. Team members should be trained on any new procedures or tools that have been introduced. This training should be practical and hands-on, allowing team members to become familiar with the updated processes and tools before they are needed in a real incident. Ongoing training and development are essential for ensuring that the incident response team remains equipped with the latest knowledge and skills.

Monitoring Outcomes

The final step in the continuous improvement process is monitoring the outcomes of the implemented changes. This step is critical for ensuring that the changes are effective and that the organization is better prepared for future incidents. Monitoring outcomes involves tracking the performance of the updated incident response processes and tools, as well as gathering feedback from the incident response team and other stakeholders.

One way to monitor outcomes is by conducting regular drills and exercises that simulate different types of incidents. These simulations allow the organization to test the effectiveness of the updated incident response plan and procedures in a controlled environment. The results of these exercises can provide valuable insights into how well the changes are working and whether any further adjustments are needed.

In addition to simulations, organizations should also track key performance metrics related to incident response. These metrics might include the time to detect, respond, and recover from incidents, as well as the number of incidents that occur over a given period. By tracking these metrics, organizations can assess whether the changes have led to improvements in incident response performance.

Feedback from the incident response team and other stakeholders is also an important part of monitoring outcomes. Team members should be encouraged to provide feedback on how well the updated processes and tools are working in practice. This feedback can help identify any issues or challenges that were not anticipated during the implementation phase.

If the monitoring process reveals that the implemented changes are not achieving the desired outcomes, further adjustments may be necessary. Continuous improvement is an iterative process, and it may take several rounds of analysis, implementation, and monitoring to fully optimize the incident response process.

The Continuous Improvement Loop

The continuous improvement process can be visualized as a loop, where the steps of analyzing incidents, identifying lessons learned, implementing changes, and monitoring outcomes are repeated on an ongoing basis. This loop ensures that the incident response process is constantly evolving and improving, rather than remaining static.

The continuous improvement loop is driven by the organization's commitment to maintaining a high standard of incident response. It requires a proactive approach, where the organization is always looking for ways to improve and adapt to new challenges. This proactive approach is essential in today's rapidly changing threat landscape, where new vulnerabilities and attack vectors are constantly emerging.

Organizations that embrace the continuous improvement loop are better equipped to respond to incidents quickly and effectively, minimizing the impact on their operations and reputation. They are also more likely to stay ahead of regulatory requirements and industry best practices, ensuring that their incident response process remains compliant and aligned with broader risk management strategies.

Case Study: Continuous Improvement in Incident Response

A leading global technology company provides an excellent example of the continuous improvement loop in action. The company faced a significant cybersecurity incident that exposed vulnerabilities in its incident response process. In response, the company conducted a comprehensive analysis of the incident, identifying several areas for improvement, including detection capabilities, communication protocols, and response procedures.

Based on the lessons learned, the company implemented a series of changes to its incident response plan. These changes included upgrading its detection tools, streamlining communication channels, and introducing new procedures for handling specific types of incidents. The company also invested in training and development for its incident response team, ensuring that team members were familiar with the updated processes and tools.

To monitor the outcomes of these changes, the company conducted regular drills and exercises, simulating different types of incidents. The results of these exercises revealed significant improvements in the company's incident response performance, including faster detection and response times and more effective communication and coordination during incidents.

Over time, the company continued to refine its incident response process, incorporating new lessons learned and adapting to emerging threats. This commitment to continuous improvement has enabled the company to maintain a high standard of incident response, protecting its operations and reputation in the face of evolving cyber threats.

Continuous improvement in incident response is not a one-time effort but an ongoing process that requires a structured framework for analyzing incidents, identifying lessons learned, implementing changes, and monitoring outcomes. By embracing this process, organizations can ensure that their incident response capabilities remain effective and responsive to the ever-changing threat landscape. This proactive approach to incident response not only enhances the organization's ability to manage incidents but also supports its broader risk management strategy, ensuring long-term resilience and success.

Incident Response and Organizational Culture

Organizational culture plays a pivotal role in shaping the effectiveness of incident response. A culture that promotes open communication, collaboration, and proactive risk management is crucial for ensuring that incidents are detected early, reported promptly, and managed efficiently. In an environment where transparency is encouraged, employees are more likely to report incidents without fear of blame, allowing for swift and effective responses. This culture of openness is essential for addressing incidents before they escalate into more significant issues.

Collaboration across departments is another critical element of a robust incident response culture. When teams from IT, legal, communications, and other areas work together, they can address incidents from multiple angles, ensuring comprehensive solutions. For example, a financial services firm that faced a phishing attack was able to mitigate the damage quickly due to the collaborative efforts of its IT and communications teams, who worked together to inform customers and secure compromised accounts. This kind of cross-functional teamwork is vital for managing incidents effectively and minimizing their impact.

A proactive approach to risk management further strengthens incident response capabilities. Organizations that encourage employees to anticipate and identify potential risks are better positioned to prevent incidents before they occur. In the healthcare industry, a hospital network that had previously suffered a ransomware attack implemented continuous monitoring and regular vulnerability assessments as part of its proactive strategy. This shift in culture helped the organization detect threats early and respond swiftly, preventing further disruptions to patient care.

Leadership's Role in Fostering a Positive Incident Response Culture

Leadership is instrumental in shaping and reinforcing a culture that supports effective incident response. When leaders prioritize incident management and provide the necessary resources, they signal its importance to the entire organization. This commitment from the top encourages employees to take incident response seriously and ensures that the necessary tools and training are in place.

For instance, in the technology sector, a software company faced a zero-day vulnerability in one of its products. The leadership's immediate response, which included allocating additional resources to the incident response team and communicating openly with customers, set a strong example for the rest of the organization. This proactive and transparent approach not only mitigated the impact of the vulnerability but also reinforced the importance of incident response within the company.

Leaders who actively support and participate in incident response efforts create an environment where continuous improvement is valued. By regularly reviewing incident outcomes and making adjustments based on lessons learned, they help the organization stay ahead of potential threats and ensure that incident response practices evolve alongside emerging risks.

Fostering a Culture of Proactive Incident Response

A proactive culture of incident response is essential for organizations to stay ahead of potential threats and manage incidents effectively. This culture is characterized by continuous vigilance, where employees are encouraged to identify and report potential risks before they escalate. Regular training and awareness programs are key to fostering this mindset, equipping employees with the knowledge and skills needed to recognize and respond to threats.

In the retail sector, a large e-commerce company implemented a proactive incident response strategy after experiencing a significant data breach. By conducting regular security assessments and fostering a culture of transparency, the company was able to prevent further breaches and maintain customer trust. This proactive approach ensured that the organization was prepared to respond quickly and effectively to any future incidents, minimizing potential disruptions to its operations.

In the telecommunications industry, a major provider faced a DDoS attack that threatened to disrupt services nationwide. Thanks to a culture of proactive risk management, the company's incident response team had already developed and tested response plans for such scenarios. The team was able to activate these plans immediately, mitigating the attack's impact and restoring service quickly. This example highlights the importance of being prepared for potential threats and having a well-practiced response plan in place.

Organizational Culture and Incident Response: A Comprehensive Checklist

Assessing the Culture of Open Communication

- **Evaluate Reporting Mechanisms:** Ensure that there are clear, accessible channels for reporting incidents. Employees should know how and where to report issues without fearing repercussions. Regularly review these mechanisms to ensure they remain effective and user-friendly.

- **Promote Transparency:** Assess whether your organizational culture promotes transparency. Are employees encouraged to speak up about potential risks and incidents? Conduct anonymous surveys to gather honest feedback from employees about their comfort level in reporting issues.

- **Remove Fear of Blame:** Examine if there is a fear of blame culture within the organization. Evaluate how incidents are discussed in team meetings—focus on solutions and learning rather than assigning blame. Consider implementing a no-blame policy to encourage reporting.

- **Leadership Communication:** Review how leaders communicate about incidents. Leaders should model transparency by sharing information about incidents openly with the entire organization. This includes discussing what happened, how it was handled, and what the organization learned from the incident.

Enhancing Cross-Departmental Collaboration

- **Cross-Functional Teams:** Check if your incident response involves collaboration across departments such as IT, legal, communications, and operations. Create dedicated cross-functional incident response teams that include members from all relevant departments.

- **Information Sharing:** Assess the effectiveness of information sharing across departments during an incident. Establish protocols for regular updates and information exchange among teams. Encourage the use of collaborative tools that allow seamless communication and sharing of relevant data.

- **Joint Training Exercises:** Organize joint training exercises that involve multiple departments to simulate incident scenarios. These exercises should test the organization's ability to work together across functions and refine collaborative efforts.

- **Role Clarity:** Ensure that each department understands its role during an incident. Develop and document clear roles and responsibilities for all departments involved in incident response to avoid confusion and ensure a coordinated effort.

Fostering Proactive Risk Management

- **Regular Risk Assessments:** Review the frequency and thoroughness of risk assessments conducted within the organization. Implement regular, comprehensive risk assessments that cover all potential threats, including cybersecurity, operational, and compliance risks.

- **Continuous Monitoring:** Evaluate whether the organization has effective continuous monitoring systems in place to detect potential threats early. Invest in technologies that provide real-time data and alerts for anomalies that could indicate an incident.

- **Employee Vigilance:** Encourage a culture of vigilance where employees are constantly aware of potential risks. This can be achieved through regular awareness programs, internal campaigns, and reminders of the importance of reporting suspicious activities.

- **Preemptive Measures:** Assess the organization's approach to preemptive risk management. Are there strategies in place to mitigate risks before they become incidents? For example, regular system updates, patch management, and vulnerability scanning should be routine practices.

Leadership's Role in Incident Response Culture

- **Leadership Engagement:** Evaluate the level of leadership engagement in incident response. Leaders should not only support the process but also actively participate in incident reviews and decision-making during a crisis.

- **Resource Allocation:** Assess whether leadership allocates sufficient resources—both financial and human—to incident response. This includes investing in the latest technology, hiring skilled personnel, and providing ongoing training.

- **Leading by Example:** Review how leaders respond to incidents. Leaders should demonstrate a commitment to transparency and continuous improvement by openly discussing the outcomes of incidents and how the organization plans to improve its response capabilities.

- **Leadership Training:** Provide specific training for leaders on incident management and communication. This ensures that they are equipped to handle their responsibilities during an incident effectively and can lead the organization through a crisis with confidence.

Building and Maintaining a Proactive Incident Response Culture

- **Training and Awareness Programs:** Assess the comprehensiveness of training programs. Regularly update and conduct training sessions for all employees, focusing on the latest threats, best practices in incident response, and the importance of their role in maintaining security.

- **Scenario-Based Drills:** Implement scenario-based drills that reflect realistic incident scenarios. These drills should test the organization's preparedness and help identify any gaps in the incident response plan.

- **Continuous Learning:** Encourage continuous learning and improvement in incident response practices. After every incident or drill, conduct a detailed review to learn what worked, what didn't, and how processes can be improved.

- **Culture of Improvement:** Review how your organization integrates lessons learned from incidents. A culture that emphasizes continuous improvement will consistently refine incident response strategies based on real-world experiences and emerging threats.

Aligning Incident Response with Organizational Strategies

- **Integration with Business Continuity Plans:** Assess whether the incident response plan is integrated with the broader business continuity and disaster recovery plans. This ensures that incident response efforts support the overall goal of maintaining business operations during a crisis.

- **Consistency Across Policies:** Evaluate the consistency of incident response policies with other organizational strategies, such as cybersecurity policies, data protection regulations, and compliance requirements. Ensure that there is alignment across all relevant policies.

- **Regular Policy Reviews:** Implement a schedule for regular reviews of the incident response plan to ensure it remains aligned with organizational goals and external regulations. This should include updates to reflect changes in the business environment, technology, or regulatory landscape.

Measuring and Monitoring the Effectiveness of Incident Response Culture

- **KPIs and Metrics:** Establish Key Performance Indicators (KPIs) and metrics to measure the effectiveness of the incident response culture. Metrics might include time to detection, time to response, employee participation in drills, and frequency of incident reports.

- **Feedback Loops:** Create feedback loops where employees can provide input on the incident response process. This can be done through surveys, anonymous feedback tools, or during post-incident reviews.

- **Performance Reviews:** Incorporate incident response performance into regular employee and leadership performance reviews. This reinforces the importance of incident response as part of the overall organizational performance.

- **Third-Party Audits:** Consider engaging third-party auditors to assess the effectiveness of your incident response culture. External audits can provide an unbiased perspective and identify areas for improvement that may not be apparent internally.

Continuous Improvement and Adaptation

- **Post-Incident Reviews:** After every incident, conduct a thorough post-incident review to identify lessons learned. Document these lessons and incorporate them into the incident response plan.

- **Adaptive Strategies:** Ensure that the incident response plan is flexible and can adapt to new threats and challenges. Regularly update the plan based on new information, emerging threats, and changes in the organizational structure or technology.

- **Employee Involvement:** Encourage ongoing employee involvement in the continuous improvement process. Employees at all levels should be engaged in identifying potential risks and suggesting improvements to the incident response plan.

- **Long-Term Commitment:** Foster a long-term commitment to improving incident response capabilities. This includes ongoing investment in technology, training, and process refinement to keep pace with evolving threats.

Incident Response in Cloud-Based ML Systems

Incident response in cloud-based machine learning (ML) systems is a complex process that requires careful consideration of the unique challenges posed by cloud environments. Unlike traditional on-premises systems, cloud-based ML systems operate in distributed, dynamic environments where the responsibility for security and incident management is shared between the cloud service provider and the organization using the service. To effectively manage incidents in this context, organizations must understand the shared responsibility model, leverage cloud-native tools, and develop robust strategies that integrate both internal and external resources.

Understanding the Shared Responsibility Model

The shared responsibility model is a fundamental concept in cloud security, defining the division of responsibilities between the cloud service provider and the organization. In cloud-based ML systems, the provider is typically responsible for the security of the cloud infrastructure, which includes the physical security of data centers, the underlying hardware, and the core cloud services. On the other hand, the organization is responsible for securing its applications, data, user access, and configurations within the cloud environment.

This division of responsibilities has significant implications for incident response. Organizations must clearly understand their role in securing their ML systems and ensure that their incident response plans reflect this understanding. For example, while the cloud provider may be responsible for mitigating threats to the infrastructure, the organization must handle incidents related to its data and ML models, such as unauthorized access or data breaches. This requires a coordinated approach where both parties collaborate to address incidents effectively.

In practice, organizations should establish clear communication channels with their cloud providers to facilitate quick responses during incidents. For instance, in the event of a security breach affecting the underlying infrastructure, the cloud provider would need to notify the organization immediately so that they can take appropriate action, such as suspending operations or isolating affected resources. This collaboration is crucial for minimizing the impact of incidents and ensuring that both parties fulfill their respective responsibilities.

Leveraging Cloud-Native Tools for Incident Detection and Response

Cloud-native tools and services offer powerful capabilities for monitoring, detecting, and responding to incidents in cloud-based ML systems. These tools are designed to integrate seamlessly with cloud environments, providing organizations with the visibility and control they need to manage incidents effectively.

AWS CloudWatch is a prime example of a cloud-native monitoring service that can be leveraged for incident response in ML systems. CloudWatch allows organizations to collect and track metrics, set alarms, and automatically respond to changes in the environment. For instance, an organization using AWS to host its ML models can set up CloudWatch alarms to detect unusual spikes in CPU usage or network traffic, which might indicate a security breach or a malfunctioning model. Once an anomaly is detected, CloudWatch can trigger automated responses, such as scaling up resources to handle increased load or initiating a Lambda function to shut down compromised instances.

A financial services company using AWS CloudWatch detected a significant increase in latency for its ML-powered fraud detection service. The CloudWatch alarm triggered an automated investigation, which revealed that the increased latency was due to a distributed denial-of-service (DDoS) attack targeting the service's endpoints. Thanks to the automated response mechanisms in place, the company was able to mitigate the attack by rerouting traffic through AWS Shield, a managed DDoS protection service, and scaling up resources to absorb the additional load. This rapid response minimized service disruption and protected the integrity of the fraud detection model.

Azure Monitor is another cloud-native tool that provides comprehensive monitoring and analytics for cloud-based ML systems. Azure Monitor collects data from various sources, including applications, operating systems, and network components, and uses this data to generate real-time insights into the system's health and performance. For ML systems, Azure Monitor can be used to track the performance of models in production, detect anomalies in data inputs, and monitor the overall health of the infrastructure supporting the ML workloads.

A healthcare provider using Azure Monitor to manage its ML models for patient data analysis detected a significant drop in prediction accuracy. The monitoring system flagged this issue, and the incident response team quickly discovered that the drop in accuracy was due to concept drift, where the statistical properties of the input data had changed over time. By integrating Azure Monitor with Azure Machine Learning, the team was able to automate the retraining of the models using updated data, restoring the accuracy of the predictions and ensuring the continued reliability of the healthcare provider's services.

These examples illustrate how cloud-native tools like AWS CloudWatch and Azure Monitor can enhance incident detection and response in cloud-based ML systems. By providing real-time insights, automating responses, and integrating with other cloud services, these tools help organizations quickly identify and address incidents, reducing the potential for damage and ensuring that ML models continue to operate effectively.

Incident Containment and Remediation in Cloud-Based Environments

When an incident occurs in a cloud-based ML system, containing the threat and mitigating its impact is critical. This process often requires close collaboration with the cloud service provider, especially when the incident involves the underlying infrastructure or managed services provided by the cloud platform.

For example, in the case of a security breach affecting a cloud-hosted ML model, the organization might need to isolate the affected resources to prevent the spread of the threat. This could involve temporarily suspending the compromised instances, revoking access permissions, or reconfiguring network security groups. Cloud service providers typically offer tools and services to assist with these tasks. For instance, **AWS Security Hub** provides a comprehensive view of the organization's security posture, integrating alerts from various AWS services and third-party tools. It can be used to quickly identify affected resources and take appropriate containment measures.

A retail company using AWS to host its customer recommendation engine discovered unauthorized access to its database containing sensitive customer information. Using AWS Security Hub, the company's incident response team quickly identified the compromised instances and isolated them by revoking access keys and modifying security groups. The team then worked with AWS support to apply necessary patches and reconfigure security settings to prevent further unauthorized access. This coordinated effort between the company and AWS minimized data exposure and allowed the company to restore its services securely.

Once the threat is contained, the organization must focus on remediation, which may involve applying patches, reconfiguring systems, or restoring data from backups. In cloud-based environments, this often requires working closely with the cloud provider to ensure that all necessary steps are taken to secure the environment and prevent a recurrence of the incident.

Google Cloud Security Command Center (SCC) is another cloud-native tool that can be instrumental in the remediation process. SCC provides visibility into an organization's assets, vulnerabilities, and threats across Google Cloud, helping incident response teams identify and prioritize remediation efforts. In one case, a technology company using Google Cloud to run its ML workloads discovered vulnerabilities in its cloud storage configurations that could potentially expose sensitive data. By using SCC, the company was able to quickly identify the misconfigured storage buckets, apply security patches, and update access controls to secure the data. The comprehensive insights provided by SCC enabled the company to address the vulnerabilities before they could be exploited, enhancing the overall security of its cloud-based ML system.

Effective Communication and Documentation

Effective communication is critical during an incident, especially in cloud-based environments where multiple stakeholders, including cloud service providers, internal teams, and possibly third-party vendors, may be involved. Clear, timely communication ensures that everyone is aware of their responsibilities and that actions are coordinated effectively to resolve the incident.

Organizations should establish predefined communication protocols as part of their incident response plan. These protocols should specify who needs to be informed about the incident, what information should be shared, and how it should be communicated. For example, during a major incident affecting a cloud-based ML system, the organization might need to communicate with the cloud provider's support team, the internal security team, and possibly external auditors or regulators. Each of these stakeholders will require different types of information, and it is crucial to ensure that the communication is clear, accurate, and timely.

ServiceNow, a popular cloud-based platform for IT service management, offers integrated incident management capabilities that can streamline communication during an incident. ServiceNow allows organizations to automate the notification process, ensuring that the right people are informed as soon as an incident is detected. It also provides a centralized platform for tracking the incident response, documenting actions taken, and communicating with stakeholders. This ensures that everyone involved has access to the same information, reducing the risk of miscommunication and improving the overall efficiency of the response.

A large financial institution using ServiceNow for incident management was able to effectively coordinate its response to a security breach affecting its cloud-based ML systems. The platform enabled the institution to quickly notify its cloud provider, internal teams, and external auditors, ensuring that everyone was on the same page. The centralized documentation of the incident response also facilitated a post-incident review, allowing the institution to identify areas for improvement and update its incident response plan accordingly.

Documentation is another critical aspect of incident response in cloud-based environments. Detailed records of the incident, including the actions taken, decisions made, and communications sent, are essential for both internal review and external compliance requirements. Proper documentation helps organizations learn from each incident, improving their response to future incidents and ensuring compliance with industry regulations and standards.

Case Studies: Successful Implementation of Cloud-Native Tools

The following case studies illustrate how organizations across various industries have successfully implemented cloud-native tools to enhance their incident response capabilities in cloud-based ML systems.

Case Study 1: Financial Services

A global financial services firm implemented AWS CloudWatch and AWS Security Hub to manage its cloud-based ML systems. When the firm detected an anomaly in its fraud detection model, CloudWatch automatically triggered an alarm, prompting the security team to investigate. Using Security Hub, the team quickly identified that the anomaly was caused by a misconfigured security group that allowed unauthorized access to the ML model's training data. The firm was able to isolate the affected resources, apply the necessary patches, and restore normal operations within hours, minimizing the impact on its services and customers.

Case Study 2: Healthcare

A healthcare provider using Azure Monitor and Azure Security Center to manage its cloud-based patient data and ML models for diagnostics encountered a significant issue when one of its models began producing inaccurate results. Azure Monitor detected unusual patterns in the model's predictions and triggered an alert. The incident response team quickly used Azure Security Center to conduct a thorough investigation, revealing that the inaccuracy was due to a data integrity issue caused by a recent update to the data pipeline. The team isolated the affected data, rolled back the changes, and retrained the model using verified data. The healthcare provider's swift action, enabled by cloud-native tools, ensured that patient care was not compromised and the issue was resolved before it could lead to incorrect diagnoses.

Case Study 3: E-Commerce

An e-commerce company that heavily relied on cloud-based ML models for product recommendations faced a critical incident when a sudden spike in demand caused its systems to slow down, leading to customer dissatisfaction. Leveraging Google Cloud's Operations Suite (formerly Stackdriver) and Security Command Center, the company's incident response team was able to identify that the slowdown was due to a combination of increased traffic and a misconfigured autoscaling policy. By quickly reconfiguring the autoscaling settings and using Security Command Center to ensure that all security parameters were in place, the company was able to restore normal service levels and mitigate the risk of a security breach during the high-traffic period. The use of these cloud-native tools not only resolved the immediate issue but also provided valuable insights for optimizing the company's infrastructure for future peak periods.

Developing and Implementing a Cloud-Based Incident Response Plan

Given the complexity and unique challenges of managing incidents in cloud-based ML systems, organizations must develop robust incident response plans that are specifically tailored to cloud environments. These plans should integrate cloud-native tools and services to ensure a comprehensive and effective response.

A well-structured cloud-based incident response plan begins with a clear understanding of the shared responsibility model. Organizations need to delineate the roles and responsibilities of both their internal teams and the cloud service provider. This includes establishing protocols for how incidents will be managed, who will be responsible for what actions, and how communication will be handled during an incident.

The plan should also include detailed procedures for incident detection and monitoring, leveraging the cloud-native tools discussed earlier. These tools should be configured to provide real-time alerts for any anomalies or security issues, with automated responses where possible to contain threats quickly. Organizations should ensure that their incident response teams are trained in the use of these tools and understand how to interpret the alerts and data they provide.

Once an incident is detected, the plan should outline clear steps for containment and remediation. This often involves close collaboration with the cloud provider to isolate affected resources, apply patches, or reconfigure security settings. The plan should also specify how to document the incident response process thoroughly, ensuring that all actions are recorded for future review and compliance purposes.

The incident response plan should include procedures for post-incident analysis and continuous improvement. After an incident is resolved, a detailed review should be conducted to identify what went well, what could be improved, and how the incident response plan should be updated. This process of continuous improvement is critical for adapting to new threats and ensuring that the organization remains prepared for future incidents.

The Importance of Continuous Training and Drills

To ensure that the incident response plan remains effective, organizations must invest in continuous training and regular drills for their incident response teams. These exercises help to reinforce the skills and knowledge needed to manage incidents in a cloud-based environment and provide an opportunity to test the plan's effectiveness.

Training should cover all aspects of the incident response process, including the use of cloud-native tools, communication protocols, and collaboration with the cloud provider. Teams should also be trained on the specific nuances of the cloud platform they are using, as different providers have different tools, services, and processes.

Regular drills, including simulated incidents and tabletop exercises, are essential for testing the incident response plan in a controlled environment. These drills should mimic real-world scenarios as closely as possible, challenging the team to respond effectively under pressure. The outcomes of these drills should be reviewed in detail, with lessons learned incorporated into the incident response plan.

Integrating Incident Response with Broader Organizational Strategies

Incident response in cloud-based ML systems should not be viewed in isolation but rather as a critical component of the organization's broader risk management and business continuity strategies. Ensuring that the incident response plan aligns with these strategies is essential for maintaining overall business resilience.

For example, the incident response plan should be integrated with the organization's disaster recovery and business continuity plans. This ensures that in the event of a major incident, such as a data breach or system failure, the organization can quickly recover its critical operations and minimize disruption to its services. The plan should also align with the organization's cybersecurity strategy, ensuring that all security measures are coordinated and that incident response efforts are supported by robust preventive controls.

Organizations should also consider regulatory requirements when developing their incident response plans. Many industries are subject to strict regulations regarding data security and incident reporting, such as the GDPR in Europe or HIPAA in the healthcare sector. The incident response plan must ensure compliance with these regulations, including the timely reporting of incidents to the appropriate authorities.

Governance and Oversight

Risk Management Committee

The Risk Management Committee (RMC) is pivotal in overseeing the risks associated with machine learning (ML) systems. It is tasked with identifying, assessing, and managing potential risks that could impact the integrity, security, and functionality of ML systems. This committee's primary objective is to establish a governance framework that ensures ML systems operate within defined risk parameters.

The committee is typically composed of cross-functional members, including data scientists, ML engineers, legal experts, compliance officers, and senior management. This diverse membership ensures that all aspects of risk—technical, legal, operational, and ethical—are comprehensively addressed. The RMC is responsible for establishing policies and procedures for risk management, regularly reviewing and updating risk management practices, and ensuring compliance with regulatory requirements. It also oversees the development and implementation of risk mitigation strategies and reports on the risk status to the board of directors or executive management.

One of the committee's primary functions is the identification of risks across the ML lifecycle. This includes risks associated with data quality, model accuracy, algorithmic bias, and potential cybersecurity threats. The committee should establish a systematic approach to risk identification, such as regular risk assessments, scenario analysis, and vulnerability testing.

Once risks are identified, the committee assesses the likelihood and potential impact of each risk. This involves a thorough analysis of how risks could affect the organization's ML operations and overall business objectives. Risks are then prioritized based on their severity, enabling the committee to focus on the most critical threats.

After prioritizing risks, the committee develops and implements strategies to mitigate them. This could involve revising data collection practices, enhancing model validation processes, implementing security measures, or ensuring compliance with regulations. The RMC must ensure that these strategies are both effective and cost-efficient.

The RMC establishes mechanisms for continuous monitoring of identified risks and the effectiveness of mitigation strategies. Regular reports are generated to inform stakeholders of the current risk landscape and the performance of risk management initiatives. These reports are crucial for maintaining transparency and accountability.

The RMC also plays a critical role in incident response. It is responsible for ensuring that the organization has a robust incident response plan in place and that all relevant personnel are trained to execute the plan effectively. The committee oversees the investigation of incidents, assesses the impact, and ensures that corrective actions are taken to prevent future occurrences.

The RMC's activities should be integrated with the organization's broader enterprise risk management (ERM) framework. This integration ensures that ML-related risks are considered within the context of the organization's overall risk appetite and that there is alignment between ML governance and broader business objectives.

The RMC must evolve continuously to address new and emerging risks in ML systems. This requires staying updated on the latest developments in ML technology, regulatory changes, and best practices in risk management. The committee should regularly review and update its policies, procedures, and risk management strategies to ensure they remain effective.

Interaction with Other Governance Bodies

The Risk Management Committee (RMC) interacts closely with other governance bodies, such as audit committees and data privacy offices, to ensure a cohesive approach to risk oversight. The RMC collaborates with the audit committee to align risk management practices with the broader financial and operational risks identified by auditors. This collaboration ensures that any risks associated with machine learning (ML) systems, such as potential financial losses due to model inaccuracies or cybersecurity breaches, are considered within the organization's overall risk management strategy. Regular joint meetings between the RMC and the audit committee allow for the sharing of risk assessments and mitigation strategies, ensuring that both financial and technological risks are addressed in a unified manner.

The interaction between the RMC and the data privacy office is crucial for managing risks related to data security and compliance with data protection regulations. The RMC relies on the data privacy office to provide insights into emerging privacy risks, particularly those associated with the collection, storage, and processing of data used in ML systems. By working together, these bodies can develop comprehensive strategies to mitigate risks, such as implementing stricter access controls, anonymizing data, and ensuring that ML models comply with regulations like GDPR or CCPA. This collaboration is essential for maintaining the integrity of data and safeguarding against privacy breaches that could lead to significant reputational and financial damage.

Effective communication between the RMC and other governance bodies, such as the audit committee and data privacy office, is facilitated through established communication channels and protocols. The RMC must ensure that its findings and decisions are communicated clearly and regularly to these bodies, enabling a coordinated response to identified risks. This communication is often formalized through joint reports, shared dashboards, and integrated risk management platforms that allow for real-time updates and collaboration. By maintaining open lines of communication, the RMC can ensure that all governance bodies are informed of the latest developments in risk management, leading to a more resilient and comprehensive risk oversight framework.

Effective Communication Strategies Between Governance Bodies

One effective communication strategy between the RMC and other governance bodies is the establishment of regular cross-functional meetings where key stakeholders from the RMC, audit committee, and data privacy office come together to discuss ongoing and emerging risks. These meetings should be structured to allow for the exchange of information and insights, with each body presenting updates on their respective areas of oversight. For instance, the audit committee might present findings from recent financial audits that have implications for ML risk, while the data privacy office could highlight new regulatory developments that need to be integrated into the RMC's risk management strategy. Such meetings foster a collaborative environment where risks are addressed from multiple perspectives, ensuring that no critical issue is overlooked.

Another communication strategy is the integration of shared digital platforms that allow real-time access to risk management data and insights across governance bodies. These platforms can be designed to aggregate data from various sources, providing a comprehensive view of the organization's risk landscape. For example, a shared dashboard could display key risk indicators, audit findings, compliance status, and incident reports, accessible to members of the RMC, audit committee, and data privacy office. This real-time access to information ensures that all governance bodies are operating with the same data, enabling quicker decision-making and more coordinated responses to potential risks.

Clear documentation and reporting protocols are also essential for effective communication between the RMC and other governance bodies. The RMC should establish standardized formats for reporting its findings and recommendations, ensuring that these documents are easily understood by non-technical stakeholders. For instance, when reporting on the risks associated with an ML system, the RMC should provide context on how these risks could impact the organization's financial health, compliance status, and overall strategic goals. By providing clear, actionable reports, the RMC can ensure that its insights are integrated into the broader risk management efforts of the organization, leading to a more holistic approach to governance and risk oversight.

Compliance Audits

Compliance audits are essential in ensuring that ML systems adhere to relevant laws, regulations, and ethical standards. These audits involve a thorough review of the organization's ML operations to verify that they comply with internal policies, industry standards, and regulatory requirements. The scope of ML compliance audits can vary depending on the industry, regulatory environment, and specific organizational needs. Audits typically cover data governance practices, model development processes, algorithmic transparency, and the ethical implications of ML applications. The scope must be clearly defined to ensure that all relevant aspects of ML governance are thoroughly examined.

Preparing for compliance audits in machine learning (ML) systems involves meticulous planning and organization to ensure that the audit process runs smoothly and yields favorable outcomes. One of the most crucial steps is maintaining thorough and up-to-date documentation. Organizations should ensure that all relevant documents, including data governance policies, model validation records, algorithmic transparency reports, and incident response plans, are current and easily accessible.

This documentation serves as the primary evidence during an audit, demonstrating that the organization adheres to regulatory requirements and industry standards. Maintaining a centralized repository where all compliance-related documents are stored and regularly updated is essential for an efficient audit process.

Training staff on compliance requirements and the specific procedures for audits is another vital aspect of preparation. Employees across the organization, especially those involved in ML operations, should be aware of the importance of compliance and their role in ensuring it. Regular training sessions should be conducted to familiarize staff with audit protocols, the organization's compliance policies, and the types of questions or documentation that auditors might request. Well-prepared staff can provide accurate information and documentation during the audit, reducing the likelihood of delays or misunderstandings. Training also reinforces a culture of compliance, where every team member understands their responsibility in maintaining regulatory standards.

Conducting pre-audit checks is a proactive strategy that can significantly enhance audit readiness. These checks involve internal reviews of the organization's ML systems and processes to identify potential areas of non-compliance before the official audit. Pre-audit checks should include an assessment of data privacy practices, model validation procedures, bias detection methods, and documentation completeness. By identifying and addressing any gaps or weaknesses in advance, organizations can mitigate the risk of negative audit findings. Pre-audit checks not only help in refining processes but also build confidence that the organization is fully prepared for the compliance audit.

The audit process for machine learning (ML) systems begins with a detailed planning phase, where auditors establish the objectives, scope, and methodology of the audit. This phase is critical as it sets the foundation for the entire audit process, ensuring that all relevant aspects of the ML system are examined. During the fieldwork phase, auditors gather evidence through interviews with key personnel, reviewing relevant documentation, and inspecting the systems in place. The collected data is then analyzed to identify any gaps in compliance or potential risks that need to be addressed.

Audits often uncover areas where organizations fail to meet compliance standards, revealing issues such as inadequate data privacy protections, a lack of transparency in model operations, insufficient documentation, and the absence of regular bias audits. These findings highlight the vulnerabilities in the organization's ML governance framework. Identifying these deficiencies is the first step toward remediation, as it allows the organization to understand where improvements are needed to enhance compliance and reduce risk.

Upon identifying non-compliance issues, organizations must act swiftly to implement corrective measures. This may involve revising existing policies, introducing new controls, or conducting additional training sessions for staff to ensure they are aware of and adhere to compliance requirements. The primary goal is to address the identified issues comprehensively to prevent them from recurring, thereby strengthening the organization's overall compliance posture.

Once the audit is completed, auditors compile a comprehensive report detailing their findings, including any non-compliance issues, associated risks, and recommendations for improvement. This report is typically presented to senior management and the board of directors, who are responsible for overseeing the implementation of the recommended corrective actions. The audit report serves as a crucial tool for guiding the organization's efforts to enhance its compliance and risk management strategies.

Continuous auditing offers a proactive approach to compliance management by enabling the regular and ongoing assessment of compliance in ML systems. Unlike traditional audits, which are conducted at set intervals, continuous auditing allows organizations to monitor compliance in real time, quickly identifying and addressing issues before they escalate. Integrating audits into the ML development lifecycle—from data collection and model training to deployment and monitoring—ensures that compliance considerations are embedded in every stage of the process. This comprehensive approach not only mitigates the risk of non-compliance but also supports the continuous improvement of the organization's ML governance framework.

Common Pitfalls and Strategies to Avoid Them

One of the most common pitfalls during compliance audits is insufficient documentation. Organizations often fall short in maintaining comprehensive records of their ML operations, including data handling practices, model development processes, and compliance activities. This lack of documentation can lead to unfavorable audit outcomes, as auditors may not have enough evidence to verify compliance with regulations. To avoid this pitfall, organizations should implement rigorous documentation practices, ensuring that every aspect of their ML operations is recorded and regularly updated. This includes maintaining detailed logs of model training sessions, data usage, algorithmic decisions, and any changes made to ML systems over time.

Inadequate bias audits are another significant challenge that organizations face during compliance audits. Bias in ML models can lead to unfair or discriminatory outcomes, which can have serious legal and reputational consequences. Many organizations fail to conduct regular bias audits or do not employ comprehensive methods to detect and mitigate bias in their ML models. To address this issue, organizations should establish robust bias auditing procedures as part of their overall risk management strategy. This includes using advanced techniques to identify bias in training data, model predictions, and decision-making processes. Regular bias audits should be conducted, with results documented and corrective actions taken where necessary to ensure compliance with ethical standards and regulations.

Another common pitfall is the lack of cross-functional collaboration during the audit preparation process. Compliance audits often require input from various departments, including IT, legal, data science, and risk management. When these departments do not communicate effectively, there can be gaps in the audit process, such as incomplete documentation, inconsistent procedures, or conflicting responses to auditor inquiries. To avoid this, organizations should foster strong collaboration across departments, ensuring that all stakeholders are aligned and working together toward a successful audit. Establishing regular communication channels, joint meetings, and shared responsibilities can help bridge any gaps and create a more cohesive approach to compliance. This collaboration is key to ensuring that all aspects of the organization's ML operations are compliant and that the audit process is smooth and efficient.

Stakeholder Engagement

Stakeholder engagement is a critical aspect of ML governance. Identifying key stakeholders, including data scientists, ML engineers, business leaders, regulators, and end users, is the first step in ensuring that all relevant parties are involved in governance processes. Each stakeholder group has unique insights and concerns that must be addressed to achieve effective governance.

Stakeholders play a vital role in shaping ML governance policies and practices. For example, data scientists and ML engineers are responsible for developing and maintaining models, while business leaders provide strategic direction and ensure that ML initiatives align with organizational goals. Regulators and compliance officers ensure that ML operations meet legal and ethical standards, while end users provide feedback on the real-world impact of ML applications.

Transparency is key to building trust with stakeholders. Organizations must be open about how ML models are developed, how data is used, and how decisions are made. This includes providing clear explanations of model outcomes, addressing concerns about bias, and being transparent about any limitations or risks associated with ML systems.

Effective communication is essential for stakeholder engagement. Organizations should establish clear communication channels for sharing information, soliciting feedback, and addressing concerns. This could include regular reports, meetings, webinars, and online portals. Communication strategies should be tailored to the needs of different stakeholder groups, ensuring that technical details are provided to those who need them, while high-level summaries are available for non-technical audiences.

Stakeholder feedback is invaluable for improving ML governance practices. Organizations should actively solicit feedback from stakeholders at all stages of the ML lifecycle, from data collection to model deployment and monitoring. This feedback should be used to refine policies, improve model performance, and address any concerns or issues raised by stakeholders.

One of the challenges of stakeholder engagement is balancing competing interests. For example, data scientists may prioritize model accuracy, while regulators may be more concerned with transparency and fairness. Organizations must find a balance that meets the needs of all stakeholders while ensuring that ML systems are effective, ethical, and compliant.

Stakeholders should be involved in the risk management process. This includes participating in risk identification and assessment, contributing to the development of risk mitigation strategies, and being informed about any incidents or risks that arise. Involving stakeholders in risk management helps ensure that all potential risks are considered and that mitigation strategies are comprehensive and effective.

Stakeholder engagement is particularly important during incidents. Organizations must communicate openly with stakeholders about the nature of the incident, its impact, and the steps being taken to resolve it. This includes providing regular updates, being transparent about any challenges or delays, and addressing stakeholder concerns promptly and effectively.

Governance bodies, such as the Risk Management Committee and Compliance Audit teams, play a key role in stakeholder engagement. These bodies are responsible for ensuring that stakeholder concerns are addressed in governance policies and practices, that stakeholders are kept informed about governance activities, and that feedback is incorporated into decision-making processes.

Organizations should measure the effectiveness of their stakeholder engagement efforts. This could involve conducting surveys, holding focus groups, and analyzing feedback to assess how well stakeholder needs and concerns are being addressed. Regularly reviewing and improving stakeholder engagement strategies helps ensure that stakeholders remain engaged and that their input is valued and acted upon.

Training and education are essential for ensuring that stakeholders understand the complexities of ML systems and the importance of governance. This could include providing training on data privacy, algorithmic transparency, and ethical considerations. Educating stakeholders helps build their capacity to contribute to governance processes and ensures that they are equipped to make informed decisions.

Stakeholders should be involved in the development of ML governance policies. This includes participating in the drafting of policies, providing feedback on proposed policies, and helping to ensure that policies are practical, effective, and aligned with the organization's goals. Involving stakeholders in policy development helps ensure that policies are well-informed and widely accepted.

Stakeholder engagement is not a one-time activity but an ongoing process. Organizations should establish long-term engagement strategies that involve regular communication, continuous feedback loops, and sustained collaboration. This helps build strong relationships with stakeholders and ensures that they remain committed to supporting ML governance efforts.

Stakeholder engagement should occur across the entire ML lifecycle, from the initial planning and design phases to model deployment, monitoring, and decommissioning. By involving stakeholders at every stage, organizations can ensure that their ML systems are designed and operated in a way that meets the needs of all stakeholders and that governance practices are effective and comprehensive.

While stakeholder engagement presents challenges, such as managing competing interests and ensuring effective communication, it also offers significant opportunities. Engaging stakeholders can lead to better governance practices, more effective ML systems, and stronger relationships with key partners. By embracing these opportunities, organizations can enhance their ML governance efforts and achieve better outcomes for all stakeholders.

Governance and oversight in machine learning systems are essential to managing risks, ensuring compliance, and engaging stakeholders effectively. Each of these elements—risk management, compliance audits, and stakeholder engagement—plays a critical role in building trust and ensuring that ML systems are used responsibly and ethically. As organizations continue to develop and deploy ML technologies, robust governance frameworks will be key to navigating the complex challenges and opportunities that arise.

Compliance Monitoring and Auditing

Compliance monitoring and auditing are critical components of ML Data Governance, ensuring that machine learning models, data pipelines, and processes adhere to relevant regulations, standards, and best practices. The key elements of compliance monitoring and auditing in ML Data Governance include:

Regulatory Compliance Monitoring

Framework Alignment and Regulatory Compliance in ML Systems

Ensuring that machine learning (ML) models and data processes comply with industry-specific regulations is a critical component of responsible AI deployment. In today's regulatory landscape, organizations must adhere to frameworks such as the General Data Protection Regulation (GDPR), Health Insurance Portability and Accountability Act (HIPAA), and California Consumer Privacy Act (CCPA), among others. Regular checks are necessary to guarantee that ML models align with these legal requirements and do not inadvertently violate any rules. Compliance isn't just about avoiding penalties; it's also about fostering trust with users and stakeholders who expect their data to be handled ethically and legally.

This process of alignment requires continuous monitoring, frequent audits, and updates to both the models and data management practices as regulations evolve. For instance, GDPR mandates that organizations ensure data privacy, requiring explicit consent from individuals for their data to be used. Similarly, HIPAA requires that healthcare-related data is protected and not used in ways that could compromise patient confidentiality. Regular checks and audits can help ensure that the use of data in ML models does not inadvertently breach these regulations.

The alignment process should include a thorough understanding of the regulatory environment, where ML models are deployed, and the specific requirements that apply to different types of data and applications. For example, in the financial sector, ML models must comply with anti-money laundering regulations, which require regular checks to ensure that data used in these models is not being misused. Additionally, as data flows through various stages of ML pipelines, organizations must ensure that each step complies with the relevant regulations, requiring robust governance structures that include legal, compliance, and technical experts who can interpret and apply regulatory standards. This alignment is not a one-time effort but an ongoing process that requires organizations to stay updated with regulatory changes and ensure that their ML systems evolve accordingly.

Policy Adherence and Ethical AI Governance

Organizations must monitor adherence to their internal policies related to data privacy, security, and the ethical use of AI. These policies are designed to ensure that ML models are developed and deployed responsibly, reflecting the organization's commitment to ethical practices. Monitoring policy adherence involves regular reviews of the processes and practices used in developing and deploying ML models, ensuring they align with the organization's stated values and ethical guidelines. This includes verifying that data privacy policies are strictly followed, with personal and sensitive data protected against unauthorized access or misuse.

Security policies must be enforced to safeguard ML models and the data they rely on from cyber threats, which could compromise the integrity of the models and lead to data breaches. Ethical AI governance also requires that ML models are transparent and explainable, allowing stakeholders to understand how decisions are made and ensuring that models are not biased or discriminatory. Regular audits should be conducted to check whether these policies are being followed, with any deviations identified and addressed promptly.

The development of ML models should be guided by clear ethical principles that prioritize fairness, accountability, and transparency. For example, organizations should ensure that their models do not perpetuate or exacerbate biases against certain groups or individuals. This requires careful consideration of the data used to train models and the algorithms themselves, which must be regularly evaluated for fairness and accuracy. Organizations should also have policies in place that guide the ethical use of AI, ensuring that models are used in ways that benefit society and do not cause harm. Adhering to these policies helps organizations avoid reputational damage and legal risks, while also fostering trust with customers and other stakeholders who expect AI to be used responsibly.

Automated Alerts and Real-Time Compliance Monitoring

The complexity of modern ML systems and the vast amounts of data they process make it challenging for organizations to manually monitor compliance with regulations and internal policies. To address this, many organizations are implementing automated systems that can detect and flag non-compliance issues in real time. These automated alerts are crucial for identifying potential risks before they escalate into major problems, such as data breaches, unauthorized access, or deviations from predefined ethical standards. For instance, automated systems can monitor data access patterns and detect anomalies that suggest unauthorized access, triggering alerts for further investigation. Similarly, these systems can monitor the behavior of ML models in production, identifying any deviations from expected performance or ethical guidelines.

Automated alerts are particularly valuable in detecting data privacy violations, where even a minor breach can have significant legal and reputational consequences. By implementing real-time monitoring systems, organizations can respond more quickly to potential issues, reducing the risk of non-compliance and ensuring that corrective actions are taken promptly. These systems can also be programmed to check for compliance with specific regulations, such as GDPR or HIPAA, by monitoring how data is used and ensuring that consent requirements are met.

Automated alerts can help organizations maintain ethical AI practices by detecting when ML models produce biased or discriminatory outcomes, allowing for immediate intervention. The implementation of these systems requires careful planning and configuration to ensure they are effective and do not generate false positives, which could lead to unnecessary disruptions. Moreover, organizations must continuously refine and update these systems to keep pace with evolving regulations and ethical standards. Automated alerts are not a replacement for human oversight but rather a tool that enhances the ability of organizations to manage compliance and ethical risks in real time. They provide an additional layer of security and governance, helping organizations maintain control over their ML systems and ensuring that they operate within the bounds of both legal and ethical guidelines.

Model Risk Management

Model risk management is a critical aspect of machine learning (ML) governance, focusing on the identification, assessment, and mitigation of risks associated with ML models. As organizations increasingly rely on ML models for decision-making across various domains, the importance of robust model risk management practices cannot be overstated. These practices ensure that ML models operate reliably, produce accurate outcomes, and comply with relevant regulations and standards. Effective model risk management involves several key components, including model validation and verification, model drift monitoring, and security audits. Each of these elements plays a vital role in safeguarding the integrity, reliability, and security of ML models.

Model Validation and Verification

Model validation and verification are foundational components of model risk management, aimed at ensuring that ML models perform as intended and meet the required standards for accuracy, fairness, and compliance. Validation involves assessing a model's performance against predefined benchmarks and evaluating its ability to generalize to new data. Verification, on the other hand, involves checking that the model's outputs align with its intended purpose and that it operates within the established parameters. Together, these processes help ensure that ML models produce reliable outcomes and do not pose undue risks to the organization or its stakeholders.

The importance of model validation and verification cannot be understated, as ML models can sometimes behave unpredictably or produce biased results if not thoroughly tested. Regular audits of ML models are essential to validate their performance and verify their compliance with regulatory requirements. These audits involve rigorous testing of the model against a variety of datasets, including those that were not part of the training process, to ensure that it can generalize effectively. Additionally, organizations should establish clear validation criteria that align with their business objectives and regulatory obligations, ensuring that models meet these criteria before they are deployed in production environments.

In practice, model validation and verification require the use of specialized tools and techniques, such as cross-validation, stress testing, and sensitivity analysis. These methods help identify potential weaknesses in the model and provide insights into how it might perform under different conditions. Moreover, validation and verification should be an ongoing process, with models regularly re-evaluated to ensure they continue to meet performance benchmarks as new data becomes available. This continuous validation process is particularly important in dynamic environments where data distributions can change over time, potentially leading to model drift and degradation in performance.

Model validation and verification are essential for managing the risks associated with ML models. By conducting regular audits and thorough testing, organizations can ensure that their models produce accurate, fair, and compliant outcomes. This not only helps mitigate the risks of deploying ML models in critical decision-making processes but also supports the broader goal of building trust in AI systems. As the use of ML continues to expand, the importance of robust model validation and verification practices will only continue to grow, making them a key focus area for organizations committed to responsible AI development.

Model Drift Monitoring

Model drift monitoring is a crucial aspect of model risk management, focusing on the ongoing evaluation of ML models to detect changes in their performance over time. Model drift occurs when the statistical properties of the input data or the relationship between the input and output variables change, leading to a decline in the model's accuracy and reliability. If left unaddressed, model drift can result in non-compliance with initial validation standards, potentially leading to biased or incorrect predictions. Therefore, continuous monitoring for model drift is essential to ensure that ML models remain effective and compliant throughout their lifecycle.

The causes of model drift are varied and can include changes in the underlying data distribution, shifts in user behavior, or the emergence of new patterns in the data that were not present during the initial training phase. To effectively monitor for model drift, organizations must implement robust monitoring systems that track model performance metrics over time and compare them to the benchmarks established during validation. These systems should be capable of detecting subtle changes in model behavior and triggering alerts when performance metrics fall below acceptable thresholds, indicating potential drift.

Organizations should establish procedures for regularly retraining ML models to address detected drift. This may involve updating the model with new data, adjusting its parameters, or even redesigning the model architecture to better align with the current data environment. Retraining should be conducted systematically, with careful consideration given to the potential risks and benefits of updating the model. In some cases, organizations may also need to implement version control and rollback mechanisms to ensure that if retraining does not produce the desired improvements, they can revert to a previous version of the model that performed better.

Model drift monitoring is not just about detecting changes in performance but also about understanding the underlying causes of drift and taking proactive steps to mitigate its impact. This requires a deep understanding of the data environment, the model's design, and the context in which it operates. By continuously monitoring for drift and implementing regular retraining procedures, organizations can ensure that their ML models remain accurate, reliable, and compliant over time. As the complexity and scale of ML deployments continue to grow, the importance of effective model drift monitoring will become increasingly critical to managing model risk and maintaining trust in AI systems.

Security Audits

Security audits are a vital component of model risk management, focusing on ensuring that ML models are protected against unauthorized access, tampering, and other security threats. As ML models become more integral to organizational decision-making, they also become attractive targets for malicious actors seeking to exploit vulnerabilities for personal or financial gain. Security audits help identify and mitigate these risks by evaluating the effectiveness of the security measures in place and ensuring that they meet the required standards for protecting sensitive data and model integrity.

The first step in conducting a security audit is to assess the access controls in place for the ML models and the data they rely on. This involves evaluating who has access to the models and data, what level of access they have, and whether there are any unnecessary or excessive permissions that could be exploited. Organizations should implement strict access controls, including role-based access, multifactor authentication, and encryption, to ensure that only authorized individuals can access and modify the models and data. Regular audits should be conducted to ensure that these controls remain effective and that any changes in personnel or roles are promptly reflected in the access control policies.

Security audits should also assess the vulnerability of ML models to external threats, such as adversarial attacks, data poisoning, or model inversion. These types of attacks can compromise the integrity of the model, leading to incorrect predictions or the leakage of sensitive information. To protect against these threats, organizations should implement robust security measures, such as adversarial training, anomaly detection, and regular penetration testing. These measures help identify potential vulnerabilities and provide insights into how the model might be exploited by malicious actors.

Security audits should also evaluate the overall security posture of the ML infrastructure, including the data storage, processing, and deployment environments. This involves assessing the encryption methods used to protect data at rest and in transit, the security of the cloud or on-premises infrastructure, and the effectiveness of the incident response plans in place to address potential security breaches. By conducting comprehensive security audits, organizations can ensure that their ML models are protected against a wide range of security threats, reducing the risk of unauthorized access, data breaches, and other security incidents.

Security audits are an essential component of model risk management, helping organizations identify and mitigate potential security risks that could compromise the integrity and reliability of their ML models. By regularly auditing access controls, vulnerability to external threats, and the overall security posture of their ML infrastructure, organizations can ensure that their models remain secure and compliant with relevant regulations. As the use of ML continues to expand, the importance of robust security audits will only continue to grow, making them a key focus area for organizations committed to responsible AI development.

Ethical and Legal Compliance

Ethical and legal compliance in machine learning (ML) systems is essential for ensuring that these technologies are used responsibly and align with both societal values and legal frameworks. As ML systems become more pervasive across various sectors, the importance of establishing robust mechanisms for ethical and legal compliance cannot be overstated. Ethical AI audits and legal liability assessments are two critical components of this compliance framework, each addressing different aspects of the potential risks associated with ML models. These practices are vital for building trust in AI systems, safeguarding against misuse, and ensuring that organizations uphold their ethical and legal responsibilities.

Ethical AI Audits

Ethical AI audits are a crucial element of ensuring that ML models adhere to ethical principles, such as fairness, accountability, and transparency. These audits involve evaluating ML models against established ethical guidelines to ensure that they do not perpetuate biases, discriminate against certain groups, or produce outcomes that could be considered unethical. The primary goal of ethical AI audits is to align ML models with societal values and organizational ethics, ensuring that their use contributes positively to society and does not result in unintended harm.

The process of conducting an ethical AI audit begins with defining the ethical standards and principles that the ML model must adhere to. These standards may be based on industry best practices, regulatory requirements, or the organization's own ethical guidelines. Once these standards are established, the audit process involves a thorough evaluation of the ML model's design, data sources, and outputs to identify potential ethical risks. This includes assessing whether the model's training data is representative and free from biases, whether the model's decision-making processes are transparent and explainable, and whether the model's outputs align with ethical principles.

Ethical AI audits also involve assessing the broader impact of the ML system on society and its alignment with organizational values. This includes considering the potential societal consequences of deploying the model, such as its impact on vulnerable populations or its potential to exacerbate existing inequalities. Organizations may also engage with external stakeholders, such as ethicists, community groups, or regulatory bodies, to ensure that their ethical AI practices are robust and aligned with broader societal expectations.

Ethical AI audits are essential for ensuring that ML models are used in a way that is consistent with ethical principles and societal values. By conducting regular ethical audits, organizations can identify and address potential ethical risks before they result in harm, thereby maintaining trust in their AI systems and upholding their ethical responsibilities. As the use of AI continues to expand, the importance of ethical AI audits will only grow, making them a key component of responsible AI governance.

Legal Liability Assessments

Legal liability assessments are a critical aspect of ensuring that ML models comply with relevant legal frameworks and do not expose organizations to potential legal risks. These assessments involve evaluating the potential legal liabilities associated with the use of ML models, including issues related to discrimination, privacy violations, and unintended harmful consequences. The primary goal of legal liability assessments is to identify and mitigate potential legal risks, ensuring that ML models are used in a manner that is compliant with the law and does not result in legal disputes or penalties.

The process of conducting a legal liability assessment begins with identifying the legal frameworks that apply to the ML model in question. This may include data protection laws, such as the General Data Protection Regulation (GDPR) or the California Consumer Privacy Act (CCPA), as well as anti-discrimination laws, such as the Fair Housing Act or the Equal Opportunity Employment Commission (EEOC) guidelines. Once the relevant legal frameworks have been identified, the assessment process involves evaluating the ML model's compliance with these laws, including assessing whether the model's data processing activities adhere to data protection requirements, whether the model's decision-making processes are free from discriminatory biases, and whether the model's outputs could result in unintended harm.

Legal liability assessments also involve considering the broader legal implications of deploying the model. This includes assessing the potential for legal disputes arising from the model's outputs, such as cases where individuals or groups may claim that they were unfairly treated or harmed as a result of the model's decisions. Organizations must also consider the potential reputational damage that could result from legal disputes or regulatory penalties and take steps to mitigate these risks by ensuring that their ML models are transparent, explainable, and compliant with the law.

Legal liability assessments are essential for protecting organizations from the potential legal risks associated with the use of ML models. By conducting regular legal assessments, organizations can identify and address potential legal liabilities before they result in legal disputes or penalties, thereby safeguarding their legal and financial interests. As the use of ML continues to expand, the importance of legal liability assessments will only grow, making them a key component of responsible AI governance.

Integration of Ethical and Legal Compliance

The integration of ethical and legal compliance into the development and deployment of ML models is crucial for ensuring that these technologies are used responsibly and in accordance with societal values and legal frameworks. Ethical AI audits and legal liability assessments are complementary processes that together provide a comprehensive approach to managing the ethical and legal risks associated with ML models. By integrating these processes into their AI governance frameworks, organizations can ensure that their ML models are both ethically sound and legally compliant, thereby building trust in their AI systems and avoiding potential ethical and legal pitfalls.

The integration process begins with establishing clear ethical and legal standards that ML models must adhere to and ensuring that these standards are reflected in the organization's AI governance policies and procedures. This includes defining the ethical principles that guide the development and deployment of ML models, as well as identifying the legal frameworks that apply to the organization's use of AI. Once these standards are established, organizations can integrate ethical AI audits and legal liability assessments into their model development lifecycle, ensuring that these processes are conducted regularly and that any identified risks are promptly addressed.

Organizations must also ensure that their AI governance frameworks include mechanisms for ongoing monitoring and evaluation. This includes conducting regular ethical and legal audits to ensure that ML models continue to comply with ethical principles and legal frameworks as they evolve over time. Organizations should also establish processes for engaging with external stakeholders, such as ethicists, legal experts, and regulatory bodies, to ensure that their ethical and legal practices remain aligned with broader societal expectations and legal requirements.

The integration of ethical and legal compliance into AI governance is essential for ensuring that ML models are used in a way that is consistent with both ethical principles and legal frameworks. By conducting regular ethical AI audits and legal liability assessments and by integrating these processes into their AI governance frameworks, organizations can identify and address potential ethical and legal risks before they result in harm, thereby building trust in their AI systems and avoiding potential legal disputes or penalties. As the use of AI continues to expand, the importance of integrating ethical and legal compliance into AI governance will only grow, making it a key focus area for organizations committed to responsible AI development.

Case Study: Ethical AI Audit in a Healthcare Technology Firm

A healthcare technology firm, specializing in AI-driven diagnostic tools, decided to conduct an ethical AI audit to ensure that its machine learning models were not only legally compliant but also aligned with ethical standards. The firm had recently come under scrutiny due to concerns about the fairness and transparency of its AI models, particularly regarding how they impacted patient care decisions. To address these concerns and strengthen trust in its technology, the firm initiated a comprehensive ethical AI audit.

Audit Process

The ethical AI audit began with the firm defining clear ethical principles that the AI models needed to adhere to, focusing on fairness, transparency, and accountability. These principles were aligned with both industry best practices and the firm's internal ethical guidelines. The audit process involved a detailed examination of the AI models, particularly their training data, algorithms, and decision-making processes. The auditors assessed whether the training data was representative of the diverse patient populations served by the firm and whether the models could explain their diagnostic decisions in a clear and understandable manner.

The auditors also conducted a series of interviews with the firm's data scientists, engineers, and legal advisors to understand the decision-making frameworks that were in place. This phase of the audit was crucial for identifying potential biases in the data and ensuring that the models' decision-making processes were transparent and accountable. The audit also included a technical review of the models' outputs to determine if there were any systematic biases affecting certain demographic groups.

Findings

The audit revealed several key areas where the firm's AI models were falling short of ethical standards. Notably, the training data was found to be skewed, with an underrepresentation of certain minority groups, leading to biased diagnostic outcomes. The models were also found to lack sufficient transparency, as their decision-making processes were not easily interpretable by healthcare providers or patients. This opacity raised concerns about the accountability of the AI system in clinical settings, where clear and explainable decisions are critical.

Lessons Implemented

Based on the audit findings, the firm took immediate action to address the identified issues. The data used to train the AI models was revised to ensure a more balanced representation of all patient demographics, thereby reducing the risk of biased outcomes. The firm also integrated advanced explainability tools into their models, such as SHAP (SHapley Additive exPlanations), to enhance the transparency of the diagnostic decisions. This allowed healthcare providers to better understand the AI-driven recommendations and make more informed clinical decisions.

The firm also established a continuous monitoring process to regularly assess the ethical performance of their AI models. This process included periodic ethical audits and the involvement of an external ethics board to review and guide the firm's AI practices. By implementing these changes, the firm not only improved the fairness and transparency of its AI systems but also reinforced its commitment to ethical AI practices, ultimately strengthening trust with both healthcare providers and patients.

Challenges and Best Practices in Ethical and Legal Compliance

While the importance of ethical and legal compliance in ML systems is widely recognized, organizations often face significant challenges in implementing these practices effectively. One of the key challenges is the complexity of ML models, which can make it difficult to assess their compliance with ethical principles and legal frameworks. For example, ML models may be opaque or "black box" in nature, making it challenging to determine whether they are producing biased or discriminatory outcomes. Similarly, the rapid pace of technological innovation in AI can make it difficult for organizations to keep up with evolving ethical and legal standards.

To address these challenges, organizations must adopt best practices for ethical and legal compliance in ML systems. This includes implementing transparency and explainability measures, such as model interpretability techniques, to ensure that ML models are understandable and their decision-making processes can be scrutinized. Organizations should also invest in continuous learning and capacity-building initiatives to ensure that their teams are up to date with the latest ethical and legal standards in AI. Additionally, organizations should establish cross-functional teams that include ethicists, legal experts, and AI practitioners to ensure that ethical and legal compliance is considered at every stage of the ML model development lifecycle.

Another best practice is to engage with external stakeholders, such as regulatory bodies, civil society organizations, and affected communities, to ensure that the organization's ethical and legal practices are aligned with broader societal expectations and legal requirements. This can involve participating in industry-wide initiatives, such as ethical AI working groups or legal compliance forums, to share best practices and learn from the experiences of other organizations. By engaging with external stakeholders, organizations can also gain valuable insights into emerging ethical and legal challenges and ensure that their AI governance frameworks remain robust and responsive to changing societal needs.

The challenges associated with ethical and legal compliance in ML systems can be effectively addressed through the adoption of best practices and a commitment to continuous improvement. By implementing transparency and explainability measures, investing in continuous learning and capacity-building, and engaging with external stakeholders, organizations can ensure that their ML models are both ethically sound and legally compliant. As the use of AI continues to expand, the importance of ethical and legal compliance will only grow, making it a key focus area for organizations committed to responsible AI development.

Governance and Accountability

Governance Structure

Establishing a robust governance structure is essential for effective management and oversight of machine learning (ML) data systems. Central to this structure are governance bodies, such as Data Governance Boards or AI Ethics Committees, which play a pivotal role in overseeing the entire ML data governance framework. These bodies are tasked with setting high-level policies, reviewing compliance, and addressing strategic issues that arise within the scope of ML operations. The primary function of these governance bodies is to ensure that ML systems operate within the bounds of regulatory requirements, organizational policies, and ethical guidelines. They are responsible for approving data management strategies, reviewing model performance and compliance reports, and making critical decisions regarding the deployment and usage of ML models. By having a dedicated committee or board, organizations can ensure that governance is not only top-down but also embedded within the strategic framework of the business, aligning with long-term objectives and regulatory expectations.

Governance bodies typically consist of a diverse group of stakeholders, including senior executives, compliance officers, data scientists, and external advisors. This diversity ensures that multiple perspectives are considered when making decisions about ML governance. For instance, senior executives bring a strategic viewpoint, while data scientists provide technical insights, and compliance officers focus on regulatory adherence. The inclusion of external advisors or ethicists can also bring an impartial perspective on emerging ethical issues or new regulations. These governance bodies are often supported by subcommittees or working groups that focus on specific areas such as data privacy, model ethics, or risk management. By leveraging the expertise of these various stakeholders, governance bodies can effectively address complex issues related to ML data governance and maintain a balance between innovation and regulatory compliance.

Regular meetings and reviews are integral to the function of governance bodies. These meetings serve as platforms for discussing ongoing issues, reviewing compliance reports, and making strategic decisions. During these sessions, governance bodies assess the effectiveness of current policies, evaluate any incidents or non-compliance issues, and discuss potential improvements to the governance framework. The frequency and format of these meetings can vary depending on the organization's size and the complexity of its ML systems. However, it is crucial that these meetings are well documented, with clear agendas and action items to ensure accountability and follow-through on decisions made. By maintaining a structured approach to governance, organizations can foster a culture of transparency and continuous improvement in their ML data practices.

Ultimately, the role of governance bodies is to provide oversight and direction, ensuring that ML systems are managed in a way that aligns with organizational goals and regulatory requirements. Their decisions and policies shape the overall governance framework, guiding the organization's approach to data management, model ethics, and risk mitigation. By establishing effective governance bodies, organizations can enhance their ability to manage ML data responsibly, address compliance issues proactively, and maintain public trust in their technological practices.

Roles and Responsibilities

Defining clear roles and responsibilities within the governance structure is crucial for ensuring effective oversight and management of ML systems. Each stakeholder involved in ML data governance has specific duties that contribute to the overall integrity and compliance of the system. Data scientists, for instance, are responsible for developing and implementing ML models, ensuring that they meet performance benchmarks and adhere to data privacy regulations. Their role involves selecting appropriate algorithms, tuning model parameters, and validating model outputs to ensure accuracy and fairness. Data engineers, on the other hand, focus on the technical aspects of data management, including data collection, preprocessing, and storage. They ensure that data pipelines are robust and secure and that data used in models is accurate, consistent, and free from bias.

Compliance officers play a critical role in ensuring that ML systems adhere to regulatory requirements and organizational policies. Their responsibilities include monitoring for compliance with laws such as GDPR, HIPAA, and CCPA, as well as internal data protection policies. They conduct regular audits, assess risks, and work with other stakeholders to address any compliance issues that arise. Compliance officers also stay abreast of changes in regulations and advise on necessary adjustments to governance practices. Business leaders, including executives and managers, are responsible for aligning ML data governance with the organization's strategic objectives. They ensure that ML initiatives support business goals, allocate resources effectively, and provide oversight on high-level decisions related to data management and model deployment.

Each role within the governance structure must be clearly defined to avoid overlaps and gaps in responsibilities. This clarity ensures that all aspects of ML data governance are addressed comprehensively and that stakeholders understand their specific contributions to the governance framework. For example, while data scientists and engineers focus on technical and operational aspects, compliance officers and business leaders provide oversight and strategic direction. Effective communication and coordination among these roles are essential for maintaining a cohesive governance framework. Regular meetings and updates among stakeholders help ensure that everyone is informed about current issues, ongoing projects, and any changes in policies or regulations.

Organizations must also establish processes for accountability and performance evaluation. This includes setting performance metrics for each role, conducting regular performance reviews, and providing feedback to ensure that stakeholders are meeting their responsibilities effectively. By establishing clear roles, responsibilities, and accountability mechanisms, organizations can enhance their governance framework and ensure that ML systems are managed in a responsible and compliant manner. This structured approach helps in maintaining the integrity of ML data practices and supports the overall success of the organization's data governance initiatives.

Policy Framework

Establishing comprehensive data management policies is fundamental to ensuring effective oversight and governance of machine learning (ML) systems. These policies cover the entire lifecycle of data, from collection and storage to processing and usage. Data management policies are designed to ensure that data is handled in a way that maintains its quality, security, and privacy while adhering to industry standards and regulatory requirements. The development of these policies typically involves defining protocols for data collection, which includes specifying the sources of data, methods for data acquisition, and consent requirements. Ensuring that data is collected in a manner that respects individuals' privacy and meets legal standards is crucial for building trust and compliance.

Storage policies are equally important and should address how data is securely stored and protected. This includes guidelines on data encryption, access controls, and backup procedures to prevent unauthorized access and data breaches. Policies should outline how data should be classified based on its sensitivity and the appropriate levels of protection required for different data types. For instance, personal or sensitive data might require more stringent security measures compared to less sensitive data. These policies help organizations manage their data assets effectively, reduce the risk of data loss or unauthorized access, and ensure compliance with data protection regulations such as GDPR or CCPA.

Processing and usage policies are also vital components of data management. These policies define how data should be processed to ensure its accuracy, consistency, and relevance. They also cover the use of data in ML models, including guidelines for data preprocessing, transformation, and integration. Ensuring that data is processed according to established standards helps maintain the quality and integrity of the data used in ML systems. Additionally, policies should specify how data can be used, shared, or transferred, ensuring that data usage aligns with legal requirements and organizational objectives.

Regular reviews and updates of data management policies are necessary to address emerging challenges and changes in regulatory landscapes. As new technologies and data protection regulations evolve, policies must be adjusted to reflect these changes and continue to meet compliance standards. Organizations should establish a process for periodic policy reviews and updates, involving key stakeholders such as data protection officers, IT professionals, and legal experts. This proactive approach helps ensure that data management practices remain effective and compliant with current regulations.

Data management policies are crucial for governing the lifecycle of data in ML systems. They ensure that data is collected, stored, processed, and used in a manner that upholds quality, security, and privacy. By aligning policies with industry standards and regulatory requirements, organizations can effectively manage their data assets and mitigate risks associated with data handling. Regular reviews and updates of these policies help maintain compliance and adapt to evolving data governance challenges.

Ethical Guidelines

Developing ethical guidelines for AI and ML systems is a critical aspect of ensuring responsible and fair use of technology. These guidelines are designed to address principles such as fairness, transparency, and accountability, which are essential for maintaining public trust and achieving ethical outcomes in ML applications. Ethical guidelines provide a framework for evaluating the impact of ML models on individuals and society, ensuring that the technology is used in ways that align with ethical standards and societal values.

Fairness is a core principle in ethical AI guidelines. It involves ensuring that ML models do not discriminate against individuals or groups based on characteristics such as race, gender, age, or socio-economic status. Fairness guidelines typically require organizations to implement measures for identifying and mitigating biases in data and algorithms. This includes conducting regular audits to detect biases, using diverse and representative datasets, and applying fairness-enhancing techniques in model development. By addressing fairness, organizations can help prevent unintended harm and ensure that ML systems benefit all users equitably.

Transparency is another key principle of ethical guidelines. It involves making the operations and decisions of ML models understandable and accessible to stakeholders. Transparency guidelines often require organizations to provide explanations of how models work, how decisions are made, and how data is used. This can be achieved through techniques such as model interpretability, which provides insights into the decision-making process of ML models, and clear communication of model limitations and potential risks. Transparency helps build trust with users and enables stakeholders to hold organizations accountable for their AI practices.

Accountability is a crucial component of ethical AI guidelines, ensuring that organizations are responsible for the outcomes of their ML systems. Accountability guidelines typically include provisions for monitoring model performance, addressing any issues that arise, and taking corrective actions when necessary. This involves establishing mechanisms for reporting and addressing ethical concerns, as well as creating processes for reviewing and updating ethical guidelines in response to new developments or feedback. Accountability ensures that organizations take responsibility for the impact of their ML models and are committed to maintaining ethical standards.

Ethical guidelines should be integrated into the overall governance framework for ML systems, involving various stakeholders such as ethicists, data scientists, and legal experts. These guidelines should be regularly reviewed and updated to reflect changes in societal values, technological advancements, and regulatory requirements. By establishing and adhering to ethical guidelines, organizations can ensure that their ML systems are developed and used responsibly, addressing potential ethical issues and fostering trust and accountability in AI technologies.

Ethical guidelines are essential for governing the use of ML systems in a way that aligns with fairness, transparency, and accountability principles. These guidelines help organizations navigate the ethical challenges associated with AI and ensure that technology is used responsibly. By integrating ethical guidelines into their governance frameworks and regularly reviewing and updating them, organizations can promote ethical practices and build trust with stakeholders.

Accountability Mechanisms

Transparency in machine learning (ML) processes is crucial for ensuring accountability and fostering trust among stakeholders. Transparency involves documenting the methodologies, data sources, and model outcomes used in ML systems. This documentation provides a clear understanding of how models are developed, the data they use, and how decisions are made. Detailed records of these aspects allow stakeholders to assess the validity and reliability of ML systems and ensure that the systems operate within ethical and regulatory boundaries.

One of the key components of transparency is documenting the data sources used in ML models. This includes specifying the origin of the data, how it was collected, and any preprocessing steps applied. Transparent documentation of data sources helps in verifying data quality, understanding potential biases, and ensuring that the data complies with regulatory requirements such as GDPR and CCPA. Additionally, it supports reproducibility by allowing others to replicate the data collection and preprocessing steps, which is essential for validating model results and maintaining scientific rigor.

The methodologies employed in developing ML models also need to be thoroughly documented. This involves detailing the algorithms used, parameter settings, training procedures, and any feature engineering techniques applied. Documenting these methodologies provides insights into how models are constructed and the rationale behind choosing specific techniques. This information is valuable for assessing the appropriateness of the methods used and for identifying any potential areas where improvements or adjustments may be needed.

Regular reporting is another crucial aspect of transparency. Providing stakeholders with periodic updates on system performance, compliance, and any issues encountered helps maintain trust and ensures accountability. These reports should include key performance indicators (KPIs), compliance status, and any deviations from expected outcomes. By regularly updating stakeholders, organizations demonstrate their commitment to transparency and provide a basis for ongoing dialogue about system performance and governance.

In addition to performance reports, it is important to communicate how ML systems are aligned with ethical standards and regulatory requirements. Reports should address how models adhere to fairness, privacy, and security principles and outline any measures taken to address potential ethical or compliance issues. This comprehensive reporting approach ensures that stakeholders are fully informed about the ethical and legal aspects of ML systems, contributing to a culture of accountability and responsible AI use.

Transparency and reporting mechanisms should be designed to facilitate easy access to information and foster open communication with stakeholders. This includes establishing channels for stakeholders to request additional information or raise concerns about ML systems. By creating an environment where information is readily available and communication is encouraged, organizations can build trust and ensure that accountability is maintained throughout the lifecycle of ML systems.

Technology and Tools

Governance Tools

Data management platforms are among the fundamental tools used in ML governance. These platforms help organizations manage the entire data lifecycle, from collection and storage to processing and analysis. They facilitate the organization, classification, and accessibility of data, ensuring that it is maintained in a secure and compliant manner. For instance, data management platforms can automate the tagging of data according to privacy regulations, making it easier to enforce data protection policies and facilitate data audits. By offering a centralized system for data management, these platforms help ensure that data is handled consistently and in accordance with governance standards.

Collibra and Informatica are two prominent examples of such platforms. Collibra is known for its data governance capabilities, offering features like data cataloging, data stewardship, and automated workflows that help organizations maintain control over their data assets. It provides a centralized platform that facilitates the tagging and classification of data according to regulatory requirements, making it easier to enforce data privacy and security policies. Informatica, on the other hand, is renowned for its robust data integration and data quality tools. It offers comprehensive data management solutions that support data governance by ensuring data consistency, accuracy, and compliance throughout its lifecycle. Both platforms enable organizations to maintain high standards of data governance by providing the necessary tools to manage, monitor, and secure their data effectively.

Compliance monitoring systems are another crucial category of governance tools. These systems continuously track and evaluate ML models and data processes to ensure adherence to regulatory requirements and internal policies. They can provide real-time alerts for potential compliance breaches, such as unauthorized access or deviations from established data handling procedures. Compliance monitoring systems often include features like automated reporting, which helps organizations stay informed about their compliance status and address issues proactively. These tools are essential for maintaining an ongoing oversight of compliance, especially in complex environments where manual monitoring would be impractical.

Compliance monitoring systems are designed to ensure that ML systems adhere to regulatory standards and internal policies. MetricStream and OneTrust are two widely used platforms in this space. MetricStream provides an integrated risk management and compliance solution that continuously monitors processes and data for potential compliance breaches. It offers real-time alerts, automated compliance reporting, and a centralized view of compliance risks, making it easier for organizations to manage their regulatory obligations proactively. OneTrust is another key player, offering tools focused on privacy management, third-party risk management, and data governance. It provides organizations with a comprehensive solution for managing compliance with global data protection regulations, such as GDPR and CCPA, by automating compliance workflows and enabling continuous monitoring. Both platforms are essential for organizations that need to ensure ongoing compliance in dynamic and complex environments.

Auditing software is also vital for governance, as it provides detailed insights into the operations and performance of ML systems. Auditing tools can track changes to data, model configurations, and access logs, enabling organizations to conduct thorough audits and reviews. These tools help identify discrepancies, assess the effectiveness of governance practices, and ensure that all activities are documented accurately. By providing comprehensive audit trails, auditing software supports accountability and transparency, making it easier for organizations to demonstrate compliance and address any issues that arise.

ACL and Galvanize are two leading auditing software solutions that help organizations conduct thorough audits of their ML systems. ACL offers advanced analytics and audit management capabilities, enabling auditors to track changes, analyze data, and generate detailed reports on system performance and compliance. It supports continuous auditing by automating the collection and analysis of audit data, making it easier to identify discrepancies and ensure that governance practices are effective. Galvanize, now part of Diligent, provides a similar suite of tools with a strong focus on risk management, compliance, and internal controls. Its platform integrates auditing functions with broader governance and risk management processes, helping organizations maintain a comprehensive view of their compliance landscape. Both tools are invaluable for ensuring that all aspects of ML governance are properly documented and that any issues are identified and addressed promptly.

The integration of these governance tools into existing systems is essential for maximizing their effectiveness. For example, integrating data management platforms with compliance monitoring systems can streamline data governance processes and ensure that data-related compliance issues are addressed promptly. Similarly, linking auditing software with data management platforms can enhance visibility into data handling practices and facilitate more comprehensive audits. By leveraging technology to integrate and automate governance processes, organizations can improve efficiency, reduce manual workloads, and maintain a higher level of oversight.

Governance tools are critical for managing and overseeing ML systems effectively. Data management platforms, compliance monitoring systems, and auditing software each play a unique role in supporting governance efforts. By integrating these tools and leveraging their capabilities, organizations can enhance their governance practices, ensure compliance with regulatory requirements, and maintain oversight of ML systems. The strategic use of technology in governance not only streamlines processes but also supports accountability, transparency, and effective management of ML systems.

Table 8-2 organizes some key differences that define various governance tools.

Table 8-2. *Comparison of governance tools*

Tool	Category	Key Features	Use Cases
Collibra	Data management platform	Data cataloging, data stewardship, automated workflows	Centralized data management, regulatory compliance, data privacy and security enforcement
Informatica	Data management platform	Data integration, data quality, comprehensive data management	Ensuring data consistency and accuracy, supporting data governance across large datasets
MetricStream	Compliance monitoring	Integrated risk management, real-time alerts, automated reporting	Continuous monitoring of regulatory compliance, proactive risk management
OneTrust	Compliance monitoring	Privacy management, third-party risk management, data governance	Managing compliance with global data protection regulations, automating compliance workflows
ACL	Auditing tool	Advanced analytics, audit management, continuous auditing	Detailed audits of ML systems, ensuring transparency and accountability in governance practices
Galvanize	Auditing Tool	Risk management, compliance, internal controls	Integrating auditing with governance and risk management processes, comprehensive compliance oversight

Automation and Integration

Automation and integration are transformative elements in enhancing governance practices within machine learning (ML) systems. Leveraging automation allows organizations to streamline various aspects of governance, including compliance checks, data quality monitoring, and model performance management. Integration, on the other hand, ensures that these automated processes work seamlessly with existing systems, creating a cohesive and efficient governance framework.

One of the primary benefits of automation in ML governance is the ability to perform compliance checks efficiently. Automated compliance tools can continuously monitor ML models and data processes to ensure adherence to regulatory requirements and internal policies. For instance, automated systems can check for data privacy compliance by verifying that data handling practices align with regulations such as GDPR and CCPA. By automating these checks, organizations can quickly identify and address potential compliance issues without relying on manual inspections, which can be time-consuming and prone to errors.

Automation also plays a crucial role in monitoring data quality. High-quality data is essential for the accuracy and reliability of ML models, and automated tools can help maintain data integrity by detecting anomalies, inconsistencies, or errors. For example, automated data quality tools can perform routine checks on data accuracy, completeness, and consistency, ensuring that data meets governance standards before it is used for model training or analysis. These tools can also provide real-time alerts when data quality issues are detected, allowing organizations to take corrective actions promptly.

Automation enhances model performance management. Automated performance monitoring tools can track key performance indicators (KPIs) for ML models and alert stakeholders to any deviations from expected performance. For instance, if a model's accuracy drops below a predefined threshold, the automated system can trigger an alert and initiate a review process. This proactive approach helps organizations address performance issues before they impact business operations or decision-making.

Integration of automation tools with existing systems is essential for creating an efficient and cohesive governance framework. For example, integrating automated compliance tools with data management platforms ensures that compliance checks are conducted as part of the data handling process. Similarly, linking automated performance monitoring tools with model management systems allows for seamless tracking of model performance and quick response to any issues. Integration also facilitates the centralization of governance activities, enabling organizations to manage and oversee ML systems more effectively.

Automation and integration are key to enhancing governance practices in ML systems. By automating compliance checks, data quality monitoring, and model performance management, organizations can improve efficiency, reduce manual workloads, and maintain a higher level of oversight. Integration ensures that these automated processes work together seamlessly, creating a cohesive governance framework that supports accountability, transparency, and effective management of ML systems. As technology continues to evolve, the adoption of advanced automation and integration strategies will be critical for maintaining robust governance practices in ML environments.

Case Studies

Automation and Integration in Healthcare

In the healthcare industry, the accuracy and reliability of data are paramount, particularly when it comes to machine learning models used for predictive analytics and patient care. A large healthcare provider in the United States faced significant challenges in maintaining data quality and ensuring compliance with stringent healthcare regulations, such as the Health Insurance Portability and Accountability Act (HIPAA). The provider's existing manual processes for data management and compliance checks were not only time-consuming but also prone to errors, which could have severe consequences in a highly regulated environment.

To address these challenges, the healthcare provider implemented an automated data management and compliance monitoring system. The organization chose to integrate Informatica's data management platform with MetricStream's compliance monitoring system. Informatica was used to manage the entire data lifecycle, ensuring data accuracy, completeness, and consistency across various sources, including electronic health records (EHRs) and laboratory information systems. The platform's automation capabilities allowed the healthcare provider to perform routine data quality checks automatically, identifying and correcting data anomalies before they could impact patient care or model accuracy.

MetricStream's compliance monitoring system was integrated with Informatica to ensure continuous adherence to HIPAA regulations. This system automatically tracked and evaluated data handling practices, providing real-time alerts in case of any potential compliance breaches. By automating these processes, the healthcare provider was able to significantly reduce the time and resources required for compliance checks, while also minimizing the risk of human error. The integration between data management and compliance monitoring systems ensured that data quality and regulatory compliance were maintained consistently, without the need for manual intervention.

The outcomes of this implementation were profound. The healthcare provider reported a 30% reduction in the time spent on compliance audits and a 25% improvement in data quality metrics. These improvements not only enhanced the reliability of the ML models used for patient care but also ensured that the organization remained compliant with healthcare regulations. The success of this automation and integration strategy highlighted the importance of leveraging advanced tools to streamline governance processes in the healthcare sector.

Automation and Integration in Finance

In the finance industry, where compliance with regulatory standards is critical, automation and integration play a pivotal role in governance. A global financial institution faced challenges in managing the compliance requirements set forth by various regulatory bodies, including the General Data Protection Regulation (GDPR) and the Sarbanes-Oxley Act (SOX). The institution's manual processes for auditing and compliance monitoring were inefficient and prone to delays, making it difficult to respond quickly to regulatory changes and ensure ongoing compliance.

To overcome these challenges, the financial institution implemented a comprehensive governance solution by integrating Collibra's data governance platform with OneTrust's compliance monitoring tools and ACL's auditing software. Collibra was chosen for its robust data governance capabilities, which included data cataloging, stewardship, and automated workflows for managing data across different departments and regions. The platform ensured that all data within the organization was consistently classified and managed according to regulatory requirements.

OneTrust's compliance monitoring system was integrated with Collibra to provide continuous oversight of data handling practices, ensuring compliance with GDPR and other regulatory standards. This system automatically generated compliance reports and alerted the compliance team to any deviations from established policies. The integration allowed the financial institution to maintain a real-time view of its compliance status, reducing the risk of non-compliance and associated penalties.

ACL's auditing software was also integrated into this governance framework to provide comprehensive audit trails and insights into the institution's financial and operational data. The software's automated auditing capabilities enabled the institution to conduct regular internal audits efficiently, ensuring that all financial transactions and data handling processes adhered to regulatory requirements. The integration of auditing tools with data governance and compliance monitoring systems created a seamless governance process, where all aspects of data management, compliance, and auditing were interconnected and automated.

The results of this integrated approach were significant. The financial institution saw a 40% reduction in the time required to complete internal audits and a 50% decrease in compliance-related incidents. These improvements not only enhanced the organization's ability to meet regulatory requirements but also increased operational efficiency and reduced costs associated with manual compliance checks. The successful implementation of automation and integration in the financial institution demonstrated the value of using advanced governance tools to maintain compliance and improve overall governance processes.

Performance Measurement

Performance measurement in the context of ML data governance is essential for ensuring the effectiveness of governance frameworks and maintaining accountability.

Governance Metrics

Governance metrics play a pivotal role in this process by providing quantifiable indicators of how well governance practices are functioning. These metrics help organizations assess compliance, evaluate risk mitigation efforts, and monitor model performance. Defining and regularly reviewing these metrics are crucial for maintaining robust governance practices and ensuring that ML systems operate effectively and ethically.

Compliance Rates: One of the primary governance metrics is compliance rates. Compliance rates measure the extent to which ML systems adhere to relevant regulations, internal policies, and industry standards. This metric includes compliance with data protection laws such as GDPR, CCPA, and HIPAA, as well as adherence to organizational policies on data privacy and security. Tracking compliance rates involves monitoring various aspects of data handling, model development, and deployment to ensure that all processes align with established guidelines. A high compliance rate indicates that governance practices are effective and that the organization is successfully meeting regulatory requirements. Conversely, a low compliance rate may signal gaps in governance practices that need to be addressed.

A multinational financial services firm, for example, tracks its compliance with data protection laws like GDPR and CCPA by monitoring the percentage of systems that comply with these regulations. This metric is calculated by auditing the data handling processes, model development workflows, and deployment procedures against the regulatory requirements. Compliance rates are monitored through automated systems that flag any deviations from established guidelines. When compliance rates dip below a certain threshold, the firm reviews its governance practices to identify gaps and implement necessary improvements.

Risk Mitigation Outcomes: Risk mitigation outcomes are another critical governance metric. This metric evaluates the effectiveness of risk management strategies in reducing or eliminating identified risks associated with ML systems. Risk mitigation outcomes can be assessed through various indicators, such as the number of incidents successfully resolved, the severity of residual risks, and the effectiveness of risk response measures. For example, if a risk management strategy involves implementing specific controls to mitigate data breaches, the outcome can be measured by tracking the number of breaches that occur and the effectiveness of the controls in preventing or minimizing the impact of these breaches. Effective risk mitigation leads to fewer incidents and lower residual risks, demonstrating the success of governance practices in managing potential threats.

A large healthcare provider, for instance, implemented a series of controls to mitigate risks associated with patient data breaches. These controls included encryption protocols, access management systems, and regular security audits. The provider measures the success of these controls by tracking the number of data breaches that occur over time and the impact of these breaches on patient data security. By analyzing these outcomes, the healthcare provider can determine the effectiveness of its risk management strategies and make informed decisions about adjusting or enhancing its controls to better protect sensitive data.

Model Performance Indicators: Model performance indicators are essential metrics for evaluating the effectiveness of ML models in meeting their intended objectives. These indicators include various performance measures such as accuracy, precision, recall, F1 score, and AUC-ROC. Regular monitoring of these indicators helps ensure that ML models operate as expected and provide reliable outcomes. For instance, if a model's accuracy drops significantly, it may indicate that the model is not performing well and may require retraining or adjustment. Tracking model performance indicators also helps identify potential issues related to data drift, model degradation, or other factors that may impact the model's effectiveness. Consistently reviewing these indicators ensures that models remain effective and aligned with their intended goals.

A retail company, for example, uses these indicators to evaluate the performance of its demand forecasting models. The accuracy of the model's predictions is regularly monitored, and any significant drops in performance trigger a review of the data inputs and model parameters. The company calculates these metrics by comparing the model's predictions with actual sales data, allowing it to identify any performance drift that may require retraining or adjustments. Continuous monitoring of these indicators ensures that the models remain aligned with the company's business goals and continue to provide reliable outcomes.

Audit Findings: Audit findings provide valuable insights into the effectiveness of governance practices by highlighting areas where improvements are needed. Regular audits assess various aspects of ML systems, including data handling, model development, and compliance with policies and regulations. The findings from these audits can reveal discrepancies, non-compliance issues, and areas of potential improvement. Key audit findings may include instances of data misuse, gaps in compliance with regulations, or inefficiencies in risk management processes. By analyzing audit findings, organizations can identify weaknesses in their governance frameworks and take corrective actions to address these issues. Monitoring audit findings over time helps track improvements and ensure that governance practices evolve to address emerging challenges.

A technology firm conducts regular audits of its ML systems to ensure compliance with industry standards and internal policies. The audit process involves reviewing the documentation of data handling practices, model development, and deployment procedures. Key findings from these audits might include instances of non-compliance with data privacy regulations or inefficiencies in the model validation process. The firm uses these findings to identify weaknesses in its governance framework and implements corrective actions to address these issues. Over time, monitoring audit findings helps the firm track its progress in improving governance practices and maintaining compliance.

Incident Resolution Times: Incident resolution times measure the efficiency of the organization's response to incidents related to ML systems. This metric tracks the time taken to identify, address, and resolve incidents such as data breaches, compliance violations, or model performance issues. Shorter resolution times indicate that the organization has effective incident response processes in place and can quickly address issues as they arise. Conversely, longer resolution times may suggest inefficiencies in the incident response process or a need for improved procedures. Monitoring incident resolution times helps organizations assess their ability to manage and mitigate the impact of incidents and ensures that governance practices are effective in addressing emerging challenges.

A global e-commerce platform, for example, tracks the time taken to resolve incidents involving unauthorized access to customer data. The platform calculates this metric by logging the time from when an incident is first detected to when it is fully resolved. Shorter resolution times indicate that the platform has effective incident response processes in place, allowing it to quickly address and mitigate the impact of incidents. Monitoring this metric helps the platform assess the effectiveness of its incident response strategies and ensure that it is prepared to handle emerging challenges swiftly and efficiently.

Data Quality Metrics: Data quality metrics assess the accuracy, completeness, and consistency of the data used in ML systems. These metrics are crucial for ensuring that ML models are based on reliable and high-quality data, which directly impacts their performance and effectiveness. Data quality metrics may include measures such as the percentage of missing or incorrect data, the frequency of data validation checks, and the results of data quality assessments. Regularly monitoring these metrics helps identify and address data quality issues, ensuring that data remains reliable and meets governance standards. High data quality contributes to the overall effectiveness of ML models and supports the integrity of governance practices.

A financial institution relies heavily on accurate data for its credit risk assessment models. To maintain high data quality, the institution monitors metrics such as the percentage of missing values, data duplication rates, and the frequency of data validation errors. These metrics are calculated by running automated data quality checks during the data ingestion process. When data quality metrics fall below acceptable thresholds, the institution triggers data cleansing procedures to rectify the issues. This proactive approach ensures that the data feeding into the ML models is reliable, which is critical for producing accurate credit risk predictions.

Ethical Compliance Metrics: Ethical compliance metrics evaluate how well ML systems adhere to ethical principles such as fairness, transparency, and accountability. These metrics assess whether ML models are designed and used in a way that aligns with ethical guidelines and societal values. For example, ethical compliance metrics may include measures of bias detection and mitigation, transparency in model decision-making, and adherence to ethical AI principles. Monitoring these metrics helps ensure that ML systems are developed and deployed responsibly and that any ethical issues are addressed promptly. Ethical compliance metrics are essential for maintaining public trust and ensuring that ML systems operate in a manner that is aligned with organizational values and societal expectations.

A social media company implements ethical compliance metrics to ensure that its ML algorithms do not perpetuate biases or make discriminatory decisions. The company uses bias detection tools to measure the fairness of its recommendation systems across different demographic groups. For instance, the tool calculates the disparity in recommendation rates between male and female users for certain content types. If the metric shows significant disparities, the company investigates the root causes and adjusts the algorithms to reduce bias. Monitoring these metrics helps the company align its ML practices with ethical standards and societal expectations.

Resource Utilization: Resource utilization metrics assess the efficiency of resource allocation and management in ML governance. These metrics include measures of how effectively resources such as time, budget, and personnel are used to support governance activities. For example, resource utilization metrics may track the amount of time spent on compliance monitoring, the budget allocated for risk management, and the number of staff involved in governance tasks. Efficient resource utilization helps ensure that governance practices are implemented effectively and that resources are allocated to areas where they are most needed. Monitoring resource utilization metrics also helps identify opportunities for improving efficiency and optimizing the use of resources in governance efforts.

A healthcare provider tracks resource utilization metrics to assess the efficiency of its ML governance activities. These metrics include the percentage of the budget spent on compliance monitoring tools, the number of personnel dedicated to data governance, and the time allocated to auditing activities. By analyzing these metrics, the provider can identify areas where resources are under- or overutilized. For instance, if too much budget is being spent on manual auditing processes, the provider might invest in automation tools to streamline these tasks, thereby optimizing resource allocation.

Model Performance Drift: Model performance drift metrics track changes in model performance over time to identify potential issues related to model degradation or data drift. These metrics include measures of shifts in model accuracy, precision, recall, and other performance indicators. Monitoring model performance drift helps ensure that models continue to operate effectively and meet their intended objectives. If significant performance drift is detected, it may indicate that the model requires retraining or adjustments to address changes in data or underlying conditions. Regularly reviewing model performance drift metrics helps maintain the effectiveness of ML models and supports the overall success of governance practices.

An e-commerce company uses model performance drift metrics to monitor changes in the performance of its recommendation engine over time. The company tracks metrics such as changes in the click-through rate (CTR) and conversion rate as indicators of model drift. These metrics are calculated by comparing the current performance with historical benchmarks. If significant drift is detected, the company investigates whether the underlying data distribution has changed or if the model needs retraining. Addressing performance drift promptly helps maintain the effectiveness of the recommendation engine, ensuring that it continues to drive customer engagement.

Compliance Incident Frequency: Compliance incident frequency metrics measure the number of compliance-related incidents that occur within a given period. These metrics provide insights into the effectiveness of compliance measures and highlight areas where improvements may be needed. For example, compliance incident frequency metrics may track the number of data breaches, regulatory violations, or policy breaches reported. Monitoring these metrics helps identify trends and patterns in compliance incidents and assess the effectiveness of risk management and compliance strategies. A high frequency of compliance incidents may indicate that additional measures are needed to address underlying issues and improve adherence to regulatory requirements.

A telecommunications company tracks the frequency of compliance incidents, such as unauthorized data access or breaches of customer privacy policies. This metric is calculated by logging each incident and categorizing it by severity. The company uses this data to identify patterns and trends, such as whether certain departments or processes are more prone to compliance breaches. If the frequency of incidents is high, the company may review and strengthen its compliance measures, such as enhancing access controls or providing additional training to staff.

Audit Trail Completeness: Audit trail completeness metrics assess the comprehensiveness and accuracy of audit trails maintained for ML systems and governance activities. These metrics include measures of the extent to which audit trails capture relevant information about data handling, model development, and compliance efforts. Ensuring that audit trails are complete and accurate is essential for maintaining transparency and accountability in governance practices. Monitoring audit trail completeness helps identify gaps in documentation and ensures that all relevant activities are recorded and accessible for review. Comprehensive audit trails support effective auditing and contribute to the overall success of governance efforts.

A retail company maintains audit trails for all its ML model deployments, tracking every change made to model parameters, data inputs, and access logs. The completeness of these audit trails is assessed by verifying that all relevant activities are recorded and that the records are accurate. If any gaps or inconsistencies are found in the audit trails, the company takes corrective actions to ensure that all future activities are fully documented. Comprehensive audit trails support transparency and accountability, making it easier for the company to conduct thorough audits and address any issues that arise.

Stakeholder Satisfaction: Stakeholder satisfaction metrics evaluate the level of satisfaction among stakeholders regarding governance practices and their impact on ML systems. These metrics include measures of stakeholder perceptions, feedback, and overall satisfaction with governance processes. For example, stakeholder satisfaction metrics may include survey results, feedback forms, and interviews with data scientists, compliance officers, and other stakeholders. Monitoring these metrics helps ensure that governance practices meet the needs and expectations of stakeholders and identify areas where improvements may be needed. High levels of stakeholder satisfaction indicate that governance practices are effective and aligned with stakeholder expectations.

A government agency measures stakeholder satisfaction with its ML governance practices through surveys and feedback sessions. Stakeholders, including data scientists, compliance officers, and end users, are asked to rate their satisfaction with the transparency, fairness, and effectiveness of the agency's ML models. The agency analyzes these satisfaction metrics to identify areas where stakeholders feel improvements are needed, such as in the clarity of model explanations or the responsiveness of governance processes. High levels of stakeholder satisfaction indicate that the agency's governance practices are aligned with stakeholder expectations and are effectively managing risks.

Data Privacy Metrics: Data privacy metrics assess how well ML systems adhere to data privacy regulations and policies. These metrics include measures of data protection compliance, data access controls, and the effectiveness of privacy safeguards. For example, data privacy metrics may track the number of data access requests processed, the effectiveness of data anonymization techniques, and compliance with data retention policies. Monitoring data privacy metrics helps ensure that data privacy requirements are met and that data is protected against unauthorized access or misuse. Effective data privacy practices contribute to the overall success of governance efforts and help maintain public trust in ML systems.

A global technology company tracks data privacy metrics to ensure compliance with data protection regulations like GDPR and CCPA. These metrics include the number of data access requests processed within the required timeframe, the effectiveness of data anonymization techniques, and the frequency of data privacy training sessions conducted for employees. The company calculates these metrics by reviewing its data processing logs and privacy impact assessments. Regular monitoring helps the company identify potential weaknesses in its data privacy practices and take steps to strengthen its compliance efforts.

Resource Allocation Efficiency: Resource allocation efficiency metrics assess how effectively resources are allocated to support governance activities and address ML system needs. These metrics include measures of the utilization of financial, human, and technological resources. For example, resource allocation efficiency metrics may track the budget allocated for compliance activities, the number of staff dedicated to governance tasks, and the effectiveness of technological tools used in governance. Monitoring these metrics helps ensure that resources are used efficiently and that governance practices are supported adequately. Effective resource allocation contributes to the overall success of governance efforts and helps optimize the use of available resources.

A pharmaceutical company monitors resource allocation efficiency metrics to evaluate how effectively its resources are being used to support ML governance activities. Metrics include the proportion of the budget allocated to compliance tools versus other governance activities and the efficiency of resource utilization in model development and monitoring. The company assesses these metrics to ensure that resources are being deployed where they are most needed and to optimize spending on governance-related activities. If inefficiencies are identified, such as excessive spending on low-impact activities, the company reallocates resources to higher-priority areas.

Model Explainability: Model explainability metrics assess how well ML models can be understood and interpreted by stakeholders. These metrics include measures of the transparency of model decision-making processes and the availability of explanations for model outputs. For example, model explainability metrics may track the use of explainability tools, the clarity of model explanations provided, and stakeholder understanding of model behavior. Monitoring these metrics helps ensure that models are transparent and that stakeholders can understand and trust the outcomes produced by ML systems. Effective model explainability practices contribute to the overall success of governance efforts and support ethical AI use.

An insurance company tracks model explainability metrics to assess how well its ML models can be understood and interpreted by non-technical stakeholders, such as underwriters and regulators. The company uses tools like SHAP (SHapley Additive exPlanations) to generate explanations for model predictions and evaluates the clarity and accessibility of these explanations through user testing. Metrics include the percentage of stakeholders who report that they understand the model's decision-making process and the ease with which explanations can be generated. Ensuring high model explainability supports the company's goals of transparency and accountability, particularly in regulatory compliance.

Governance Training Effectiveness: Governance training effectiveness metrics assess the impact of training programs on stakeholders' understanding and adherence to governance practices. These metrics include measures of training participation rates, knowledge retention, and the application of governance principles. For example, governance training effectiveness metrics may track the number of training sessions conducted, the results of knowledge assessments, and the implementation of governance practices by trained stakeholders. Monitoring these metrics helps ensure that training programs are effective and that stakeholders are well-equipped to support governance efforts. Effective training programs contribute to the overall success of governance practices and support stakeholder engagement.

A large financial institution measures the effectiveness of its governance training programs through metrics such as employee participation rates, knowledge retention scores from post-training assessments, and the application of governance principles in daily operations. The institution collects data on these metrics by tracking attendance at training sessions, conducting follow-up quizzes, and reviewing governance practices in routine audits. If the training effectiveness metrics indicate gaps in knowledge or application, the institution revises its training content and delivery methods to better meet the needs of its staff.

Ethical Adherence Metrics: Ethical adherence metrics evaluate how well ML systems adhere to ethical principles and guidelines. These metrics include measures of fairness, transparency, and accountability in model development and use. For example, ethical adherence metrics may track the results of bias assessments, the clarity of model decision-making processes, and the responsiveness to ethical concerns. Monitoring these metrics helps ensure that ML systems operate in a manner that aligns with ethical standards and societal values. Effective ethical adherence practices contribute to the overall success of governance efforts and support responsible AI use.

A healthcare analytics company monitors ethical adherence metrics to evaluate how well its ML systems align with ethical guidelines related to patient care and data usage. These metrics include measures of fairness in treatment recommendations, the transparency of patient data usage, and the responsiveness to ethical concerns raised by stakeholders. The company tracks these metrics by conducting regular ethics reviews and bias audits, ensuring that its ML systems do not disproportionately affect any patient group or violate ethical standards. Maintaining strong ethical adherence metrics helps the company build trust with patients and healthcare providers, reinforcing its commitment to ethical AI practices.

Continuous Improvement

Continuous improvement is a fundamental aspect of effective ML data governance, driving ongoing enhancements in governance practices based on performance metrics, stakeholder feedback, and lessons learned from audits and incidents. Implementing a continuous improvement process ensures that governance practices evolve to address emerging challenges, adapt to changing regulations, and meet the needs of stakeholders. By regularly reviewing and refining governance practices, organizations can maintain high standards of effectiveness, compliance, and accountability in their ML systems.

The continuous improvement process begins with a thorough review of performance metrics. This review involves analyzing governance metrics such as compliance rates, risk mitigation outcomes, model performance indicators, and audit findings. By evaluating these metrics, organizations can identify areas where governance practices are performing well and areas where improvements are needed. For example, if compliance rates are lower than expected, it may indicate the need for enhanced monitoring or updated policies. Regularly reviewing performance metrics provides valuable insights into the effectiveness of governance practices and helps guide improvement efforts.

Integrating stakeholder feedback into the continuous improvement process is essential for ensuring that governance practices meet the needs and expectations of those directly affected by ML systems. Stakeholder feedback can be collected through surveys, interviews, and feedback forms, providing insights into the strengths and weaknesses of governance practices. For example, feedback from data scientists may reveal challenges related to data access or model transparency, while feedback from end users may highlight issues related to fairness or usability. By incorporating this feedback into governance practices, organizations can address concerns, enhance stakeholder satisfaction, and improve the overall effectiveness of governance efforts.

Lessons learned from audits and incidents are crucial for driving continuous improvement in governance practices. Audits provide insights into compliance with regulations, internal policies, and best practices, while incidents highlight areas where governance practices may need to be strengthened. For example, an audit may reveal gaps in data handling procedures, while an incident such as a data breach may uncover weaknesses in risk management strategies. By analyzing lessons learned from audits and incidents, organizations can identify root causes, implement corrective actions, and refine governance practices to prevent similar issues in the future. This iterative approach helps ensure that governance practices remain effective and responsive to emerging challenges.

Updating policies and procedures is a key component of the continuous improvement process. Based on the insights gained from performance metrics, stakeholder feedback, and lessons learned, organizations may need to revise their governance policies and procedures to address identified gaps and enhance effectiveness. For example, if a new regulation is introduced, organizations may need to update their data management policies to ensure compliance. Similarly, if feedback indicates a need for improved transparency, organizations may need to enhance their model explainability practices. Regularly updating policies and procedures helps ensure that governance practices remain current, relevant, and aligned with evolving standards and requirements.

Training and capacity building are essential for supporting continuous improvement in governance practices. Providing ongoing training and professional development opportunities for stakeholders helps ensure that they are equipped with the knowledge and skills needed to adhere to updated policies and procedures. For example, training programs may focus on new regulatory requirements, changes in governance practices, or emerging best practices in ML governance. By investing in training and capacity building, organizations can enhance stakeholder competence, support effective implementation of governance practices, and foster a culture of continuous improvement.

Leveraging advancements in technology and tools is a critical aspect of continuous improvement in ML governance. As technology evolves, new tools and technologies become available to support governance practices, such as advanced data management platforms, automated compliance monitoring systems, and sophisticated auditing software. By adopting and integrating these technologies, organizations can enhance their governance capabilities, streamline processes, and improve oversight. For example, implementing automation tools can help streamline compliance checks and data quality monitoring, while integrating new technologies can improve model performance management and risk mitigation. Staying abreast of technological advancements and incorporating them into governance practices supports continuous improvement and enhances overall governance effectiveness.

Developing and refining performance measurement frameworks is essential for continuous improvement in governance practices. These frameworks provide a structured approach to measuring and evaluating governance effectiveness, helping organizations track progress and identify areas for enhancement. Performance measurement frameworks may include a range of metrics, such as compliance rates, risk mitigation outcomes, model performance indicators, and audit findings. By regularly reviewing and updating these frameworks, organizations can ensure that they capture relevant data, align with evolving standards, and provide actionable insights for continuous improvement.

Effective change management is crucial for implementing continuous improvement initiatives in governance practices. Change management involves planning, executing, and monitoring changes to governance processes, policies, and technologies to ensure successful implementation and adoption. For example, when updating policies or introducing new technologies, organizations must manage the transition effectively to minimize disruption and ensure that stakeholders are informed and prepared. By applying change management principles, organizations can facilitate smooth transitions, address challenges, and support the successful implementation of continuous improvement efforts.

Benchmarking and adopting best practices are valuable strategies for driving continuous improvement in governance practices. By comparing their governance practices against industry standards and best practices, organizations can identify areas where they excel and areas where improvements are needed. Benchmarking may involve analyzing the governance practices of leading organizations, participating in industry forums, or reviewing relevant literature. Incorporating best practices helps organizations enhance their governance frameworks, adopt innovative approaches, and stay aligned with industry standards. Regular benchmarking ensures that governance practices remain competitive and effective in a rapidly evolving landscape.

Evaluating the impact and effectiveness of continuous improvement initiatives is essential for assessing progress and ensuring that governance practices achieve their intended outcomes. This evaluation involves measuring the results of implemented changes, assessing the impact on governance metrics, and determining the effectiveness of new policies, procedures, and technologies. For example, organizations may evaluate the impact of updated policies on compliance rates, assess the effectiveness of new technologies in enhancing governance processes, and measure stakeholder satisfaction with improved practices. By evaluating impact and effectiveness, organizations can ensure that continuous improvement efforts are successful and contribute to the overall success of ML data governance.

Continuous Improvement Framework

To establish a robust continuous improvement framework, organizations must follow a structured approach that includes assessing current practices, identifying gaps, setting improvement goals, implementing changes, and evaluating outcomes. This process helps organizations maintain high standards of governance, compliance, and accountability while continuously adapting to new challenges and opportunities.

Assessing Current Practices

The first phase of continuous improvement involves a comprehensive assessment of current governance practices. Organizations need to review existing policies, procedures, and tools to determine their effectiveness in managing ML systems. This assessment includes analyzing governance metrics such as compliance rates, risk mitigation outcomes, and model performance indicators. The goal is to identify areas where governance practices are working well and areas where they may be falling short. For instance, if compliance rates are lower than expected, this could indicate issues with the monitoring processes or the need for updated policies. By thoroughly assessing current practices, organizations can establish a baseline understanding of their governance performance.

Identifying Gaps

After assessing current practices, the next phase involves identifying gaps in governance processes. This step is critical for pinpointing specific areas that require improvement. Organizations should use a combination of performance metrics, stakeholder feedback, and lessons learned from audits and incidents to identify these gaps. For example, if audit findings reveal weaknesses in data handling procedures or if stakeholder feedback highlights concerns about model transparency, these issues represent clear areas for improvement. Identifying gaps provides a focused direction for the continuous improvement process, ensuring that efforts are targeted where they are most needed.

Setting Improvement Goals

Once gaps have been identified, organizations should set clear and measurable improvement goals. These goals should be specific, actionable, and aligned with the organization's overall governance objectives. For example, if a gap in data privacy compliance is identified, an improvement goal could be to achieve a specific compliance rate within a defined timeframe. Setting improvement goals provides a roadmap for the changes that need to be implemented and helps ensure that all stakeholders are aligned with the organization's governance priorities. These goals also serve as benchmarks against which progress can be measured.

Implementing Changes

With improvement goals in place, the next phase is to implement the necessary changes. This may involve updating governance policies, revising procedures, adopting new technologies, or enhancing training programs. Effective change management is crucial during this phase to ensure that changes are implemented smoothly and that stakeholders are adequately prepared. For instance, if a new automated compliance monitoring system is introduced, organizations must provide training to ensure that employees can use the system effectively. Implementing changes requires careful planning and execution to minimize disruptions and ensure that improvements are successfully integrated into the governance framework.

Evaluating Outcomes

The final phase of the continuous improvement process involves evaluating the outcomes of the implemented changes. Organizations should measure the impact of these changes on governance metrics and assess whether the improvement goals have been achieved. For example, if a goal was to improve compliance rates, organizations should analyze post-implementation data to determine if there has been a measurable increase in compliance. Evaluation also involves gathering feedback from stakeholders to assess their satisfaction with the changes. By evaluating outcomes, organizations can determine the effectiveness of their continuous improvement efforts and identify any additional adjustments that may be needed.

Checklist for Continuous Improvement

To help organizations implement continuous improvement effectively, a checklist and toolkit can be used to guide each phase of the process.

Assessment

- Review Existing Governance Frameworks

    - Evaluate current governance policies, procedures, and tools.

    - Check alignment with regulatory requirements and industry standards.

    - Analyze recent governance reports, audit findings, and compliance results.

    - Identify key stakeholders and their roles in governance.

- Analyze Performance Metrics

    - Gather and review data on compliance rates (e.g., GDPR, CCPA compliance).

    - Assess risk mitigation outcomes, including the number and severity of incidents.

    - Evaluate model performance indicators (e.g., accuracy, bias, transparency).

    - Analyze audit findings for recurring issues or compliance gaps.

- Collect Stakeholder Feedback

    - Distribute surveys to gather input from data scientists, compliance officers, and end users.

    - Conduct interviews with key stakeholders to identify challenges and areas of concern.

    - Review feedback from past incidents or audit results to identify patterns.

Gap Identification

- Identify Governance Gaps

    - Compare current practices against best practices and industry benchmarks.

    - Use data analysis tools to identify discrepancies in data handling, model accuracy, or compliance.

    - Review past incidents and audit findings to pinpoint weaknesses in current governance practices.

    - Assess the effectiveness of communication channels between governance teams.

- Prioritize Gaps

    - Rank identified gaps based on their potential impact on compliance, risk, and model performance.

    - Focus on gaps that pose the highest risk to the organization or have the greatest potential for improvement.

    - Consider both short-term fixes and long-term strategic improvements.

Goal Setting

- Define Improvement Goals

    - Set specific, measurable, achievable, relevant, and time-bound (SMART) goals for addressing identified gaps.

    - Align goals with the organization's overall governance strategy and objectives.

    - Establish clear milestones and deadlines for each goal.

- Communicate Goals

  - Inform all relevant stakeholders of the improvement goals and their roles in achieving them.

  - Ensure that goals are documented and accessible to the entire governance team.

  - Assign responsibility for each goal to specific teams or individuals.

- Allocate Resources

  - Determine the resources required (e.g., budget, personnel, tools) to achieve each goal.

  - Ensure that sufficient resources are allocated and that teams are equipped to meet their objectives.

  - Plan for potential roadblocks and establish contingency plans.

Implementation

- Develop an Implementation Plan

  - Create a detailed plan outlining the steps needed to achieve each improvement goal.

  - Include timelines, responsible parties, and key milestones.

  - Integrate the implementation plan into the organization's broader governance framework.

- Execute the Plan

  - Roll out changes systematically, starting with high-priority areas.

  - Implement new governance policies, procedures, or technologies as needed.

  - Ensure all stakeholders are trained on new practices or tools.

  - Monitor progress against milestones and adjust the plan as necessary.

- Manage Change Effectively

  - Communicate regularly with stakeholders to keep them informed of progress and any changes.

  - Address any resistance to change by providing support and resources.

  - Ensure that all changes are documented and that relevant governance records are updated.

Evaluation

- Monitor and Measure Outcomes

  - Reassess the governance metrics post-implementation to determine the effectiveness of changes.

  - Measure the impact on compliance rates, risk mitigation, and model performance.

  - Evaluate whether the improvement goals have been met according to the defined success criteria.

- Collect and Analyze Feedback

  - Solicit feedback from stakeholders on the implemented changes.

  - Identify any areas where further improvements are needed based on stakeholder input.

  - Document lessons learned during the implementation process.

- Continuous Monitoring and Reporting

  - Establish ongoing monitoring processes to ensure sustained improvement.

  - Create a schedule for regular reviews of governance practices and outcomes.

  - Report on progress to senior management and relevant committees to maintain transparency and accountability.

- Document and Share Best Practices

  - Record successful strategies and approaches for future reference.

  - Share best practices within the organization to foster a culture of continuous improvement.

  - Review and update the continuous improvement framework periodically to incorporate new insights and developments.

Summary

This chapter delves into the essential role of regulatory compliance and risk management in the deployment and operation of machine learning (ML) systems. It highlights how adherence to legal frameworks like GDPR, CCPA, and HIPAA is crucial for ensuring that ML systems function within ethical and legal limits. It underscores that these regulations impose strict requirements related to data protection, privacy, and the prevention of bias in automated decision-making, necessitating their integration into ML practices.

A significant focus is placed on how organizations must incorporate these regulatory demands into their ML workflows to maintain compliance and build stakeholder trust. This chapter discusses the necessity of adopting ethical AI practices—such as transparency, fairness, and accountability—to ensure alignment with societal values and to avoid potential legal repercussions. The discussion also introduces frameworks like the AJL Model Bias Framework, which are instrumental in mitigating algorithmic bias and ensuring non-discriminatory outcomes from ML systems.

This chapter emphasizes the ongoing need for continuous monitoring and regular updates to compliance practices, especially as regulations and industry standards evolve. It encourages organizations to engage actively with legal and compliance experts and to participate in industry forums to remain informed about changes in the regulatory landscape. This proactive approach is portrayed as critical to maintaining a robust compliance posture in the face of an ever-changing regulatory environment.

The chapter highlights how robust compliance efforts not only mitigate risks such as data privacy breaches and algorithmic bias but also contribute to operational efficiency. By embedding compliance into the core of ML operations, organizations can create a culture of accountability and transparency, which in turn strengthens their overall governance framework. This proactive stance on compliance is depicted as a key factor in fostering trust with both internal and external stakeholders.

The chapter discusses the broader benefits of a strong compliance framework, noting that it enhances an organization's reputation and provides a competitive edge in the marketplace. Organizations that prioritize regulatory compliance and ethical AI practices are better positioned to navigate the complexities of modern data governance and to build long-term trust with customers, partners, and regulators. This comprehensive approach to compliance is presented as a vital component of sustainable ML governance.

CHAPTER 9

Organizational Culture and Change Management

Organizational culture and change management are critical components in ensuring the successful implementation of machine learning (ML) data governance frameworks. As organizations increasingly rely on ML systems to drive business decisions, the need for robust governance becomes paramount. This governance ensures that ML models are developed, deployed, and maintained in a manner that aligns with ethical standards, regulatory requirements, and organizational objectives. However, the effectiveness of ML data governance is deeply influenced by the culture of the organization and its ability to manage change.

Organizational culture refers to the shared values, beliefs, and norms that shape the behavior and decision-making processes within an organization. It is the foundation upon which governance frameworks are built and sustained. In the context of ML data governance, a strong culture of ethics, accountability, and transparency is essential. Such a culture fosters an environment where governance practices are not only implemented but also embraced by all members of the organization.

Change management, on the other hand, involves the processes and strategies that guide an organization through transitions, such as the adoption of new technologies or governance frameworks. Effective change management ensures that these transitions occur smoothly and that the organization remains resilient in the face of challenges. When implementing ML data governance, change management is crucial for aligning the organization's practices with new governance requirements and ensuring that employees are adequately prepared and motivated to adhere to these changes.

In this chapter, we will explore the interplay between organizational culture, change management, and ML data governance. We will examine how these elements work together to create a governance framework that not only meets regulatory requirements but also drives ethical and effective use of ML systems.

© Aditya Nandan Prasad 2024
A. Nandan Prasad, *Introduction to Data Governance for Machine Learning Systems*,
https://doi.org/10.1007/979-8-8688-1023-7_9

The Role of Organizational Culture in ML Data Governance

Organizational culture plays a critical role in shaping the framework and execution of machine learning (ML) data governance within any institution. It forms the bedrock of how data is perceived, valued, and utilized across the organization. A culture that prioritizes data integrity, transparency, and ethical use of data naturally fosters a robust governance environment. When an organization's culture emphasizes the importance of data as a strategic asset, it inherently supports the establishment of stringent data governance policies that ensure the ethical and compliant use of ML models. Conversely, in an organization where data is seen merely as a byproduct, ML data governance might be relegated to a mere compliance checkbox, leading to potential risks and ethical challenges. The alignment between organizational culture and ML data governance is vital because it influences how policies are developed, communicated, and adhered to, shaping the overall success of the governance framework.

Embedding Ethics in AI

The adoption and implementation of ethical AI practices are deeply influenced by the culture within an organization. When ethical considerations are deeply ingrained in the organizational culture, ML data governance frameworks are more likely to prioritize fairness, accountability, and transparency. This cultural foundation supports the development of AI systems that are not only technically proficient but also ethically sound, ensuring that ML models are designed with a strong awareness of potential biases, data privacy concerns, and the broader social impact of their deployment.

A strong organizational culture that emphasizes ethics in AI fosters the creation of governance policies that extend beyond mere compliance, encouraging innovation while mitigating ethical risks. This commitment to ethical practices not only shields the organization from legal and reputational threats but also strengthens trust among stakeholders, including customers, regulators, and employees. By embedding ethics into the core of AI practices, organizations can ensure that their AI systems align with both technical excellence and societal values.

Leadership's Role in ML Data Governance

Leadership plays a pivotal role in cultivating a culture that supports effective ML data governance. Leaders set the tone for the organization's values and priorities, influencing how data is managed and governed. In organizations where leadership actively promotes a data-driven culture, there is a clear emphasis on data quality, accuracy, and responsible use. Leaders in such environments encourage continuous learning about data governance, ensuring that all employees understand the importance of their role in maintaining data integrity. They also advocate for the integration of data governance into all business processes, recognizing its importance in the success of ML initiatives. Leadership commitment to data governance ensures that it is not just an IT or compliance function but a core business imperative that is supported and upheld across the organization.

Effective leaders in data governance set clear objectives that align with the organization's overall strategy, ensuring that governance policies are not just theoretical but are actionable and measurable. By establishing specific goals, such as improving data quality or ensuring compliance with regulatory standards, leaders provide a roadmap that guides the organization toward better data management practices. Furthermore, leaders play a crucial role in communicating the importance of data ethics, making it a visible and ongoing part of the organizational dialogue. This communication helps to embed data ethics into the company culture, ensuring that employees at all levels understand why ethical considerations are vital and how they relate to the organization's success.

In addition to setting objectives and communicating values, effective leaders lead by example in adhering to governance policies. When leaders consistently follow the same rules and guidelines that they expect from their teams, it reinforces the importance of these policies and encourages employees to take them seriously. This behavior fosters a culture of accountability, where everyone from the top down is responsible for maintaining data integrity and ethical practices. Leaders who prioritize data governance and demonstrate their commitment through their actions help to build a strong foundation for a governance culture that supports the ethical and responsible use of data across the organization.

Employee Engagement in ML Data Governance

For ML data governance to be effective, it requires the active engagement and buy-in of all employees. Organizational culture plays a significant role in determining the level of employee engagement in governance practices. In a culture that values collaboration, transparency, and accountability, employees are more likely to take ownership of data governance responsibilities. They understand that their actions directly impact the quality and integrity of data used in ML models, and they are motivated to adhere to governance policies. This buy-in is crucial because even the most well-designed governance frameworks can fail if employees are not committed to following them. By fostering a culture of inclusion and shared responsibility, organizations can ensure that every employee, from data scientists to business analysts, plays a role in maintaining the integrity of the ML data governance framework.

To effectively engage employees in data governance, organizations should implement specific training programs that reinforce the importance of these practices. Regular workshops and seminars focused on data governance principles can help employees understand the role they play in maintaining data integrity and ethical AI practices. These sessions can cover topics such as data privacy, ethical AI development, and the potential consequences of poor data governance. Additionally, interactive training modules that simulate real-world scenarios can be particularly effective in helping employees grasp the complexities of data governance and the impact of their decisions.

Organizations can also adopt methods like peer learning and mentorship programs to reinforce governance practices. Pairing less experienced employees with seasoned data professionals can help embed a culture of data stewardship throughout the organization. This mentorship approach allows for the transfer of knowledge and best practices in a more personalized and context-specific manner, ensuring that employees are not only trained but also supported in their ongoing adherence to governance policies. By integrating these specific training programs and methods, organizations can cultivate a workforce that is both knowledgeable and committed to upholding the principles of ML data governance.

Training and Development for Data Governance

Training and development are critical in cultivating an organizational culture that upholds ML data governance. Regular and structured training programs empower employees to thoroughly understand the principles of data governance, the ethical implications of AI, and the necessity of adhering to regulatory standards. These programs are instrumental in embedding the organization's core values, such as data accuracy, privacy, and fairness, into the daily practices of its workforce. By consistently emphasizing these elements, training initiatives help establish a strong foundation for a culture that supports robust data governance.

A culture that emphasizes continuous learning is vital for equipping employees with the skills and knowledge required to manage the complexities of ML data governance effectively. When employees are regularly exposed to updated training on governance practices, they become more adept at handling the intricacies of data management and the ethical considerations associated with AI. This continuous education ensures that the workforce remains competent and confident in their ability to maintain high standards of data governance, which is crucial for the integrity of ML models.

Ongoing training fosters an environment of innovation within the organization. As employees stay informed about the latest advancements in AI and data governance, they are better positioned to contribute to the organization's ability to adapt to new challenges and seize opportunities in the ever-changing ML landscape. This commitment to continuous learning not only strengthens the organization's governance framework but also encourages a proactive approach to evolving industry standards, ensuring long-term success in a dynamic field.

Organizational Structure and Data Governance

The organizational structure plays a significant role in shaping its data governance culture. In hierarchical organizations, the clear delineation of responsibility and accountability for data governance can facilitate the enforcement of compliance and ensure that governance policies are consistently applied across the board. This clarity can be beneficial for maintaining order and adherence to established protocols. However, such a rigid structure may also inhibit innovation, making it challenging for the organization to adapt swiftly to the fast-evolving ML landscape.

Conversely, organizations with a decentralized or flat structure often benefit from a culture that promotes innovation and collaboration. These structures allow for more flexible and responsive governance practices, enabling teams to experiment and adapt to new developments in ML more readily. However, the lack of a clear hierarchy can lead to difficulties in ensuring that governance policies are uniformly applied across different teams and departments, potentially leading to inconsistencies in data management and compliance.

Given these dynamics, it is crucial for organizations to carefully assess how their structure aligns with their data governance goals. Adjustments may be necessary to strike a balance between fostering innovation and maintaining consistent governance practices. By aligning their organizational structure with their data governance objectives, organizations can create an environment that both supports effective ML data governance and encourages adaptability in a rapidly changing field.

Balancing Innovation and ML Data Governance

One of the key challenges organizations face is finding the right balance between fostering innovation and adhering to the requirements of ML data governance. In cultures that heavily emphasize innovation, there can be a natural inclination to explore the full potential of machine learning and AI technologies. However, this push for cutting-edge advancements sometimes risks overshadowing the importance of governance and compliance. When innovation becomes the primary focus, it may lead to shortcuts or lapses in data management practices, potentially compromising the integrity and ethical use of data.

Despite this tension, it is possible to cultivate a governance culture that harmonizes with innovation. Organizations that excel in this area are those that integrate governance principles directly into their innovation processes. By embedding considerations of ethics and regulatory compliance into the development of new ideas and technologies, these organizations ensure that their pursuit of innovation does not come at the cost of governance standards. This integrated approach allows them to develop ML models that are not only groundbreaking but also responsibly managed and ethically sound.

Fostering a culture that equally values innovation and governance creates an environment where employees feel empowered to explore new ideas and take calculated risks, knowing that there is a robust framework in place to guide their efforts. This balance enables teams to push the boundaries of what is possible with AI and ML, while still adhering to the necessary governance protocols. In such a culture, innovation does not exist in a vacuum but is supported by a foundation of responsible data use and ethical considerations.

By promoting a culture that respects both innovation and governance, organizations can achieve the dual goals of advancing technology and maintaining the highest standards of data integrity and ethics. This approach not only protects the organization from potential risks associated with non-compliance but also positions it as a leader in responsible AI development. In doing so, the organization can foster sustainable innovation that benefits both the company and its broader community of stakeholders.

Compliance and Regulatory Adherence

Compliance with regulations such as GDPR, CCPA, and HIPAA is a crucial element of ML data governance, and the organizational culture significantly influences how these regulations are prioritized and implemented. In organizations where compliance is seen as a strategic priority, there is a strong commitment to understanding and adhering to regulatory requirements. This approach fosters the creation of robust data governance policies that ensure compliance, while also encouraging employees to stay informed about any changes or updates to regulations.

On the other hand, when compliance is viewed as a burden or an afterthought, the organization is at a greater risk of regulatory violations, which can lead to severe legal and financial repercussions. Therefore, cultivating a culture that values compliance is essential for minimizing risk and ensuring that ML data governance practices are fully aligned with legal and regulatory standards. This proactive approach not only protects the organization from potential penalties but also reinforces its commitment to ethical data management.

Data Privacy and Security in ML Data Governance

Data privacy is a crucial aspect of ML data governance, and the organizational culture significantly influences how privacy is handled. In organizations where data privacy is a priority, there is a strong commitment to safeguarding individuals' personal information and ensuring it is used responsibly. This commitment is evident in the organization's data governance policies, which enforce strict guidelines for data collection, storage, and sharing. Employees are trained to understand the importance of data privacy and are encouraged to adhere to best practices for protecting sensitive information.

A culture that values data privacy not only ensures compliance with ethical standards and legal requirements but also fosters transparency with customers and stakeholders. This transparency helps build trust and confidence in the organization's ML systems, as stakeholders feel assured that their data is being handled with the utmost care. By embedding a strong privacy-focused culture, organizations can reduce the risk of data breaches and other privacy-related incidents, thereby reinforcing their commitment to responsible data governance.

Accountability in ML Data Governance

Accountability is a key principle of ML data governance, and organizational culture plays a crucial role in promoting accountability at all levels of the organization. In a culture of accountability, employees are encouraged to take ownership of their actions and decisions, particularly when it comes to data management and the use of ML models. This culture is supported by clear policies and procedures that outline the roles and responsibilities of different stakeholders in the governance process. Employees are held accountable for following these policies and for ensuring that their work aligns with the organization's governance objectives. A culture of accountability also encourages transparency, as employees are expected to document their actions and decisions and to report any issues or concerns. This culture helps to create a strong foundation for ML data governance, ensuring that governance practices are consistently applied and that any deviations are quickly identified and addressed.

Promoting fairness and diversity in ML is a key aspect of data governance, and organizational culture plays a significant role in achieving this goal. In a culture that values fairness and diversity, there is a strong commitment to developing ML models that are free from bias and that reflect the diversity of the organization's customers and stakeholders. This culture is reflected in the organization's data governance policies, which include guidelines for ensuring that data used in ML models is representative and that models are regularly audited for bias. Employees in such organizations are encouraged to consider the ethical implications of their work and to take steps to address any potential biases in their models. A culture of fairness and diversity also promotes collaboration and inclusivity, as employees from different backgrounds and perspectives are encouraged to contribute to the development and governance of ML models. By fostering a culture that values fairness and diversity, organizations can ensure that their ML data governance practices are aligned with ethical principles and that their models are equitable and inclusive.

Data security is a critical component of ML data governance, and organizational culture plays a key role in determining how seriously security is taken. In a culture that prioritizes data security, there is a strong commitment to protecting data from unauthorized access, breaches, and other security threats. This culture is reflected in the organization's data governance policies, which include strict guidelines for data encryption, access control, and incident response. Employees in such organizations are trained to recognize and respond to security threats, and they are encouraged to follow best practices for data protection. A culture of security also promotes vigilance and proactivity, as employees are expected to stay informed about emerging threats and to take steps to mitigate risks. By fostering a culture that values data security, organizations can ensure that their ML data governance practices are aligned with best practices and that their data is protected from threats.

Responsible AI and Continuous Improvement

Responsible AI use is a key aspect of ML data governance, and organizational culture plays a significant role in promoting responsible AI practices. In a culture that values responsibility, there is a strong commitment to developing and deploying ML models in a way that aligns with ethical principles and societal values. This culture is reflected in the organization's data governance policies, which include guidelines for ensuring that ML models are used in a way that is fair, transparent, and accountable. Employees in such organizations are encouraged to consider the potential impact of their work on society and to take steps to mitigate any negative consequences. A culture of responsibility also promotes continuous learning and improvement, as employees are encouraged to stay informed about the latest developments in AI ethics and to incorporate best practices into their work. By fostering a culture that values responsible AI use, organizations can ensure that their ML data governance practices are aligned with ethical standards and that their models are used in a way that benefits society.

Continuous improvement is a key principle of ML data governance, and organizational culture plays a crucial role in promoting a culture of continuous improvement. In a culture that values continuous improvement, there is a strong commitment to regularly reviewing and refining data governance practices to ensure that they remain effective and aligned with organizational goals. This culture is reflected in the organization's data governance policies, which include guidelines for monitoring and assessing the performance of ML models and for making adjustments as needed. Employees in such organizations are encouraged to provide feedback on governance practices and to suggest improvements based on their experiences. A culture of continuous improvement also promotes innovation, as employees are encouraged to experiment with new approaches to governance and to stay informed about emerging trends and technologies. By fostering a culture that values continuous improvement, organizations can ensure that their ML data governance practices remain effective and that they are able to adapt to changing circumstances.

Change Management in ML Data Governance

Managing change is a significant challenge in ML data governance, and the effectiveness of this management is deeply influenced by organizational culture. In organizations that prioritize adaptability and resilience, there is a strong focus on handling change in ways that minimize disruption while ensuring the continuity of governance practices. This commitment to smooth transitions is embedded in the organization's data governance policies, which provide clear guidelines for managing various changes, whether they involve updates to regulations, the adoption of new technologies, or shifts in organizational structure.

In such a culture, employees are encouraged to be flexible and embrace change as a part of their professional responsibilities. They are provided with the necessary training and resources to adapt to new circumstances, which empowers them to maintain high standards in data governance even as the environment evolves. This proactive approach to change management helps ensure that employees are not only prepared for change but are also active participants in navigating it, which enhances the overall resilience of the governance framework.

A culture that values change management emphasizes the importance of communication and collaboration. Employees are encouraged to work together, share their experiences, and offer insights on how best to handle changes in the ML governance landscape. This collaborative environment fosters a sense of collective responsibility and ensures that the organization can respond effectively to new challenges and opportunities. By embedding these values into the organizational culture, companies can maintain robust ML data governance practices that are resilient in the face of change.

Data Stewardship and Transparency

Data stewardship is a key component of ML data governance, and organizational culture plays a crucial role in promoting effective data stewardship practices. In a culture that values data stewardship, there is a strong commitment to the responsible management and protection of data throughout its lifecycle. This culture is reflected in the organization's data governance policies, which include guidelines for data collection, storage, and sharing, as well as for ensuring data quality and integrity. Employees in such organizations are trained to understand the importance of data stewardship and are encouraged to take ownership of their role in maintaining data quality and security. A culture of data stewardship also promotes collaboration and accountability, as employees are expected to work together to ensure that data is managed in a way that aligns with organizational goals and ethical principles. By fostering a culture that values data stewardship, organizations can ensure that their ML data governance practices are aligned with best practices and that their data is managed in a way that supports their strategic objectives.

Transparency is a key principle of ML data governance, and organizational culture plays a significant role in promoting transparency in governance practices. In a culture that values transparency, there is a strong commitment to open communication and information sharing, both within the organization and with external stakeholders. This culture is reflected in the organization's data governance policies, which include guidelines for documenting and reporting on data governance practices, such as how data is collected, processed, and used in ML models. Employees in such organizations are encouraged to be transparent in their work, and they are provided with the tools and resources they need to share information effectively. A culture of transparency also promotes trust and accountability, as employees are expected to be honest and open about their actions and decisions. By fostering a culture that values transparency, organizations can ensure that their ML data governance practices are aligned with ethical standards and that they are able to build trust with stakeholders.

Data-Driven Decision-Making and Collaboration

Data-driven decision-making is a key aspect of ML data governance, and organizational culture plays a significant role in promoting data-driven decision-making practices. In a culture that values data-driven decision-making, there is a strong commitment to using data as the basis for making informed and objective decisions. This culture is reflected in the organization's data governance policies, which include guidelines for ensuring that data used in decision-making is accurate, reliable, and up-to-date. Employees in such organizations are encouraged to use data to inform their decisions, and they are provided with the tools and resources they need to access and analyze data effectively.

A culture of data-driven decision-making also promotes innovation and agility, as employees are encouraged to experiment with new approaches and to use data to identify opportunities and risks. By fostering a culture that values data-driven decision-making, organizations can ensure that their ML data governance practices are aligned with best practices and that their decisions are based on sound data and analysis.

Collaboration is a key principle of ML data governance, and organizational culture plays a significant role in promoting collaboration in governance practices. In a culture that values collaboration, there is a strong commitment to working together across departments and teams to achieve common goals. This culture is reflected in the organization's data governance policies, which include guidelines for cross-functional collaboration, such as how to share data and insights, how to coordinate on data governance initiatives, and how to resolve conflicts. Employees in such organizations are encouraged to collaborate with others, and they are provided with the tools and resources they need to work together effectively. A culture of collaboration also promotes innovation and creativity, as employees are encouraged to share ideas and perspectives and to work together to solve complex problems. By fostering a culture that values collaboration, organizations can ensure that their ML data governance practices are aligned with best practices and that they are able to leverage the collective expertise of their employees.

Building a Governance-Focused Organizational Culture

Building a governance-focused organizational culture is a deliberate process that requires a clear vision, strategic planning, and ongoing effort. It involves creating an environment where governance is valued, understood, and integrated into the daily operations of the organization. A governance-focused culture not only supports the implementation of ML data governance but also ensures its sustainability over time.

The first step in building a governance-focused culture is to define the core values that support governance and accountability in ML systems. These values should reflect the organization's commitment to ethical AI use, data privacy, security, and compliance with regulatory requirements. Once these values are established, they should be communicated clearly to all members of the organization and incorporated into the organization's mission statement, policies, and practices. Defining and promoting these values helps create a shared understanding of the importance of governance and sets the foundation for a culture that prioritizes it.

Effective communication is essential for fostering a governance-focused culture. This involves developing communication strategies that raise awareness of ML data governance and its significance within the organization. Communication should be ongoing and should include information about governance policies, best practices, and the role of employees in upholding governance standards. Regular updates, newsletters, and town hall meetings can be used to keep governance top of mind and ensure that all employees are informed about the latest developments and expectations related to ML data governance.

Training and education are critical components of building a governance-focused culture. Employees need to be equipped with the knowledge and skills necessary to adhere to governance policies and contribute to the organization's governance efforts. This can be achieved through targeted training programs that cover topics such as data privacy, ethical AI use, compliance, and risk management. Training should be tailored to the specific roles and responsibilities of employees, ensuring that they understand how governance applies to their work. In addition to formal training, ongoing education and development opportunities should be provided to keep employees informed about new governance trends and best practices.

Incentivizing compliance and adherence to governance policies is an effective way to reinforce a governance-focused culture. Reward systems can be implemented to recognize and reward employees who demonstrate a commitment to governance and contribute to the organization's governance objectives. This could include recognition programs, bonuses, or other incentives that motivate employees to prioritize governance in their work. By linking rewards to governance outcomes, organizations can encourage employees to take ownership of governance practices and strive for continuous improvement.

In some cases, building a governance-focused culture may require a cultural transformation. This involves a comprehensive effort to shift the organization's culture from one that may not prioritize governance to one that does. Cultural transformation requires strong leadership, clear communication, and a commitment to change at all levels of the organization. It may involve redefining the organization's values, revising policies and procedures, and implementing new practices that support governance. Cultural transformation is a long-term process, but it is essential for creating a culture that supports and sustains ML data governance.

Framework for Building a Governance-Focused Organizational Culture

Creating a governance-focused organizational culture requires a systematic and intentional approach. The following framework provides a detailed, step-by-step guide to building such a culture, ensuring that governance is deeply embedded in the organization's values, communication, training, incentives, and overall practices. This approach will help organizations develop and maintain a culture that supports the effective and sustainable implementation of ML data governance.

Step 1: Define Core Values that Support Governance

1. **Identify Key Governance Values:** Start by identifying the core values that align with the principles of governance and accountability in ML systems. These should include ethical AI use, data privacy, security, compliance with regulatory requirements, transparency, and accountability.

2. **Align Values with Organizational Mission:** Ensure that these governance values are aligned with the broader mission and vision of the organization. This alignment will help in integrating these values seamlessly into the organization's culture.

3. **Document the Values:** Clearly document these values in formal organizational documents such as the mission statement, code of conduct, and strategic plans. This documentation will serve as a reference point for all governance-related initiatives.

4. **Incorporate Values into Policies:** Embed these values into the organization's policies and procedures. For example, create policies that mandate ethical AI practices, strict data privacy measures, and adherence to regulatory standards.

Step 2: Communicate the Values Organization Wide

1. **Develop a Communication Plan:** Create a comprehensive communication plan to disseminate the governance values throughout the organization. The plan should include key messages, communication channels, and a timeline for rolling out the values.

2. **Use Multiple Channels:** Leverage multiple communication channels such as emails, newsletters, intranet, town hall meetings, and workshops to communicate these values. This multichannel approach ensures that all employees, regardless of their role or location, receive the message.

3. **Leadership Involvement:** Ensure that leadership is actively involved in communicating these values. When leaders consistently talk about and demonstrate these values, it reinforces their importance and encourages employees to adopt them.

4. **Ongoing Communication:** Maintain regular communication to keep governance values top of mind. Regular updates, reminders, and discussions about the importance of governance should be a continuous part of organizational communication.

Step 3: Integrate Values into Everyday Practices

1. **Embed Values in Decision-Making:** Incorporate governance values into daily decision-making processes. For example, when developing new ML models, ensure that ethical considerations and data privacy are key factors in the decision-making process.

2. **Governance Committees:** Establish governance committees or working groups that are responsible for ensuring that these values are upheld in all organizational practices. These committees should have representatives from different departments to ensure a holistic approach.

3. **Policy Implementation:** Ensure that the policies reflecting governance values are implemented across all levels of the organization. This might involve revising existing procedures to align with the new governance-focused policies.

4. **Integrate into Performance Metrics:** Align performance metrics with governance values. For example, include adherence to governance policies as a key performance indicator (KPI) in employee evaluations.

Step 4: Provide Training and Education

1. **Conduct a Training Needs Assessment:** Begin by assessing the current knowledge and skills of employees regarding ML data governance. This will help in identifying gaps and tailoring training programs to meet specific needs.

2. **Develop Role-Specific Training Programs:** Create targeted training programs that are specific to different roles within the organization. For example, data scientists might need training on ethical AI practices, while legal teams might require training on compliance with data protection regulations.

3. **Use Varied Training Methods:** Employ a variety of training methods such as workshops, e-learning modules, webinars, and simulations to cater to different learning preferences and ensure comprehensive coverage of governance topics.

4. **Provide Ongoing Education:** Implement continuous education opportunities to keep employees updated on the latest governance trends, best practices, and regulatory changes. This could include regular workshops, access to online courses, and subscriptions to relevant publications.

5. **Evaluate Training Effectiveness:** Regularly assess the effectiveness of training programs through feedback, assessments, and performance evaluations. Use this data to make necessary adjustments and improvements to the training content and delivery methods.

Step 5: Incentivize Compliance and Adherence to Governance Policies

1. **Design a Recognition Program:** Develop a recognition program that rewards employees who consistently adhere to governance policies and contribute to the organization's governance objectives. This could include awards, public recognition, or special incentives.

2. **Link Rewards to Governance Outcomes:** Tie bonuses, promotions, and other rewards to the achievement of governance-related goals. For example, employees who successfully implement data privacy measures or improve compliance might receive financial incentives.

3. **Create a Governance Champion Role:** Identify and empower governance champions within the organization who advocate for governance practices and help motivate their peers to adhere to these standards. These champions can play a crucial role in driving cultural change from within.

4. **Encourage Peer Recognition:** Foster a culture where employees recognize and appreciate each other's contributions to governance. Peer recognition programs can be a powerful tool in reinforcing governance-focused behaviors.

Step 6: Implement Cultural Transformation Initiatives

1. **Assess Current Culture:** Begin by conducting a thorough assessment of the current organizational culture to understand how governance is currently perceived and practiced. This could involve surveys, focus groups, and interviews with employees across different levels.

2. **Engage Leadership in Transformation:** Secure commitment from leadership to drive the cultural transformation. Leaders should be visible champions of the change, consistently demonstrating and reinforcing governance values in their actions and decisions.

3. **Redefine Organizational Values:** If necessary, redefine the organization's values to place a stronger emphasis on governance. This might involve revising the mission statement, updating the code of conduct, and communicating these changes organization wide.

4. **Revise Policies and Procedures:** Review and update existing policies, procedures, and processes to align with the new governance-focused values. This might include introducing new policies, revising outdated ones, and ensuring that all procedures reflect the organization's commitment to governance.

5. **Implement Change Management Strategies:** Use change management strategies to facilitate the cultural transformation. This could include creating a change management team, developing a detailed change management plan, and providing support to employees as they adapt to new ways of working.

6. **Monitor and Sustain the Transformation:** Establish mechanisms to monitor the progress of the cultural transformation and sustain the momentum over time. This might involve regular check-ins, cultural audits, and adjustments to the transformation strategy based on feedback and observed outcomes.

Change Management Strategies for ML Data Governance

Change management is a critical component of any successful machine learning (ML) data governance initiative. The rapid evolution of ML technologies, coupled with the increasing complexity of data ecosystems, requires organizations to be agile in adapting to new challenges and opportunities. Effective change management strategies ensure that these adaptations occur smoothly, minimizing disruptions while maximizing the benefits of the changes. In the context of ML data governance, change management involves aligning people, processes, and technology to meet the evolving needs of the organization. This includes addressing changes in regulatory requirements, data governance frameworks, and ML model deployment practices. A well-executed change management strategy helps organizations maintain control over their data governance processes, even as they navigate the complexities of an ever-changing technological landscape. The success of ML data governance depends not only on the robustness of the governance framework but also on the organization's ability to manage change effectively.

Leadership is a pivotal factor in driving successful change management in ML data governance. Leaders set the vision and direction for the organization, and their commitment to change management can inspire and motivate employees to embrace new processes and technologies. In the context of ML data governance, leadership must actively promote the importance of governance practices and the need for change to maintain compliance, enhance data quality, and ensure ethical AI use. Leaders are responsible for communicating the strategic importance of these changes, explaining how they align with the organization's goals and values. Additionally, leadership must model the behaviors they expect from employees, demonstrating a commitment to data governance and a willingness to adapt to new methodologies. By fostering a culture of openness and collaboration, leaders can create an environment where change is not feared but embraced as an opportunity for growth and improvement in ML data governance.

Employee engagement is a crucial aspect of change management in ML data governance. For change to be effective, employees at all levels must be involved in the process, understanding not only what changes are being made but also why these changes are necessary. Engaging employees early in the change process helps to build trust and reduce resistance, as it gives them a sense of ownership and responsibility for the success of the initiative. In the context of ML data governance, this might involve including data scientists, analysts, and IT professionals in the development and implementation of new governance policies. Providing training and resources to help employees understand and adapt to new data governance practices is also essential. When employees feel supported and empowered, they are more likely to embrace change and contribute positively to the organization's ML data governance objectives. Effective employee engagement strategies ensure that the transition to new governance practices is smooth and that the organization can maintain its commitment to data integrity and ethical AI use.

Clear and consistent communication is vital to the success of change management initiatives in ML data governance. Organizations must develop communication strategies that effectively convey the rationale behind changes, the benefits these changes will bring, and the impact they will have on employees and business processes. Communication should be tailored to different audiences within the organization, ensuring that each group receives the information most relevant to their roles and responsibilities. For example, data scientists may need detailed information about changes to data handling protocols, while executives may require high-level overviews of the strategic benefits of enhanced data governance. Additionally, communication should be ongoing, with regular updates provided throughout the change process to keep employees informed and engaged. Utilizing a variety of communication channels, such as meetings, emails, and internal portals, can help ensure that the message reaches everyone in the organization. Effective communication not only helps to reduce uncertainty and resistance but also fosters a sense of collaboration and shared purpose in achieving the organization's ML data governance goals.

Training and development are essential components of change management in ML data governance. As organizations implement new governance frameworks and processes, it is crucial to provide employees with the skills and knowledge they need to adapt to these changes. This includes training on new data governance policies, tools, and technologies, as well as ongoing education on emerging trends and best practices in ML and AI. Training programs should be designed to address the specific needs of different employee groups, ensuring that everyone, from data scientists to business analysts, understands their role in the new governance framework. In addition to formal training sessions, organizations should also provide opportunities for hands-on learning and practical application of new skills. This could involve workshops, simulations, or pilot projects that allow employees to gain experience with new governance practices in a controlled environment. By investing in training and development, organizations can ensure that their employees are well-prepared to navigate the complexities of ML data governance and that they have the confidence to embrace change.

Stakeholder involvement is critical to the success of change management in ML data governance. Stakeholders, both internal and external, have a vested interest in the organization's data governance practices and must be considered in the change management process. Internal stakeholders include employees across various departments, such as IT, data science, compliance, and management, all of whom play a role in data governance. External stakeholders may include customers, partners, regulators, and shareholders, who rely on the organization to manage data responsibly and comply with relevant regulations. Engaging stakeholders in the change management process involves soliciting their input and feedback, addressing their concerns, and keeping them informed about progress and outcomes. This collaborative approach helps to build trust and ensures that the changes being implemented align with the expectations and requirements of all stakeholders. Effective stakeholder involvement is key to ensuring that ML data governance changes are accepted and supported by all parties, ultimately contributing to the success of the organization's governance framework.

Resistance to change is a common challenge in any change management process, and ML data governance is no exception. Employees may resist changes to governance practices for a variety of reasons, including fear of the unknown, concerns about job security, or discomfort with new technologies and processes. To manage resistance effectively, organizations must first understand the root causes of this resistance and address them directly. This may involve providing additional training, offering reassurances about job stability, or involving employees more closely in the change process to reduce fear and uncertainty. Clear communication is also critical in managing resistance, as it helps to clarify the reasons for the change and the benefits it will bring. Additionally, organizations should identify and engage change champions—employees who are supportive of the changes and can help to influence their peers positively. By proactively managing resistance to change, organizations can reduce disruptions and ensure that their ML data governance initiatives are implemented successfully.

Technology plays a central role in ML data governance, and its adoption is a key aspect of change management. As organizations implement new data governance tools and technologies, they must ensure that employees are able to effectively use these tools to manage data and comply with governance policies. This requires a comprehensive approach to technology adoption, including the selection of user-friendly tools, integration with existing systems, and providing training and support to employees. Organizations should also consider the scalability and flexibility of new technologies, ensuring that they can adapt to future changes in the data governance landscape. Additionally, technology adoption should be aligned with the organization's broader strategic goals, ensuring that it supports not only governance objectives but also business growth and innovation. By carefully managing the adoption of new technologies, organizations can enhance their ML data governance capabilities while minimizing disruptions and ensuring that employees are fully equipped to leverage these tools.

A well-structured change management plan is essential for the successful implementation of ML data governance initiatives. This plan should outline the objectives of the change, the steps required to achieve these objectives, and the roles and responsibilities of all stakeholders involved. It should also include a timeline for implementation, with clear milestones and checkpoints to monitor progress. Risk management is another critical component of the change management plan, as it helps to identify potential challenges and develop strategies to mitigate them. The plan should also address communication and training needs, ensuring that all employees are informed and prepared for the changes ahead. Additionally, the change management plan should include metrics for evaluating the success of the initiative, allowing the organization to assess the impact of the changes and make adjustments as needed. By developing a comprehensive change management plan, organizations can ensure that their ML data governance initiatives are implemented smoothly and effectively, with minimal disruption to business operations.

Flexibility is a crucial attribute of successful change management in ML data governance. Given the dynamic nature of the ML landscape, organizations must be prepared to adapt their change management strategies as new challenges and opportunities arise. This requires a flexible approach to planning and implementation, allowing for adjustments to be made as needed. For example, if an unexpected regulatory change occurs during the implementation of a new data governance framework, the organization must be able to quickly reassess its approach and make the necessary modifications. Flexibility also involves being open to feedback from employees and stakeholders and being willing to make changes to the plan based on this input. By adopting a flexible approach to change management, organizations can better navigate the complexities of ML data governance and ensure that their initiatives remain relevant and effective in a rapidly changing environment.

Measuring the success of change management initiatives is essential for ensuring that ML data governance goals are met. Organizations should establish clear metrics and key performance indicators (KPIs) that align with their governance objectives, such as improvements in data quality, compliance rates, or employee adoption of new governance practices. These metrics should be monitored throughout the change process to assess progress and identify areas that may require additional attention. Regular evaluations and feedback loops can help organizations to make data-driven decisions about the effectiveness of their change management strategies and to adjust their approach as needed. Additionally, organizations should consider conducting postimplementation reviews to evaluate the overall impact of the change management initiative and to identify lessons learned for future projects. By systematically measuring the success of their change management efforts, organizations can ensure that their ML data governance initiatives deliver the desired outcomes and contribute to the long-term success of the organization.

Sustaining change is a critical aspect of change management in ML data governance. Once new governance practices have been implemented, organizations must ensure that these changes are maintained and integrated into the daily operations of the business.

This requires ongoing support and reinforcement, including continued training, regular communication, and monitoring of compliance with new policies. Organizations should also establish mechanisms for continuous improvement, allowing them to refine and enhance their governance practices over time. This might involve setting up a governance committee or task force to oversee the implementation and sustainability of the changes, as well as to address any issues that arise. Additionally, organizations should recognize and reward employees who contribute to the success of the change initiative, helping to reinforce positive behaviors and ensure long-term adherence to new governance practices. By focusing on sustaining change, organizations can ensure that their ML data governance initiatives continue to deliver value and support the organization's strategic goals.

Organizational culture plays a significant role in the success of change management for ML data governance. The culture of an organization influences how employees perceive and respond to change, and it can either facilitate or hinder the adoption of new governance practices. In organizations with a culture that values innovation, collaboration, and continuous improvement, change management initiatives are more likely to be successful, as employees are generally more open to new ideas and approaches. Conversely, in organizations with a more rigid or risk-averse culture, resistance to change may be higher, making it more challenging to implement new governance practices. To address cultural considerations, organizations should assess their current culture and identify any potential barriers to change. This might involve promoting a culture of learning and adaptability, encouraging open communication, and fostering a sense of shared responsibility for the success of the change initiative. By aligning change management strategies with the organization's culture, organizations can increase the likelihood of successful implementation of ML data governance initiatives.

Continuous improvement is a key principle of effective change management in ML data governance. Organizations must recognize that change is not a one-time event but an ongoing process that requires regular review and refinement. This involves continuously monitoring the effectiveness of governance practices, gathering feedback from employees and stakeholders, and making adjustments as needed. Continuous improvement also involves staying informed about emerging trends and best practices in ML and data governance, ensuring that the organization remains at the forefront of the field. By fostering a culture of continuous improvement, organizations can ensure that their data governance practices evolve in response to new challenges and opportunities, and that they continue to support the organization's strategic objectives. This proactive approach to change management helps organizations to maintain agility and resilience in a rapidly changing technological landscape, ensuring long-term success in ML data governance.

Change Management Frameworks

Implementing such governance practices often requires significant changes to organizational structures, processes, and cultures. Change management frameworks like Kotter's 8-Step Change Model, the ADKAR model, and Lewin's Change Management Model provide structured approaches to managing these changes.

Kotter's 8-Step Change Model

John Kotter's 8-Step Change Model is one of the most widely recognized frameworks for managing organizational change. It provides a step-by-step approach to leading change, from creating urgency to anchoring new practices in the organization's culture.

Step 1: Create a Sense of Urgency

The first step in Kotter's model is to create a sense of urgency about the need for change. In the context of ML data governance, this might involve highlighting the risks of poor data governance, such as regulatory penalties, data breaches, or biased AI models that can damage the organization's reputation.

A financial institution might conduct a risk assessment revealing vulnerabilities in its data governance processes, such as inadequate encryption or lack of audit trails. By presenting these findings to leadership and emphasizing the potential consequences, the organization can create a sense of urgency that motivates stakeholders to take immediate action.

Step 2: Build a Guiding Coalition

The second step is to build a coalition of influential leaders and stakeholders who support the change. For ML data governance, this coalition should include representatives from data science, IT, legal, compliance, and business leadership.

A healthcare organization might form a data governance task force that includes data scientists, compliance officers, and senior executives. This group would be responsible for developing governance policies, advocating for their implementation, and ensuring that all departments are aligned with the new governance standards.

Step 3: Develop a Vision and Strategy

A clear vision and strategy are essential for guiding the change effort. For ML data governance, this might involve defining what effective governance looks like, such as ensuring data accuracy, enhancing data security, and maintaining transparency in AI models.

An e-commerce company might establish a vision to become a leader in ethical AI by implementing a comprehensive data governance framework that prioritizes customer privacy and data security. The strategy would include specific actions like adopting new encryption technologies, conducting regular audits, and training employees on data governance best practices.

Step 4: Communicate the Vision

Effective communication is critical to ensuring that all stakeholders understand and support the change. This step involves regularly communicating the vision and strategy for ML data governance through various channels.

The e-commerce company could launch a communication campaign that includes newsletters, webinars, and town hall meetings to explain the importance of data governance, outline the new policies, and answer employees' questions.

Step 5: Empower Broad-Based Action

Removing obstacles and empowering employees to take action is crucial for making progress. In ML data governance, this might involve providing the necessary tools, resources, and training to employees so they can adhere to governance policies effectively.

The organization might implement a data governance platform that automates compliance checks and provides users with real-time insights into data usage. Additionally, training sessions could be held to ensure employees are comfortable using the new tools and understand their role in maintaining data governance.

Step 6: Generate Short-Term Wins

Achieving short-term wins helps to build momentum and demonstrate the value of the change. For ML data governance, these wins might include successfully implementing a new data governance policy or achieving compliance with a regulatory requirement.

The organization might roll out a pilot project in one department to implement a new data classification system. If successful, this short-term win can be showcased to the rest of the organization as proof of the benefits of robust data governance.

Step 7: Consolidate Gains and Produce More Change

To sustain momentum, it's important to build on the short-term wins and continue driving change. This might involve expanding the data governance initiative to other departments or refining policies based on feedback and lessons learned.

After the successful pilot, the organization could expand the data classification system across all departments, integrating additional features such as automated data masking for sensitive information. Continuous improvement efforts would ensure the governance framework evolves with the organization's needs.

Step 8: Anchor New Approaches in the Culture

For change to be sustained, it must become part of the organizational culture. In ML data governance, this means embedding governance practices into the everyday operations of the organization and ensuring that all employees understand and value the importance of governance.

The organization could include data governance metrics in employee performance reviews, recognize teams that excel in adhering to governance standards, and continuously reinforce the importance of governance through ongoing training and communication.

The ADKAR Model

The ADKAR model, developed by Prosci, is another popular change management framework that focuses on individual change. It outlines five key elements—awareness, desire, knowledge, ability, and reinforcement—that are essential for successful change at the individual level. This model is particularly useful in ML data governance initiatives, where individual behavior and adherence to policies play a crucial role.

Awareness

The first step is to create awareness of the need for change. In ML data governance, this involves educating employees about the importance of data governance, the risks of non-compliance, and the benefits of following best practices.

An organization might launch an awareness campaign highlighting recent data breaches in the industry and the potential consequences of poor data governance, thus making employees aware of the stakes involved.

Desire

Once awareness is established, it's important to build the desire for change. This can be achieved by addressing employees' concerns, demonstrating the personal and professional benefits of adopting governance practices, and gaining their buy-in.

The organization could engage employees in discussions about how data governance can protect their work, reduce errors, and enhance the overall quality of ML models, thus creating a desire to participate in the change.

Knowledge

Providing employees with the knowledge they need to implement change is crucial. For ML data governance, this involves training programs, workshops, and access to resources that teach employees how to apply governance policies effectively.

The organization might develop a series of training modules on topics like data privacy laws, ethical AI practices, and how to use new governance tools, ensuring that all employees have the necessary knowledge to comply with governance standards.

Ability

Knowledge alone is not enough; employees must also have the ability to implement change. This means ensuring they have the right tools, support, and resources to put their knowledge into practice.

The organization could provide hands-on workshops where employees can practice using governance tools in real-world scenarios, thereby building their ability to apply what they've learned.

Reinforcement

Finally, reinforcement is necessary to ensure that the change sticks. This involves recognizing and rewarding employees who adhere to governance practices, providing continuous feedback, and addressing any issues that arise.

The organization could establish a recognition program that rewards teams who consistently demonstrate compliance with data governance policies, reinforcing the importance of these practices over time.

Lewin's Change Management Model

Kurt Lewin's Change Management Model is one of the oldest and most well-known change management frameworks. It describes change as a three-step process: unfreeze, change, and refreeze. This model is particularly useful in ML data governance because it emphasizes the need to prepare the organization for change, implement the change, and then solidify the change to make it permanent.

Unfreeze

The first stage, unfreeze, involves preparing the organization for change by breaking down the existing status quo. This might involve challenging existing beliefs, behaviors, and practices that are not aligned with effective ML data governance.

In an organization where data governance has not been a priority, the unfreeze stage might involve highlighting the risks of non-compliance and the potential benefits of a strong governance framework. This could include presentations, workshops, and discussions that encourage employees to think critically about the current state of data governance and why change is necessary.

Change

The change stage involves implementing the new practices, processes, and behaviors that are aligned with effective ML data governance. This might involve introducing new data governance policies, tools, and technologies, as well as providing training and support to employees.

The organization could roll out a new data governance policy that includes stricter controls over data access, improved data classification systems, and regular audits of ML models. Training sessions would be held to ensure that all employees understand the new policies and how to implement them.

Refreeze

The final stage, refreeze, involves solidifying the changes to make them a permanent part of the organization's culture. This might involve embedding data governance practices into everyday operations, providing ongoing support and resources, and recognizing and rewarding employees who adhere to the new practices.

To refreeze the changes, the organization might integrate data governance metrics into employee performance reviews, establish a data governance committee to oversee ongoing compliance, and provide continuous training and updates on governance best practices. Recognizing and rewarding employees who excel in data governance would also help to reinforce the new behaviors.

Applying Change Management Frameworks to ML Data Governance Initiatives

Implementing ML data governance is a complex process that requires careful planning and execution. Change management frameworks like Kotter's 8-Step Change Model, the ADKAR model, and Lewin's Change Management Model provide structured approaches to managing the changes required to establish and maintain effective data governance.

Example 1: Implementing a New Data Classification System

An organization that wants to implement a new data classification system as part of its ML data governance initiative could use Kotter's 8-Step Change Model to guide the process. The organization would start by creating a sense of urgency around the need for better data classification to reduce errors and improve compliance. A guiding coalition would be formed, including representatives from IT, data science, legal, and compliance. The organization would then develop a clear vision and strategy for the new data classification system, communicate the vision to all stakeholders, and empower employees to take action by providing the necessary tools and training. Short-term wins, such as successful pilot projects, would be used to build momentum, and the system would be gradually expanded across the organization. Finally, the new data classification practices would be anchored in the organization's culture through ongoing training, communication, and recognition.

Example 2: Enhancing Data Privacy Practices

An organization that wants to enhance its data privacy practices as part of its ML data governance initiative could use the ADKAR model. The organization would start by creating awareness of the importance of data privacy and the risks of non-compliance with regulations like GDPR. Next, the organization would build the desire for change by engaging employees in discussions about the benefits of enhanced data privacy practices, such as protecting customer trust and reducing the risk of data breaches.

Employees would then be provided with the knowledge they need to implement the changes, including training on data privacy laws and best practices. The organization would ensure that employees have the ability to implement the changes by providing the necessary tools, support, and resources. Finally, the organization would reinforce the changes by recognizing and rewarding employees who demonstrate a commitment to data privacy.

Example 3: Transforming Data Governance Culture

An organization that wants to transform its data governance culture could use Lewin's Change Management Model. The organization would start by unfreezing the current culture, challenging existing beliefs and behaviors that do not align with effective data governance. This might involve highlighting the risks of non-compliance, the benefits of strong data governance, and the need for a cultural shift. The organization would then implement the change by introducing new data governance policies, tools, and technologies, as well as providing training and support to employees. Finally, the organization would refreeze the changes by embedding data governance practices into everyday operations, providing ongoing support and resources, and recognizing and rewarding employees who adhere to the new practices.

Integrating ML Data Governance into Organizational Processes

Integrating machine learning (ML) data governance into organizational processes is an essential step for ensuring the responsible and effective use of data. This integration starts with understanding the core principles of data governance, which include data quality, privacy, security, and compliance. Organizations must first establish a robust framework that aligns with their strategic objectives and regulatory requirements. This framework should address data stewardship, roles, and responsibilities, ensuring that data governance is not just a theoretical concept but a practical, actionable strategy. Effective data governance involves creating policies and procedures that govern data collection, storage, processing, and dissemination. This ensures that data is accurate, accessible, and secure while meeting legal and ethical standards. For ML models, which rely heavily on data, the governance framework must also account for data integrity and bias, as well as the impact of data-driven decisions on stakeholders. By embedding data governance into organizational processes, companies can mitigate risks, enhance decision-making, and build trust with customers and regulators.

The first step in integrating ML data governance into organizational processes is to establish a data governance committee or team. This team is responsible for developing and enforcing data policies and ensuring that data governance practices are adhered to across the organization. The committee should include stakeholders from various departments, including IT, legal, compliance, and data science, to provide a holistic view of data management needs. It is crucial to define clear roles and responsibilities for each member of the team, ensuring accountability and effective oversight. The committee should also be tasked with regularly reviewing and updating data governance policies to adapt to new regulations, technologies, and business needs. This collaborative approach helps to foster a culture of data stewardship within the organization, ensuring that data governance is integrated into everyday business processes.

Another critical aspect of integrating ML data governance is the implementation of data quality management practices. High-quality data is fundamental to the success of ML models, as poor data quality can lead to inaccurate predictions and biased outcomes. Organizations should establish processes for data validation, cleansing, and enrichment to maintain data accuracy and completeness. This involves setting up automated data quality checks, monitoring data sources, and addressing data anomalies promptly. Additionally, organizations should invest in data lineage tools that track the origin and transformation of data throughout its lifecycle. By maintaining high data quality standards, organizations can ensure that their ML models are trained on reliable data, leading to more accurate and actionable insights.

Data privacy and security are paramount when integrating ML data governance into organizational processes. Organizations must implement stringent measures to protect sensitive data from unauthorized access and breaches. This includes using encryption, access controls, and secure data storage solutions. Additionally, organizations should adhere to data protection regulations such as the General Data Protection Regulation (GDPR) and the California Consumer Privacy Act (CCPA). These regulations require organizations to implement privacy-by-design principles, which involve incorporating data protection measures into the development of ML models and systems. Regular security audits and risk assessments should be conducted to identify and address potential vulnerabilities. By prioritizing data privacy and security, organizations can build trust with customers and avoid legal and financial repercussions.

Compliance with regulatory requirements is a critical component of ML data governance. Organizations must stay informed about relevant laws and regulations that impact data management and ML practices. This includes understanding requirements related to data protection, fairness, and transparency. For instance, the Fair Housing Act and Equal Opportunity Employment Commission regulations mandate non-discrimination in decision-making processes, including those driven by ML models. Organizations should establish processes for ensuring that their ML models comply with these regulations, including conducting regular bias and fairness audits. It is also essential to document compliance efforts and maintain records to demonstrate adherence to regulatory requirements. By integrating compliance into data governance practices, organizations can mitigate legal risks and ensure ethical use of data.

Training and awareness programs are vital for successfully integrating ML data governance into organizational processes. Employees across all levels of the organization need to understand the importance of data governance and their role in maintaining data quality and compliance. Training programs should cover topics such as data privacy, security, and ethical use of data. Additionally, employees should be educated about the specific data governance policies and procedures relevant to their roles. Regular workshops, seminars, and e-learning modules can help keep staff updated on best practices and emerging trends in data governance. By fostering a culture of data literacy and accountability, organizations can enhance their data governance efforts and ensure that employees are equipped to manage data responsibly.

The integration of ML data governance also involves establishing clear data ownership and stewardship practices. Data ownership defines who is responsible for managing and safeguarding data within the organization, while data stewardship involves the ongoing care and maintenance of data assets. Organizations should assign data stewards or custodians to oversee specific data domains and ensure that data governance policies are followed. These individuals should have a deep understanding of data management practices and be empowered to make decisions regarding data usage and access. Clear documentation of data ownership and stewardship roles helps to prevent data misuse and ensures that data governance responsibilities are clearly defined. This structure supports effective data management and enhances the overall quality of data governance practices.

Implementing data governance tools and technologies is another important aspect of integrating ML data governance into organizational processes. Various tools are available to support data governance efforts, including data cataloging, data lineage, and data quality management solutions. These tools help organizations manage and monitor data throughout its lifecycle, providing visibility into data sources, transformations, and usage. Additionally, data governance platforms can automate compliance checks and generate reports to facilitate regulatory adherence. By leveraging these technologies, organizations can streamline their data governance processes and improve the efficiency and effectiveness of their data management practices.

Monitoring and auditing are essential components of an effective ML data governance strategy. Organizations should establish processes for continuously monitoring data usage and model performance to ensure compliance with governance policies. This includes conducting regular audits of data quality, security, and privacy practices. Audits help identify potential issues and areas for improvement, enabling organizations to address them proactively. Additionally, monitoring tools can provide real-time insights into data-related activities and alert organizations to any anomalies or violations. Regular reviews and audits help maintain the integrity of data governance practices and ensure that data management processes remain aligned with organizational goals and regulatory requirements.

The integration of ML data governance into organizational processes also requires a focus on ethical considerations. Organizations should establish ethical guidelines for the use of data and ML models, addressing issues such as fairness, transparency, and accountability. This involves developing policies that promote responsible data use and ensure that ML models do not perpetuate biases or discrimination. Ethical considerations should be integrated into the development and deployment of ML models, with mechanisms in place to address potential ethical dilemmas. By prioritizing ethical data practices, organizations can build trust with stakeholders and demonstrate their commitment to responsible data management.

Data governance in the context of ML also involves establishing clear communication channels and reporting mechanisms. Effective communication ensures that data governance policies and procedures are well understood and followed throughout the organization. Organizations should develop reporting mechanisms for employees to raise concerns or report issues related to data governance. This includes creating channels for reporting data breaches, security incidents, and compliance violations. Transparent communication and reporting processes help to address issues promptly and maintain a strong data governance framework. By fostering open dialogue and encouraging feedback, organizations can enhance their data governance practices and improve overall data management.

Incorporating feedback and continuous improvement is a crucial aspect of integrating ML data governance into organizational processes. Organizations should regularly solicit feedback from employees, stakeholders, and external auditors to identify areas for improvement in their data governance practices. This feedback should be used to refine policies, procedures, and technologies to better align with organizational needs and regulatory requirements. Continuous improvement involves staying updated on emerging trends and best practices in data governance and incorporating them into existing processes. By adopting a culture of continuous improvement, organizations can enhance their data governance efforts and adapt to changing data management challenges.

Integrating ML data governance into organizational processes requires a strategic approach that aligns data governance with overall business objectives. Data governance should be viewed as a strategic enabler that supports the organization's mission and goals. This involves aligning data governance practices with key business priorities, such as customer satisfaction, operational efficiency, and innovation. Organizations should develop a data governance strategy that supports their long-term vision and provides a framework for achieving strategic goals. By integrating data governance into the broader business strategy, organizations can ensure that data management practices contribute to overall success and drive value across the organization.

Integrating ML data governance into organizational processes is a multifaceted endeavor that requires careful planning and execution. By establishing a data governance framework, implementing data quality management practices, prioritizing data privacy and security, and ensuring regulatory compliance, organizations can effectively manage their data assets and mitigate risks. Training and awareness programs, clear data ownership and stewardship, and the use of data governance tools are also essential for successful integration. Continuous monitoring, ethical considerations, and strategic alignment further support the integration of ML data governance into organizational processes. Through a comprehensive approach, organizations can enhance their data management practices, build trust with stakeholders, and drive value through responsible and effective use of data.

Strategies for Integrating ML Data Governance into Organizational Processes

Aligning ML Data Governance with Business Objectives

Strategic Importance of Data Governance

ML data governance should be closely aligned with the organization's broader business objectives to ensure that data practices contribute to achieving strategic goals. Rather than being treated as an isolated function, data governance must be embedded within the organization's strategic framework. This alignment ensures that the governance practices support key business outcomes, such as enhancing customer satisfaction, driving innovation, improving operational efficiency, and maintaining regulatory compliance.

Data governance plays a strategic role by ensuring that the data used for ML is accurate, consistent, and secure. This, in turn, enhances the effectiveness and reliability of ML models, which are often critical to business operations and decision-making. When data governance is aligned with business objectives, it ensures that data-related decisions contribute directly to the organization's success.

Customizing Governance Policies to Business Priorities

Governance policies should be designed to reflect the unique needs and priorities of the organization. This involves setting standards for data quality, defining access controls, and establishing compliance protocols that align with the organization's strategic goals. For instance, organizations that prioritize customer experience may focus on data accuracy and privacy, while those in highly regulated industries may emphasize data security and compliance.

By tailoring data governance policies to business priorities, organizations can ensure that data governance supports their strategic initiatives. This approach also helps in addressing specific challenges that the organization faces, such as regulatory compliance in financial services or data privacy concerns in healthcare.

Leadership Involvement in Governance

Involving business leaders in data governance decision-making is essential for ensuring that governance practices are aligned with organizational objectives. Business leaders bring a strategic perspective that helps in aligning data governance with the company's overall goals. Their involvement ensures that governance initiatives receive the necessary support and resources, making it easier to integrate these practices into the organization's processes.

Leadership involvement also ensures that data governance is seen as a critical business function rather than just an IT responsibility. By championing data governance, business leaders can drive cultural change, making data governance a core part of the organization's strategic planning and execution.

Creating Cross-Functional Teams for Data Governance

Necessity of Cross-Functional Collaboration

Effective ML data governance requires collaboration across various departments, including IT, data science, compliance, legal, and business units. Cross-functional teams are crucial in developing and implementing governance practices that address the diverse needs of the organization. These teams bring together different perspectives and expertise, ensuring that data governance is comprehensive and aligned with the organization's objectives.

Cross-functional collaboration helps in identifying and addressing gaps in data governance, such as inconsistencies in data management practices across departments. It also facilitates the development of governance policies that are practical and applicable across the organization, ensuring that all departments adhere to the same standards.

Defining Roles and Responsibilities

For cross-functional teams to be effective, it is essential to establish clear roles and responsibilities. Each team member should have a specific role, such as data steward, compliance officer, or IT manager, with defined responsibilities related to data governance. Clear roles help in ensuring accountability and make it easier to track progress on governance initiatives.

Well-defined roles also help in avoiding overlaps and conflicts between team members, which can slow down the governance process. By assigning specific responsibilities, organizations can ensure that each aspect of data governance, from data quality management to regulatory compliance, is effectively managed.

Enhancing Communication and Collaboration

Effective communication is critical to the success of cross-functional data governance teams. Regular meetings, shared platforms, and collaborative tools help in ensuring that team members stay aligned and can quickly address any governance challenges. Transparent communication also helps in building trust and fostering a collaborative environment, which is essential for the successful implementation of governance initiatives.

Collaboration tools, such as project management platforms and data governance software, can facilitate communication and coordination among team members. These tools allow teams to track progress, share updates, and collaborate on governance tasks in real time, ensuring that governance initiatives are implemented efficiently.

Using Integrated Data Governance Platforms to Streamline Processes

Importance of Technology in Data Governance

Technology plays a crucial role in streamlining ML data governance processes, particularly in organizations that deal with large volumes of data across multiple systems. Integrated data governance platforms offer a centralized solution for managing data governance tasks, such as data quality management, metadata management, and compliance monitoring. These platforms help in automating governance processes, reducing manual effort, and ensuring consistency in data governance practices across the organization.

Integrated platforms also provide a single source of truth for data governance, making it easier to track and manage data assets. This is particularly important in large organizations with complex data ecosystems, where maintaining data consistency and quality can be challenging. By centralizing data governance tasks, these platforms help in ensuring that governance practices are applied uniformly across the organization.

Streamlining Governance Processes

Integrated data governance platforms can streamline various governance processes, such as data classification, access management, and compliance monitoring. These platforms offer tools for automating data governance tasks, such as identifying and classifying sensitive data, setting access controls, and monitoring compliance with regulations. Automation reduces the risk of human error and ensures that governance practices are consistently applied.

For example, a data governance platform can automate the process of classifying data based on sensitivity levels, ensuring that sensitive data is properly protected. It can also monitor data access in real time, alerting the organization to any unauthorized access attempts. By automating these tasks, organizations can improve the efficiency and effectiveness of their data governance processes.

Ensuring Compliance with Regulations

Compliance with regulations such as GDPR, CCPA, and HIPAA is a critical aspect of ML data governance. Integrated data governance platforms help organizations ensure compliance by providing tools for tracking and managing regulatory requirements. These platforms can automate compliance monitoring, generate audit reports, and ensure that data governance practices align with regulatory standards.

For instance, a data governance platform can track data lineage, ensuring that the organization can demonstrate how data has been processed and used in compliance with regulations. It can also automate the generation of compliance reports, making it easier for the organization to meet regulatory requirements. By providing these tools, integrated platforms help organizations maintain compliance with data protection regulations.

Embedding Data Governance into Organizational Culture

Fostering a Culture of Accountability

Integrating ML data governance into organizational processes requires fostering a culture of accountability. Employees at all levels of the organization should understand the importance of data governance and be held accountable for adhering to governance practices. This involves providing training on data governance, setting clear expectations, and recognizing and rewarding compliance.

Accountability is essential for ensuring that data governance practices are consistently applied across the organization. When employees understand their role in maintaining data governance and are held accountable for their actions, it helps in embedding governance practices into the organizational culture.

Providing Training and Education

Training and education are critical for integrating ML data governance into organizational processes. Employees need to be equipped with the knowledge and skills necessary to adhere to governance practices and contribute to the organization's governance efforts. This involves providing training on data governance principles, regulatory requirements, and the use of governance tools.

Ongoing education is also important for keeping employees informed about new governance trends and best practices. Regular training sessions, workshops, and webinars can help in ensuring that employees stay up-to-date with the latest developments in data governance.

Recognizing and Rewarding Compliance

Recognizing and rewarding employees who adhere to data governance practices is essential for fostering a culture of compliance. This could involve recognizing teams that consistently demonstrate compliance, offering incentives for meeting governance targets, or including data governance metrics in performance reviews.

By recognizing and rewarding compliance, organizations can motivate employees to take data governance seriously and ensure that governance practices are consistently applied. This also helps in reinforcing the importance of data governance within the organizational culture.

Challenges in Integrating ML Data Governance

Resistance to Change

One of the primary challenges in integrating ML data governance into organizational processes is resistance to change. Employees may be resistant to adopting new governance practices, particularly if they perceive them as burdensome or unnecessary. Overcoming this resistance requires clear communication about the importance of data governance, providing adequate training, and involving employees in the governance process.

Resistance to change can be mitigated by demonstrating the benefits of data governance, such as improved data quality, reduced risk of regulatory breaches, and enhanced decision-making. By highlighting these benefits, organizations can help employees see the value of data governance and encourage them to embrace governance practices.

Balancing Governance with Innovation

Another challenge is balancing the need for data governance with the need for innovation. Strict governance practices can sometimes be perceived as stifling innovation, particularly in fast-paced environments where agility is key. Organizations need to strike a balance between ensuring data governance and allowing for flexibility and innovation.

This balance can be achieved by adopting a risk-based approach to data governance, where governance practices are tailored to the level of risk associated with different types of data. For example, stricter governance practices might be applied to sensitive data, while more flexible practices are applied to non-sensitive data. This approach allows organizations to maintain data governance without stifling innovation.

Ensuring Consistency Across the Organization

Ensuring consistency in data governance practices across the organization can be challenging, particularly in large organizations with multiple departments and data sources. Inconsistent governance practices can lead to data quality issues, compliance risks, and inefficiencies.

To address this challenge, organizations should establish standardized governance policies and procedures that apply across the organization. Cross-functional teams can help in ensuring that these policies are consistently applied, while integrated data governance platforms can provide the tools needed to monitor and enforce consistency.

Case Studies: Successful Cultural and Change Management in ML Data Governance

Examining real-world case studies of organizations that have successfully implemented cultural and change management strategies for ML data governance provides valuable insights and lessons that can be applied to other organizations. These case studies highlight the challenges faced, the strategies used to overcome them, and the outcomes achieved.

Case Study 1: Cultural Transformation

XYZ Corp, a global technology company, recognized the need to strengthen its ML data governance practices in response to increasing regulatory scrutiny and ethical concerns. The company embarked on a cultural transformation initiative to align its organizational culture with governance objectives. This involved redefining the company's values to prioritize ethics, accountability, and transparency in AI use. Leadership played a key role in driving this transformation, with senior executives championing the importance of governance and modeling the desired behaviors. The company also implemented a comprehensive training program to educate employees about governance and provide them with the skills needed to adhere to governance policies. As a result of these efforts, XYZ Corp successfully embedded governance into its organizational culture, leading to improved compliance, reduced risk, and enhanced trust among stakeholders.

The Challenge: Navigating Regulatory and Ethical Pressures

XYZ Corp was increasingly under pressure from external regulators and internal stakeholders to ensure that its ML models adhered to high standards of ethics and compliance. The company faced several key challenges:

1. **Regulatory Scrutiny:** Governments and regulatory bodies worldwide were tightening regulations around data privacy, AI ethics, and transparency. XYZ Corp needed to ensure that its ML practices complied with a complex web of global regulations, including GDPR in Europe and CCPA in California. Failure to comply could result in significant legal and financial penalties, as well as damage to the company's reputation.

2. **Ethical Concerns:** There were growing concerns about the ethical implications of AI, particularly around bias and transparency. XYZ Corp's stakeholders, including customers, partners, and employees, were increasingly demanding that the company take a proactive stance on ethical AI. This required a shift from viewing data governance as a compliance exercise to embracing it as a core organizational value.

3. **Cultural Misalignment:** The existing organizational culture at XYZ Corp was heavily focused on innovation and speed, sometimes at the expense of governance and accountability. This cultural misalignment posed a significant challenge to implementing effective data governance, as employees were not fully aligned with the importance of these practices.

Solution: A Strategic Cultural Transformation

To address these challenges, XYZ Corp embarked on a comprehensive cultural transformation initiative aimed at aligning the organization's culture with its data governance objectives. This transformation involved several key strategies:

1. **Redefining Core Values:** The first step in the transformation was to redefine the company's core values to emphasize ethics, accountability, and transparency. XYZ Corp's leadership recognized that without a strong ethical foundation, any governance efforts would be superficial and unsustainable. The new values were communicated across the organization, with a clear message that ethical AI and data governance were non-negotiable priorities.

   Leadership Commitment: Senior executives at XYZ Corp took an active role in championing the new values. They modeled the desired behaviors by consistently emphasizing the importance of governance in their decision-making processes. Leadership's visible commitment to these values was critical in driving cultural change, as it signaled to employees that data governance was not just an IT issue but a strategic priority for the entire organization.

2. **Comprehensive Training Programs:** Recognizing that cultural change requires education and skill-building, XYZ Corp implemented a comprehensive training program focused on ML data governance. The training was designed to equip employees with the knowledge and skills needed to adhere to governance policies and to understand the ethical implications of their work.

   Tailored Training Modules: The training program included tailored modules for different roles within the organization. For example, data scientists received training on ethical AI practices and how to mitigate bias in ML models, while compliance officers were trained on the latest regulatory requirements. This role-specific approach ensured that all employees understood how governance applied to their work.

Ongoing Education: In addition to initial training, XYZ Corp committed to ongoing education through regular workshops, webinars, and updates on governance best practices. This continuous learning approach helped to reinforce the importance of governance and kept employees informed about new developments in the field.

3. **Embedding Governance into Daily Operations:** To ensure that the cultural transformation was sustainable, XYZ Corp worked to embed governance practices into the organization's daily operations. This involved integrating governance metrics into performance evaluations, establishing governance committees, and creating clear accountability structures.

    Performance Metrics: Governance-related metrics were included in employee performance evaluations to ensure that adherence to governance policies was recognized and rewarded. This helped to align individual incentives with the company's governance objectives, encouraging employees to take ownership of these practices.

    Governance Committees: XYZ Corp established cross-functional governance committees responsible for overseeing the implementation of governance policies and addressing any issues that arose. These committees included representatives from various departments, ensuring that governance was approached holistically and that all aspects of the organization were aligned.

    Accountability Structures: Clear accountability structures were put in place to ensure that governance practices were consistently applied across the organization. This included defining roles and responsibilities for governance at every level of the organization, from executives to frontline employees.

Outcome: A Successful Cultural Transformation

The cultural transformation initiative at XYZ Corp yielded significant results, addressing the challenges the company faced and leading to a more robust governance framework:

1. **Improved Compliance:** As a result of the cultural transformation, XYZ Corp achieved higher levels of compliance with regulatory requirements. The company was able to navigate the complex regulatory landscape more effectively, reducing the risk of legal and financial penalties.

2. **Enhanced Ethical Standards:** The emphasis on ethics and transparency led to the development of more ethical AI models. XYZ Corp was able to address concerns about bias and transparency in its ML practices, building trust among stakeholders and enhancing its reputation as a responsible technology company.

3. **Reduced Risk:** By embedding governance into its organizational culture, XYZ Corp was able to reduce risks associated with data breaches, regulatory violations, and ethical lapses. The company's proactive approach to governance helped to mitigate potential threats and protect its long-term interests.

4. **Increased Stakeholder Trust:** The cultural transformation led to enhanced trust among XYZ Corp's stakeholders, including customers, partners, and employees. The company's commitment to governance and ethics was recognized and appreciated by stakeholders, leading to stronger relationships and greater brand loyalty.

5. **Sustainable Governance Practices:** The cultural transformation ensured that governance practices were not only implemented but also sustained over time. By embedding governance into the company's values, operations, and accountability structures, XYZ Corp created a foundation for ongoing compliance and ethical AI use.

Case Study 2: Change Management

ABC Inc., a financial services firm, faced significant challenges in implementing ML data governance due to resistance from employees and a lack of alignment between governance policies and existing processes. To address these challenges, the company developed a change management plan that focused on stakeholder engagement, communication, and support. The plan involved involving key stakeholders in the development of governance policies, providing clear and consistent communication about the reasons for the change, and offering ongoing support to employees through training and resources. The company also used technology to facilitate the integration of governance practices into existing processes, such as automating compliance checks and monitoring model performance. Through these efforts, ABC Inc. successfully overcame resistance and achieved a smooth transition to a governance-focused organization.

Challenges Faced by ABC Inc.

Employee Resistance

One of the most pressing challenges ABC Inc. faced was resistance from employees. The implementation of ML data governance required changes in how data was handled, processed, and managed across the organization. Employees, particularly those who had been with the company for a long time, were accustomed to established processes and were reluctant to adopt new practices. This resistance was partly due to a fear of increased workloads, a perceived threat to autonomy, and uncertainty about the implications of these changes on their daily responsibilities.

Lack of Process Alignment

Another significant challenge was the misalignment between the new governance policies and ABC Inc.'s existing operational processes. The firm's pre-existing systems and workflows were not designed with stringent data governance in mind, which led to friction when trying to overlay new governance policies onto old processes. This misalignment created inefficiencies, confusion, and even errors, further exacerbating employee resistance and complicating the implementation of governance practices.

Solutions Implemented by ABC Inc.

To address these challenges, ABC Inc. developed a multifaceted change management plan that focused on stakeholder engagement, clear communication, ongoing support, and the strategic use of technology. This plan was crucial in overcoming resistance and ensuring a smooth transition to a governance-focused organization.

Stakeholder Engagement

ABC Inc. recognized that the success of its governance initiative depended heavily on the buy-in from key stakeholders. To achieve this, the company involved these stakeholders in the development of governance policies from the outset. By including data scientists, IT personnel, compliance officers, and business leaders in the policy development process, ABC Inc. ensured that the governance framework was practical, relevant, and aligned with the operational realities of each department. This inclusive approach not only reduced resistance by addressing stakeholder concerns early on but also empowered these stakeholders to become champions of the change within their respective teams.

Clear and Consistent Communication

Effective communication was another cornerstone of ABC Inc.'s change management strategy. The company made it a priority to communicate the reasons for implementing ML data governance clearly and consistently across all levels of the organization. This communication included detailed explanations of the regulatory requirements driving the change, the ethical implications of poor data governance, and the long-term benefits of adopting robust governance practices. By framing the change in terms of both compliance and business value, ABC Inc. was able to shift the narrative from one of burden to one of opportunity. Regular updates, Q&A sessions, and town hall meetings were used to maintain transparency and address any ongoing concerns or questions from employees.

Ongoing Support and Training

Recognizing that resistance often stems from a lack of understanding or fear of the unknown, ABC Inc. invested heavily in training and ongoing support for its employees. The company developed comprehensive training programs tailored to different roles within the organization, ensuring that each employee had the knowledge and skills necessary to comply with the new governance policies. These training sessions included hands-on workshops, online modules, and one-on-one coaching, providing multiple avenues for employees to learn and ask questions. Additionally, ABC Inc. established a support network of governance experts who were available to assist employees as they navigated the new processes. This support network was crucial in alleviating fears, building confidence, and fostering a culture of continuous learning.

Leveraging Technology for Integration

To further facilitate the transition, ABC Inc. turned to technology to integrate governance practices into its existing processes. One of the key technological solutions implemented was the automation of compliance checks and the monitoring of ML model performance. By leveraging automated tools, the company was able to reduce the manual burden on employees, minimize errors, and ensure that governance practices were consistently applied across the organization. These tools also provided real-time insights into data governance metrics, enabling proactive management and quick adjustments when necessary. The use of technology not only streamlined the implementation of governance practices but also helped to overcome some of the resistance by demonstrating the efficiency and effectiveness of the new processes.

Outcome: A Smooth Transition to Governance-Focused Operations

Through the implementation of a well-structured change management plan, ABC Inc. successfully navigated the challenges of employee resistance and process misalignment. The focus on stakeholder engagement, clear communication, ongoing support, and the strategic use of technology allowed the company to overcome these obstacles and achieve a smooth transition to a governance-focused organization. As a result, ABC Inc. not only strengthened its ML data governance practices but also built a more resilient, compliant, and ethically aware organization, better equipped to navigate the complexities of the financial services industry.

Lessons Learned

These case studies highlight several key lessons for organizations seeking to implement cultural and change management strategies for ML data governance. First, leadership is critical in driving cultural transformation and ensuring that governance is prioritized across the organization. Second, engaging stakeholders in the change process is essential for building buy-in and reducing resistance. Third, communication is key to ensuring that employees understand the reasons for the change and are motivated to adhere to governance policies. Finally, technology plays a vital role in facilitating the integration of governance practices into organizational processes and ensuring that they are consistently applied.

While these case studies demonstrate successful cultural and change management efforts, they also highlight the challenges that organizations may face. These challenges include resistance to change, misalignment between governance policies and existing processes, and the need for continuous support and reinforcement. Solutions to these challenges include involving employees in the change process, providing ongoing communication and training, and using technology to streamline governance practices. By addressing these challenges proactively, organizations can increase the likelihood of successful implementation and sustainability of ML data governance.

The cultural and change management efforts described in these case studies had a significant impact on the organizations' governance outcomes. Both XYZ Corp and ABC Inc. experienced improved compliance with regulatory requirements, reduced risk, and enhanced trust among stakeholders as a result of their efforts. These outcomes demonstrate the importance of cultural and change management in ensuring the success of ML data governance. When organizations prioritize governance and effectively manage the change process, they are better positioned to achieve their governance objectives and build a culture of accountability and transparency.

Measuring the Success of Cultural and Change Management Initiatives

Ensuring the success of cultural and change management initiatives related to ML data governance is vital for organizations striving to achieve robust governance practices. These initiatives are complex and multifaceted, requiring careful planning, execution, and, crucially, measurement. To evaluate the effectiveness of these efforts, organizations can leverage a variety of tools, including key performance indicators (KPIs), employee feedback, governance metrics, and continuous improvement processes. By tracking and analyzing these elements, organizations can gain a comprehensive understanding of the impact of their initiatives, identify areas for improvement, and ensure that governance practices are sustainably embedded into the organizational culture.

Key Performance Indicators (KPIs) in ML Data Governance

KPIs are essential tools for measuring the success of cultural and change management initiatives in ML data governance. They provide quantitative metrics that allow organizations to assess the effectiveness of their governance efforts over time. Specific KPIs relevant to ML data governance include compliance rates, the number of governance-related incidents, employee adherence to governance policies, and the effectiveness of governance training programs.

Compliance Rates

Compliance rates are a critical KPI for assessing the effectiveness of governance initiatives. These rates can be tracked by monitoring the organization's adherence to relevant regulations, such as GDPR, CCPA, or industry-specific standards. For instance, a financial institution might track the percentage of data audits that result in full compliance with regulatory requirements. A high compliance rate indicates that the organization's governance practices are effective and that employees are following the required protocols.

Employee Adherence to Governance Policies

Tracking employee adherence to governance policies provides insight into how well governance practices are being integrated into the organization's day-to-day operations. This can be measured through audits, self-assessments, and reports of policy violations. For example, an organization might monitor the frequency of data breaches or unauthorized data access incidents as a KPI. A reduction in such incidents over time would suggest that employees are increasingly adhering to governance policies, reflecting the success of the cultural and change management initiatives.

Effectiveness of Governance Training Programs

The effectiveness of governance training programs can also be evaluated as a KPI. This can be measured by assessing the knowledge retention of employees after training sessions through tests, quizzes, or practical assessments. Additionally, the application of this knowledge in real-world scenarios, such as the proper classification of data or correct use of data management tools, can serve as indicators of training effectiveness. Organizations might set benchmarks, such as achieving a certain percentage of employees who pass post-training assessments, to evaluate the success of their training programs.

By regularly tracking these KPIs, organizations can assess the impact of their cultural and change management efforts on ML data governance. Trends in these KPIs can help identify areas where further improvements are needed, such as additional training or adjustments to governance policies.

Employee Feedback as a Measure of Success

Employee feedback is a crucial qualitative measure for assessing the success of cultural and change management initiatives. Gathering feedback through surveys, focus groups, and interviews provides organizations with valuable insights into employees' perceptions of the organizational culture, the change process, and their understanding and adherence to governance policies.

Surveys and Polls

Surveys and polls are commonly used tools for collecting employee feedback on ML data governance initiatives. These tools can measure employees' awareness of governance policies, their confidence in applying them, and their overall satisfaction with the governance training they have received. For example, an organization might survey its employees to determine their level of understanding of new data privacy regulations and how confident they feel in complying with these regulations in their daily work. High scores in such surveys indicate that employees are well-informed and comfortable with the governance practices, whereas lower scores may signal the need for further education and support.

Focus Groups and Interviews

Focus groups and interviews offer more in-depth insights into the employee experience with ML data governance initiatives. These methods allow employees to express their thoughts, concerns, and suggestions in a more open-ended format. For instance, during a focus group, employees might discuss the challenges they face in adhering to governance policies, such as difficulties in accessing necessary resources or unclear guidelines. This feedback can be invaluable for identifying specific areas where the change management process may need to be adjusted or where additional support is required.

Addressing Feedback

The information gathered from employee feedback should be used to inform adjustments to cultural and change management strategies. For example, if feedback reveals that employees are struggling with the practical application of governance policies, the organization might introduce more hands-on training sessions or provide additional resources to assist employees in meeting governance standards. By continuously gathering and acting on employee feedback, organizations can create a more supportive environment that fosters compliance and adherence to governance practices.

Governance Metrics and Continuous Improvement

Governance metrics provide a direct measure of the effectiveness of ML data governance practices and are crucial for evaluating the success of cultural and change management initiatives. These metrics might include the number of compliance violations, the accuracy and fairness of ML models, the timeliness of data updates, and the effectiveness of risk management practices.

Compliance Violations

Tracking the number of compliance violations is a straightforward way to measure the impact of governance initiatives. A decrease in violations over time suggests that governance policies are being effectively implemented and that employees are following the necessary protocols. For example, an organization might track the number of incidents where data is mishandled or where data privacy regulations are breached. A downward trend in these incidents would indicate that cultural and change management efforts are successfully embedding governance practices into the organization's operations.

Model Accuracy and Fairness

In the context of ML data governance, metrics related to the accuracy and fairness of ML models are also critical. These can be tracked by regularly auditing ML models to ensure they are performing as expected and do not exhibit biased outcomes. For instance, an organization might monitor the performance of its credit scoring models to ensure they do not unfairly disadvantage certain demographic groups. Improvements in these metrics suggest that governance practices are effectively mitigating risks associated with biased or inaccurate models.

Timeliness of Data Updates

Another important metric is the timeliness of data updates, which reflects the organization's ability to maintain up-to-date and accurate data for its ML models. This can be tracked by measuring the time taken to update data repositories or by monitoring the currency of data used in ML models. Timely data updates are essential for ensuring the accuracy and reliability of ML outputs, and improvements in this metric indicate that governance practices are being effectively implemented.

Continuous Improvement and Long-Term Sustainability

Continuous improvement is a vital component of successful cultural and change management initiatives in ML data governance. It involves regularly reviewing the effectiveness of governance practices, identifying areas for improvement, and implementing changes to address any issues. This process should be informed by KPIs, employee feedback, and governance metrics and should involve input from all relevant stakeholders.

Regular Review Processes

Organizations should establish regular review processes to evaluate the effectiveness of their governance practices. These reviews might involve quarterly audits, annual compliance assessments, or ongoing monitoring of governance metrics. The insights gained from these reviews can help organizations identify trends, spot potential issues early, and make data-driven decisions about how to enhance their governance practices.

Stakeholder Involvement

Engaging stakeholders in the continuous improvement process is essential for ensuring that governance practices remain aligned with organizational goals and regulatory requirements. Stakeholders, including employees, compliance officers, and business leaders, should be involved in reviewing KPIs, providing feedback, and suggesting improvements to governance practices. Their input is crucial for making informed adjustments that address real-world challenges and support long-term sustainability.

Adapting to Change

Continuous improvement also involves adapting to changes in the external environment, such as new regulations, technological advancements, or shifts in market conditions. For example, as new data privacy laws are introduced, organizations may need to update their governance policies, revise training programs, or implement new technologies to ensure compliance. By staying proactive and responsive to these changes, organizations can maintain effective governance practices that support their strategic objectives.

Summary

The chapter discusses the critical role of organizational culture and change management in the successful implementation of machine learning (ML) data governance frameworks. As ML systems increasingly drive business decisions, robust governance is essential to ensure that these systems align with ethical standards, regulatory requirements, and organizational objectives. It highlights how organizational culture— the shared values, beliefs, and norms within a company—forms the foundation for these governance frameworks. A strong culture of ethics, accountability, and transparency fosters an environment where governance practices are embraced and sustained.

It also delves into the importance of change management, which involves guiding an organization through transitions, such as adopting new governance frameworks. Effective change management ensures that these transitions are smooth and that the organization remains resilient amidst challenges. It emphasizes that leadership plays a pivotal role in setting the tone for governance by promoting a data-driven culture and integrating governance into all business processes.

Employee engagement is underscored as crucial for ML data governance. A culture that values collaboration, transparency, and accountability motivates employees to take ownership of governance responsibilities, ensuring the integrity of ML data. The chapter suggests implementing specific training programs to reinforce governance practices and developing a continuous learning environment to adapt to the evolving ML landscape.

The chapter discusses the impact of organizational structure on data governance. Hierarchical structures may facilitate compliance but can stifle innovation, while decentralized structures promote flexibility but may pose challenges in consistent policy application. The chapter concludes by stressing the need for a balance between fostering innovation and maintaining strong governance practices, ensuring that ethical considerations are embedded in ML systems and that the organization adapts effectively to changes.

CHAPTER 10

Future Trends and Emerging Challenges

Machine learning (ML) has rapidly transformed from a niche technology to a critical tool that powers various aspects of business operations and decision making. With this transformation, the governance of data, particularly within the context of ML, has emerged as a vital concern for organizations. Data governance in ML refers to the processes, policies, and practices that ensure the quality, privacy, security, and ethical use of data across the entire lifecycle of an ML model. As the dependence on data-driven technologies grows, the importance of robust ML data governance frameworks cannot be overstated. These frameworks not only help in managing data-related risks but also ensure compliance with an increasingly complex web of regulations and standards.

In the current landscape, organizations are already grappling with challenges such as data quality management, bias mitigation, and compliance with privacy laws like the GDPR. However, as ML technology continues to evolve, so too will the challenges and opportunities related to data governance. The future of ML data governance is likely to be shaped by trends such as the rise of decentralized data sources, the increasing use of AI for automating governance tasks, and the growing demand for ethical AI practices. These trends will require organizations to rethink and innovate their governance strategies to stay ahead of potential risks and leverage new opportunities.

Moreover, emerging challenges such as the scalability of governance frameworks, cross-border data flows, and the management of diverse data types will test the resilience and adaptability of current data governance practices. As organizations navigate these challenges, they will need to develop more sophisticated, agile, and proactive approaches to ML data governance. This essay explores the future trends that are expected to influence ML data governance and delves into the emerging challenges

© Aditya Nandan Prasad 2024
A. Nandan Prasad, *Introduction to Data Governance for Machine Learning Systems*,
https://doi.org/10.1007/979-8-8688-1023-7_10

that organizations will face. By understanding these trends and challenges, organizations can better prepare for the future, ensuring that their data governance practices remain effective and relevant in an ever-changing technological landscape.

Existing Challenges in Implementing ML Data Governance

Despite advancements in ML data governance frameworks, organizations still encounter considerable challenges in their implementation. One significant hurdle is the complexity and variety of data sources used in ML models. Organizations frequently rely on data from diverse origins, including internal databases, external third-party providers, and publicly accessible datasets. Managing the quality, consistency, and compliance of this data presents a formidable challenge, particularly when dealing with unstructured or semi-structured formats.

Integrating ML data governance into existing organizational processes is another major challenge. Many organizations struggle to align their data governance practices with broader business goals, resulting in gaps and inconsistencies in data management. This misalignment can hinder the effectiveness of data governance frameworks and complicate efforts to ensure data integrity and compliance.

The rapidly evolving regulatory landscape further complicates ML data governance. As new data protection laws and AI regulations are introduced, organizations must continuously update their governance practices to stay compliant. This involves not only staying informed about legal changes but also adapting internal processes and systems to align with new requirements. The dynamic nature of regulations requires a proactive approach to ensure ongoing compliance.

Another challenge organizations face is the transparency and explainability of ML models. Many ML models, especially those based on deep learning, are inherently complex and difficult to interpret. Ensuring that these models are understandable and trustworthy to stakeholders is a crucial aspect of data governance. However, achieving this transparency and explainability in practice remains a significant difficulty.

The integration of complex ML models into existing data governance frameworks also poses challenges. Organizations must ensure that their governance practices can accommodate the unique needs of advanced ML models while maintaining overall data quality and compliance. This requires a nuanced approach to governance that addresses the specific requirements of different types of ML models.

Organizations must address the issue of data stewardship and accountability. Effective data governance requires clear roles and responsibilities for managing data across its lifecycle. Ensuring that data stewards are properly trained and that accountability mechanisms are in place is essential for maintaining data quality and compliance.

Another key challenge is managing the data lifecycle, including data acquisition, storage, and usage. Organizations must develop robust processes for handling data throughout its lifecycle to ensure that it remains accurate, secure, and compliant with regulations. This involves implementing appropriate data management practices and technologies to support ongoing data governance efforts.

The need for continuous improvement in data governance practices is crucial. Organizations must regularly review and refine their data governance frameworks to address emerging challenges and adapt to changing requirements. This ongoing process of evaluation and enhancement is essential for maintaining effective data governance and supporting the successful deployment of ML models.

Future Trends in ML Data Governance

As the field of machine learning (ML) continues to evolve, the governance of data within this context is also expected to undergo significant changes. The future of ML data governance will be shaped by various trends, including technological advancements, regulatory developments, and shifts in societal expectations. These trends will present both opportunities and challenges for organizations, requiring them to adapt their data governance strategies to stay ahead of the curve.

Data Decentralization and Edge Computing

One of the most significant trends in ML data governance is the decentralization of data, driven by the rise of edge computing. Traditionally, data governance has focused on managing data within centralized data centers or cloud environments. However, as more organizations adopt edge computing technologies, data is increasingly being generated, processed, and stored at the edge of the network, closer to where it is created. This shift presents new challenges for data governance, as organizations must now manage data across a more distributed and decentralized environment.

Edge computing offers several advantages, including reduced latency, improved performance, and enhanced privacy by keeping data closer to its source. However, it also complicates data governance, as organizations must ensure that data remains consistent, secure, and compliant across multiple edge locations. This requires the development of new data governance frameworks that can operate effectively in decentralized environments. Additionally, organizations must address the challenges of data integration and interoperability, as data generated at the edge must often be combined with data from other sources to train ML models.

To govern data in decentralized environments, organizations will need to adopt more flexible and adaptive data governance practices. This may include the use of distributed ledger technologies, such as blockchain, to ensure data integrity and traceability across the edge network. Organizations will also need to invest in edge-specific data management tools that can handle the unique challenges of edge computing, such as data synchronization, replication, and conflict resolution. By developing robust data governance practices for decentralized environments, organizations can unlock the full potential of edge computing while ensuring that their data remains well governed.

AI-Driven Data Governance

As organizations collect and process increasing amounts of data, traditional manual approaches to data governance are becoming less feasible. AI-driven data governance offers a solution by automating routine governance tasks, such as data classification, policy enforcement, and compliance monitoring. This not only improves the efficiency of data governance efforts but also enables organizations to manage larger and more complex data environments.

AI-driven data governance leverages advanced analytics and machine learning algorithms to monitor and manage data in real time. For example, AI can be used to automatically classify data based on its sensitivity, apply encryption to protect sensitive data, and monitor data access patterns to detect potential security threats. Additionally, AI can help organizations identify and mitigate data quality issues by automatically detecting anomalies, inconsistencies, and errors in data. By automating these tasks, AI-driven data governance reduces the burden on human data stewards and allows organizations to focus on more strategic governance activities.

One of the key benefits of AI-driven data governance is its ability to provide predictive insights into data governance risks. By analyzing historical data and patterns, AI can predict potential compliance violations, data breaches, or other governance issues before they occur. This enables organizations to take proactive measures to address these risks, rather than reacting to them after the fact. Additionally, AI can help organizations optimize their data governance strategies by identifying areas for improvement and recommending best practices based on data-driven insights. As AI technology continues to advance, its role in data governance is expected to grow, offering new opportunities for organizations to enhance their governance efforts.

Enhanced Data Privacy Regulations

Data privacy is a cornerstone of any data governance strategy, and this focus is set to intensify as global regulations become more stringent. With increasing public awareness and concern over data privacy, governments worldwide are introducing more comprehensive and stringent data protection laws. These regulations are shaping how organizations collect, store, process, and share data, particularly in the context of ML, where data is the lifeblood of algorithms.

The General Data Protection Regulation (GDPR) in the European Union set a global benchmark for data privacy, influencing legislation in other regions. Since its implementation, other jurisdictions have followed suit, introducing their own regulations with similar or even stricter provisions. For example, the California Consumer Privacy Act (CCPA) and the more recent California Privacy Rights Act (CPRA) in the United States impose significant requirements on organizations regarding data transparency, consumer rights, and the handling of personal data.

Looking ahead, this trend is expected to continue, with more countries and regions adopting rigorous data privacy laws. Additionally, existing regulations are likely to be updated and expanded to address new challenges posed by emerging technologies like ML and AI. For instance, the GDPR is expected to evolve to cover aspects such as automated decision making and AI transparency, which are directly relevant to ML data governance. The increasing complexity of these regulations will require organizations to continuously adapt their data governance frameworks to remain compliant.

Enhanced data privacy regulations will also necessitate more sophisticated data governance practices. Organizations will need to implement stronger data access controls, encryption, and anonymization techniques to protect personal data. Additionally, they will need to develop mechanisms to ensure that ML models comply with privacy regulations throughout their lifecycle, from training to deployment. This includes implementing privacy-by-design principles, where privacy considerations are integrated into the development of ML models from the outset.

The concept of data sovereignty, where data must be stored and processed within the geographic boundaries of a specific country or region, is gaining traction. This trend is likely to complicate data governance, particularly for multinational organizations that need to manage data across multiple jurisdictions. Organizations will need to navigate the complexities of cross-border data flows and ensure that their ML models comply with diverse regional data privacy laws.

In response to these challenges, organizations are increasingly adopting privacy-enhancing technologies (PETs) such as homomorphic encryption, differential privacy, and federated learning. These technologies allow organizations to perform ML on sensitive data without compromising privacy, enabling them to comply with data privacy regulations while still leveraging the power of ML. As these technologies mature, they are expected to play a critical role in the future of ML data governance.

Ethical AI and Responsible Data Use

The ethical implications of AI and ML are becoming a central concern for organizations and regulators alike. As ML models are increasingly used to make decisions that impact individuals and society, there is growing recognition of the need to ensure that these models are fair, transparent, and accountable. This has led to the rise of ethical AI and responsible data use as key trends in ML data governance.

Ethical AI refers to the practice of designing, developing, and deploying AI and ML systems in a way that aligns with ethical principles, such as fairness, transparency, and respect for human rights. In the context of ML data governance, this means ensuring that data used to train ML models is representative, free from bias, and handled in a manner that respects privacy and autonomy. It also involves implementing governance mechanisms to ensure that ML models are transparent, explainable, and accountable.

One of the primary challenges in ethical AI is addressing bias in ML models. Bias can enter the ML pipeline at various stages, from the selection of training data to the design of algorithms and the interpretation of model outputs. If left unchecked, bias can lead to unfair and discriminatory outcomes, particularly when ML models are used in high-stakes domains such as hiring, lending, law enforcement, and healthcare. As a result, organizations are increasingly focusing on bias detection and mitigation as part of their ML data governance efforts.

Transparency is another critical aspect of ethical AI. Stakeholders, including regulators, customers, and the public, are demanding greater transparency in how ML models make decisions. This requires organizations to develop explainable AI (XAI) techniques that allow them to provide clear and understandable explanations of model decisions. This is particularly important in regulated industries, where organizations must demonstrate that their ML models comply with legal and ethical standards.

Accountability is also a key concern in ethical AI. As ML models are increasingly used to automate decision making, it is essential to ensure that there is human oversight and that organizations are held accountable for the outcomes of their models. This includes implementing governance mechanisms to monitor and audit ML models throughout their lifecycle, ensuring that they remain aligned with ethical principles and do not drift over time.

In response to these challenges, organizations are developing ethical AI frameworks and guidelines that outline best practices for responsible data use and AI development. These frameworks often include principles such as fairness, transparency, accountability, and privacy, as well as specific guidelines for implementing these principles in practice. As ethical AI becomes a more prominent concern, organizations will need to integrate these frameworks into their broader ML data governance strategies.

Data Interoperability and Standardization

As ML and AI technologies become more widespread, the need for data interoperability and standardization is becoming increasingly important. Data interoperability refers to the ability of different systems, platforms, and organizations to exchange and use data seamlessly, while standardization involves the development of common data formats, protocols, and guidelines that facilitate this interoperability. Both of these trends are critical for effective ML data governance, particularly in complex, multicloud, and hybrid environments.

One of the key drivers of the need for data interoperability is the growing use of cloud computing and the proliferation of cloud service providers. Many organizations now operate in multicloud environments, where data is stored and processed across multiple cloud platforms. This creates challenges for data governance, as organizations must ensure that data remains consistent, secure, and compliant as it moves between different environments. Data interoperability is essential for enabling organizations to manage and govern their data effectively in these complex environments.

Standardization plays a crucial role in achieving data interoperability. By adopting common data formats, protocols, and guidelines, organizations can ensure that their data can be easily exchanged and used across different systems and platforms. This is particularly important for ML, where data from multiple sources often needs to be combined to train models. Standardization also facilitates data integration, allowing organizations to break down data silos and create unified data environments that support more effective ML.

Standardization plays a crucial role in ensuring data quality and consistency. By implementing standardized data governance practices, organizations can guarantee that their data remains accurate, complete, and reliable, no matter where it is stored or processed. This is vital for maintaining the quality and effectiveness of ML models, as they depend on high-quality data to generate accurate and dependable predictions.

Standardization is also important for compliance with data protection regulations. Many data protection laws, such as GDPR, require organizations to implement standardized data protection measures, such as encryption and access controls, to protect personal data. By adopting standardized data governance practices, organizations can ensure that they comply with these regulations and avoid the risks of non-compliance.

As the need for data interoperability and standardization continues to grow, industry bodies and standards organizations are working to develop common frameworks and guidelines for ML data governance. For example, the ISO (International Organization for Standardization) and the IEEE (Institute of Electrical and Electronics Engineers) are developing standards for AI and ML, including guidelines for data governance. These standards are expected to play a critical role in shaping the future of ML data governance, providing organizations with the tools and guidance they need to manage their data effectively in a rapidly changing technological landscape.

Integration of Blockchain for Data Governance

Blockchain technology, originally developed as the underlying technology for cryptocurrencies, is gaining traction as a potential tool for enhancing data governance in ML. Blockchain's decentralized and immutable nature makes it an attractive solution for ensuring data integrity, traceability, and transparency, all of which are critical components of effective data governance. As organizations seek to improve their ML data governance practices, the integration of blockchain technology is emerging as a significant trend.

One of the key benefits of blockchain in the context of ML data governance is its ability to provide a tamper-proof record of data transactions. By recording every data transaction on a blockchain, organizations can create an immutable audit trail that tracks the entire lifecycle of the data, from its creation to its use in ML models. This audit trail can be used to verify the authenticity and integrity of the data, ensuring that it has not been tampered with or altered. This is particularly important for industries where data integrity is critical, such as finance, healthcare, and supply chain management.

Blockchain also offers the potential to enhance data transparency and accountability in ML. By using smart contracts, which are self-executing contracts with the terms of the agreement directly written into code, organizations can automate and enforce data governance policies. For example, a smart contract could automatically enforce data access controls, ensuring that only authorized users can access sensitive data. This not only improves data security but also ensures that data governance policies are consistently applied across the organization.

Blockchain can also facilitate data sharing and collaboration in ML. Many organizations are hesitant to share data due to concerns about data security and privacy. However, blockchain's decentralized nature allows organizations to share data in a secure and controlled manner without the need for a central intermediary. By using blockchain, organizations can create a shared, decentralized data ecosystem where data can be securely exchanged and used for ML while still maintaining control over their data.

Despite its potential, the integration of blockchain into ML data governance is not without challenges. One of the primary challenges is the scalability of blockchain technology. Traditional blockchain networks, such as Bitcoin and Ethereum, are known for their limited scalability, which can hinder their ability to handle large volumes of data transactions. However, new developments in blockchain technology, such as layer 2 solutions and alternative consensus mechanisms, are addressing these scalability issues, making blockchain more viable for data governance applications.

Another challenge is the complexity of integrating blockchain with existing data governance frameworks. Blockchain is still a relatively new technology, and many organizations may lack the expertise and resources to implement it effectively. Additionally, the regulatory landscape for blockchain is still evolving, and organizations must navigate a complex web of regulations to ensure that their blockchain-based data governance practices comply with legal requirements.

As blockchain technology continues to mature, it is expected to play an increasingly important role in ML data governance. Organizations that successfully integrate blockchain into their data governance frameworks will be better equipped to ensure data integrity, transparency, and accountability, positioning themselves for success in a data-driven future.

As ML continues to transform industries and redefine how organizations operate, the importance of robust data governance cannot be overstated. The future of ML data governance will be shaped by a complex interplay of technological advancements, regulatory developments, and evolving organizational needs. Organizations that proactively address emerging trends and challenges in ML data governance will be better positioned to harness the full potential of ML while minimizing risks and ensuring compliance.

Key trends such as the rise of automated governance, enhanced data privacy regulations, ethical AI, data interoperability, and the integration of blockchain technology are poised to reshape the landscape of ML data governance. These trends will require organizations to adopt new technologies, frameworks, and best practices to ensure that their data governance strategies remain effective in an increasingly complex and dynamic environment.

At the same time, organizations must remain vigilant to emerging challenges, such as the scalability of blockchain, the complexity of integrating new technologies, and the evolving regulatory landscape. By staying ahead of these challenges and continuously adapting their data governance practices, organizations can ensure that they remain compliant, competitive, and resilient in the face of rapid technological change.

Emerging Challenges

As ML has grown into a critical component of business operations, existing challenges and hurdles emerge. We'll discuss some of these challenges and discuss the frameworks needed to overcome them.

Data Quality and Integrity

Data quality and integrity are foundational to the success of any ML initiative. Poor data quality can lead to inaccurate models, biased predictions, and ultimately, flawed decision making. As organizations increasingly rely on ML, the challenge of maintaining high data quality becomes more pronounced.

One of the primary issues is the sheer volume and variety of data that modern organizations must manage. Data is generated from numerous sources, including IoT devices, social media, transactional systems, and more. This diversity makes it difficult to maintain consistency, accuracy, and completeness across datasets. Furthermore, data can quickly become outdated, leading to "data decay," where the information used to train ML models no longer reflects current realities.

Data integrity is another critical concern. Ensuring that data remains unaltered and trustworthy throughout its lifecycle is essential for building reliable ML models. However, the increasing complexity of data environments, coupled with the growing threat of cyberattacks, makes it challenging to protect data integrity. For example, unauthorized data manipulation could result in ML models making incorrect predictions, with potentially severe consequences in areas like finance, healthcare, and security.

Organizations need to implement rigorous data governance practices such as data cleansing, validation, and regular audits to effectively tackle these challenges. Leveraging automated tools can enhance monitoring by identifying data quality issues in real-time, enabling proactive problem resolution before they affect ML outcomes. Moreover, maintaining transparency and accountability requires robust data lineage tracking, which allows organizations to trace the origins and transformations of data utilized in ML models.

Ethical and Social Implications of ML

The ethical and social implications of ML are becoming increasingly prominent as these technologies are deployed in critical areas such as criminal justice, healthcare, hiring, and finance. ML models have the potential to perpetuate and even exacerbate existing biases and inequalities, leading to unfair outcomes.

One of the major challenges is algorithmic bias, where ML models unintentionally produce discriminatory outcomes due to biased training data. This can occur when historical data reflects societal biases, such as gender or racial discrimination, which the ML model then learns and perpetuates. The challenge for organizations is to identify and mitigate these biases before deploying ML models in real-world applications.

Another ethical concern is the lack of transparency in ML models, often referred to as the "black box" problem. Many ML models, particularly deep learning models, are highly complex and difficult to interpret. This opacity makes it challenging for organizations to explain how decisions are made, which can undermine trust and accountability. This is especially problematic in regulated industries, where organizations must demonstrate compliance with legal and ethical standards.

The social implications of ML also extend to issues of privacy and surveillance. The widespread use of ML in areas like facial recognition, social media monitoring, and predictive policing raises concerns about privacy and civil liberties. Organizations must navigate these concerns while balancing the potential benefits of ML with the need to protect individual rights.

Organizations need to adopt ethical AI frameworks that emphasize fairness, transparency, and accountability. This involves conducting bias audits, applying explainable AI techniques, and actively engaging with stakeholders to ensure that ML models are aligned with societal values.

Regulatory Compliance and Governance

As ML technologies evolve, so too does the regulatory landscape. Governments and regulatory bodies worldwide are increasingly scrutinizing the use of ML and AI, particularly in sensitive areas such as data privacy, consumer protection, and financial regulation. Staying compliant with these evolving regulations presents a significant challenge for organizations.

One of the key challenges is navigating the complex and often conflicting regulatory requirements across different jurisdictions. For instance, the GDPR in Europe imposes stringent data protection requirements, including the right to explanation, which impacts how ML models must be designed and deployed. Meanwhile, other regions, such as the United States and China, have their own regulatory frameworks, which may differ significantly from the GDPR. Organizations operating globally must ensure that their ML models comply with the relevant regulations in each jurisdiction, which can be a complex and resource-intensive process.

Another regulatory challenge is the increasing demand for transparency and explainability in ML models. Regulators are beginning to require organizations to provide clear explanations of how their ML models make decisions, particularly in areas like credit scoring, hiring, and healthcare. This demand for explainability can be difficult to meet, especially for complex models like deep neural networks, which are inherently opaque.

Regulatory compliance also extends to data governance practices, including data security, privacy, and auditability. Organizations must implement robust data governance frameworks that can withstand regulatory scrutiny and demonstrate compliance with legal requirements. This includes maintaining detailed records of data processing activities, implementing data protection measures, and conducting regular audits to ensure that ML models remain compliant over time.

Talent and Expertise Gap

The rapid advancement of ML technologies has outpaced the availability of skilled professionals who can effectively manage and govern these technologies. This talent and expertise gap is a significant challenge for organizations looking to integrate ML into their operations.

One of the primary issues is the scarcity of data scientists and ML engineers with the necessary skills to develop, deploy, and govern ML models. These professionals must not only have a deep understanding of ML algorithms and techniques but also possess expertise in data governance, ethics, and regulatory compliance. The demand for these skills far exceeds the current supply, leading to intense competition for talent.

Organizations also need professionals with expertise in data governance and compliance. These individuals play a crucial role in ensuring that ML models are developed and deployed in a manner that aligns with regulatory requirements and ethical standards. However, there is a shortage of professionals with this combination of skills, which can hinder organizations' ability to effectively govern their ML initiatives.

Organizations must invest in upskilling their current workforce and promoting a culture of continuous learning. This approach involves offering training programs in ML, data governance, and ethics while also fostering collaboration among data scientists, ML engineers, and governance professionals. Exploring alternative talent strategies, such as partnerships with academic institutions or engaging external consultants, may also be necessary to address the talent gap.

Data Silos and Integration

Data silos are a persistent challenge for organizations, particularly as they seek to leverage ML for data-driven decision making. Data silos occur when different departments or business units within an organization maintain separate, isolated data systems, making it difficult to share and integrate data across the organization.

In the context of ML, data silos can hinder the development of accurate and comprehensive models. ML models often require large and diverse datasets to produce reliable predictions. When data is siloed, it limits the availability of relevant data, leading to suboptimal models that may not fully capture the complexity of the problem at hand.

Data integration is another challenge closely related to data silos. Even when data is accessible across the organization, integrating data from different sources can be difficult due to differences in data formats, structures, and standards. Inconsistent data can lead to errors and discrepancies, undermining the accuracy and reliability of ML models.

To overcome these challenges, organizations must implement data governance strategies that promote data sharing and integration. This includes breaking down organizational silos, standardizing data formats and practices, and implementing data integration tools that facilitate seamless data exchange. Additionally, organizations must foster a culture of collaboration and data-driven decision making, where data is viewed as a shared asset rather than a departmental resource.

ML Model Drift and Maintenance

ML models are not static; they require ongoing maintenance to ensure they remain accurate and relevant over time. Model drift occurs when the performance of an ML model degrades due to changes in the underlying data or the environment in which it operates. This is a significant challenge for organizations, as it can lead to incorrect predictions and decisions.

There are two main types of model drift: data drift and concept drift. Data drift occurs when the distribution of the input data changes over time, while concept drift occurs when the relationship between the input data and the target variable changes. Both types of drift can significantly impact the performance of ML models, leading to inaccurate predictions and decisions.

Organizations must implement monitoring and maintenance practices as part of their ML data governance framework to overcome challenges of model drift. This includes regularly retraining models on new data, monitoring model performance, and implementing automated systems that can detect and respond to drift in real-time. Additionally, organizations must establish clear governance processes for updating and validating models, ensuring that they remain accurate and compliant with regulatory requirements.

Cost and Resource Management

The deployment and governance of ML models require significant investments in terms of cost and resources. From acquiring and processing large datasets to developing and maintaining ML models, the costs associated with ML initiatives can quickly escalate. Managing these costs effectively is a critical challenge for organizations.

One of the primary cost drivers in ML is the need for computing power and storage. Training large ML models, particularly deep learning models, requires substantial computational resources, often involving expensive GPUs or cloud-based services. Additionally, storing and processing the vast amounts of data required for ML can also be costly, particularly when dealing with high-velocity data streams or large-scale datasets.

Resource management is another challenge, as organizations must allocate sufficient resources to support their ML initiatives without overextending their budgets. This includes not only computational resources but also human resources, as skilled professionals are needed to develop, deploy, and maintain ML models.

To manage costs and resources effectively, organizations must adopt a strategic approach to ML deployment. This includes prioritizing ML initiatives based on their potential value, optimizing resource allocation, and leveraging cost-effective solutions such as cloud computing or automated ML tools. Additionally, organizations should implement governance processes that monitor and control ML-related expenditures, ensuring that resources are used efficiently and aligned with business objectives.

Data Security and Cyber Threats

As ML becomes increasingly integrated into organizational processes, the importance of data security and protection from cyber threats cannot be overstated. ML systems are particularly vulnerable to various forms of cyberattacks, which can compromise data integrity, disrupt operations, and expose organizations to significant risks.

One of the primary security challenges is the potential for adversarial attacks, where malicious actors manipulate input data to deceive ML models. For example, in image recognition systems, adversarial attacks can involve adding subtle perturbations to images that cause the model to misclassify them. These attacks can have serious consequences, particularly in security-sensitive applications such as facial recognition or autonomous vehicles.

Another concern is the security of the data used to train ML models. If an attacker gains access to the training data, they can introduce malicious data points that skew the model's predictions, a tactic known as data poisoning. Protecting the integrity of training data is therefore crucial to ensuring the reliability of ML models.

ML systems are also vulnerable to traditional cyber threats, such as data breaches, malware, and ransomware. As organizations collect and process increasing amounts of data for ML, they become attractive targets for cybercriminals. Ensuring the security of this data is essential to safeguarding the organization and maintaining the trust of customers and stakeholders.

Implementing robust data protection measures—such as encryption, access controls, and secure data storage—is critical for overcoming these security challenges. Incorporating ML-specific security practices, like adversarial training, which enhances model resilience by exposing them to adversarial examples during training, is equally important. Furthermore, conducting regular security audits and updates is essential for staying ahead of the evolving cyber threat landscape.

Interpretability and Explainability of ML Models

The interpretability and explainability of ML models are critical challenges, particularly as these models are increasingly used in high-stakes decision-making processes. Stakeholders, including regulators, customers, and employees, often demand clear explanations of how ML models arrive at their predictions and decisions.

Interpretability refers to the ability to understand the internal workings of an ML model, while explainability relates to the ability to articulate the model's decision-making process in a way that is comprehensible to non-experts. Both are essential for ensuring transparency, trust, and accountability in ML applications.

One of the main challenges is that many ML models, especially complex ones like deep learning models, are inherently difficult to interpret. These models often involve thousands or even millions of parameters, making it challenging to pinpoint how specific inputs influence the model's output. This lack of interpretability can be a significant barrier to the adoption of ML in critical areas such as healthcare, finance, and legal systems, where stakeholders need to understand the reasoning behind decisions.

Prioritizing the development and deployment of interpretable and explainable ML models is essential in overcoming this challenge. This might include opting for simpler models that are inherently easier to understand or employing post-hoc explanation techniques like LIME (Local Interpretable Model-agnostic Explanations) or SHAP (SHapley Additive exPlanations), which offer insights into the decision-making process of a model. Additionally, it's important to design explainability strategies with different stakeholders in mind, ensuring that the explanations are appropriately tailored to the audience's level of expertise.

Ethical AI Governance

As organizations increasingly rely on ML, the need for robust ethical AI governance frameworks has become more urgent. Ethical AI governance involves establishing policies, practices, and standards to ensure that ML models are developed and deployed in a manner that aligns with ethical principles and societal values.

One of the key challenges in ethical AI governance is the alignment of ML models with ethical principles such as fairness, accountability, and transparency. Ensuring that ML models do not perpetuate bias, discrimination, or other unethical practices requires a comprehensive approach to governance that spans the entire ML lifecycle, from data collection to model deployment.

Another challenge is the need for cross-disciplinary collaboration in ethical AI governance. Ethical considerations often intersect with legal, technical, and social issues, requiring input from diverse stakeholders, including ethicists, legal experts, data scientists, and community representatives. Coordinating this collaboration can be complex, particularly in large organizations with diverse teams and interests.

Establishing clear ethical AI governance frameworks that outline the guiding principles and standards for ML development and deployment is crucial in overcoming these challenges. This involves conducting regular ethical audits, implementing strategies to mitigate bias, and ensuring active stakeholder engagement throughout the governance process. Cultivating a culture of ethical awareness and accountability, where ethical considerations are embedded into daily decision making, is also essential for long-term success.

Scalability and Operationalization of ML Models

As organizations scale their ML initiatives, they face significant challenges in operationalizing and managing ML models at scale. The process of deploying, monitoring, and maintaining ML models in production environments can be complex and resource-intensive, particularly as the number of models increases.

One of the primary challenges is ensuring the scalability of ML models. As the volume of data and the number of models grow, organizations must implement scalable infrastructure and processes that can handle the increased complexity. This includes adopting cloud-based platforms, containerization, and orchestration tools such as Kubernetes to manage and scale ML workloads efficiently.

Operationalization, or "MLOps," is another critical challenge. MLOps involves the integration of ML models into the broader IT and business processes, ensuring that models can be deployed, monitored, and updated seamlessly. This requires collaboration between data scientists, ML engineers, and IT operations teams, as well as the implementation of automated tools and processes to streamline model deployment and monitoring.

To address these challenges, organizations must invest in building scalable and resilient ML infrastructure that can support their growth. This includes adopting cloud-based solutions, implementing MLOps practices, and ensuring that the necessary tools and resources are in place to manage ML models at scale. Additionally, organizations should prioritize the continuous monitoring and maintenance of ML models, ensuring that they remain accurate, reliable, and aligned with business objectives.

Case Studies

To further examine how data governance practices and frameworks can meet the inherent challenges of machine learning models, let's take a look at a few case studies.

Healthcare Industry

Healthcare data is inherently complex, comprising diverse sources such as electronic health records (EHRs), medical imaging, genetic data, and patient-reported outcomes. The quality of this data is paramount, as inaccuracies or inconsistencies can lead to incorrect diagnoses or treatment recommendations. For instance, if an ML model is trained on incomplete or erroneous patient data, it may generate biased predictions that could harm patient outcomes.

Data Quality Challenges

Healthcare organizations must implement stringent data governance practices to handle such challenges. This includes regular data cleansing processes to remove duplicates, correct errors, and fill in missing information. Additionally, standardizing data formats and terminology across different systems is essential for ensuring consistency and interoperability. For example, adopting common standards like HL7 (Health Level Seven) and SNOMED CT (Systematized Nomenclature of Medicine-Clinical Terms) can help harmonize data from various sources, improving overall data quality.

Ethical Implications and Algorithmic Bias

The ethical implications of using ML in healthcare are profound, particularly concerning algorithmic bias. Bias in ML models can arise from the data used to train them, which may reflect historical disparities in healthcare access and treatment. For instance, if a model is trained predominantly on data from a specific demographic, it may not perform well for other populations, leading to unequal treatment outcomes.

Healthcare organizations must take proactive steps to identify and mitigate bias in ML models. This includes conducting bias audits, where models are tested across different demographic groups to ensure fairness. Additionally, incorporating diverse datasets that represent the full spectrum of patient populations can help reduce the risk of biased predictions. For example, a study by Obermeyer et al. (2019) highlighted the need for careful consideration of socioeconomic factors in predictive models to avoid perpetuating health disparities.

The "black box" nature of some ML models poses a challenge in healthcare, where transparency and explainability are crucial for clinical decision making. Physicians and patients need to understand the rationale behind a model's recommendations to make informed decisions. To address this, healthcare organizations can adopt explainable AI techniques, such as decision trees or rule-based models, which provide clear and interpretable outputs. Alternatively, post-hoc explanation methods like LIME can be used to generate explanations for more complex models, ensuring that clinicians can trust and verify the model's predictions.

Financial Services

Financial institutions operate in a highly regulated environment, where compliance with laws such as the General Data Protection Regulation (GDPR) and the Dodd-Frank Act is mandatory. The use of ML in decision-making processes, such as loan approvals or credit scoring, must align with these regulations to ensure fairness and transparency.

Regulatory Compliance Challenges

One of the key challenges is ensuring that ML models comply with the "right to explanation" under GDPR, which requires organizations to provide clear explanations for automated decisions. This can be particularly difficult for complex ML models, such as deep learning networks, which are often opaque and difficult to interpret. To address this, financial institutions can use interpretable models or employ techniques like SHAP values to explain model decisions in a way that satisfies regulatory requirements.

Another regulatory challenge is managing data privacy and ensuring that customer data is handled in accordance with legal standards. Financial institutions must implement robust data governance frameworks that include data encryption, access controls, and audit trails to protect sensitive information. For instance, adopting data masking techniques can help anonymize personal information, reducing the risk of data breaches and ensuring compliance with privacy regulations.

Data Security and Cyber Threats

The financial services industry is a prime target for cyberattacks, given the high value of the data it handles. ML models themselves can be vulnerable to various forms of cyber threats, such as data poisoning, where attackers manipulate the training data to skew model predictions, or adversarial attacks, where malicious inputs cause models to make incorrect decisions.

To mitigate these risks, financial institutions must adopt advanced cybersecurity measures as part of their ML data governance strategy. This includes implementing multi-factor authentication, intrusion detection systems, and continuous monitoring of ML models for signs of tampering or abnormal behavior. Additionally, adopting adversarial training techniques can help improve the resilience of ML models against attacks by exposing them to malicious inputs during the training phase.

Financial institutions must also ensure that their ML models are regularly updated and retrained to account for changes in the threat landscape. For example, as fraud tactics evolve, fraud detection models must be continuously refined to stay effective. This requires a combination of automated model monitoring tools and human oversight to detect and respond to emerging threats in real time.

Retail Industry

Retail organizations often struggle with data silos, where customer, sales, inventory, and supply chain data are stored in separate systems, making it difficult to create a unified view of the business. This fragmentation can limit the effectiveness of ML models, which rely on comprehensive and integrated datasets to generate accurate predictions.

Data Silos and Integration

To overcome data silos, retail organizations must invest in data integration platforms that facilitate the seamless exchange of information across different systems. For example, adopting cloud-based data lakes can help centralize data from various sources, enabling ML models to access the full spectrum of relevant data. Additionally, implementing data governance practices that standardize data formats and ensure consistent data quality across departments is essential for improving the accuracy and reliability of ML models.

Scalability and Operationalization of ML Models

As retail organizations expand their use of ML, they face challenges in scaling and operationalizing their models across multiple stores, regions, and channels. Ensuring that ML models can handle large volumes of data and operate efficiently in real-time environments is critical for maintaining a competitive edge.

To address scalability challenges, retail organizations can adopt cloud-based ML platforms that offer the flexibility to scale computing resources as needed. For instance, using serverless computing services allows organizations to dynamically allocate resources based on demand, ensuring that ML models can handle peak loads during busy shopping periods. Additionally, containerization technologies like Docker can help package and deploy ML models consistently across different environments, reducing the complexity of managing large-scale ML deployments.

Operationalizing ML models, or MLOps, is another critical aspect of scalability. Retail organizations must establish robust MLOps practices that integrate ML workflows into their existing IT and business processes. This includes automating model deployment, monitoring, and retraining processes to ensure that models remain accurate and up-to-date. For example, implementing continuous integration and continuous deployment (CI/CD) pipelines for ML models can streamline the process of updating models as new data becomes available, reducing the time to market for new ML-driven initiatives.

Manufacturing Sector

The manufacturing sector faces a significant talent gap when it comes to implementing and managing ML initiatives. While there is growing interest in using ML to drive operational improvements, there is a shortage of professionals with the necessary skills to develop, deploy, and govern ML models in manufacturing environments.

Talent and Expertise Gap

To counter this gap, manufacturing organizations must invest in upskilling their existing workforce and fostering a culture of continuous learning. This includes providing training programs in ML, data science, and data governance, as well as encouraging cross-disciplinary collaboration between data scientists, engineers, and production managers. For example, partnerships with academic institutions and industry consortia can help develop tailored training programs that address the specific needs of the manufacturing sector.

Manufacturing organizations may need to explore alternative talent strategies, such as leveraging external consultants or adopting automated ML tools that simplify model development and deployment. Automated ML platforms can help bridge the talent gap by enabling domain experts with limited ML expertise to build and deploy models, democratizing access to ML capabilities across the organization.

Model Drift and Maintenance

Model drift is a significant challenge in manufacturing, where changing conditions on the production floor or in the supply chain can impact the accuracy of ML models. For example, an ML model used to predict equipment failures may become less accurate as new machinery is introduced or as operating conditions change over time.

To address model drift, manufacturing organizations must implement robust monitoring and maintenance practices as part of their ML data governance framework. This includes continuously monitoring model performance and retraining models on new data to ensure they remain accurate and relevant. For instance, predictive maintenance models can be retrained using the latest sensor data to account for changes in equipment behavior or environmental conditions.

Manufacturing organizations can also implement automated model monitoring tools that detect signs of drift in real-time and trigger alerts or corrective actions. For example, if a predictive maintenance model starts to exhibit decreased accuracy, the system could automatically retrain the model using the latest data or escalate the issue to human operators for further investigation.

Manufacturing organizations should establish clear governance processes for updating and validating ML models. This includes defining criteria for when a model should be retrained, as well as ensuring that all updates are thoroughly tested before being deployed in production environments. By maintaining rigorous oversight of ML models, manufacturing organizations can mitigate the risks associated with model drift and ensure that their ML initiatives continue to deliver value over time.

Public Sector

Transparency is a fundamental requirement for ML applications in the public sector, where decisions can have far-reaching impacts on citizens' lives. For example, ML models used in areas such as social services, law enforcement, or tax collection must be transparent to ensure that they are fair, unbiased, and accountable to the public.

Transparency and Public Accountability

One of the key challenges is ensuring that ML models are explainable to both policymakers and the general public. Public sector organizations must adopt explainable AI techniques that allow stakeholders to understand how models make decisions and to identify any potential biases or errors. For example, using decision trees or rule-based models can provide clear and interpretable outputs, making it easier to communicate model decisions to non-experts.

Public sector organizations must also establish clear governance processes that involve stakeholders in the development and oversight of ML models. This includes conducting public consultations, engaging with civil society organizations, and providing transparency reports that detail how ML models are used in public services. By fostering open and inclusive governance practices, public sector organizations can build public trust and ensure that ML applications align with societal values.

Security and Data Privacy

Security and data privacy are critical concerns in the public sector, where ML models often rely on sensitive personal information, such as health records, financial data, or social security numbers. Protecting this data from cyber threats and ensuring that it is used responsibly is essential for maintaining public confidence in ML-driven public services.

Public sector organizations must implement robust data protection measures as part of their ML data governance framework. This includes encrypting data at rest and in transit, implementing access controls to restrict data access to authorized personnel, and conducting regular security audits to identify and mitigate vulnerabilities. For example, adopting advanced encryption techniques and multi-factor authentication can help safeguard sensitive data from unauthorized access.

Public sector organizations must also ensure that ML models comply with data privacy regulations, such as the GDPR or the California Consumer Privacy Act (CCPA). This includes implementing data anonymization techniques to protect individuals' identities and ensuring that data is collected and used in accordance with legal and ethical standards. For instance, when using ML for predictive policing, organizations must carefully consider the implications of using personal data and take steps to prevent potential misuse or discrimination.

Public sector organizations can also leverage privacy-enhancing technologies, such as differential privacy, which allows for the analysis of data while preserving individual privacy. By adopting these technologies, public sector organizations can balance the need for data-driven insights with the obligation to protect citizens' privacy and rights.

Energy Sector

Optimizing Data Utilization for Predictive Maintenance

In the energy sector, predictive maintenance using ML has the potential to drastically reduce operational costs by predicting equipment failures before they occur, thus avoiding unplanned downtime. However, the effectiveness of these models hinges on the quality and utilization of the data collected from various sources, such as sensors, operational logs, and environmental data.

One of the key challenges is the sheer volume of data generated by energy systems. For instance, a single wind turbine can generate terabytes of data daily, covering aspects like temperature, vibration, and power output. This data must be efficiently processed, stored, and analyzed to provide actionable insights. Energy companies must therefore invest in scalable data infrastructure, such as cloud-based storage solutions and distributed computing systems, to handle these large datasets.

Integrating data from various sources, such as different types of sensors, historical maintenance records, and weather data, is crucial for improving the accuracy of predictive maintenance models. This requires robust data governance practices, including data normalization and standardization, to ensure that all data sources are compatible and can be effectively utilized by ML models.

Energy companies must also address the challenge of real-time data processing, which is essential for predicting imminent equipment failures. Implementing edge computing solutions, where data processing occurs close to the source, can help reduce latency and enable faster decision making. For example, installing edge devices on wind turbines can allow real-time monitoring and prediction of potential faults, enabling timely maintenance interventions.

Ethical Concerns and Decision Making in Predictive Maintenance

The use of ML in predictive maintenance raises ethical concerns, particularly in terms of decision making and the impact on human workers. For example, reliance on ML models to predict equipment failures could lead to decisions that prioritize cost savings over safety, potentially putting workers and the environment at risk.

Energy companies must therefore establish ethical guidelines as part of their ML data governance framework. This includes ensuring that predictive maintenance models are transparent and that the decision-making process is well documented and subject to human oversight. For instance, while ML models can recommend maintenance actions, the final decision should involve human operators who can consider additional factors, such as safety protocols and environmental regulations.

There is a need to address the potential impact of predictive maintenance on the workforce. As ML models become more prevalent, there may be concerns about job displacement or changes in job roles. Energy companies should engage with workers and labor unions to address these concerns and provide training and upskilling opportunities to help workers adapt to new technologies. For example, maintenance workers could be trained in data analysis and ML to take on more advanced roles in predictive maintenance.

The use of predictive maintenance models must also consider the broader environmental and social impact. For instance, decisions made by ML models should align with the company's sustainability goals and environmental regulations. This could involve incorporating environmental impact assessments into the predictive maintenance process to ensure that decisions do not inadvertently harm the environment.

Energy companies can also explore the use of AI ethics frameworks, such as the Ethical AI Principles developed by organizations like the World Economic Forum, to guide the development and deployment of predictive maintenance models. By integrating ethical considerations into their ML data governance practices, energy companies can ensure that their use of ML aligns with societal values and contributes to sustainable and responsible energy production.

Telecommunications
Data Privacy Concerns

Telecommunications companies handle vast amounts of personal data, including call records, location data, and internet usage patterns. This data is invaluable for building ML models that can optimize network performance, personalize customer experiences, and identify potential churn. However, the use of such data raises significant privacy concerns, particularly in light of stringent data protection regulations such as the GDPR and the CCPA.

One of the primary challenges is ensuring that customer data is collected, processed, and stored in a manner that complies with privacy laws. This requires telecommunications companies to implement robust data governance frameworks that include data anonymization, encryption, and access control measures. For example, using techniques like differential privacy can help anonymize customer data while still allowing for meaningful analysis and ML model development.

Another challenge is obtaining informed consent from customers for the use of their data in ML models. Telecommunications companies must be transparent about how customer data will be used and ensure that customers have the ability to opt out of data collection if they choose. This involves clear communication through privacy notices and providing customers with easy-to-use tools for managing their data preferences.

Telecommunications companies must also address the risk of data breaches, which can compromise customer privacy and result in significant legal and financial penalties. Implementing advanced cybersecurity measures, such as intrusion detection systems and regular security audits, is essential for protecting customer data and ensuring compliance with data protection regulations.

Ensuring Fairness in Customer Segmentation

ML models used for customer segmentation in telecommunications can offer significant business value by enabling companies to tailor their services to different customer groups. However, there is a risk that these models may inadvertently perpetuate or exacerbate existing biases, leading to unfair treatment of certain customer segments.

For example, if an ML model is trained on historical data that reflects discriminatory practices, such as offering better deals to certain demographic groups, it may continue to replicate these biases in its predictions. This can result in some customers receiving less favorable treatment based on factors like age, gender, or ethnicity.

To ensure fairness in customer segmentation, telecommunications companies must implement bias detection and mitigation strategies as part of their ML data governance framework. This includes conducting regular audits of ML models to identify and address any potential biases. For instance, using fairness metrics, such as demographic parity or equalized odds, can help assess whether a model's predictions are biased against specific groups.

Telecommunications companies should also consider the broader ethical implications of their customer segmentation strategies. For example, decisions based on ML models should not lead to the exclusion or marginalization of certain customer groups. To address this, companies can adopt an inclusive design approach, where customer segmentation models are developed with input from diverse stakeholders, including representatives from different demographic groups.

In addition to technical measures, telecommunications companies should also establish clear governance processes for reviewing and approving ML models used for customer segmentation. This includes setting up ethics committees or review boards that can provide oversight and ensure that models align with the company's values and ethical standards.

Transparency is crucial for maintaining customer trust. Telecommunications companies should provide customers with clear explanations of how their data is used and how segmentation decisions are made. This can involve offering customers access to tools that allow them to see how they have been segmented and provide feedback if they believe they have been treated unfairly.

Transportation and Logistics

In transportation and logistics, data is generated from a wide range of sources, including GPS systems, IoT sensors, traffic management systems, and supply chain management software. Ensuring that this data can be effectively integrated and utilized by ML models is a significant challenge, particularly given the diversity of data formats, protocols, and standards used across the industry.

Data Interoperability Challenges

One of the primary challenges is achieving data interoperability, where different systems and devices can seamlessly exchange and use data. For example, a logistics company may need to integrate data from its fleet of trucks, warehouse management systems, and third-party suppliers to optimize delivery routes and manage inventory. However, if these systems use incompatible data formats or protocols, it can be difficult to create a unified view of the supply chain.

To address this challenge, transportation and logistics companies must adopt data governance practices that promote standardization and interoperability. This includes implementing industry standards, such as the Electronic Data Interchange (EDI) or the Open Data Protocol (OData), to facilitate data exchange between different systems. Additionally, companies can use data integration platforms that offer APIs and connectors to link disparate systems, enabling seamless data flow across the supply chain.

The use of cloud-based platforms can help centralize data from various sources, making it easier for ML models to access and analyze the information. For example, using a cloud-based transportation management system (TMS) can provide a single source of truth for all transportation-related data, improving the accuracy of ML-driven route optimization models.

Managing the Risks of Autonomous Systems

The development of autonomous systems, such as self-driving vehicles and drones, is one of the most exciting applications of ML in the transportation and logistics industry. However, these systems also pose significant risks, particularly in terms of safety and liability.

One of the key challenges is ensuring that autonomous systems are safe and reliable in real-world environments. ML models used in autonomous systems must be trained on vast amounts of data to account for the myriad of scenarios they may encounter, from varying weather conditions to unpredictable human behavior. However, even with extensive training, there is always a risk that an autonomous system may make incorrect decisions, leading to accidents or other safety incidents.

To mitigate these risks, transportation and logistics companies must implement rigorous testing and validation processes for autonomous systems. This includes conducting extensive simulations and real-world testing to ensure that the ML models can handle a wide range of scenarios.

Real Estate

ML has the potential to transform market analysis in real estate by providing predictive insights into property values, market trends, and investment opportunities. For instance, ML models can analyze historical sales data, property features, and economic indicators to predict future property values and identify emerging market trends.

Leveraging ML for Market Analysis

One of the key challenges in leveraging ML for market analysis is ensuring that models are accurate and reliable. Real estate markets are influenced by a wide range of factors, including economic conditions, local regulations, and demographic trends, which can vary significantly across different regions. To address this challenge, real estate professionals must use diverse and comprehensive datasets that capture these factors and train ML models accordingly.

For example, incorporating data from multiple sources, such as property listings, sales records, and economic reports, can help create a more complete picture of market conditions. Additionally, using advanced ML techniques, such as ensemble methods or deep learning, can improve the accuracy of market predictions by combining insights from different models and accounting for complex interactions between variables.

Another challenge is ensuring that ML models can adapt to changing market conditions. Real estate markets are dynamic and can be affected by sudden shifts in economic or regulatory factors. To address this, real estate professionals must implement regular model updates and retraining processes to ensure that models remain relevant and accurate over time. For instance, using a continuous learning approach, where models are periodically updated with new data, can help maintain their predictive power and relevance.

Addressing Ethical Concerns

The use of ML in real estate raises ethical concerns, particularly in terms of fairness and transparency. For example, ML models used for property valuation or investment recommendations can inadvertently perpetuate biases or reinforce existing inequalities in the housing market.

One of the key ethical concerns is ensuring that ML models do not discriminate against certain demographic groups or perpetuate biases based on factors such as race, gender, or socioeconomic status. To address this, real estate professionals must implement fairness and bias detection techniques as part of their ML data governance framework. This includes conducting regular audits of ML models to identify and mitigate any potential biases, as well as using fairness metrics to assess the impact of models on different demographic groups.

Transparency is crucial for maintaining trust in ML-driven real estate decisions. Real estate professionals should provide clear explanations of how ML models are used and how decisions are made, ensuring that clients and stakeholders understand the factors influencing property valuations and investment recommendations. This can involve creating user-friendly dashboards or reports that highlight key model inputs and outputs.

Ethical considerations also include ensuring that ML models respect privacy and data protection regulations. Real estate professionals must handle personal and financial data responsibly, implementing measures such as data anonymization and secure storage to protect client information. For example, when using ML to analyze property buyers' preferences, ensuring that personal data is anonymized and aggregated can help maintain privacy and comply with data protection laws.

Insurance

ML has the potential to enhance risk assessment in insurance by providing more accurate predictions of claim likelihood, policyholder behavior, and potential losses. For example, ML models can analyze historical claims data, customer profiles, and external factors to assess the risk associated with different insurance policies.

Improving Risk Assessment

One of the key challenges in improving risk assessment is ensuring that ML models are accurate and reliable. Insurance companies must use comprehensive datasets that capture a wide range of factors influencing risk, including historical claims, policyholder information, and external variables such as economic conditions or weather events.

To address this challenge, insurers can invest in data collection and analysis tools that provide a holistic view of risk factors. For example, integrating data from multiple sources, such as telematics devices in vehicles or health monitoring systems, can help improve the accuracy of risk predictions. Additionally, using advanced ML techniques, such as ensemble methods or deep learning, can enhance the predictive power of risk assessment models.

Managing Model Interpretability

Model interpretability is a critical challenge in the insurance industry, where decisions based on ML models can have significant impacts on policyholders and claimants. Ensuring that ML models are interpretable and transparent is essential for maintaining trust and compliance with regulatory requirements.

One of the primary challenges is providing clear explanations of how ML models make decisions and assess risk. For example, if an ML model is used to determine insurance premiums or claim payouts, policyholders should be able to understand the factors influencing these decisions and verify that they are fair and accurate.

To address this challenge, insurers can use explainable AI techniques that provide insights into model decision-making processes. For example, techniques such as SHAP (SHapley Additive exPlanations) or LIME (Local Interpretable Model-agnostic Explanations) can help explain the contributions of different features to model predictions. Additionally, insurers can provide transparency reports that detail the methodologies and data used in ML models, ensuring that stakeholders have access to relevant information.

Insurers must also consider the ethical implications of ML-based risk assessments. For example, models should be designed to avoid discriminatory practices and ensure fairness in underwriting and claim processing. Implementing fairness audits and bias detection techniques can help identify and mitigate any potential biases in ML models, ensuring that decisions are equitable and aligned with regulatory standards.

Summary

The chapter explores the evolving landscape of machine learning (ML) data governance. It begins by emphasizing how ML has transitioned from a niche technology to a critical component of business operations, making robust data governance increasingly vital. It highlights the need for comprehensive frameworks that ensure data quality, privacy, security, and ethical use throughout the ML model lifecycle. It delves into existing challenges such as managing diverse data sources, integrating governance into organizational processes, and staying compliant with an ever-changing regulatory landscape. Transparency and explainability of ML models are also significant hurdles, particularly as these models become more complex.

The chapter then shifts focus to future trends that are likely to shape ML data governance, including the rise of decentralized data sources due to edge computing, the integration of AI for automating governance tasks, and the growing demand for ethical AI practices. It stresses the importance of adapting governance strategies to these trends to manage emerging challenges like the scalability of governance frameworks, cross-border data flows, and diverse data types.

The chapter concludes by discussing the need for organizations to develop more sophisticated, agile, and proactive approaches to ML data governance to stay ahead of potential risks and capitalize on new opportunities in the rapidly evolving technological landscape.

APPENDIX A

Frameworks and Standards in Machine Learning Data Governance

Data Governance Frameworks

Several frameworks have been developed to guide organizations in establishing effective data governance practices. While not all are specifically tailored to ML, they provide foundational principles that can be adapted to the unique needs of ML projects.

The DAMA-DMBOK Framework

The Data Management Body of Knowledge (DAMA-DMBOK) is one of the most comprehensive frameworks for data management and governance. Developed by the Data Management Association (DAMA), it outlines key areas of data management, including data governance, data quality, data security, and data architecture.

Key Components Relevant to ML

- Data Quality Management: Ensures that data used in ML models is accurate, consistent, and reliable.

- Data Security Management: Protects data used in ML from unauthorized access and breaches.

- Data Architecture Management: Defines the structure and organization of data used in ML, ensuring it is accessible and well structured.

711

© Aditya Nandan Prasad 2024
A. Nandan Prasad, *Introduction to Data Governance for Machine Learning Systems*,
https://doi.org/10.1007/979-8-8688-1023-7

Application to ML

The DAMA-DMBOK framework provides a solid foundation for managing the data lifecycle in ML projects. By ensuring that data governance practices are in place from data acquisition to model deployment, organizations can mitigate risks and enhance the reliability of their ML models.

COBIT (Control Objectives for Information and Related Technologies)

COBIT, developed by ISACA, is a framework for developing, implementing, monitoring, and improving information technology (IT) governance and management practices. Although originally designed for IT governance, COBIT's principles can be applied to data governance in ML.

Key Components Relevant to ML

- Alignment of Data Governance with Business Goals: Ensures that data governance supports the overall objectives of the organization, including those related to ML.

- Risk Management: Identifies and mitigates risks associated with data governance in ML, including data breaches and compliance failures.

- Performance Management: Monitors the effectiveness of data governance practices in achieving desired outcomes, such as improved model performance.

Application to ML

COBIT's focus on aligning data governance with business objectives is particularly relevant for ML projects, where the goal is often to derive actionable insights and drive business value. By applying COBIT principles, organizations can ensure that their ML initiatives are well governed and aligned with strategic goals.

ISO/IEC 38500:2015—IT Governance

The ISO/IEC 38500:2015 standard provides principles for the effective, efficient, and acceptable use of IT within organizations. While it is not exclusively focused on data governance, its principles are highly relevant for governing data in ML projects.

Key Principles Relevant to ML

- Responsibility: Clearly define roles and responsibilities for data governance in ML.

- Strategy: Ensure that data governance strategies support the overall business and IT strategies, including ML initiatives.

- Acquisition: Govern the acquisition and use of data in ML to ensure it is appropriate and ethical.

Application to ML

ISO/IEC 38500:2015 provides a high-level framework for governance that can be adapted to the specific needs of ML projects. Its emphasis on responsibility, strategy, and acquisition is particularly relevant for ensuring that ML models are built on a solid foundation of well-governed data.

The CMMI (Capability Maturity Model Integration) Framework

CMMI is a process level improvement training and appraisal program that helps organizations improve their processes. It includes specific guidance on data management, which is crucial for ML projects.

Key Components Relevant to ML

- Process Management: Ensures that data governance processes are well defined and consistently applied across ML projects.

- Performance Measurement: Monitors the effectiveness of data governance practices in supporting ML model development and deployment.

- Continuous Improvement: Encourages ongoing refinement of data governance practices to adapt to evolving ML requirements.

Application to ML

CMMI's focus on process management and continuous improvement is highly relevant for ML projects, where data governance practices must evolve to keep pace with new technologies and methodologies. By adopting CMMI principles, organizations can ensure that their data governance practices remain effective and aligned with best practices.

Industry Standards for Data Privacy

Data privacy is a critical concern in ML, particularly when models are trained on personal or sensitive data. Several industry standards provide guidelines for protecting data privacy, ensuring compliance with regulations, and safeguarding individual rights.

General Data Protection Regulation (GDPR)

The GDPR is a comprehensive data protection regulation enacted by the European Union (EU) in 2018. It sets stringent requirements for the collection, processing, and storage of personal data, with significant implications for ML projects.

Key Requirements Relevant to ML

- Consent: Obtain explicit consent from individuals before collecting their data for ML purposes.

- Data Minimization: Collect only the data necessary for the specific ML task.

- Right to Explanation: Ensure that ML models are interpretable and that individuals have the right to understand how decisions affecting them are made.

- Data Subject Rights: Implement processes to allow individuals to access, correct, or delete their data.

Application to ML

Compliance with GDPR is essential for any ML project that involves personal data. This includes not only data collected within the EU but also data from EU citizens regardless of where the processing occurs. Organizations must implement robust data governance practices to ensure that their ML models are GDPR-compliant, including mechanisms for obtaining consent, minimizing data collection, and providing explanations for automated decisions.

California Consumer Privacy Act (CCPA)

The CCPA is a state statute intended to enhance privacy rights and consumer protection for residents of California, USA. It is similar to GDPR in its emphasis on data privacy and consumer rights but has some distinct provisions.

Key Requirements Relevant to ML

- Disclosure: Inform consumers about the types of personal data being collected and the purposes for which it is used in ML models.

- Opt-Out: Provide consumers with the option to opt-out of the sale of their personal data, which may include data used in ML models.

- Data Access and Deletion: Allow consumers to access and request the deletion of their personal data.

Application to ML

Like GDPR, CCPA compliance is critical for ML projects that handle personal data from California residents. Organizations must establish data governance practices that enable them to meet CCPA requirements, including transparent data practices, consumer opt-out options, and mechanisms for data access and deletion.

Health Insurance Portability and Accountability Act (HIPAA)

HIPAA is a US law that sets standards for protecting sensitive patient health information. It is particularly relevant for ML projects in the healthcare sector that use patient data for model training and analysis.

Key Requirements Relevant to ML

- Protected Health Information (PHI) Safeguards: Implement safeguards to protect PHI used in ML models.

- De-identification: Ensure that data is de-identified when used for ML purposes to prevent the identification of individual patients.

- Compliance with Privacy Rule: Adhere to the Privacy Rule, which governs the use and disclosure of PHI.

Application to ML

ML projects in healthcare must comply with HIPAA to protect patient data and avoid legal penalties. This includes implementing robust data governance practices that safeguard PHI, ensure data is de-identified, and maintain compliance with HIPAA regulations.

NIST Privacy Framework

The National Institute of Standards and Technology (NIST) Privacy Framework is a voluntary framework designed to help organizations manage privacy risks. It provides guidelines for identifying and mitigating privacy risks, which are particularly relevant for ML projects.

Key Components Relevant to ML

- Privacy Risk Assessment: Identify and assess privacy risks associated with the use of personal data in ML models.

- Data Protection Strategies: Implement strategies to protect personal data, including encryption and access controls.

- Transparency: Ensure that data practices related to ML are transparent and that individuals are informed about how their data is used.

Application to ML

The NIST Privacy Framework provides a structured approach to managing privacy risks in ML projects. By following its guidelines, organizations can ensure that their ML models are designed and deployed with privacy in mind, minimizing risks and building trust with users.

AI Ethics Frameworks

As ML and AI technologies become more pervasive, ethical considerations are increasingly important. AI ethics frameworks provide guidelines for developing and deploying ML models in a way that is fair, transparent, and accountable.

The European Commission's Ethics Guidelines for Trustworthy AI

The European Commission has developed Ethics Guidelines for Trustworthy AI, which outline principles for ensuring that AI systems, including ML models, are ethical and trustworthy.

Key Principles Relevant to ML

- Human Agency and Oversight: Ensure that ML models support human decision-making and do not undermine human autonomy.

- Fairness: Avoid biases in ML models that could lead to unfair treatment of individuals or groups.

- Transparency: Make ML models and their decision-making processes transparent and understandable to users.

Application to ML

These guidelines are particularly relevant for organizations deploying ML models in areas with significant ethical implications, such as healthcare, finance, and criminal justice. By adhering to these principles, organizations can build ML models that are not only effective but also ethical and trustworthy.

The IEEE Global Initiative on Ethics of Autonomous and Intelligent Systems

The IEEE Global Initiative has developed a series of standards and guidelines for the ethical development and deployment of autonomous and intelligent systems, including ML.

Key Components Relevant to ML

- Ethically Aligned Design: Ensure that ML models are designed with ethical considerations in mind, including fairness, accountability, and transparency.

- Bias and Fairness: Implement strategies to identify and mitigate biases in ML models.

- Data Governance: Establish data governance practices that ensure the ethical use of data in ML models.

Application to ML

The IEEE guidelines provide a comprehensive approach to ethical AI, with a strong emphasis on bias and fairness. Organizations can use these guidelines to ensure that their ML models are ethically sound and aligned with best practices in AI ethics.

The Montreal Declaration for Responsible AI

The Montreal Declaration for Responsible AI is a set of principles aimed at guiding the development and deployment of AI technologies, including ML, in a way that respects human rights and promotes social well-being.

Key Principles Relevant to ML

- Well-being: Ensure that ML models contribute to the well-being of individuals and society.

- Respect for Autonomy: Design ML models that respect individuals' autonomy and decision-making capabilities.

- Non-discrimination: Prevent discrimination and bias in ML models to ensure fair and equitable treatment of all individuals.

Application to ML

The Montreal Declaration provides a human-centered approach to AI ethics, emphasizing the need for ML models to promote well-being and respect individual rights. Organizations can use these principles to guide the ethical development of their ML models, ensuring they align with broader societal values.

ML-Specific Governance Guidelines

In addition to general data governance frameworks and AI ethics guidelines, several ML-specific governance guidelines have been developed to address the unique challenges of ML model development and deployment.

Model Governance Frameworks

Model governance refers to the processes and controls that ensure ML models are developed, deployed, and monitored in a way that is consistent with organizational policies and regulatory requirements.

Key Components Relevant to ML

- Model Development: Establish guidelines for the development of ML models, including data selection, model training, and validation.

- Model Deployment: Implement controls to ensure that ML models are deployed in a way that is secure, compliant, and aligned with business goals.

- Model Monitoring and Maintenance: Develop processes for monitoring ML models in production, including performance tracking and bias detection.

Application to ML

Model governance is critical for ensuring that ML models are reliable, compliant, and aligned with organizational objectives. By establishing a robust model governance framework, organizations can mitigate risks and ensure that their ML models deliver value while adhering to ethical and regulatory standards.

The FAIR Principles for ML

The FAIR Principles (Findability, Accessibility, Interoperability, and Reusability) were originally developed for data management but have been adapted for ML to ensure that ML models and their associated data are well governed and accessible.

Key Components Relevant to ML

- Findability: Ensure that ML models and their data are easily findable within the organization.

- Accessibility: Make ML models and their data accessible to those who need them, while protecting sensitive information.

- Interoperability: Design ML models and their data to be interoperable with other systems and technologies.

- Reusability: Ensure that ML models and their data can be reused and repurposed for future projects.

Application to ML

The FAIR Principles provide a practical framework for managing ML models and their data throughout their lifecycle. By adhering to these principles, organizations can enhance the governance of their ML projects, making models more accessible, interoperable, and reusable.

Responsible AI Governance

Responsible AI governance refers to the practices and guidelines that ensure AI systems, including ML models, are developed and deployed in a responsible manner, taking into account ethical, legal, and societal considerations.

Key Components Relevant to ML

- Ethical AI Design: Incorporate ethical considerations into the design and development of ML models.

- Compliance Monitoring: Monitor ML models for compliance with relevant regulations and ethical guidelines.

- Stakeholder Engagement: Engage stakeholders, including users, regulators, and affected communities, in the governance of ML models.

Application to ML

Responsible AI governance is essential for ensuring that ML models are developed and used in a way that is ethical, legal, and socially responsible. By adopting responsible AI governance practices, organizations can build trust in their ML models and mitigate potential risks.

Checklists and Templates: Practical Tools for Implementing ML Data Governance Practices

The rapid adoption of machine learning (ML) across industries has created a demand for robust governance frameworks to ensure that data used in ML models is managed, secured, and aligned with ethical standards. Machine Learning Data Governance (MLDG) encompasses the practices, policies, and frameworks that govern how data is handled throughout the ML lifecycle—from data collection and preprocessing to model deployment and monitoring. As organizations seek to establish MLDG, practical tools such as checklists and templates become invaluable. These tools provide structured guidance for implementing governance practices, creating policies, and adapting frameworks to specific organizational needs.

Checklists for Implementing ML Data Governance Practices

Checklists are essential tools that help organizations ensure they cover all necessary steps when implementing MLDG practices. The following are checklists tailored to various aspects of MLDG.

Data Governance Framework Implementation Checklist

Define Data Governance Objectives and Goals

- Identify Business Objectives: Clearly define how data governance will support the organization's business goals.

- Set Specific Goals: Establish specific, measurable, achievable, relevant, and time-bound (SMART) goals for data governance.

- Align with Strategic Initiatives: Ensure that data governance goals align with the broader strategic initiatives of the organization.

- Communicate Objectives: Communicate the data governance objectives and goals across the organization to ensure alignment.

Establish a Data Governance Team

- Identify Key Stakeholders: Identify the key stakeholders who will be involved in data governance, including executives, data stewards, IT, and legal teams.

- Assign Roles and Responsibilities: Clearly define the roles and responsibilities of each stakeholder in the data governance process.

- Create a Data Governance Council: Form a data governance council or steering committee to oversee the implementation and ongoing management of the framework.

- Appoint Data Stewards: Assign data stewards responsible for managing data assets and ensuring data quality within specific domains.

Develop Data Governance Policies and Standards

- Create Data Management Policies: Develop policies that cover data creation, usage, sharing, storage, and disposal.

- Establish Data Quality Standards: Define standards for data accuracy, completeness, consistency, and timeliness.

- Implement Data Security Policies: Establish policies for data access, encryption, and compliance with data privacy regulations.

- Develop Data Privacy Policies: Create policies to ensure compliance with data protection laws such as GDPR or CCPA.

- Define Data Classification Guidelines: Implement guidelines for classifying data based on sensitivity and criticality.

Design the Data Governance Framework

- Define the Framework Structure: Outline the structure of the data governance framework, including governance layers, decision-making processes, and reporting lines.

- Select a Governance Model: Choose an appropriate governance model (centralized, decentralized, or hybrid) based on the organization's size, structure, and complexity.

- Create Data Governance Workflows: Develop workflows for key data governance processes, such as data quality management, data access requests, and issue resolution.

- Integrate with Existing Processes: Ensure that the data governance framework is integrated with existing business processes and IT systems.

Implement Data Governance Tools and Technologies

- Select Data Governance Tools: Choose tools for data cataloging, data lineage, data quality monitoring, and metadata management.

- Implement Data Quality Monitoring: Set up tools for real-time monitoring and reporting of data quality metrics.

- Deploy Data Lineage Tools: Implement tools to track data flow, transformations, and usage across the organization.

- Integrate with Data Management Platforms: Ensure that data governance tools are integrated with data management platforms, such as data warehouses or data lakes.

- Test Tools and Systems: Conduct thorough testing of data governance tools and systems before full-scale implementation.

Develop Data Governance Metrics and KPIs

- Define Key Metrics: Identify key metrics for measuring the effectiveness of data governance, such as data quality scores, compliance rates, and data usage metrics.

- Set KPIs: Establish Key Performance Indicators (KPIs) to track progress against data governance goals.

- Implement Reporting Dashboards: Develop dashboards for real-time tracking of data governance metrics and KPIs.

- Monitor and Review Metrics: Regularly monitor and review data governance metrics to identify areas for improvement.

- Adjust Metrics as Needed: Continuously refine metrics and KPIs based on evolving business needs and data governance maturity.

Conduct Data Governance Training and Awareness Programs

- Develop Training Programs: Create training programs for all levels of the organization, from executives to data stewards and end-users.

- Focus on Data Governance Best Practices: Ensure that training programs cover data governance best practices, policies, and standards.

- Promote Data Literacy: Implement programs to enhance data literacy across the organization, ensuring that all employees understand the importance of data governance.

- Provide Ongoing Training: Offer continuous training and refresher courses to keep employees updated on new data governance practices and tools.

- Assess Training Effectiveness: Evaluate the effectiveness of training programs through assessments and feedback surveys.

Implement Data Governance Processes and Workflows

- Define Key Processes: Clearly define key data governance processes, such as data quality management, data access control, and data issue resolution.

- Document Workflows: Document workflows for each process, detailing steps, roles, and responsibilities.

- Automate Where Possible: Implement automation for repetitive and routine data governance tasks to improve efficiency and reduce human error.

- Ensure Consistency: Standardize processes across the organization to ensure consistency in data governance practices.

- Monitor Process Effectiveness: Regularly review and refine processes to ensure they are effective and aligned with data governance goals.

Establish Data Governance Communication and Change Management

- Develop a Communication Plan: Create a plan for communicating data governance initiatives, policies, and changes to all stakeholders.

- Engage Stakeholders: Actively engage stakeholders in the data governance process, seeking their input and feedback.

- Manage Change: Implement change management strategies to ensure smooth adoption of data governance practices across the organization.

- Promote a Data-Driven Culture: Foster a culture where data governance is seen as a critical component of business success.

- Celebrate Successes: Recognize and celebrate successes in data governance implementation to build momentum and support.

Monitor and Review Data Governance Implementation

- Conduct Regular Audits: Perform regular audits of data governance processes, policies, and tools to ensure compliance and effectiveness.

- Review Data Governance Framework: Periodically review and update the data governance framework to reflect changes in business needs, technology, or regulations.

- Address Gaps and Issues: Identify and address any gaps or issues in data governance implementation.

- Seek Continuous Improvement: Continuously seek opportunities to improve data governance practices, processes, and tools.

- Report Progress to Leadership: Regularly report on data governance progress and outcomes to organizational leadership.

Ensure Regulatory Compliance

- Identify Applicable Regulations: Identify all relevant data protection and privacy regulations that apply to the organization.

- Implement Compliance Measures: Ensure that data governance practices are designed to meet regulatory requirements, including data retention, consent management, and data access controls.

- Conduct Compliance Audits: Regularly audit data governance practices to ensure ongoing compliance with regulations.

- Maintain Documentation: Keep thorough documentation of compliance measures, audit results, and corrective actions.

- Stay Updated on Regulatory Changes: Monitor changes in regulations and adjust data governance practices as needed to maintain compliance.

Evaluate and Refine Data Governance Framework

- Assess Framework Effectiveness: Evaluate the effectiveness of the data governance framework against established goals and KPIs.

- Gather Stakeholder Feedback: Collect feedback from stakeholders on the usability and effectiveness of the data governance framework.

- Identify Areas for Improvement: Identify areas where the framework can be improved or refined to better meet organizational needs.

- Update Framework as Needed: Make necessary updates to the data governance framework to reflect feedback and evolving business requirements.

- Document Lessons Learned: Document lessons learned during the implementation process to inform future data governance initiatives.

Sustain and Scale Data Governance

- Plan for Scalability: Ensure that the data governance framework is scalable to accommodate future growth in data volume, complexity, and organizational needs.

- Sustain Governance Practices: Develop strategies to sustain data governance practices over the long term, including continuous training, regular audits, and stakeholder engagement.

- Integrate with New Initiatives: Ensure that data governance is integrated into new business initiatives, projects, and technology implementations.

- Measure Long-Term Impact: Track the long-term impact of data governance on business outcomes, including data-driven decision-making, risk management, and regulatory compliance.

- Adapt to Emerging Trends: Stay informed about emerging trends in data governance, technology, and regulations, and adapt the framework as necessary.

Prepare for Future Challenges and Opportunities

- Anticipate Future Challenges: Identify potential future challenges in data governance, such as increasing data volume, new regulations, or advances in technology.

- Develop Contingency Plans: Create contingency plans to address potential challenges, ensuring that the organization is prepared to respond effectively.

- Explore New Technologies: Stay informed about new technologies that could enhance data governance, such as artificial intelligence, machine learning, and blockchain.

- Foster Innovation: Encourage innovation in data governance practices, exploring new approaches to data management, quality, and security.

- Position for Competitive Advantage: Leverage data governance as a strategic asset to gain a competitive advantage in the marketplace.

Data Quality Checklist

Data Collection and Acquisition

Data Identification and Planning

- Define Data Requirements: Clearly specify the type, format, and quantity of data needed for the ML project.

- Identify Data Sources: Determine potential sources of data, including internal databases, external datasets, and data collection methods.

- Evaluate Data Quality: Assess the quality of available data sources in terms of accuracy, completeness, consistency, timeliness, and relevance.

- Consider Data Costs: Evaluate the costs associated with acquiring, storing, and processing data.

- Develop Data Collection Plan: Create a comprehensive plan outlining the steps involved in data collection, acquisition, and storage.

Data Acquisition Methods

- Internal Data Sources: Utilize existing internal databases, data warehouses, or data lakes.

- External Data Sources: Acquire data from third-party vendors, public datasets, or open-source repositories.

- Data Collection Tools: Employ data collection tools (e.g., web scraping, APIs, sensors) to gather relevant data.

- Data Partnerships: Collaborate with other organizations to share or exchange data.

- Data Generation: Create synthetic data or simulate real-world scenarios to augment existing datasets.

Data Collection and Storage

- Data Collection Protocols: Establish protocols for data collection, including data collection methods, data formats, and data quality standards.

- Data Storage: Ensure secure and efficient storage of collected data using appropriate data storage solutions.

- Data Privacy and Security: Implement measures to protect data privacy and security throughout the collection and storage process.

- Data Labeling and Annotation: Label or annotate data as needed for supervised learning tasks.

- Data Versioning and Tracking: Maintain data versioning and tracking to ensure data integrity and traceability.

Data Quality Checks

- Data Validation: Validate collected data against predefined rules, constraints, or patterns.

- Data Cleaning: Identify and address data quality issues such as missing values, outliers, and inconsistencies.

- Data Standardization: Ensure data is consistent in terms of format, units, and encoding.

- Data Quality Metrics: Calculate data quality metrics to assess the overall quality of the collected data.

- Data Profiling: Generate summary statistics to understand the characteristics of the collected data.

Data Preprocessing

Data Cleaning and Handling Missing Values

- Handle Missing Values: Identify and address missing values using appropriate techniques (e.g., imputation, deletion).

- Remove Outliers: Detect and remove or handle outliers that may skew the data distribution.

- Correct Errors: Identify and correct any errors or inconsistencies in the data.

- Data Imputation: Use appropriate imputation techniques (e.g., mean, median, mode, interpolation) to fill missing values.

- Data Standardization: Standardize numerical features to a common scale (e.g., z-score normalization).

Data Transformation and Feature Engineering

- Feature Engineering: Create new features from existing ones to improve model performance.

- Feature Selection: Select the most relevant features to reduce dimensionality and improve model efficiency.

- Data Normalization: Normalize numerical features to a common range (e.g., min–max scaling).

- Data Transformation: Apply transformations (e.g., log transformations, square root) to improve model performance.

- Handle Categorical Data: Convert categorical variables into numerical representations (e.g., one-hot encoding, label encoding).

Data Augmentation

- Data Augmentation: Generate additional training data by applying transformations to existing data (e.g., rotations, flips, scaling).

- Synthetic Data Generation: Create synthetic data to augment the dataset and improve model robustness.

Data Quality Checks

- Data Profiling: Generate summary statistics to understand the characteristics of the data.

- Data Consistency Checks: Verify data consistency across different sources and formats.

- Data Validation Rules: Define and apply validation rules to ensure data meets specific criteria.

- Data Quality Metrics: Calculate data quality metrics to assess the overall quality of the preprocessed data.

Data Validation and Integrity Checks

Data Validation

- Data Quality Metrics: Calculate data quality metrics (e.g., completeness rate, accuracy rate, consistency rate) to assess overall data quality.

- Data Profiling: Generate summary statistics (e.g., counts, means, medians, modes) to understand data characteristics and identify anomalies.

- Data Consistency Checks: Verify data consistency across different sources and formats to ensure data integrity.

- Data Validation Rules: Define and apply validation rules to ensure data meets specific criteria (e.g., range checks, format checks, uniqueness checks).

- Data Anomaly Detection: Identify and address data anomalies (e.g., outliers, inconsistencies) that may impact model performance.

Data Integrity Checks

- Data Lineage: Track the origin and transformation of data throughout its lifecycle to ensure data integrity and traceability.

- Data Security: Implement measures to protect data from unauthorized access, modification, or deletion.

- Data Privacy: Comply with relevant data privacy regulations (e.g., GDPR, CCPA) to ensure data is handled securely and responsibly.

- Data Versioning: Maintain data versioning and tracking to manage changes to data and ensure data integrity.

- Data Backup and Recovery: Implement robust backup and recovery procedures to protect data from loss and ensure data availability.

- Data Quality Monitoring: Continuously monitor data quality metrics and identify potential issues that may affect model performance.

- Data Drift Detection: Detect changes in data distribution over time that may impact model accuracy.

- Data Consistency Checks: Verify data consistency across different sources and formats to ensure data integrity.

- Data Validation Rules: Define and apply validation rules to ensure data meets specific criteria (e.g., range checks, format checks, uniqueness checks).

- Data Anomaly Detection: Identify and address data anomalies (e.g., outliers, inconsistencies) that may impact model performance.

Continuous Monitoring and Evaluation

- Data Quality Monitoring: Continuously monitor data quality metrics and identify potential issues that may affect model performance.

- Data Drift Detection: Detect changes in data distribution over time that may impact model accuracy.

- Model Performance Evaluation: Assess model performance on new, unseen data to identify data quality issues.

- Feedback Loops: Incorporate feedback from data users to improve data quality and governance processes.

- Regular Audits: Conduct regular audits to assess data integrity and compliance with governance standards.

Data Quality Assessment

Data Completeness

- Missing Values: Identify and quantify missing values in the dataset.

- Completeness Rate: Calculate the percentage of non-missing values for each attribute.

- Missing Value Patterns: Analyze patterns of missing values to determine potential causes.

- Impact of Missing Values: Assess the impact of missing values on model performance and accuracy.

- Imputation Strategies: Consider appropriate imputation techniques to handle missing values (e.g., mean, median, mode, imputation models).

Data Accuracy

- Data Validation: Verify data against predefined rules, constraints, or patterns.

- Accuracy Rate: Calculate the percentage of correct values compared to the total number of values.

- Error Detection: Identify and correct errors or inconsistencies in the data.

- Data Quality Metrics: Use data quality metrics (e.g., error rate, accuracy rate) to assess accuracy.

- Data Cleaning Techniques: Employ data cleaning techniques (e.g., outlier detection, anomaly detection) to improve accuracy.

Data Consistency

- Data Duplicates: Identify and remove duplicate records to avoid redundancy.

- Data Standardization: Ensure consistency in data formats, units, and encoding.

- Data Reconciliation: Reconcile data from multiple sources to ensure consistency.

- Data Integrity Checks: Perform data integrity checks to verify data consistency and reliability.

- Data Quality Metrics: Calculate data quality metrics (e.g., duplicate rate, consistency rate) to assess consistency.

Data Timeliness

- Data Age: Determine the age of the data and its relevance to the current context.

- Data Refresh Frequency: Assess the frequency at which data is updated.

- Data Latency: Measure the time lag between data generation and availability.

- Timeliness Metrics: Calculate timeliness metrics (e.g., data age, update frequency) to evaluate timeliness.

- Data Archival: Implement data archival policies to preserve historical data.

Data Relevance

- Data Alignment with Business Goals: Ensure data aligns with the organization's business objectives.

- Data Relevance to ML Model: Verify that data is relevant to the specific ML model being developed.

- Data Bias: Identify and address potential biases in the data that could affect model performance.

- Data Quality Metrics: Calculate data quality metrics (e.g., relevance rate, bias score) to assess relevance.

- Data Augmentation: Consider data augmentation techniques to improve data relevance and diversity.

Data Uniformity

- Data Format: Ensure consistency in data formats and structures.

- Data Granularity: Verify that data is at the appropriate level of granularity for the analysis.

- Data Completeness: Ensure that all relevant attributes are included in the dataset.

- Data Uniformity Metrics: Calculate data uniformity metrics (e.g., format consistency rate, granularity level) to assess uniformity.

- Data Transformation: Apply data transformations to ensure uniformity (e.g., normalization, standardization).

Data Reliability

- Data Source Credibility: Evaluate the credibility and reliability of data sources.

- Data Provenance: Track the origin and history of data.

- Data Trustworthiness: Assess the trustworthiness of data based on its quality and accuracy.

- Data Reliability Metrics: Calculate data reliability metrics (e.g., source credibility score, data provenance completeness) to evaluate reliability.

- Data Validation: Implement data validation processes to ensure data reliability.

Data Versioning and Traceability

Version Control System (VCS) Setup

- Implement a robust VCS for data, models, and code (e.g., Git, DVC).

- Ensure the VCS supports branching and merging to track different versions.

- Define naming conventions for datasets, models, and code to maintain consistency.

- Establish policies for committing changes, including review processes and approval workflows.

Data Metadata Management

- Capture detailed metadata for each dataset version, including creation date, source, format, and schema.

- Store metadata in a centralized repository accessible to all relevant stakeholders.

- Use automated tools to extract and update metadata, ensuring it stays current.

- Include metadata on data quality metrics, preprocessing steps, and usage history.

Data Provenance Tracking

- Record the origin and lineage of all data used in ML models, including sources and transformations.

- Use automated tools to track data lineage across the entire data pipeline.

- Document all intermediate processing steps and transformations applied to the data.

- Ensure that provenance information is linked to version control and metadata systems.

Data Reproducibility

- Maintain detailed records of all preprocessing and feature engineering steps to enable reproducibility.

- Store configuration files, scripts, and environment settings used during model training.

- Validate that models can be re-trained and produce consistent results using stored data versions.

- Implement regular checks to ensure that pipelines and models remain reproducible over time.

Data Versioning Policies

- Establish clear guidelines for when new versions of data and models should be created.

- Define roles and responsibilities for managing version control and ensuring compliance.

- Document and communicate versioning policies across teams to ensure alignment.

- Include criteria for archiving outdated versions and processes for restoring archived data if needed.

Audit Trails and Compliance

- Implement audit trails to track all changes made to data, models, and code.

- Ensure audit logs are secure, immutable, and accessible for compliance purposes.

- Regularly review audit trails to identify any anomalies or unauthorized changes.

- Align audit trails with regulatory requirements, ensuring all relevant data governance standards are met.

Data Quality Monitoring and Maintenance

Data Quality Metrics Definition

- Determine the key data attributes that directly impact the performance and outcomes of ML models.

- Prioritize attributes based on their influence on model accuracy, fairness, and interpretability.

- Ensure that the data quality metrics align with the overall business goals and objectives of the ML projects.

- Consider how data quality affects decision-making, compliance, and customer satisfaction.

- Define baseline metrics for data quality such as completeness, accuracy, consistency, timeliness, and uniqueness.

- Use historical data to set realistic and achievable benchmarks for these metrics.

- Tailor data quality metrics to address the specific needs of ML models, such as the requirement for balanced data distributions or low noise levels.

- Include metrics that account for model-specific factors like data drift, feature importance, and input–output correlations.

- Integrate metrics that ensure data quality compliance with relevant regulations, such as GDPR or industry-specific standards.

- Include measures to track adherence to data privacy, security, and ethical guidelines.

- Set clear and quantifiable thresholds for each data quality metric to determine acceptable levels of quality.

- Establish criteria for flagging data that falls below these thresholds and requires corrective action.

- Schedule regular reviews of the data quality metrics to ensure they remain relevant as the ML models and business requirements evolve.

- Update metrics as needed to reflect changes in data sources, regulatory requirements, or technological advancements.

Automated Data Quality Checks

- Determine the key dimensions of data quality that are essential for your ML models, such as accuracy, completeness, consistency, timeliness, and validity.

- Prioritize these dimensions based on their impact on model performance and business outcomes.

- Choose data quality management tools that support automation, such as data profiling, data validation, and anomaly detection.

- Ensure the tools integrate seamlessly with your existing data pipelines and ML workflows.

- Establish specific validation rules for each data quality dimension, such as range checks, format validation, and duplicate detection.

- Implement these rules within your data quality tools to automatically flag and correct errors.

- Set up real-time monitoring for data quality metrics, enabling immediate detection and resolution of data issues as they arise.

- Use automated alerts and notifications to inform relevant stakeholders of any data quality breaches.

- Integrate automated data quality checks into your ETL (Extract, Transform, Load) or ELT (Extract, Load, Transform) processes.

- Ensure that data is validated and cleaned during ingestion, before it is used in ML model training or analysis.

- Schedule automated batch audits to assess data quality over time, identifying trends, recurring issues, and potential areas for improvement.

- Compare results against historical benchmarks to evaluate data quality progress.

- Configure your data quality tools to perform automated corrections where possible, such as filling in missing values or normalizing data formats.

- Establish protocols for handling corrections that require human intervention, ensuring accountability and traceability.

- Keep a detailed log of all automated data quality checks, including detected issues, corrective actions taken, and the outcomes.

- Use this log to analyze the effectiveness of your data quality management strategies and make informed adjustments.

Data Profiling and Anomaly Detection

- Identify the critical attributes and features of your data, such as data types, distributions, and patterns, that need to be profiled.

- Establish thresholds and benchmarks for these characteristics to distinguish between normal and anomalous behavior.

- Choose tools that offer robust data profiling capabilities, including the ability to generate summary statistics, identify outliers, and detect missing or inconsistent data.

- Ensure the tools integrate well with your ML pipelines and support automated profiling.

- Develop and deploy algorithms that can automatically detect anomalies in the data, such as unusual patterns, outliers, or data drift.

- Customize these algorithms to suit the specific requirements of your ML models and data governance framework.

- Schedule regular data profiling activities, both at the initial data ingestion stage and periodically throughout the data lifecycle.

- Use automated profiling to continuously monitor data quality and detect any deviations from expected norms.

- Regularly review and update the rules and thresholds used for profiling and anomaly detection based on new insights, changing data characteristics, or evolving ML model needs.

- Document changes to ensure transparency and maintain a history of profiling criteria adjustments.

Data Cleansing and Transformation Procedures

- Develop standardized procedures for data cleansing and transformation to address identified data quality issues.

- Create scripts and tools for automating data cleansing processes, such as removing duplicates, handling missing values, and correcting errors.

741

- Validate the results of data cleansing to ensure that data quality improvements meet established thresholds.

- Maintain a log of all data cleansing activities, including changes made and the rationale behind them.

Data Quality Monitoring Dashboard

- Create a centralized dashboard to monitor data quality metrics in real-time.

- Ensure the dashboard provides visualizations and reports that are easily interpretable by different stakeholders.

- Include drill-down capabilities to investigate specific data quality issues at a granular level.

- Regularly update the dashboard to reflect new data sources, metrics, and business requirements.

Data Quality Maintenance Schedule

- Establish a regular maintenance schedule for data quality monitoring, including periodic reviews and updates.

- Assign responsibilities for monitoring and maintaining data quality to specific teams or individuals.

- Include data quality checks as part of the regular model retraining and deployment process.

- Review and refine data quality processes based on feedback from stakeholders and the performance of ML models.

Data Quality Issue Resolution Process

- Develop a standardized process for identifying, prioritizing, and resolving data quality issues.

- Implement a ticketing system to track the progress of data quality issue resolution.

- Define roles and responsibilities for resolving data quality issues, including escalation procedures for critical issues.

- Document and communicate the resolution of data quality issues to relevant stakeholders.

Handling Data Quality Issues

Identifying Data Quality Issues

- Regularly perform data profiling to assess the data's structure, content, and quality.

- Conduct audits on data sources to identify discrepancies, missing values, and inconsistencies.

- Implement automated data quality assessment tools to continuously monitor and report issues.

- Utilize exploratory data analysis (EDA) techniques to detect anomalies, outliers, and data distribution irregularities.

- Ensure that data profiling includes checks for data completeness, accuracy, timeliness, and consistency.

- Establish a process for stakeholders to report data quality issues, including a clear communication channel.

- Regularly collect feedback from data consumers (e.g., data scientists, analysts) on data quality concerns.

- Integrate issue reporting into data governance tools to track and prioritize data quality issues.

- Create a centralized repository for documenting reported issues and the actions taken to resolve them.

- Engage with data stewards to validate and assess the impact of reported issues.

Categorizing and Prioritizing Data Quality Issues

- Categorize data quality issues based on their severity (e.g., critical, high, medium, low).

- Assess the impact of each issue on ML model performance, business processes, and decision-making.

- Prioritize issues that affect key datasets used in high-stakes ML models or regulatory compliance.

- Evaluate the potential risks associated with unresolved data quality issues, including financial, reputational, and legal risks.

- Align prioritization criteria with the organization's data governance policies and strategic goals.

- Classify data quality issues into categories such as accuracy, completeness, consistency, validity, and timeliness.

- Identify root causes for each category, such as data entry errors, system integration failures, or outdated data.

- Use issue classification to tailor specific remediation strategies for different types of data quality problems.

- Document classification criteria and ensure they are consistently applied across the organization.

- Involve data owners and stewards in the classification process to ensure accuracy and relevance.

Remediating Data Quality Issues

- Develop automated data cleansing procedures to correct common errors, such as misspellings, incorrect formats, and duplicate entries.

- Implement data validation rules to enforce accuracy and prevent the introduction of erroneous data.

- Apply imputation techniques to handle missing data, selecting methods that are appropriate for the data type and use case.

- Use advanced analytics or machine learning models to predict and correct data anomalies.

- Document all cleansing and correction actions taken, ensuring they are traceable and reproducible.

- Enhance data quality by enriching datasets with external sources or additional attributes.

- Use data augmentation techniques to generate synthetic data that fills gaps or diversifies the dataset.

- Validate enriched or augmented data to ensure it meets quality standards and aligns with ML model requirements.

- Collaborate with data providers to obtain high-quality, authoritative data sources.

- Continuously monitor enriched data for changes in quality over time.

Preventing Future Data Quality Issues

- Define clear data quality standards that outline acceptable levels of accuracy, completeness, consistency, and timeliness.

- Develop data quality scorecards to measure and report adherence to these standards across the organization.

- Integrate data quality standards into data governance policies and ensure they are communicated to all stakeholders.

- Regularly review and update data quality standards to adapt to changing business needs and technological advancements.

- Incorporate data quality metrics into performance evaluations for data management roles.

- Implement automated data quality checks within data pipelines to detect and address issues in real-time.

- Set up alerts and notifications for data quality breaches, ensuring prompt remediation actions.

- Use AI-driven tools to predict potential data quality issues based on historical patterns and trends.

- Continuously monitor data quality metrics and KPIs, adjusting strategies as necessary to maintain high standards.

- Integrate data quality monitoring with data lineage and provenance tracking to ensure traceability.

Documenting and Reporting Data Quality Management

- Maintain a detailed log of all identified data quality issues, including their status, severity, and remediation actions taken.

- Use version control systems to document changes made to data quality processes and standards.

- Develop a centralized documentation repository for all data quality-related reports, guidelines, and procedures.

- Ensure that all data quality documentation is accessible to relevant stakeholders and updated regularly.

- Create audit trails for data quality activities, ensuring compliance with regulatory and governance requirements.

- Regularly report on data quality metrics and trends to senior management and relevant stakeholders.

- Use dashboards and visualizations to communicate the current state of data quality across the organization.

- Provide detailed reports on critical data quality issues, including root causes, impact assessments, and resolution timelines.

- Facilitate regular meetings or forums to discuss data quality challenges, successes, and areas for improvement.

- Encourage transparency and open communication about data quality issues, fostering a culture of continuous improvement.

Compliance and Ethics Checklist

Legal Compliance Checklist

Data Privacy Regulations

General Data Protection Regulation (GDPR)

- Data Subject Consent: Ensure that explicit consent is obtained from data subjects before collecting or processing their personal data.

- Right to Access: Implement mechanisms that allow data subjects to access their data and understand how it is being used.

- Right to Rectification: Provide data subjects with the ability to correct inaccurate or incomplete data.

- Right to Erasure (Right to be Forgotten): Establish processes to delete personal data upon request, provided there are no legal grounds for retaining it.

- Data Portability: Facilitate the transfer of personal data to another service provider at the request of the data subject.

- Data Protection Impact Assessment (DPIA): Conduct DPIAs for ML projects to assess the potential impact on data privacy and mitigate risks.

California Consumer Privacy Act (CCPA)

- Right to Opt-Out: Implement a system that allows consumers to opt-out of the sale of their personal data.

- Right to Know: Ensure that consumers can request information about the categories of personal data collected, the sources of that data, and the purposes for which it is used.

- Right to Delete: Provide consumers with the ability to request the deletion of their personal data.

- Non-discrimination: Ensure that consumers are not discriminated against for exercising their privacy rights under CCPA.

Health Insurance Portability and Accountability Act (HIPAA)

- Protected Health Information (PHI): Ensure that all ML models involving healthcare data comply with HIPAA regulations regarding the use, storage, and sharing of PHI.

- De-identification: Implement de-identification techniques to remove personally identifiable information from healthcare data used in ML models.

- Data Breach Notification: Establish procedures for promptly notifying affected individuals and authorities in the event of a data breach involving PHI.

Data Security Regulations
ISO/IEC 27001

- Information Security Management System (ISMS): Implement an ISMS that aligns with ISO/IEC 27001 standards to ensure the security of data used in ML models.

- Risk Assessment: Conduct regular risk assessments to identify and mitigate security risks associated with ML data.

- Access Control: Establish strict access control mechanisms to ensure that only authorized personnel can access ML data.

- Encryption: Ensure that all data used in ML models is encrypted both in transit and at rest to prevent unauthorized access.

NIST Cybersecurity Framework

- Identify: Continuously identify and assess cybersecurity risks to ML data and systems.

- Protect: Implement measures to protect ML data from cyber threats, including firewalls, anti-virus software, and secure coding practices.

- Detect: Establish systems to detect and respond to potential security incidents involving ML data.

- Respond: Develop an incident response plan for addressing and mitigating the impact of cybersecurity breaches.

- Recover: Implement procedures to recover data and restore ML systems to normal operation following a security incident.

Non-discrimination Regulations

Equal Employment Opportunity Commission (EEOC)

- Fair Hiring Practices: Ensure that ML models used in hiring processes comply with EEOC guidelines, avoiding bias or discrimination based on race, gender, age, or other protected characteristics.

- Adverse Impact Analysis: Regularly conduct adverse impact analysis to identify and mitigate potential discriminatory effects of ML models.

- Bias Mitigation: Implement bias detection and mitigation strategies in ML models to ensure fair and equitable outcomes.

Fair Housing Act

- Non-discriminatory Lending Practices: Ensure that ML models used in lending decisions do not discriminate against applicants based on race, color, religion, sex, familial status, national origin, or disability.

- Algorithmic Transparency: Provide transparency into the decision-making processes of ML models used in housing-related decisions.

- Regular Audits: Conduct regular audits of ML models to ensure compliance with Fair Housing Act regulations and to detect any potential bias.

Intellectual Property Regulations

Copyright Law

- Data Ownership: Ensure that the data used in ML models is legally acquired and that ownership rights are respected.

- Model Licensing: Obtain appropriate licenses for third-party data or algorithms used in ML models to avoid copyright infringement.

- Data Usage Rights: Clearly define and document the usage rights of data, including any limitations or restrictions imposed by data providers.

Patent Law

- Algorithm Patents: Ensure that ML algorithms developed in-house are protected by patents, if applicable.

- Infringement Avoidance: Conduct thorough patent searches to avoid infringement of existing patents when developing new ML algorithms.

- Documentation: Maintain detailed documentation of ML model development processes to support patent applications and protect intellectual property.

Ethical Compliance Checklist

Transparency and Explainability

Model Explainability

- Interpretability: Develop ML models that are interpretable, allowing stakeholders to understand how decisions are made.

- Post-Hoc Explainability: Implement post-hoc explainability techniques, such as SHAP (Shapley Additive Explanations) or LIME (Local Interpretable Model-agnostic Explanations), to provide insights into complex models.

- Documentation: Maintain comprehensive documentation that explains the rationale behind model design, training data selection, and decision-making processes.

Transparent Reporting

- Model Documentation: Provide clear and detailed documentation of ML models, including their purpose, data sources, and decision-making criteria.

- Impact Assessments: Publish the results of impact assessments conducted to evaluate the ethical implications of ML models.

- Stakeholder Communication: Communicate openly with stakeholders about the limitations, risks, and potential biases of ML models.

Fairness and Non-discrimination

Bias Detection

- Data Bias Analysis: Regularly analyze training data for biases that could influence model outcomes, ensuring that data is representative and unbiased.

- Model Bias Testing: Implement bias testing techniques to evaluate ML models for potential discriminatory behavior.

- Bias Mitigation: Employ techniques such as re-sampling, re-weighting, or adversarial debiasing to mitigate identified biases in ML models.

Inclusive Design

- Diverse Teams: Ensure that ML development teams are diverse and include individuals with different perspectives, backgrounds, and expertise.

- Stakeholder Engagement: Involve a diverse group of stakeholders in the design, development, and deployment of ML models to ensure that different viewpoints are considered.

- Cultural Sensitivity: Design ML models with cultural sensitivity in mind, avoiding decisions that could be perceived as discriminatory or insensitive.

Accountability and Responsibility

Responsible AI Principles

- Ethical Guidelines: Develop and adhere to ethical guidelines that outline the principles of responsible AI within the organization.

- Accountability Framework: Establish an accountability framework that clearly defines the roles and responsibilities of individuals involved in the ML lifecycle.

- Ethics Committees: Create ethics committees or advisory boards to oversee ML projects and ensure that ethical considerations are incorporated into decision-making.

Consequence Management

- Impact Assessments: Conduct thorough impact assessments to evaluate the potential social, economic, and environmental consequences of ML models.

- Mitigation Strategies: Develop and implement strategies to mitigate any negative consequences identified during impact assessments.

- Incident Response: Establish an incident response plan for addressing ethical concerns or issues that arise during the deployment of ML models.

Human-Centric Design

User-Centered Development

- User Feedback: Involve end-users in the development process, gathering feedback to ensure that ML models meet their needs and expectations.

- Usability Testing: Conduct usability testing to evaluate the user experience and identify areas for improvement.

- Iterative Development: Use an iterative development approach that incorporates user feedback at each stage, allowing for continuous refinement of ML models.

Human Oversight

- Decision-Making: Ensure that critical decisions made by ML models are subject to human oversight, allowing for intervention if necessary.

- Transparency: Provide end-users with clear information about how ML models work and how decisions are made, empowering them to make informed choices.

- Ethical Safeguards: Implement ethical safeguards that prevent ML models from making

Data Security Checklist

Data Classification and Sensitivity Assessment

Data Classification

Define Data Categories

- Identify Data Types: Determine the different types of data used in ML processes (e.g., personal data, financial data, proprietary data, sensor data).

- Categorize Data Based on Sensitivity: Classify data into categories such as public, internal, confidential, and restricted based on its sensitivity and the potential impact of unauthorized access or disclosure.

- Align with Regulatory Requirements: Ensure data categories align with relevant regulations (e.g., GDPR, HIPAA) and industry standards (e.g., ISO/IEC 27001).

- Document Data Classification Criteria: Establish and document the criteria used for classifying data, including factors such as legal requirements, business impact, and data subject rights.

- Include Data Source Information: Clearly define and document the sources of data, including external data providers, internal databases, and third-party services, to understand the origin and reliability of data.

- Consider Data Usage Scenarios: Categorize data based on how it is used in ML processes, such as training, validation, or prediction, to apply appropriate controls at each stage.

- Account for Data Aggregation: Consider the impact of data aggregation, where combining data from multiple sources may change its classification or increase sensitivity.

Implement Data Classification Policies

- Develop Data Classification Policies: Create policies that outline the process for classifying data, including responsibilities and procedures for classification.

- Assign Ownership and Responsibility: Designate data owners responsible for classifying data and ensuring compliance with classification policies.

- Integrate Classification into Data Lifecycle: Ensure data classification is integrated into the data lifecycle, from creation to archival or deletion.

- Regularly Review and Update Classifications: Periodically review data classifications to ensure they remain accurate and reflect current business and regulatory environments.

- Implement Data Handling Guidelines: Develop guidelines for handling data based on its classification, including access controls, encryption requirements, and storage practices.

- Conduct Training for Data Classification: Provide training to employees involved in data classification to ensure they understand the criteria and importance of accurate classification.

- Monitor Compliance with Classification Policies: Regularly audit data classification activities to ensure adherence to policies and identify areas for improvement.

Use Automated Tools for Data Classification

- Evaluate Data Classification Tools: Assess and select automated tools that can assist in identifying, categorizing, and labeling data based on predefined criteria.

- Implement Data Tagging Mechanisms: Use metadata tagging to label data according to its classification, facilitating automated processing and protection.

- Monitor and Validate Classification Accuracy: Continuously monitor the effectiveness of automated classification tools and validate their accuracy through audits and reviews.

- Integrate with ML Pipelines: Ensure that data classification is integrated into ML pipelines to apply appropriate protections and controls throughout the model lifecycle.

- Leverage Machine Learning for Classification: Consider using ML algorithms to assist in data classification by identifying patterns and anomalies in data that may influence its classification.

- Support Multi-Level Classification: Implement tools that support multi-level classification, where data can be tagged with multiple classifications to reflect its complexity and usage.

Sensitivity Assessment Checklist

Identify and Assess Sensitive Data

- Conduct a Data Inventory: Perform a comprehensive inventory of all data used in ML processes to identify sensitive data (e.g., PII, PHI, financial information).

- Assess Sensitivity Based on Context: Evaluate data sensitivity based on the context in which it is used, including the potential for harm if the data is exposed or misused.

- Classify Data by Sensitivity Level: Assign sensitivity levels to data (e.g., low, medium, high) based on the assessment, considering both the data content and its usage.

- Consider Data Aggregation Risks: Assess the risks associated with data aggregation, where combining multiple data sources could increase sensitivity or risk.

- Review Regulatory Requirements: Identify and document regulatory requirements that impact the handling of sensitive data, such as GDPR, HIPAA, or industry-specific regulations.

- Evaluate Data Retention Policies: Review data retention policies to ensure that sensitive data is not kept longer than necessary, minimizing exposure risks.

- Account for Data Access and Sharing: Consider the sensitivity of data when shared with third parties or accessed by different teams, applying necessary restrictions or controls.

Implement Sensitivity Assessment Policies

- Develop Sensitivity Assessment Guidelines: Create guidelines for conducting sensitivity assessments, including criteria and procedures for determining data sensitivity.

- Assign Responsibility for Sensitivity Assessments: Designate roles responsible for conducting and approving sensitivity assessments as part of the data governance framework.

- Incorporate Sensitivity Assessment into ML Workflows: Ensure sensitivity assessments are a mandatory step in ML project workflows, influencing decisions on data use, access, and protection.

- Regularly Update Sensitivity Assessments: Periodically review and update sensitivity assessments to reflect changes in data usage, regulatory requirements, or organizational priorities.

- Create a Sensitivity Classification Matrix: Develop a matrix that maps different data types and usage scenarios to sensitivity levels, providing a clear framework for assessment.

- Involve Stakeholders in Sensitivity Assessments: Engage data owners, legal, and compliance teams in sensitivity assessments to ensure a comprehensive evaluation.

- Document Sensitivity Assessment Outcomes: Maintain records of sensitivity assessment outcomes, including the rationale for assigned sensitivity levels and any actions taken.

Mitigate Risks Associated with Sensitive Data

- Implement Data Minimization Practices: Reduce the amount of sensitive data used in ML models by applying data minimization principles, ensuring only necessary data is processed.

- Use Data Anonymization and Pseudonymization: Apply techniques such as anonymization or pseudonymization to reduce the risk of exposure for sensitive data.

- Apply Differential Privacy Techniques: Consider implementing differential privacy to provide strong privacy guarantees when using sensitive data in ML models.

- Restrict Access to Sensitive Data: Limit access to sensitive data based on the principle of least privilege, ensuring only authorized personnel can access or process the data.

- Encrypt Sensitive Data: Ensure sensitive data is encrypted both at rest and in transit to protect against unauthorized access.

- Monitor Data Access and Usage: Implement monitoring tools to track access to sensitive data and detect any unauthorized or unusual activity.

- Conduct Regular Security Audits: Perform regular audits of data protection measures to identify and address potential vulnerabilities in handling sensitive data.

Monitor and Protect Sensitive Data

- Implement Data Protection Controls: Apply encryption, access controls, and monitoring mechanisms to protect sensitive data throughout its lifecycle.

- Conduct Regular Audits and Assessments: Regularly audit data protection measures and sensitivity assessments to ensure ongoing compliance and effectiveness.

- Monitor Data Usage and Access Patterns: Continuously monitor data usage and access patterns to detect any unauthorized or suspicious activity involving sensitive data.

- Establish Incident Response Procedures: Develop and implement incident response procedures for handling data breaches or security incidents involving sensitive data.

- Automate Sensitive Data Monitoring: Use automated tools to continuously monitor sensitive data for compliance with classification and protection policies.

- Perform Data Loss Prevention (DLP) Checks: Implement DLP technologies to detect and prevent unauthorized sharing or transmission of sensitive data.

- Review and Update Protection Controls: Regularly review data protection controls to ensure they are up-to-date and effective against emerging threats.

Access Control

Establish Access Control Policies

- Define Access Control Requirements: Clearly outline the requirements for access control based on the sensitivity of data and the roles within the organization.

- Classify Data Access Levels: Categorize data access levels (e.g., public, internal, confidential, restricted) corresponding to the data classification and sensitivity assessment.

- Develop Role-Based Access Control (RBAC): Implement RBAC, assigning permissions based on roles and responsibilities, ensuring that users have access only to the data necessary for their tasks.

- Incorporate Principle of Least Privilege: Ensure that access is granted on a need-to-know basis, limiting data access to the minimum level required for performing job functions.

- Document Access Control Policies: Create detailed documentation for access control policies, including who has access, the level of access granted, and the rationale behind these decisions.

- Establish Data Stewardship Roles: Appoint data stewards responsible for managing access control, overseeing who can access specific datasets, and ensuring adherence to policies.

- Align with Regulatory Compliance: Ensure access control policies comply with relevant regulations and industry standards (e.g., GDPR, HIPAA, CCPA) to protect sensitive data.

Implement Technical Access Controls

- Use Multi-factor Authentication (MFA): Implement MFA for accessing critical ML systems and sensitive datasets, adding an extra layer of security.

- Deploy Identity and Access Management (IAM) Tools: Utilize IAM solutions to manage user identities, enforce access controls, and monitor access to ML systems and data.

759

- Encrypt Data at Rest and in Transit: Ensure that all sensitive data is encrypted both when stored and when transmitted, protecting it from unauthorized access.

- Set Up Data Access Logs: Implement logging mechanisms to track who accessed what data, when, and for what purpose, providing a complete audit trail.

- Integrate Access Controls with ML Pipelines: Embed access control mechanisms within ML pipelines to ensure that only authorized personnel can interact with the data and models at each stage.

- Control Access to Data in Collaboration Tools: Apply access controls to data shared in collaboration tools (e.g., cloud storage, version control systems) to prevent unauthorized access.

- Implement Time-Based Access Controls: Use time-based access controls to grant temporary access to data for specific tasks or projects, automatically revoking access after the set period.

- Leverage Encryption Keys Management: Employ robust encryption key management practices to control access to encrypted data, ensuring that only authorized users can decrypt sensitive information.

Monitor and Review Access Controls

- Conduct Regular Access Reviews: Periodically review and audit access permissions to ensure that only the appropriate individuals have access to specific data and systems.

- Implement Continuous Monitoring: Use monitoring tools to continuously track access patterns and detect any unusual or unauthorized access attempts in real-time.

- Establish an Incident Response Plan: Develop and implement an incident response plan to quickly address and mitigate the impact of unauthorized access or security breaches.

- Automate Access Monitoring: Employ automated systems to monitor access controls, identifying and flagging potential security risks or compliance violations.

- Perform Access Audits: Conduct regular audits of access logs and permissions to verify that access controls are being followed and are effective in protecting data.

- Review Access Control Effectiveness: Continuously assess the effectiveness of access control mechanisms and make necessary adjustments to improve security and compliance.

- Implement Role Recertification: Regularly recertify user roles and associated access permissions, particularly after organizational changes such as promotions, transfers, or departures.

- Track and Report Access Violations: Maintain a log of access violations and report them to relevant stakeholders, including data stewards, compliance officers, and senior management.

Manage Access Control for Third-Party Collaborations

- Establish Third-Party Access Policies: Develop and enforce policies for granting third-party access to ML systems and data, ensuring that external partners comply with your access control standards.

- Use Data Sharing Agreements: Formalize data sharing arrangements with third parties through legal agreements that specify access levels, responsibilities, and data protection requirements.

- Limit Third-Party Access: Restrict third-party access to only the data necessary for their role, applying the principle of least privilege and time-bound access.

- Monitor Third-Party Access: Continuously monitor access by third parties to detect any unauthorized or unusual access patterns, ensuring compliance with access control policies.

- Conduct Third-Party Audits: Regularly audit third-party access to verify adherence to your access control policies and agreements.

- Revoke Access Post-Project: Immediately revoke access for third parties upon completion of a project or termination of a partnership to prevent unauthorized future access.

- Review Third-Party Security Practices: Evaluate the security practices of third parties to ensure they meet your organization's standards for data protection and access control.

- Integrate Third-Party Access with IAM: Utilize IAM tools to manage third-party access, ensuring consistent application of access control policies across internal and external users.

Data Encryption

Encryption Strategy and Policy Development

- Define Encryption Requirements: Identify the specific encryption needs based on data sensitivity, regulatory requirements, and organizational policies.

- Develop an Encryption Policy: Establish a formal encryption policy that outlines the standards, protocols, and tools to be used for encrypting data in the ML lifecycle.

- Determine Encryption Scope: Specify the scope of encryption, including data at rest, data in transit, and data in use, ensuring comprehensive protection across all stages.

- Select Encryption Algorithms: Choose appropriate encryption algorithms (e.g., AES-256, RSA) that meet industry standards and provide strong security.

- Establish Key Management Practices: Develop a robust key management strategy that includes key generation, distribution, rotation, storage, and destruction practices.

- Incorporate Compliance Requirements: Ensure the encryption policy aligns with relevant legal and regulatory frameworks, such as GDPR, HIPAA, and industry-specific standards.

- Document Encryption Procedures: Create detailed documentation for encryption procedures, including configuration settings, approved algorithms, and key management practices.

- Designate Roles and Responsibilities: Assign clear roles and responsibilities for encryption, key management, and compliance monitoring to appropriate personnel.

Data Encryption Implementation

- Encrypt Data at Rest: Implement encryption for all sensitive data stored in databases, file systems, and cloud storage to protect it from unauthorized access.

- Encrypt Data in Transit: Use Transport Layer Security (TLS) or similar protocols to encrypt data during transmission between clients, servers, and cloud services.

- Encrypt Data in Use: Explore encryption technologies that protect data while it is being processed, such as homomorphic encryption or secure multiparty computation.

- Use End-to-End Encryption: Implement end-to-end encryption for communication channels to ensure data remains encrypted from the source to the destination.

- Apply Encryption to ML Models: Encrypt ML models, especially when deploying them to production environments, to protect intellectual property and sensitive insights.

- Enable Full-Disk Encryption: Use full-disk encryption for devices and servers handling sensitive ML data to protect against unauthorized physical access.

- Implement Database Encryption: Apply transparent database encryption (TDE) or column-level encryption for databases storing ML-related data.

- Secure Backup Data: Ensure all backups of ML data are encrypted to prevent unauthorized access in case of loss or theft.

Key Management and Security

- Implement a Key Management System (KMS): Use a centralized KMS to generate, store, and manage encryption keys securely, ensuring they are protected from unauthorized access.

- Automate Key Rotation: Set up automated key rotation policies to regularly update encryption keys, reducing the risk of key compromise.

- Use Hardware Security Modules (HSMs): Store and manage encryption keys in HSMs to provide enhanced security and compliance with stringent regulatory standards.

- Encrypt Encryption Keys: Ensure that encryption keys themselves are encrypted (key encryption keys—KEK) for additional security.

- Control Access to Keys: Limit access to encryption keys based on the principle of least privilege, ensuring only authorized personnel or systems can access them.

- Implement Multi-factor Authentication (MFA): Require MFA for accessing key management systems to enhance security.

- Monitor Key Usage: Continuously monitor and log key usage to detect any unauthorized access or unusual activity.

- Develop Key Backup and Recovery Plans: Establish secure backup and recovery procedures for encryption keys to ensure data access continuity in case of loss.

Compliance and Auditing

- Align with Regulatory Standards: Ensure encryption practices comply with industry standards and regulations such as GDPR, HIPAA, FIPS 140-2, and PCI DSS.

- Conduct Regular Encryption Audits: Perform regular audits of encryption practices to verify compliance with internal policies and external regulations.

- Monitor Encryption Effectiveness: Continuously monitor and assess the effectiveness of encryption controls and update them as necessary to address emerging threats.

- Log Encryption Activities: Maintain detailed logs of all encryption-related activities, including key management actions and access to encrypted data.

- Review and Update Encryption Policies: Periodically review encryption policies and procedures to ensure they reflect the latest security best practices and regulatory requirements.

- Provide Encryption Compliance Reports: Generate and submit compliance reports to relevant stakeholders, including auditors and regulatory bodies, as required.

- Establish an Incident Response Plan: Develop a response plan for encryption-related incidents, such as key compromise or unauthorized data access, including notification procedures and remediation steps.

Encryption in Cloud Environments

- Leverage Cloud Provider Encryption Services: Use built-in encryption services provided by cloud providers (e.g., AWS KMS, Azure Key Vault) for encrypting data stored and processed in the cloud.

- Implement Cloud Data Encryption Policies: Establish encryption policies specific to cloud environments, including requirements for data at rest and in transit.

- Control Access to Cloud Encryption Keys: Ensure that access to cloud-based encryption keys is tightly controlled and monitored, with strict IAM policies in place.

- Encrypt Cloud Storage and Databases: Apply encryption to all cloud storage services (e.g., S3 buckets, Azure Blob Storage) and databases to protect sensitive ML data.

- Use Customer-Managed Keys (CMKs): Consider using customer-managed encryption keys for cloud services to maintain control over key management and meet compliance requirements.

- Encrypt Cloud-Based ML Models: Ensure that ML models deployed in the cloud are encrypted to protect against unauthorized access and potential intellectual property theft.

- Monitor Cloud Encryption Compliance: Regularly monitor and audit cloud encryption practices to ensure they meet the organization's security policies and compliance requirements.

- Implement Hybrid Cloud Encryption Strategies: If using a hybrid cloud approach, ensure consistent encryption practices across on-premises and cloud environments.

Encryption for Data Sharing and Collaboration

- Encrypt Shared Data: Ensure that any data shared with third parties or collaborators is encrypted, both during transmission and when stored on external systems.

- Use Secure File Transfer Protocols: Implement secure file transfer protocols (e.g., SFTP, FTPS) for sharing encrypted data across networks.

- Establish Data Sharing Agreements: Include encryption requirements in data sharing agreements with third parties to ensure data is adequately protected.

- Use Encryption for API Communications: Encrypt data transmitted via APIs, especially when integrating ML models with external systems or services.

- Monitor and Log Data Sharing Activities: Keep detailed logs of all data sharing activities, including encryption status, to ensure compliance and traceability.

- Enforce Encryption for Collaborative Tools: Require encryption for data exchanged through collaboration tools, such as shared repositories or project management platforms.

- Implement Secure Collaboration Channels: Use end-to-end encrypted communication channels (e.g., encrypted email, secure messaging) for discussing sensitive ML data.

Data Masking and Anonymization

Data Masking Checklist

Determine Data Masking Requirements

- Identify Sensitive Data: List and categorize all sensitive data elements that require masking (e.g., PII, PHI, financial data, proprietary business information).

- Define Masking Scope: Determine the scope of data masking, including which datasets, tables, or fields need to be masked within the ML pipelines.

- Evaluate Regulatory Compliance: Ensure that data masking practices meet the requirements of relevant regulations and standards (e.g., GDPR, CCPA, HIPAA).

- Determine Masking Techniques: Choose appropriate masking techniques (e.g., character scrambling, nulling out, shuffling, encryption) based on the data type and use case.

- Assess Impact on ML Models: Evaluate the potential impact of data masking on ML model performance, ensuring that the masked data still supports accurate and reliable model outcomes.

- Identify Masking Triggers: Define specific triggers or conditions that necessitate data masking, such as data being shared with third parties or moved to less secure environments.

Implement Data Masking Procedures

- Develop Masking Policies: Create and document policies that outline the procedures for masking sensitive data, including roles and responsibilities.

- Automate Data Masking Processes: Implement tools or scripts to automate the data masking process, reducing the risk of human error and ensuring consistency.

- Apply Dynamic Data Masking: Consider using dynamic data masking to provide different masking levels based on user roles and access permissions.

- Test Masking Effectiveness: Conduct thorough testing of the masking process to ensure that the sensitive data cannot be reverse-engineered or deciphered.

- Monitor Masked Data Access: Track and monitor who accesses masked data to ensure that only authorized personnel have the necessary permissions.

- Document Masking Process: Keep detailed records of how data was masked, including the techniques used and the datasets affected, to support audits and compliance checks.

- Regularly Review Masking Policies: Periodically review and update masking policies to accommodate new data types, regulatory changes, or advancements in masking technology.

Manage Data Masking Tools and Techniques

- Select Appropriate Masking Tools: Evaluate and select data masking tools that best meet your organization's requirements and integrate seamlessly with your existing ML workflows.

- Ensure Compatibility with ML Pipelines: Confirm that the selected masking tools or techniques are compatible with the ML models, ensuring minimal disruption to the pipeline.

- Support for Various Data Types: Ensure the masking tools can handle various data types (e.g., text, numeric, dates) and formats (e.g., JSON, XML).

- Implement Fine-Grained Masking: Use fine-grained masking to apply different masking techniques to different parts of the data, depending on sensitivity and use cases.

- Maintain Data Integrity: Ensure that the masking process does not alter the overall data structure or introduce inconsistencies that could affect ML model outcomes.

- Train Users on Masking Tools: Provide training to relevant teams on how to use data masking tools effectively and securely.

- Monitor for Masking Gaps: Continuously monitor data masking processes to identify and address any gaps or vulnerabilities.

Data Anonymization Checklist

Identify Anonymization Requirements

- Determine the Need for Anonymization: Identify the datasets or scenarios where anonymization is required, especially when data will be used for public release, research, or shared with external partners.

- Classify Data for Anonymization: Classify data based on the level of anonymity required, such as full anonymization, pseudonymization, or differential privacy.

- Define Anonymization Techniques: Choose appropriate anonymization techniques (e.g., k-anonymity, l-diversity, t-closeness) based on the type of data and the risks associated with re-identification.

- Assess Re-identification Risks: Evaluate the risk of re-identification after anonymization, considering factors like data uniqueness, data volume, and external data sources.

- Align with Privacy Regulations: Ensure that anonymization practices
  comply with privacy regulations like GDPR, which may require
  specific measures to protect data subjects' privacy.

- Involve Legal and Compliance Teams: Consult with legal and
  compliance teams to confirm that anonymization strategies meet all
  necessary legal and regulatory requirements.

Implement Anonymization Techniques

- Develop Anonymization Policies: Document policies that define the
  process for anonymizing data, including roles, responsibilities, and
  approved techniques.

- Apply Anonymization Consistently: Ensure that anonymization is
  applied consistently across all relevant datasets to prevent accidental
  exposure of identifiable information.

- Use Automated Anonymization Tools: Implement tools to automate
  the anonymization process, reducing the risk of errors and ensuring
  consistency.

- Test Anonymization Robustness: Perform robustness testing to
  ensure that anonymized data cannot be re-identified using external
  data or advanced techniques.

- Document Anonymization Process: Keep detailed documentation of
  the anonymization process, including techniques used and datasets
  affected, to support audits and compliance efforts.

- Review Anonymization Effectiveness: Regularly review and update
  anonymization methods to reflect changes in technology, data types,
  and privacy risks.

- Anonymize Data Before Sharing: Ensure that data is fully anonymized
  before it is shared with external parties or used in environments with
  lower security controls.

Manage Anonymization Tools and Techniques

- Evaluate Anonymization Tools: Assess and select tools that can effectively anonymize your datasets while integrating with existing ML workflows.

- Support for Advanced Anonymization: Choose tools that support advanced anonymization techniques like differential privacy or synthetic data generation.

- Ensure Compatibility with ML Models: Confirm that anonymized data remains useful for ML models, balancing the trade-off between privacy and data utility.

- Monitor for Anonymization Failures: Continuously monitor the effectiveness of anonymization processes to detect and address any failures or potential re-identification risks.

- Integrate with Data Governance Framework: Ensure that anonymization practices are fully integrated into your broader data governance framework, aligning with other policies and procedures.

- Maintain Anonymization Consistency: Implement controls to ensure that anonymization is consistently applied across all datasets, regardless of their source or format.

- Train Teams on Anonymization Best Practices: Provide training on anonymization techniques, tools, and the importance of maintaining privacy to all relevant teams.

Data Backup and Recovery

Backup Strategy

- Backup Frequency: Determine the optimal frequency for backups based on data sensitivity, volume, and regulatory requirements.

- Backup Scope: Identify the specific data assets that need to be backed up, including raw data, processed data, models, and metadata.

- Backup Location: Choose a secure and reliable location for storing backups, considering factors such as accessibility, redundancy, and disaster recovery capabilities.

- Backup Retention Policy: Define a retention policy specifying how long backups should be retained based on regulatory compliance and business needs.

- Backup Media: Select appropriate backup media (e.g., tapes, disks, cloud storage) based on storage capacity, cost, and performance requirements.

Backup Implementation

- Backup Software: Choose a reliable backup software solution that supports the selected backup media and provides features like compression, encryption, and deduplication.

- Backup Scheduling: Automate backup processes using scheduling tools to ensure timely and consistent backups.

- Testing and Verification: Regularly test backup procedures to verify their effectiveness and identify any issues.

- Data Encryption: Implement strong encryption to protect sensitive data during transmission and storage.

- Backup Labeling and Documentation: Clearly label and document backups to facilitate easy identification and retrieval.

Disaster Recovery Planning

- Disaster Recovery Site: Identify a suitable location for a disaster recovery site, considering factors such as proximity to the primary site, infrastructure availability, and security.

- Recovery Procedures: Develop detailed recovery procedures outlining the steps required to restore operations in case of a disaster.

- Testing and Drills: Conduct regular disaster recovery drills to ensure that staff is trained and prepared to execute recovery plans effectively.

- Business Continuity Planning: Develop a business continuity plan to address the impact of a disaster on business operations and identify critical functions that need to be prioritized for recovery.

- Data Replication: Consider implementing data replication to maintain multiple copies of data at different locations for redundancy and disaster recovery purposes.

Backup Monitoring and Management

- Backup Monitoring: Monitor backup processes for errors, failures, and performance issues.

- Backup Verification: Regularly verify the integrity of backups to ensure that data can be restored successfully.

- Backup Rotation: Implement a backup rotation scheme to manage backup storage efficiently.

- Backup Auditing: Conduct regular audits to assess the effectiveness of backup procedures and identify areas for improvement.

- Backup Review and Updates: Periodically review and update backup plans to reflect changes in data volumes, regulatory requirements, and technology advancements.

Templates for ML Data Governance Policies
Data Governance Policy Template

Purpose

This policy establishes the framework for managing data within the organization, ensuring that data used in ML models is accurate, secure, and compliant with relevant regulations.

Scope

This policy applies to all data collected, processed, and used in ML models across the organization.

Objectives

- Ensure data quality and accuracy.

- Protect data from unauthorized access and breaches.

- Comply with legal and regulatory requirements.

- Address ethical considerations, including bias and fairness.

Governance Structure

- Data Governance Committee: Responsible for overseeing data governance practices and policies.

- Data Stewards: Assigned to manage specific datasets, ensuring compliance with governance standards.

Data Quality Standards

- Data Profiling: Regular profiling to assess data quality.

- Data Cleaning: Procedures for correcting data errors and inconsistencies.

- Data Validation: Rules for validating data accuracy.

Data Security Measures

- Access Controls: Role-based access controls to limit data access.

- Encryption: Encryption of data at rest and in transit.

- Incident Response: Plans for responding to data breaches.

Compliance and Ethics

- Regulatory Compliance: Procedures for ensuring compliance with applicable regulations.

- Ethical Considerations: Strategies for addressing bias and promoting fairness in ML models.

Review and Monitoring

- Policy Review: Regular review and updating of the data governance policy.

- Monitoring: Continuous monitoring of data governance practices.

Data Privacy and Security Policy Template

Purpose
This policy outlines the measures taken to protect data privacy and ensure the security of data used in ML models.

Scope
This policy applies to all personal and sensitive data collected, processed, and stored within the organization.

Data Privacy Principles

- Data Minimization: Collect only the data necessary for ML purposes.

- Anonymization: Anonymize personal data where possible.

- User Consent: Obtain explicit consent from users for data collection and use.

Data Security Measures

- Access Controls: Implement role-based access controls (RBAC).

- Encryption: Encrypt data at rest and in transit using industry-standard protocols.

- Data Masking: Apply data masking techniques for sensitive data.

Incident Response

- Detection: Continuous monitoring for potential security breaches.

- Containment: Immediate containment measures in the event of a breach.

- Notification: Prompt notification of affected parties as required by law.

Compliance

- Regulatory Adherence: Ensure compliance with data protection regulations (e.g., GDPR, CCPA).

- Audits: Conduct regular audits of data privacy and security practices.

Review and Monitoring

- Policy Review: Annual review of the data privacy and security policy.

- Security Monitoring: Ongoing monitoring of security controls and practices.

Model Governance Framework Template

Purpose

This framework establishes the governance standards for ML models, ensuring they are developed, deployed, and maintained in a compliant and ethical manner.

Scope

This framework applies to all ML models developed and used within the organization.

Model Development Standards

- Data Selection: Ensure data used for model training is accurate and representative.

- Bias Detection: Implement bias detection and mitigation strategies during model development.

- Explainability: Develop models with explainability in mind, ensuring stakeholders can understand model decisions.

Model Deployment Standards

- Testing: Thoroughly test models before deployment to ensure they meet performance and governance criteria.

- Approval Process: Establish an approval process for model deployment, involving relevant stakeholders.

Model Monitoring and Maintenance

- Continuous Monitoring: Monitor model performance regularly to detect and address any issues.

- Retraining: Implement retraining schedules to maintain model accuracy and relevance.

- Audits: Conduct regular audits of deployed models to ensure ongoing compliance with governance standards.

Ethical Considerations

- Fairness: Ensure models do not perpetuate bias or discrimination.

- Transparency: Maintain transparency in model development and deployment processes.

- Accountability: Assign clear accountability for model outcomes and governance.

Compliance

- Regulatory Compliance: Ensure models comply with applicable regulations and standards.

- Documentation: Maintain thorough documentation of model development, deployment, and monitoring processes.

Review and Monitoring

- Framework Review: Regularly review and update the model governance framework.

- Monitoring: Continuously monitor model governance practices and adjust as necessary.

APPENDIX C

Data Governance Tools Comparison

Comparison Criteria for ML Data Governance Tools

To evaluate ML data governance tools, we will use the following criteria:

- Feature Set: The range of functionalities provided by the tool, including data quality management, data lineage tracking, compliance management, and bias detection.

- Ease of Use: The user-friendliness of the tool, including its interface, documentation, and integration capabilities with existing ML workflows.

- Scalability: The tool's ability to handle large volumes of data and its performance in enterprise-scale environments.

- Integration Capabilities: How well the tool integrates with other ML and data management platforms, such as data lakes, data warehouses, and ML model management systems.

- Compliance and Security Features: The robustness of the tool's security measures and its ability to help organizations meet regulatory requirements.

© Aditya Nandan Prasad 2024
A. Nandan Prasad, *Introduction to Data Governance for Machine Learning Systems*,
https://doi.org/10.1007/979-8-8688-1023-7

- Cost: The pricing model of the tool, including licensing fees, subscription costs, and any additional expenses related to implementation and maintenance.

- Community and Support: The availability of community support, customer service, and the frequency of updates and new feature releases.

Using these criteria, we will compare some of the leading ML data governance tools available in the market.

Informatica Axon

Feature Set
Informatica Axon is a comprehensive data governance tool that offers features such as data cataloging, data lineage, and business glossary management. It allows organizations to manage data quality through rules-based governance and supports collaboration across teams by providing a centralized platform for data governance activities.

Ease of Use
Informatica Axon is known for its intuitive user interface and comprehensive documentation. It offers a customizable dashboard that allows users to monitor key metrics and governance activities in real-time.

Scalability
The tool is highly scalable and can handle large datasets, making it suitable for enterprises with extensive data governance needs. Informatica's cloud-native architecture further enhances its scalability.

Integration Capabilities
Informatica Axon integrates seamlessly with other Informatica products, such as Informatica Data Quality and Informatica Enterprise Data Catalog. It also supports integration with third-party tools, including popular data lakes and ML platforms.

Compliance and Security Features
Axon provides robust compliance management features, helping organizations adhere to regulations like GDPR and CCPA. It also offers data masking and encryption capabilities to enhance data security.

Cost

Informatica Axon is priced on the higher end of the spectrum, with costs varying based on the size of the organization and the specific features required. However, its comprehensive feature set justifies the investment for large enterprises.

Community and Support

Informatica has a strong community presence and offers extensive support through its customer service team, online forums, and regular webinars. The company also provides training programs to help users get the most out of their investment.

Advantages

- Comprehensive feature set tailored for large enterprises.

- Strong integration with other Informatica products.

- Robust compliance and security features.

Drawbacks

- High cost, which may be prohibitive for small to mid-sized organizations.

- Steeper learning curve due to the extensive range of features.

Collibra Data Intelligence Platform

Feature Set

Collibra Data Intelligence Platform provides a wide range of data governance functionalities, including data cataloging, data lineage, and policy management. The tool also offers advanced analytics capabilities, allowing users to gain insights into their data governance processes and make data-driven decisions.

Ease of Use

Collibra is praised for its user-friendly interface and drag-and-drop functionality, making it accessible to users with varying levels of technical expertise. The platform also offers extensive documentation and training resources.

Scalability

Collibra is designed to scale with the needs of growing organizations. Its cloud-based architecture ensures that it can handle large volumes of data without compromising performance.

Integration Capabilities

The tool offers robust integration capabilities, supporting connections with a wide range of data sources, including databases, data lakes, and ML platforms. Collibra also provides APIs for custom integrations.

Compliance and Security Features

Collibra helps organizations maintain compliance with data privacy regulations through features such as data masking, encryption, and automated compliance reporting. The tool also offers role-based access controls to enhance security.

Cost

Collibra operates on a subscription-based pricing model, with costs varying depending on the size of the organization and the number of users. While it is more affordable than Informatica Axon, it still represents a significant investment for smaller organizations.

Community and Support

Collibra has an active user community and offers various support channels, including online forums, customer service, and professional services. The company also regularly updates the platform with new features and enhancements.

Advantages

- User-friendly interface with drag-and-drop functionality.
- Strong integration capabilities with various data sources.
- Advanced analytics for data governance insights.

Drawbacks

- Subscription costs can be high for smaller organizations.
- Some advanced features may require additional customization.

Talend Data Fabric

Feature Set

Talend Data Fabric offers a comprehensive suite of data management and governance tools, including data integration, data quality, and data cataloging. The platform also provides ML-specific governance features, such as data versioning and model management.

Ease of Use

Talend is known for its ease of use, particularly for users familiar with open-source platforms. The tool offers a graphical interface for designing data workflows and includes extensive documentation and tutorials.

Scalability

Talend Data Fabric is highly scalable, with both on-premises and cloud deployment options. The tool is designed to handle large datasets and complex data governance needs.

Integration Capabilities

Talend offers a wide range of connectors for integrating with various data sources, including databases, cloud storage, and ML platforms. The platform also supports custom integrations through its API.

Compliance and Security Features

Talend provides robust compliance and security features, including data masking, encryption, and role-based access controls. The tool also offers automated compliance reporting to help organizations meet regulatory requirements.

Cost

Talend operates on a subscription-based pricing model, with different tiers depending on the features and level of support required. It is generally more affordable than Informatica and Collibra, making it a popular choice for mid-sized organizations.

Community and Support

Talend has a strong open-source community and offers extensive support through its customer service team, online forums, and training resources. The platform also benefits from regular updates and contributions from the open-source community.

Advantages

- Comprehensive suite of data management and governance tools.

- Strong support for open-source platforms.

- More affordable pricing model compared to competitors.

Drawbacks

- Some advanced features may require additional configuration.

- Integration with non-Talend tools may require custom development.

IBM Watson Knowledge Catalog

Feature Set

IBM Watson Knowledge Catalog offers a comprehensive data governance solution, including data cataloging, data lineage, and data quality management. The tool also provides AI-driven data discovery and classification features, making it easier for organizations to manage their data assets.

Ease of Use

IBM Watson Knowledge Catalog is designed with usability in mind, offering a clean and intuitive interface. The platform includes AI-driven recommendations, which help users manage data governance tasks more efficiently.

Scalability

The tool is highly scalable and can be deployed on IBM Cloud, on-premises, or in hybrid environments. It is designed to handle large datasets and complex governance requirements, making it suitable for enterprise-level organizations.

Integration Capabilities

IBM Watson Knowledge Catalog integrates seamlessly with other IBM products, such as IBM Cloud Pak for Data, as well as third-party data sources and ML platforms. The tool also supports REST APIs for custom integrations.

Compliance and Security Features

IBM Watson Knowledge Catalog offers robust compliance and security features, including data masking, encryption, and automated compliance reporting. The platform also provides AI-driven data privacy tools, which help organizations identify and manage sensitive data.

Cost

IBM Watson Knowledge Catalog is priced at a premium, reflecting its enterprise-grade feature set and IBM's brand reputation. The cost can vary significantly based on the deployment model (cloud, on-premises, or hybrid) and the specific features required. While the pricing may be on the higher side, many enterprises find the investment worthwhile due to the tool's advanced capabilities and seamless integration with other IBM products.

Community and Support

IBM provides extensive support through its customer service, online resources, and a strong community of users and developers. The platform is regularly updated with new features, and IBM offers training and professional services to help organizations maximize the value of their investment.

Advantages

- AI-driven data governance features that enhance efficiency and accuracy.

- Seamless integration with other IBM products and third-party tools.

- Strong compliance and security capabilities, including data privacy management.

Drawbacks

- High cost, which may not be suitable for smaller organizations.

- The platform's advanced features may require specialized knowledge or training to fully leverage.

Microsoft Purview

Feature Set

Microsoft Purview is a unified data governance service that offers a wide range of features, including data cataloging, data lineage, data classification, and data sharing. Purview is particularly well suited for organizations that are heavily invested in the Microsoft ecosystem, as it integrates seamlessly with Azure services and other Microsoft products.

Ease of Use

Microsoft Purview is designed to be user-friendly, with an intuitive interface that allows users to easily navigate through its various features. The platform offers guided setup wizards and extensive documentation, making it accessible even to users who are new to data governance.

Scalability

Purview is built on Azure, ensuring that it can scale to meet the needs of organizations of all sizes. Its cloud-native architecture allows it to handle large datasets and complex data governance requirements, making it ideal for enterprises with significant data assets.

Integration Capabilities

Microsoft Purview integrates seamlessly with Azure data services, as well as other Microsoft products like Power BI, Microsoft 365, and Dynamics 365. It also supports integration with third-party data sources and platforms, although its strongest integration capabilities are within the Microsoft ecosystem.

Compliance and Security Features

Purview offers robust compliance and security features, including data classification, data encryption, and automated compliance reporting. The platform is designed to help organizations meet regulatory requirements such as GDPR, HIPAA, and CCPA. It also includes features for managing sensitive data and implementing role-based access controls.

Cost

Microsoft Purview is priced competitively, especially for organizations already using other Azure services. The cost is based on usage, with pricing tiers that accommodate different levels of data governance needs. This flexible pricing model makes Purview an attractive option for organizations of all sizes, from small businesses to large enterprises.

Community and Support

Microsoft provides extensive support for Purview through its customer service, online resources, and community forums. The platform is regularly updated with new features, and Microsoft offers training and certification programs to help users get the most out of the tool.

Advantages

- Seamless integration with the Microsoft ecosystem, particularly Azure.

- Competitive pricing with flexible tiers based on usage.

- Comprehensive compliance and security features.

Drawbacks

- Best suited for organizations already using Microsoft products; integration with non-Microsoft tools may be less straightforward.

- Some advanced features may require additional configuration and customization.

Alation

Feature Set

Alation is a leading data catalog and governance platform that provides features such as data discovery, data stewardship, and data quality management. Alation is particularly known for its collaborative features, which enable users to work together to manage and govern data more effectively.

Ease of Use

Alation is designed with usability in mind, offering a user-friendly interface that facilitates collaboration across teams. The platform includes features like guided navigation, search capabilities, and AI-driven recommendations, making it easy for users to find and manage data assets.

Scalability

Alation is scalable and can be deployed in cloud, on-premises, or hybrid environments. The platform is designed to handle large datasets and is well suited for organizations with complex data governance needs.

Integration Capabilities

Alation offers robust integration capabilities, with connectors for a wide range of data sources, including databases, data lakes, and BI tools. The platform also supports APIs for custom integrations, allowing organizations to tailor the tool to their specific needs.

Compliance and Security Features

Alation provides strong compliance and security features, including data lineage tracking, data masking, and role-based access controls. The platform also supports automated compliance reporting, helping organizations meet regulatory requirements.

Cost

Alation operates on a subscription-based pricing model, with costs varying based on the size of the organization and the specific features required. While the platform is not the cheapest option on the market, its collaborative features and ease of use make it a valuable investment for many organizations.

Community and Support

Alation has an active user community and offers extensive support through its customer service, online resources, and professional services. The platform is regularly updated with new features, and Alation provides training programs to help users get the most out of the tool.

Advantages

- Strong focus on collaboration, making it easier for teams to work together on data governance tasks.

- User-friendly interface with AI-driven recommendations.

- Robust integration capabilities with a wide range of data sources.

Drawbacks

- Subscription costs can be high, particularly for organizations with extensive data governance needs.

- Some advanced features may require additional configuration and customization.

Ataccama ONE

Feature Set

Ataccama ONE is a unified data management platform that offers a wide range of data governance features, including data quality management, data cataloging, and master data management. The platform is designed to provide end-to-end data governance, with AI-driven automation to streamline processes and improve efficiency.

Ease of Use

Ataccama ONE is known for its user-friendly interface, which includes drag-and-drop functionality and guided workflows. The platform is designed to be accessible to users with varying levels of technical expertise, making it easy for organizations to implement and use.

Scalability

Ataccama ONE is highly scalable and can be deployed in cloud, on-premises, or hybrid environments. The platform is designed to handle large datasets and complex data governance requirements, making it suitable for enterprises with significant data assets.

Integration Capabilities

Ataccama ONE offers robust integration capabilities, with connectors for a wide range of data sources, including databases, data lakes, and cloud storage. The platform also supports APIs for custom integrations, allowing organizations to tailor the tool to their specific needs.

Compliance and Security Features

Ataccama ONE provides strong compliance and security features, including data masking, encryption, and role-based access controls. The platform also supports automated compliance reporting, helping organizations meet regulatory requirements such as GDPR and CCPA.

Cost

Ataccama ONE operates on a subscription-based pricing model, with costs varying based on the size of the organization and the specific features required. While the platform is more affordable than some of the other tools discussed, it still represents a significant investment for smaller organizations.

Community and Support

Ataccama has a growing community of users and offers extensive support through its customer service, online resources, and professional services. The platform is regularly updated with new features, and Ataccama provides training programs to help users get the most out of the tool.

Advantages

- AI-driven automation for improved efficiency and streamlined processes.
- User-friendly interface with drag-and-drop functionality.
- Strong integration capabilities with a wide range of data sources.

Drawbacks

- Subscription costs may be high for smaller organizations.
- Some advanced features may require additional customization and configuration.

Informatica Enterprise Data Catalog

Feature Set

Informatica Enterprise Data Catalog (EDC) is a comprehensive data governance tool that provides features such as data cataloging, metadata management, and data lineage tracking. The platform also offers AI-driven data discovery and classification, making it easier for organizations to manage their data assets.

Ease of Use

Informatica EDC is designed with usability in mind, offering an intuitive interface and guided workflows. The platform includes features like search capabilities and AI-driven recommendations, making it easy for users to find and manage data assets.

Scalability

Informatica EDC is highly scalable and can be deployed in cloud, on-premises, or hybrid environments. The platform is designed to handle large datasets and complex data governance requirements, making it suitable for enterprises with significant data assets.

Integration Capabilities

Informatica EDC offers robust integration capabilities, with connectors for a wide range of data sources, including databases, data lakes, and BI tools. The platform also supports APIs for custom integrations, allowing organizations to tailor the tool to their specific needs.

Compliance and Security Features

Informatica EDC provides strong compliance and security features, including data lineage tracking, data masking, and role-based access controls. The platform also supports automated compliance reporting, helping organizations meet regulatory requirements.

Cost

Informatica EDC operates on a subscription-based pricing model, with costs varying based on the size of the organization and the specific features required. While the platform is not the cheapest option on the market, its comprehensive feature set and scalability make it a valuable investment for many organizations.

Community and Support

Informatica has a strong community presence and offers extensive support through its customer service, online resources, and professional services. The platform is regularly updated with new features, and Informatica provides training programs to help users get the most out of the tool.

Advantages

- AI-driven data governance features that enhance efficiency and accuracy.

- Seamless integration with other Informatica products and third-party tools.

- Strong compliance and security capabilities, including data privacy management.

Drawbacks

- High cost, which may not be suitable for smaller organizations.

- The platform's advanced features may require specialized knowledge or training to fully leverage.

Sample Documentation in ML Data Governance

This appendix explores key examples of documentation used in ML data governance processes, focusing on data dictionaries, data lineage reports, and audit logs. These documents serve as vital references for data stewards, data scientists, compliance officers, and other stakeholders involved in the governance of ML data. By providing real-world examples and best practices, this appendix aims to offer practical insights that can be adapted to various organizational contexts.

Problem Definition

Data Governance Strategy Document

Project Name: [Insert Project Name]
Date: [Insert Date]
Version: [Insert Version Number]

1. Introduction

This Data Governance Strategy Document outlines the governance framework for managing data within the [Insert Project Name]. It defines the principles, standards, roles, and responsibilities that guide the collection, storage, processing, and sharing of data throughout the ML lifecycle. This strategy is designed to ensure that data is managed in a manner that is consistent with organizational goals, ethical considerations, and legal obligations.

2. Purpose and Scope

2.1 Purpose

The purpose of this document is to establish a comprehensive framework for data governance within the [Insert Project Name]. This framework ensures that data used for ML model development and deployment is accurate, secure, compliant with regulations, and ethically handled.

2.2 Scope

This strategy applies to all data-related activities within the [Insert Project Name], including data collection, storage, processing, analysis, sharing, and deletion. It covers structured, unstructured, and semi-structured data from all internal and external sources used in the project.

3. Data Governance Policies

3.1 Overview of Data Governance Policies

Data governance policies are the foundation of our data management strategy. They establish the principles, standards, and procedures that guide the use of data within the organization, ensuring consistency with organizational goals, ethical standards, and legal requirements.

3.2 Importance of Data Governance in ML

In the context of ML, data governance is critical because it ensures that the data used to train, test, and deploy models is accurate, secure, and compliant with relevant regulations. Without clear policies, there is a risk of using biased, incomplete, or non-compliant data, which can lead to inaccurate models and potentially harmful outcomes. For instance, all data used in model training must be anonymized to protect individual privacy.

3.3 Components of a Data Governance Policy

A comprehensive data governance policy for the [Insert Project Name] includes the following components:

- Purpose and Scope: A statement outlining the purpose of the policy and the scope of its application within the organization.

- Roles and Responsibilities: A description of the roles and responsibilities of individuals and teams involved in data governance, including data stewards, data scientists, and compliance officers.

- Data Quality Standards: Guidelines for ensuring the accuracy, completeness, and consistency of data used in ML processes.

- Data Security and Privacy: Policies for protecting sensitive data, including encryption, access controls, and data anonymization.

- Compliance and Ethics: Procedures for ensuring compliance with relevant regulations (e.g., GDPR) and ethical considerations in the use of data for ML.

- Data Lifecycle Management: Guidelines for managing data throughout its lifecycle, from collection and storage to processing and deletion.

- Audit and Monitoring: Procedures for regular audits and monitoring of data governance practices to ensure ongoing compliance and effectiveness.

4. Governance Framework

4.1 Governance Structure

The governance framework for the [Insert Project Name] is composed of the following elements:

- Data Governance Council: A cross-functional team responsible for overseeing the implementation of data governance policies. The council ensures alignment with organizational objectives and regulatory standards.

- Data Stewards: Individuals responsible for maintaining data quality and integrity. Data Stewards enforce data governance practices at the operational level.

- Data Custodians: IT professionals responsible for the technical management of data, including storage, access control, and security measures.

- Data Owners: Business units or individuals who have ownership over specific datasets. Data Owners are responsible for defining data access policies and ensuring data quality.

- Data Users: Analysts, data scientists, and other stakeholders who access and utilize data for ML model development and analysis.

5. Roles and Responsibilities

5.1 Data Governance Council

- Responsibilities:

  - Oversee the implementation of the data governance strategy.

  - Ensure alignment with organizational goals and regulatory requirements.

  - Review and approve data governance policies and procedures.

  - Monitor compliance and address governance issues.

- Members:

  - [Insert Role], Chief Data Officer (CDO)

  - [Insert Role], Data Privacy Officer (DPO)

  - [Insert Role], IT Security Manager

  - [Insert Role], ML Project Lead

  - [Insert Role], Business Unit Representatives

5.2 Data Stewards

- Responsibilities:

  - Monitor data quality and enforce data governance policies at the operational level.

  - Collaborate with data owners and custodians to resolve data quality issues.

  - Ensure proper documentation and metadata management for datasets.

  - Conduct regular data quality assessments and report findings to the Data Governance Council.

- Appointed Data Stewards:

  - [Insert Name], Data Steward for [Insert Data Type]

  - [Insert Name], Data Steward for [Insert Data Type]

5.3 Data Custodians

- Responsibilities:

  - Manage the technical aspects of data storage, access control, and security.

  - Implement data protection measures, including encryption and access controls.

  - Ensure that the IT infrastructure supports data governance requirements.

  - Maintain audit trails and logs for data access and usage.

- Appointed Data Custodians:

  - [Insert Name], IT Infrastructure Lead

  - [Insert Name], Database Administrator

5.4 Data Owners

- Responsibilities:

  - Define data access policies and ensure data quality for owned datasets.

  - Approve data usage requests and manage data sharing agreements.

  - Ensure that data usage complies with governance policies and regulations.

- Appointed Data Owners:

  - [Insert Name], Head of [Insert Data Type]

  - [Insert Name], Head of [Insert Data Type]

5.5 Data Users

- Responsibilities:

  - Use data responsibly and in compliance with governance policies.

  - Report any data quality issues or security concerns to Data Stewards or Custodians.

  - Ensure that data usage aligns with ethical standards and project objectives.

- Data Users:

  - [List of Data Users or Roles, e.g., Data Scientists, Analysts, etc.]

6. Data Governance Principles

The following principles guide data governance within the [Insert Project Name]:

- Accountability: Clear accountability is established for all data-related activities, ensuring that roles and responsibilities are defined and enforced.

- Transparency: Data practices are transparent, with clear documentation and communication of data-related decisions and processes.

- Data Quality: High standards of data quality are maintained to ensure that the data used in the ML project is accurate, reliable, and fit for purpose.

- Security and Privacy: Data is protected through robust security measures and privacy controls, minimizing the risk of data breaches and ensuring compliance with regulations.

- Ethical Use: Data is used ethically, with respect for individual privacy and without introducing bias or harm in the ML models.

- Compliance: All data-related activities comply with relevant legal, regulatory, and organizational requirements.

7. Data Quality Standards

All data used in ML models within the [Insert Project Name] must meet the following quality standards:

- Accuracy: Data must accurately represent the real-world entities or phenomena it describes.

- Completeness: All necessary data elements must be present and accounted for.

- Consistency: Data must be consistent across different sources and systems.

- Timeliness: Data must be up-to-date and available when needed.

- Validity: Data must conform to defined formats, standards, and rules.

8. Data Security and Privacy

Sensitive data must be protected to ensure privacy and security throughout the ML lifecycle:

- Encryption: All sensitive data must be encrypted both at rest and in transit.

- Access Controls: Data access must be restricted based on roles and responsibilities, with access granted on a need-to-know basis.

- Data Anonymization: Personally identifiable information (PII) must be anonymized or pseudonymized before being used in ML models to protect individual privacy.

- Compliance: All data practices must comply with relevant privacy regulations, such as GDPR and HIPAA.

9. Compliance and Ethics

The following procedures ensure that data governance practices adhere to legal and ethical standards:

- Regulatory Compliance: All data-related activities must comply with relevant regulations, including GDPR, HIPAA, and other industry-specific requirements.

- Ethical Considerations: Data usage must prioritize fairness, transparency, and the avoidance of bias. ML models must be regularly evaluated to ensure they do not perpetuate or amplify biases.

- Data Usage Approval: Data usage for ML model development must be approved by Data Owners and the Data Governance Council to ensure compliance with policies and ethical standards.

10. Data Lifecycle Management

Data must be managed throughout its lifecycle according to the following guidelines:

- Data Collection: Data must be collected in a lawful, transparent, and purpose-driven manner.

- Data Storage: Data must be stored securely with appropriate access controls and encryption.

- Data Processing: Data must be processed in compliance with data quality, security, and privacy standards.

- Data Retention: Data must be retained for a period specified by organizational and legal requirements, after which it must be securely deleted or archived.

- Data Deletion: Data must be securely deleted when it is no longer needed, in compliance with organizational and legal requirements.

11. Audit and Monitoring

To ensure ongoing compliance and effectiveness, the following audit and monitoring procedures are in place:

- Regular Audits: Conduct periodic audits of data governance practices to assess compliance with policies and regulations.

- Continuous Monitoring: Implement tools and processes to continuously monitor data quality, security, and access controls.

- Incident Reporting: Any deviations from this policy must be reported to the Compliance Officer immediately for investigation and remediation.

12. Training and Awareness

To ensure that all stakeholders understand and adhere to the data governance strategy:

- Training Programs: Develop and deliver training programs for all stakeholders involved in data governance, including data stewards, custodians, and users.

- Awareness Campaigns: Conduct awareness campaigns to highlight the importance of data governance and promote best practices.

- Knowledge Sharing: Facilitate knowledge-sharing sessions to discuss challenges, solutions, and best practices in data governance.

13. Document Control

- Document Owner: [Insert Name, Role]

- Approval Date: [Insert Date]

- Next Review Date: [Insert Date]

- Version History:

  - Version [Insert Version Number]—Initial Document—[Insert Date]—Approved by [Insert Name]

Data Ethics and Privacy Assessment

Project Name: [Insert Project Name]
Date: [Insert Date]
Version: [Insert Version Number]

1. Introduction

This document provides a comprehensive assessment of ethical considerations and privacy concerns related to the data used in the [Insert Project Name]. The assessment identifies potential biases, evaluates privacy risks, and outlines strategies to mitigate these issues to ensure the ethical and responsible use of data throughout the machine learning (ML) lifecycle.

2. Purpose and Scope

2.1 Purpose

The purpose of this Data Ethics and Privacy Assessment is to:

- Evaluate the ethical implications of data usage within the [Insert Project Name].

- Identify and address potential biases that may affect the ML model's fairness and accuracy.

- Assess privacy risks associated with the collection, storage, processing, and sharing of data.

- Recommend measures to mitigate identified risks and ensure compliance with legal and ethical standards.

2.2 Scope

This assessment covers all data-related activities within the [Insert Project Name], including data collection, storage, processing, analysis, and sharing. It applies to structured, unstructured, and semi-structured data from all internal and external sources used in the project.

3. Ethical Considerations

3.1 Data Bias Identification

Data bias can significantly impact the fairness and accuracy of ML models. The following types of bias have been assessed for their potential impact on the [Insert Project Name]:

- Sampling Bias: Assess whether the dataset is representative of the population or if certain groups are underrepresented, leading to biased model outcomes.

- Measurement Bias: Evaluate whether the data collection methods introduce systematic errors, such as inaccurate measurements or mislabeling.

- Confirmation Bias: Consider whether the data reflects preconceived notions or expectations, potentially leading to biased predictions.

- Algorithmic Bias: Analyze whether the algorithms used in the project might amplify existing biases in the data.

3.2 Ethical Implications

The ethical implications of using biased data or algorithms in the [Insert Project Name] could include:

- Unfair Treatment: Certain groups may be disadvantaged by biased predictions, leading to unfair outcomes in decision-making processes.

- Lack of Transparency: The opacity of ML models can make it difficult to identify and correct biases, raising concerns about accountability.

- Reinforcement of Stereotypes: Biased data may reinforce harmful stereotypes, perpetuating existing societal inequalities.

3.3 Mitigation Strategies

To address the identified ethical considerations, the following strategies will be implemented:

- Bias Detection and Correction: Implement tools and techniques for detecting and correcting biases in the dataset before model training.

- Diverse Data Collection: Ensure that data is collected from a diverse range of sources to reduce the risk of sampling bias.

- Fairness Audits: Conduct regular fairness audits to evaluate the model's performance across different demographic groups.

- Explainability Techniques: Apply explainable AI (XAI) techniques to improve model transparency and identify potential sources of bias.

4. Privacy Concerns

4.1 Data Privacy Risks

The following privacy risks have been identified in the [Insert Project Name]:

- Personally Identifiable Information (PII): Assess the presence of PII in the dataset, which could lead to privacy violations if not properly protected.

- Data Anonymization and Re-identification: Evaluate the effectiveness of data anonymization techniques and the risk of re-identification, where anonymized data could potentially be traced back to individuals.

- Data Sharing and Access: Consider the risks associated with sharing data with third parties or providing access to sensitive data within the organization.

- Data Breach Risks: Assess the likelihood and potential impact of data breaches, which could expose sensitive information to unauthorized entities.

4.2 Legal and Regulatory Compliance

The [Insert Project Name] must comply with relevant data privacy regulations, including but not limited to:

- General Data Protection Regulation (GDPR): Ensure that data processing activities comply with GDPR requirements, including obtaining informed consent, providing data access rights, and enabling data erasure upon request.

- Health Insurance Portability and Accountability Act (HIPAA): For projects involving health data, ensure compliance with HIPAA standards for protecting sensitive health information.

- California Consumer Privacy Act (CCPA): Ensure compliance with CCPA requirements for data collection, usage, and sharing, particularly for California residents.

4.3 Mitigation Strategies

To mitigate the identified privacy risks, the following strategies will be implemented:

- Data Anonymization and Pseudonymization: Apply robust anonymization and pseudonymization techniques to protect PII and reduce the risk of re-identification.

- Access Controls: Implement strict access controls to limit who can view and process sensitive data, based on roles and responsibilities.

- Encryption: Encrypt sensitive data both at rest and in transit to protect it from unauthorized access.

- Privacy Impact Assessments (PIAs): Conduct regular PIAs to assess the potential impact of data processing activities on individual privacy and to ensure compliance with legal requirements.

- Data Minimization: Limit the collection of data to only what is necessary for the ML project, reducing the amount of sensitive information handled.

5. Risk Assessment Summary

The table below provides a summary of the identified ethical and privacy risks, their potential impact, and the proposed mitigation strategies:

Risk	Potential Impact	Mitigation Strategy
Sampling Bias	Unfair treatment of underrepresented groups	Diverse Data Collection, Bias Detection, and Correction
Measurement Bias	Inaccurate predictions due to systematic errors	Data Validation, Fairness Audits
Algorithmic Bias	Amplification of existing biases	Algorithmic Fairness Techniques, Explainability Techniques
PII Exposure	Privacy violations, regulatory non-compliance	Data Anonymization, Access Controls, Encryption

(*continued*)

Risk	Potential Impact	Mitigation Strategy
Re-identification Risk	Potential re-identification of anonymized data	Robust Anonymization, Pseudonymization, Regular PIAs
Data Breach	Exposure of sensitive data to unauthorized entities	Encryption, Access Controls, Breach Response Plan

6. Recommendations

Based on the assessment, the following recommendations are made to ensure ethical and privacy-compliant data usage in the [Insert Project Name]:

- Regular Bias Audits: Conduct ongoing audits to detect and mitigate biases in the dataset and ML models.

- Enhanced Privacy Controls: Strengthen privacy controls through advanced anonymization techniques and strict access management.

- Stakeholder Involvement: Involve a diverse group of stakeholders in the development and review of data governance policies to ensure that ethical considerations are fully addressed.

- Continuous Monitoring: Implement continuous monitoring of data usage and model performance to identify and address any emerging ethical or privacy concerns.

7. Document Control

- Document Owner: [Insert Name, Role]

- Approval Date: [Insert Date]

- Next Review Date: [Insert Date]

- Version History:

    - Version [Insert Version Number]—Initial Document—[Insert Date]—Approved by [Insert Name]

Data Collection and Exploration

Data Sourcing Agreement

Project Name: [Insert Project Name]
Date: [Insert Date]
Version: [Insert Version Number]

1. Introduction

This Data Sourcing Agreement outlines the terms, conditions, and compliance requirements for sourcing data for the [Insert Project Name]. It establishes the legal and ethical framework for data acquisition, including data ownership, licensing, and the responsibilities of all parties involved. This agreement ensures that data is sourced in a manner consistent with organizational policies, regulatory requirements, and best practices.

2. Purpose and Scope

2.1 Purpose

The purpose of this Data Sourcing Agreement is to:

- Define the terms and conditions for acquiring and using data in the [Insert Project Name].

- Establish data ownership and licensing arrangements between the data provider and the organization.

- Ensure compliance with relevant legal, regulatory, and ethical standards during data acquisition and use.

- Outline the responsibilities of all parties involved in data sourcing and usage.

2.2 Scope

This agreement applies to all data sourcing activities for the [Insert Project Name], including data acquisition from external vendors, partners, and internal sources. It covers all forms of data, including structured, unstructured, and semi-structured data.

3. Definitions

- Data Provider: The entity or individual providing the data to be used in the [Insert Project Name].

- Data Recipient: The organization or project team responsible for receiving and using the data within the [Insert Project Name].

- Data Ownership: The legal rights and control over the data, including the ability to use, distribute, and modify the data.

- Licensing: The terms under which the data provider grants the data recipient the rights to use the data.

4. Data Ownership and Licensing

4.1 Data Ownership

- Ownership Rights: The data provider retains ownership of the data unless otherwise specified in this agreement. The data recipient acknowledges that the data provider holds all intellectual property rights, including copyright, trademarks, and trade secrets, associated with the data.

- Transfer of Ownership: If ownership of the data is to be transferred, the terms and conditions of such a transfer must be explicitly stated in this agreement. The data provider must agree to the transfer in writing, and all relevant documentation must be provided to the data recipient.

4.2 Licensing Terms

- License Grant: The data provider grants the data recipient a [non-exclusive/exclusive], [perpetual/limited-term] license to use the data for the purposes outlined in this agreement. The license includes the rights to access, process, and analyze the data as necessary for the [Insert Project Name].

- License Restrictions: The data recipient agrees to use the data solely for the purposes specified in this agreement. The data may not be sold, licensed, or otherwise transferred to third parties without the explicit written consent of the data provider.

- Sub-licensing: The data recipient may not grant sub-licenses to third parties unless expressly permitted in this agreement. Any sub-licensing arrangements must be approved by the data provider in writing.

- License Termination: The data provider reserves the right to terminate the license if the data recipient violates the terms of this agreement. Upon termination, the data recipient must cease all use of the data and return or destroy all copies of the data as directed by the data provider.

5. Data Usage and Compliance

5.1 Permitted Use

- Scope of Use: The data may only be used for the purposes outlined in this agreement and within the scope of the [Insert Project Name]. Any use of the data outside this scope requires prior written approval from the data provider.

- Compliance with Laws: The data recipient agrees to comply with all applicable laws and regulations related to data usage, including data protection and privacy laws (e.g., GDPR, CCPA). The data recipient must ensure that the data is used in a manner that respects individual privacy rights and does not result in unlawful discrimination or harm.

- Data Modification: The data recipient may modify or transform the data as necessary for the [Insert Project Name]. However, any such modifications must be documented, and the data provider must be informed of the nature and purpose of the modifications.

5.2 Data Security

–   Security Measures: The data recipient must implement appropriate technical and organizational measures to protect the data from unauthorized access, use, or disclosure. This includes encryption, access controls, and regular security audits.

–   Data Breach Notification: In the event of a data breach, the data recipient must notify the data provider within [Insert Timeframe] of becoming aware of the breach. The notification must include a description of the breach, the data affected, and the steps taken to mitigate the breach.

–   Data Storage: The data must be stored in a secure environment that complies with industry best practices and relevant legal requirements. The data recipient is responsible for ensuring that all storage solutions meet these standards.

6. Data Quality and Maintenance

6.1 Data Quality

–   Accuracy and Completeness: The data provider represents and warrants that the data provided is accurate, complete, and up-to-date to the best of their knowledge. The data recipient is responsible for verifying the quality of the data before use.

–   Data Validation: The data recipient may perform data validation checks to ensure the accuracy and reliability of the data. Any discrepancies or errors identified during validation must be reported to the data provider promptly.

6.2 Data Updates

–   Data Refresh: If the data requires regular updates or refreshes, the data provider agrees to provide these updates within the agreed-upon timeframe. The data recipient is responsible for incorporating these updates into the project as needed.

- Data Correction: If errors or inaccuracies are identified in the data after delivery, the data provider agrees to correct the data and provide the corrected version to the data recipient promptly.

7. Confidentiality

7.1 Confidential Information

- Definition: Confidential Information refers to any non-public information disclosed by either party in connection with this agreement, including the data itself, business practices, technical information, and any other proprietary information.

- Obligations: Both parties agree to maintain the confidentiality of all Confidential Information and to use it solely for the purposes outlined in this agreement. Confidential Information may not be disclosed to third parties without the prior written consent of the disclosing party.

7.2 Exceptions

- Public Domain: Confidential Information does not include information that is or becomes publicly available without breach of this agreement.

- Legal Requirements: If either party is required by law to disclose Confidential Information, they must notify the other party promptly and cooperate in seeking a protective order or other appropriate remedy.

8. Term and Termination

8.1 Term

- Agreement Term: This agreement shall commence on [Insert Start Date] and continue until [Insert End Date], unless terminated earlier in accordance with the terms of this agreement.

8.2 Termination

- Termination for Cause: Either party may terminate this agreement with immediate effect if the other party breaches any material term of this agreement and fails to remedy the breach within [Insert Cure Period] after receiving written notice.

- Termination for Convenience: Either party may terminate this agreement for convenience by providing [Insert Notice Period] written notice to the other party.

- Effect of Termination: Upon termination of this agreement, the data recipient must cease all use of the data and return or destroy all copies of the data as directed by the data provider. The data recipient must certify in writing that all data has been returned or destroyed.

9. Indemnification and Liability

9.1 Indemnification

- Data Provider Indemnity: The data provider agrees to indemnify, defend, and hold harmless the data recipient from and against any claims, damages, liabilities, and expenses arising out of the data provider's breach of this agreement or any representations or warranties.

- Data Recipient Indemnity: The data recipient agrees to indemnify, defend, and hold harmless the data provider from and against any claims, damages, liabilities, and expenses arising out of the data recipient's use of the data in violation of this agreement or applicable laws.

9.2 Limitation of Liability

- Liability Cap: Except in cases of gross negligence or willful misconduct, the liability of either party under this agreement shall not exceed [Insert Liability Cap Amount].

- Exclusion of Consequential Damages: Neither party shall be liable for any indirect, incidental, special, or consequential damages arising out of or related to this agreement, even if advised of the possibility of such damages.

10. Dispute Resolution

10.1 Governing Law

- Applicable Law: This agreement shall be governed by and construed in accordance with the laws of [Insert Jurisdiction], without regard to its conflict of law principles.

10.2 Dispute Resolution

- Negotiation: In the event of a dispute arising out of or relating to this agreement, the parties agree to first attempt to resolve the dispute through good faith negotiations.

- Arbitration: If the dispute cannot be resolved through negotiation within [Insert Timeframe], the parties agree to submit the dispute to binding arbitration in accordance with the rules of [Insert Arbitration Institution]. The arbitration shall take place in [Insert Location], and the language of the arbitration shall be [Insert Language].

11. Miscellaneous

11.1 Entire Agreement

- Integration: This agreement constitutes the entire agreement between the parties with respect to the subject matter hereof and supersedes all prior agreements, understandings, and representations.

11.2 Amendments

– Modification: Any amendments or modifications to this agreement must be made in writing and signed by both parties.

11.3 Assignment

– Assignment Restrictions: Neither party may assign or transfer its rights or obligations under this agreement without the prior written consent of the other party, except in the case of a merger, acquisition, or sale of substantially all of its assets.

11.4 Notices

– Notice Requirements: All notices required or permitted under this agreement shall be in writing and delivered to the addresses specified below or to such other address as may be designated by a party in writing.

11.5 Severability

– Severability Clause: If any provision of this agreement is found to be invalid or unenforceable by a court of competent jurisdiction, the remaining provisions shall remain in full force and effect.

12. Signatures

This Data Sourcing Agreement is executed by the duly authorized representatives of the parties as of the date first written above.

Data Provider:

[Insert Name]

[Insert Title]

[Insert Company Name]

[Insert Signature]

[Insert Date]

Data Recipient:

    [Insert Name]

    [Insert Title]

    [Insert Company Name]

    [Insert Signature]

    [Insert Date]

Data Lineage Documentation

Project Name: [Insert Project Name]
Date: [Insert Date]
Version: [Insert Version Number]

1. Introduction

This Data Lineage Documentation outlines the origin, flow, and transformation of data within the [Insert Project Name]. It provides a comprehensive view of how data moves through various systems and processes, ensuring traceability, accountability, and compliance with data governance policies. This documentation is crucial for understanding the data lifecycle, identifying potential data quality issues, and maintaining data integrity across the project.

2. Purpose and Scope

2.1 Purpose

The purpose of this Data Lineage Documentation is to:

- Track the origin and flow of data from its source to its final destination.

- Document all transformations, processing steps, and systems involved in the data lifecycle.

- Ensure traceability and accountability for data handling within the [Insert Project Name].

- Facilitate troubleshooting, data quality management, and regulatory compliance.

2.2 Scope

This documentation covers all data sources, systems, processes, and transformations involved in the [Insert Project Name]. It includes structured, unstructured, and semi-structured data, as well as internal and external data sources.

3. Data Lineage Overview

3.1 Data Sources

This section provides an overview of the primary data sources used in the [Insert Project Name]. Each data source is described with its origin, ownership, and any relevant metadata.

- Source Name: [Insert Data Source Name]

- Description: [Brief description of the data source]

- Origin: [Internal/External; specify the original provider or system]

- Data Type: [Structured/Unstructured/Semi-structured]

- Data Owner: [Insert Data Owner Name]

- Frequency of Updates: [e.g., Real-time, Daily, Weekly]

- Format: [e.g., CSV, JSON, XML, Database Table]

3.2 Data Ingestion

This section describes how data is ingested into the project's systems, including the tools and methods used for data extraction, loading, and storage.

- Ingestion Tool: [Insert Tool or Process Used]

- Ingestion Method: [e.g., Batch Processing, Real-time Streaming]

- Ingestion Schedule: [e.g., Daily, Hourly, On-Demand]

- Data Storage Location: [Insert Location, e.g., Data Warehouse, Data Lake, Cloud Storage]

- Data Validation: [Describe any validation processes applied during ingestion]

4. Data Transformation and Processing

4.1 Transformation Steps

This section outlines the key transformation steps that data undergoes after ingestion, including data cleaning, enrichment, and aggregation processes.

- Transformation Step Name: [Insert Transformation Name]

- Description: [Brief description of the transformation]

- Tool/Technology Used: [Insert Tool or Technology Name]

- Transformation Logic: [Describe the logic or algorithm applied, e.g., filtering, joining, aggregating]

- Input Data: [List of data fields or datasets used as input]

- Output Data: [List of data fields or datasets produced as output]

- Data Quality Checks: [Describe any checks or validations applied]

4.2 Processing Workflows

This section provides an overview of the processing workflows, detailing how data moves through different systems and processes.

- Workflow Name: [Insert Workflow Name]

- Description: [Brief description of the workflow]

- Systems Involved: [List of systems or platforms involved in the workflow]

- Data Flow: [Describe the flow of data through the workflow, including inputs, outputs, and any intermediate steps]

- Automation: [Specify whether the workflow is automated or manual]

- Error Handling: [Describe how errors are detected and managed within the workflow]

5. Data Storage and Management

5.1 Data Storage Locations

This section documents where data is stored at different stages of its lifecycle within the [Insert Project Name].

- Storage Name: [Insert Storage Name]

- Location: [Insert Physical or Cloud Location]

- Data Type: [Structured/Unstructured/Semi-structured]

- Format: [e.g., Database, File System, Cloud Storage]

- Access Controls: [Describe who has access to the storage and how access is managed]

- Data Retention: [Specify retention policies, including duration and archiving procedures]

5.2 Data Backup and Recovery

This section outlines the backup and recovery procedures to ensure data integrity and availability.

- Backup Schedule: [e.g., Daily, Weekly, Real-time]

- Backup Location: [Insert Backup Storage Location]

- Recovery Procedures: [Describe the steps for recovering data in case of loss or corruption]

- Testing Frequency: [Specify how often recovery procedures are tested]

6. Data Access and Security

6.1 Access Controls

This section details the access controls in place to protect data within the [Insert Project Name].

- Access Control Method: [e.g., Role-Based Access Control (RBAC), Attribute-Based Access Control (ABAC)]

- User Roles: [List of roles with access to the data, e.g., Data Scientists, Analysts, Administrators]

- Permissions: [Describe the permissions associated with each role, e.g., Read, Write, Modify]

- Authentication Methods: [e.g., Single Sign-On (SSO), Multi-factor Authentication (MFA)]

- Audit Trails: [Describe how access is monitored and recorded]

6.2 Data Security Measures

This section outlines the security measures in place to protect data from unauthorized access, breaches, and other threats.

- Encryption: [Describe encryption methods used for data at rest and in transit]

- Data Masking: [Describe any data masking techniques applied to sensitive data]

- Security Monitoring: [Describe the tools and processes used for monitoring data security]

- Incident Response: [Outline the steps taken in response to a data security incident]

7. Data Lineage Visualization

7.1 Data Lineage Diagrams

This section includes visual representations of data lineage within the [Insert Project Name]. These diagrams illustrate the flow of data from its source through various transformations, storage locations, and final outputs.

– Diagram Name: [Insert Diagram Name]

– Description: [Brief description of what the diagram represents]

– Key Components: [List of key components shown in the diagram, e.g., Data Sources, Transformation Steps, Storage Locations]

– Legend: [Explanation of symbols, colors, and other notations used in the diagram]

8. Data Lineage Use Cases

8.1 Data Quality Management

This section describes how data lineage is used to manage and improve data quality within the [Insert Project Name].

– Use Case Name: [Insert Use Case Name]

– Description: [Brief description of the use case]

– Objective: [Describe the goal of the use case, e.g., Identifying data quality issues, Ensuring data accuracy]

– Process: [Describe the steps involved in using data lineage for this use case]

– Outcome: [Describe the expected or actual outcome of the use case]

8.2 Compliance and Auditing

This section outlines how data lineage supports compliance with regulatory requirements and auditing processes.

– Use Case Name: [Insert Use Case Name]

– Description: [Brief description of the use case]

- Objective: [Describe the goal of the use case, e.g., Demonstrating compliance, Supporting audits]

- Process: [Describe the steps involved in using data lineage for this use case]

- Outcome: [Describe the expected or actual outcome of the use case]

9. Maintenance and Updates

9.1 Documentation Maintenance

This section outlines the procedures for maintaining and updating the data lineage documentation.

- Responsible Parties: [List of roles responsible for maintaining the documentation]

- Update Frequency: [Specify how often the documentation should be reviewed and updated]

- Version Control: [Describe the version control process, including how changes are tracked and documented]

9.2 Continuous Improvement

This section describes how the data lineage process is continuously improved to enhance data management and governance.

- Feedback Mechanisms: [Describe how feedback is collected from users and stakeholders]

- Improvement Initiatives: [List any ongoing or planned initiatives to improve data lineage processes]

- Performance Metrics: [Describe the metrics used to evaluate the effectiveness of data lineage management]

10. Document Control

- Document Owner: [Insert Name, Role]

- Approval Date: [Insert Date]

- Next Review Date: [Insert Date]

- Version History:

  - Version [Insert Version Number]—Initial Document—[Insert Date]—Approved by [Insert Name]

Data Privacy Impact Assessment (DPIA)

Project Name: [Insert Project Name]
Date: [Insert Date]
Version: [Insert Version Number]

1. Introduction

This Data Privacy Impact Assessment (DPIA) evaluates the potential impact of data processing activities on individuals' privacy within the [Insert Project Name]. The DPIA ensures that data processing complies with relevant privacy regulations, such as the General Data Protection Regulation (GDPR), and identifies measures to mitigate privacy risks. This assessment is a critical component of the project's data governance framework, designed to protect individuals' rights and maintain organizational accountability.

2. Purpose and Scope

2.1 Purpose

The purpose of this DPIA is to:

- Identify and assess the privacy risks associated with data processing activities within the [Insert Project Name].

- Ensure compliance with applicable privacy regulations, including GDPR.

- Develop and implement measures to mitigate identified privacy risks.

- Provide documentation that demonstrates the organization's commitment to protecting individuals' privacy rights.

2.2 Scope

This DPIA covers all data processing activities within the [Insert Project Name], including the collection, storage, use, sharing, and deletion of personal data. The assessment applies to structured and unstructured data, and includes data collected from internal and external sources.

3. Data Processing Overview

3.1 Description of Data Processing Activities

This section provides a detailed description of the data processing activities involved in the [Insert Project Name].

- Data Categories: [List of personal data categories being processed, e.g., Name, Email Address, Health Information]

- Processing Purpose: [Describe the purpose of data processing, e.g., Personalization of services, Data analysis, ML model training]

- Data Sources: [List of data sources, e.g., Internal databases, Third-party providers, User inputs]

- Data Flow: [Brief description of how data moves through the system, from collection to processing and storage]

3.2 Legal Basis for Processing

This section identifies the legal basis for processing personal data under GDPR, ensuring that data processing activities are lawful.

- Legal Basis: [Specify the legal basis for processing, e.g., Consent, Contractual necessity, Legitimate interests]

- Consent Management: [Describe how consent is obtained, recorded, and managed, if applicable]

- Data Subject Rights: [Outline how data subjects can exercise their rights, such as access, rectification, or deletion]

4. Privacy Risk Assessment

4.1 Risk Identification

This section identifies potential privacy risks associated with the data processing activities.

- Risk 1: [Describe the first identified risk, e.g., Unauthorized access to personal data]

- Risk 2: [Describe the second identified risk, e.g., Data breaches resulting in the exposure of sensitive information]

- Risk 3: [Describe the third identified risk, e.g., Inadequate data anonymization leading to re-identification of individuals]

4.2 Risk Evaluation

This section evaluates the identified risks based on their potential impact on individuals' privacy and the likelihood of occurrence.

Risk	Impact	Likelihood	Risk Level
Unauthorized access to personal data	High	Medium	High
Data breaches	High	Low	Medium
Inadequate data anonymization	Medium	Medium	Medium

- Impact: The potential consequences for individuals if the risk materializes.

- Likelihood: The probability of the risk occurring.

- Risk Level: A combined assessment of impact and likelihood, indicating the overall risk level.

5. Mitigation Measures

5.1 Implemented Measures

This section outlines the measures implemented to mitigate the identified privacy risks.

- Measure 1: [Describe the first mitigation measure, e.g., Implementation of role-based access control (RBAC) to prevent unauthorized data access]

- Measure 2: [Describe the second mitigation measure, e.g., Encryption of sensitive data both at rest and in transit]

- Measure 3: [Describe the third mitigation measure, e.g., Use of advanced anonymization techniques to protect individual identities]

5.2 Additional Measures

This section recommends additional measures to further reduce privacy risks.

- Measure 4: [Describe the fourth mitigation measure, e.g., Regular privacy impact assessments and audits]

- Measure 5: [Describe the fifth mitigation measure, e.g., Enhanced monitoring and logging of data access activities]

- Measure 6: [Describe the sixth mitigation measure, e.g., Training programs on data privacy for all employees involved in the project]

6. Data Subject Rights

6.1 Overview of Data Subject Rights

This section outlines the rights of data subjects under GDPR and how these rights are respected in the [Insert Project Name].

- Right to Access: [Describe how individuals can request access to their personal data]

- Right to Rectification: [Describe how individuals can request correction of inaccurate or incomplete data]

- Right to Erasure (Right to be Forgotten): [Describe how individuals can request the deletion of their personal data]

- Right to Data Portability: [Describe how individuals can request the transfer of their data to another controller]

- Right to Object: [Describe how individuals can object to the processing of their data, particularly for direct marketing]

6.2 Procedures for Exercising Rights

This section describes the procedures in place for individuals to exercise their rights.

- Request Submission: [Explain how data subjects can submit requests, e.g., through a web portal, email, or in writing]

- Verification Process: [Describe the process for verifying the identity of the data subject making the request]

- Response Timeframe: [Specify the timeframe within which the organization will respond to requests, e.g., within 30 days]

- Handling Objections: [Describe how objections to data processing are handled and resolved]

7. Compliance and Documentation

7.1 Regulatory Compliance

This section ensures that the data processing activities comply with GDPR and other relevant regulations.

- Compliance Checks: [Describe the process for regularly checking compliance with GDPR, e.g., internal audits, external reviews]

- Documentation Requirements: [List the documents required for demonstrating compliance, e.g., consent records, DPIA reports, data processing agreements]

7.2 Record-Keeping

This section outlines the record-keeping practices to ensure compliance and accountability.

- Records of Processing Activities: [Describe the records maintained for processing activities, as required by Article 30 of GDPR]

- Data Breach Records: [Describe how data breaches are documented and reported to supervisory authorities]

- DPIA Documentation: [Explain how DPIA reports and related documentation are stored and managed]

8. Review and Updates

8.1 Regular Reviews

This section describes the process for regularly reviewing and updating the DPIA to ensure its relevance and effectiveness.

- Review Frequency: [Specify how often the DPIA will be reviewed, e.g., annually, biannually]

- Responsible Parties: [List the roles responsible for conducting the review, e.g., Data Protection Officer, Compliance Team]

- Update Triggers: [Describe situations that may trigger an update to the DPIA, e.g., changes in data processing activities, new regulatory requirements]

8.2 Continuous Monitoring

This section outlines the continuous monitoring of data processing activities to identify and address emerging privacy risks.

- Monitoring Tools: [Describe the tools and technologies used for monitoring data processing activities]

- Incident Reporting: [Explain the process for reporting and responding to privacy incidents identified during monitoring]

- Feedback Mechanism: [Describe how feedback from data subjects and stakeholders is collected and used to improve privacy practices]

9. Document Control

- Document Owner: [Insert Name, Role]

- Approval Date: [Insert Date]

- Next Review Date: [Insert Date]

- Version History:

  - Version [Insert Version Number]—Initial Document—[Insert Date]—Approved by [Insert Name]

Data Preparation

Data Quality Assessment Report

Project Name: [Insert Project Name]
Date: [Insert Date]
Version: [Insert Version Number]

1. Introduction

This Data Quality Assessment Report provides a comprehensive evaluation of the data used in the [Insert Project Name]. The report examines the data's quality in terms of completeness, accuracy, consistency, and timeliness. It identifies any issues that may impact the quality of the data and outlines remediation actions to address these issues. The goal of this assessment is to ensure that the data meets the standards necessary for reliable and accurate analysis within the project.

2. Purpose and Scope

2.1 Purpose

The purpose of this Data Quality Assessment Report is to:

- – Evaluate the quality of data being used in the [Insert Project Name].

- – Identify any data quality issues that could impact the reliability and accuracy of the project's outcomes.

- – Document the completeness, accuracy, consistency, and timeliness of the data.

- – Recommend remediation actions to address identified data quality issues.

2.2 Scope

This report covers all datasets used within the [Insert Project Name], including structured, unstructured, and semi-structured data. The assessment includes data sourced from both internal and external systems.

3. Data Quality Dimensions

3.1 Completeness

Definition: Completeness refers to the extent to which all required data is present and accounted for in the dataset.

- – Assessment Criteria:
  - • Percentage of missing values in key data fields.
  - • Availability of all required data fields.
  - • Presence of mandatory records.
- – Findings:
  - • [Insert finding on missing values, e.g., "5% of records in the dataset are missing values in the 'Date of Birth' field."]

- [Insert finding on availability, e.g., "All required data fields are present in the dataset."]
- [Insert finding on mandatory records, e.g., "97% of mandatory records are available, with a 3% shortfall."]

– Remediation Actions:

- [Insert action, e.g., "Implement data validation checks at the point of data entry to reduce missing values."]
- [Insert action, e.g., "Audit data sources to identify and correct the shortfall in mandatory records."]

3.2 Accuracy

Definition: Accuracy refers to the extent to which data correctly represents the real-world entities or events it is intended to model.

– Assessment Criteria:

- Verification of data against authoritative sources.
- Error rates in data fields (e.g., incorrect entries, typos).
- Consistency between related data fields.

– Findings:

- [Insert finding on verification, e.g., "90% of the 'Postal Codes' match the authoritative source, with a 10% error rate."]
- [Insert finding on error rates, e.g., "2% of entries in the 'Employee ID' field contain typos."]
- [Insert finding on consistency, e.g., "There is a 5% inconsistency rate between 'City' and 'State' fields."]

– Remediation Actions:

- [Insert action, e.g., "Correct mismatched 'Postal Codes' by cross-referencing with the authoritative source."]
- [Insert action, e.g., "Implement automated data validation scripts to catch and correct typos at the point of entry."]

3.3 Consistency

Definition: Consistency refers to the uniformity of data across different datasets and within the same dataset, ensuring that data does not contradict itself.

- Assessment Criteria:

  - Alignment of data formats across datasets.

  - Consistency between related datasets (e.g., customer records in different systems).

  - Cross-field consistency within the same dataset.

- Findings:

  - [Insert finding on format alignment, e.g., "Date formats are inconsistent across datasets, with both 'MM/DD/YYYY' and 'YYYY-MM-DD' formats being used."]

  - [Insert finding on dataset consistency, e.g., "Customer names in the billing system do not consistently match those in the CRM system, with a 7% discrepancy rate."]

  - [Insert finding on cross-field consistency, e.g., "15% of records show inconsistent 'Order Date' and 'Shipping Date' fields."]

- Remediation Actions:

  - [Insert action, e.g., "Standardize date formats across all datasets to 'YYYY-MM-DD'."]

  - [Insert action, e.g., "Implement data reconciliation processes between the billing and CRM systems to reduce discrepancies."]

  - [Insert action, e.g., "Apply business rules to ensure 'Shipping Date' is always after 'Order Date'."]

3.4 Timeliness

Definition: Timeliness refers to the extent to which data is up-to-date and available when needed for analysis and decision-making.

- – Assessment Criteria:
  - • Frequency of data updates.
  - • Lag time between data collection and availability.
  - • Relevance of data based on its currency.
- – Findings:
  - • [Insert finding on update frequency, e.g., "Data is updated weekly, but some critical fields are only refreshed monthly."]
  - • [Insert finding on lag time, e.g., "There is a 3-day lag between data collection and its availability for analysis."]
  - • [Insert finding on data relevance, e.g., "90% of the data is relevant for current analysis, with 10% considered outdated."]
- – Remediation Actions:
  - • [Insert action, e.g., "Increase the update frequency of critical fields to align with the weekly schedule."]
  - • [Insert action, e.g., "Implement real-time data processing for critical data streams to reduce lag time."]
  - • [Insert action, e.g., "Review and archive outdated data, ensuring that only relevant data is used for analysis."]

4. Overall Data Quality Score

Methodology: The overall data quality score is calculated based on the weighted average of the scores for completeness, accuracy, consistency, and timeliness.

- – Completeness Score: [Insert Score] (e.g., 85%)
- – Accuracy Score: [Insert Score] (e.g., 90%)
- – Consistency Score: [Insert Score] (e.g., 80%)

- Timeliness Score: [Insert Score] (e.g., 95%)

- Overall Data Quality Score: [Insert Overall Score] (e.g., 88%)

Interpretation: The overall data quality score of [Insert Overall Score] indicates that the data used in the [Insert Project Name] meets [Insert Level, e.g., "a high standard"] for quality, with particular strengths in [Insert Areas of Strength, e.g., "timeliness and accuracy"]. However, improvements are needed in [Insert Areas of Improvement, e.g., "consistency and completeness"] to fully ensure the reliability of the data.

5. Remediation Plan

5.1 Immediate Actions

- Action 1: [Insert Immediate Action, e.g., "Conduct a data cleansing operation to address the identified inconsistencies in customer records."]

- Action 2: [Insert Immediate Action, e.g., "Update data validation scripts to improve accuracy during data entry."]

- Action 3: [Insert Immediate Action, e.g., "Increase the frequency of data updates for critical fields to improve timeliness."]

5.2 Long-Term Actions

- Action 1: [Insert Long-Term Action, e.g., "Implement a comprehensive data governance framework to ensure ongoing data quality management."]

- Action 2: [Insert Long-Term Action, e.g., "Train staff on data quality best practices and the importance of accurate data entry."]

- Action 3: [Insert Long-Term Action, e.g., "Develop automated tools to monitor data quality continuously and flag potential issues in real time."]

6. Recommendations

Based on the data quality assessment, the following recommendations are made to improve data quality within the [Insert Project Name]:

- Implement regular data audits: Schedule regular data quality audits to identify and address issues proactively.

- Standardize data formats: Ensure that data formats are consistent across all datasets to improve data consistency.

- Enhance data validation processes: Strengthen data validation processes at the point of entry to reduce errors and improve accuracy.

- Increase data update frequency: Review and adjust data update schedules to ensure that all critical data is timely and relevant.

7. Document Control

- Document Owner: [Insert Name, Role]

- Approval Date: [Insert Date]

- Next Review Date: [Insert Date]

- Version History:

  - Version [Insert Version Number]—Initial Document—[Insert Date]—Approved by [Insert Name]

Data Anonymization and Pseudonymization Records

Project Name: [Insert Project Name]
Date: [Insert Date]
Version: [Insert Version Number]

1. Introduction

This document provides a comprehensive record of the methods and practices used to anonymize and pseudonymize data within the [Insert Project Name]. The purpose of these techniques is to protect sensitive information, ensuring compliance with data protection regulations and maintaining the privacy of individuals whose data is used in the project.

2. Purpose and Scope

2.1 Purpose

The purpose of this document is to:

- Detail the methods used to anonymize and pseudonymize data within the [Insert Project Name].

- Ensure compliance with relevant data protection regulations (e.g., GDPR, HIPAA).

- Provide a reference for auditing and compliance purposes.

- Outline the responsibilities of individuals and teams involved in the anonymization and pseudonymization process.

2.2 Scope

This document applies to all data anonymization and pseudonymization activities conducted within the [Insert Project Name]. It covers all types of data, including structured, unstructured, and semi-structured data, originating from internal and external sources.

3. Definitions

- Anonymization: The process of removing or modifying personal identifiers in data so that individuals cannot be re-identified by any means reasonably likely to be used.

- Pseudonymization: The process of replacing personal identifiers with a pseudonym or artificial identifier, which can be reversed if the key to the pseudonym is available.

4. Anonymization Techniques

4.1 Data Masking

- Description: Data masking involves replacing sensitive data with fictional but realistic-looking data to protect personal information.

- Method Applied: [Describe the specific data masking technique used, e.g., character shuffling, randomization]

- Data Fields Affected: [List the data fields where data masking was applied, e.g., Social Security Numbers, Credit Card Numbers]

- Rationale: [Explain why data masking was chosen as the anonymization method for these fields]

4.2 Generalization

- Description: Generalization involves reducing the granularity of data to make it less specific and thereby protect individual identities.

- Method Applied: [Describe how generalization was applied, e.g., converting exact ages to age ranges, specific locations to broader regions]

- Data Fields Affected: [List the data fields where generalization was applied, e.g., Date of Birth, Geographic Location]

- Rationale: [Explain why generalization was used for these fields and how it helps in anonymization]

4.3 Suppression

- Description: Suppression involves completely removing specific data elements to prevent the identification of individuals.

- Method Applied: [Describe which data elements were suppressed, e.g., removing certain identifiers from the dataset]

- Data Fields Affected: [List the data fields where suppression was applied, e.g., Names, Phone Numbers]

- Rationale: [Explain why suppression was necessary and how it contributes to anonymization]

4.4 Data Perturbation

- Description: Data perturbation involves altering data slightly to obscure individual information while maintaining the overall dataset's utility.

- Method Applied: [Describe the perturbation method used, e.g., adding noise to numerical values, adjusting dates slightly]

- Data Fields Affected: [List the data fields where perturbation was applied, e.g., Income Data, Transaction Dates]

- Rationale: [Explain why perturbation was chosen and how it maintains data utility while protecting privacy]

5. Pseudonymization Techniques

5.1 Tokenization

- Description: Tokenization replaces sensitive data elements with a token that can be mapped back to the original data using a secure key.

- Method Applied: [Describe the tokenization process, e.g., replacing credit card numbers with randomly generated tokens]

- Data Fields Affected: [List the data fields where tokenization was applied, e.g., Credit Card Numbers, Social Security Numbers]

- Rationale: [Explain why tokenization was selected and how the mapping key is securely managed]

5.2 Reversible Encryption

- Description: Reversible encryption involves encrypting data using a key that allows the original data to be restored if necessary.

- Method Applied: [Describe the encryption method used, e.g., AES encryption with a secure key management process]

- Data Fields Affected: [List the data fields where reversible encryption was applied, e.g., Email Addresses, Phone Numbers]

- Rationale: [Explain why reversible encryption was used and how the encryption keys are stored and managed]

5.3 Pseudonym Assignment

- Description: Pseudonym assignment involves replacing personal identifiers with pseudonyms, such as a random identifier or code.

- Method Applied: [Describe the pseudonym assignment process, e.g., assigning random IDs to user accounts]

- Data Fields Affected: [List the data fields where pseudonym assignment was applied, e.g., User IDs, Customer Numbers]

- Rationale: [Explain why pseudonym assignment was used and how the pseudonymization process is reversible]

6. Compliance and Regulatory Considerations

6.1 GDPR Compliance

- Anonymization under GDPR: The anonymization techniques used comply with GDPR requirements by ensuring that personal data is rendered anonymous in such a way that the data subject is not identifiable.

- Pseudonymization under GDPR: Pseudonymization is used as an additional security measure under GDPR, reducing the risks associated with data processing while maintaining the ability to re-identify data subjects if necessary, with proper authorization.

6.2 HIPAA Compliance

- De-Identification under HIPAA: The methods used for data anonymization meet HIPAA's de-identification standards, ensuring that health information cannot be used to identify an individual.

- Limited Data Set: For data that is pseudonymized under HIPAA, the Limited Data Set standard is followed, allowing for the removal of certain identifiers while permitting the use of data for research, public health, or healthcare operations.

6.3 CCPA Compliance

- Consumer Privacy under CCPA: The data anonymization and pseudonymization techniques ensure that the data meets CCPA standards for consumer privacy, including the right to be forgotten and the right to access personal information.

7. Documentation and Audit Trails

7.1 Record Keeping

- Data Anonymization Records: All anonymization activities are logged, including the date, method applied, and the data fields affected. This ensures that there is a complete record of how and when data was anonymized.

- Pseudonymization Key Management: A secure process is in place for managing and storing pseudonymization keys, with access limited to authorized personnel only. A log is maintained for all key access and usage.

7.2 Audit Trails

- Anonymization Audit Trail: The audit trail for data anonymization includes detailed records of the anonymization process, the tools used, and any verification steps taken to confirm that the data is effectively anonymized.

– Pseudonymization Audit Trail: The audit trail for pseudonymization includes logs of pseudonym assignment, encryption processes, and key management activities. It ensures traceability and accountability for all pseudonymization actions.

8. Roles and Responsibilities

8.1 Data Protection Officer (DPO)

– Responsibilities: The DPO oversees the anonymization and pseudonymization processes, ensuring compliance with legal and regulatory requirements. The DPO is also responsible for maintaining records and audit trails.

8.2 Data Anonymization Team

– Responsibilities: The team is responsible for implementing and validating the anonymization techniques. They work closely with data owners to understand the data structure and ensure effective anonymization.

8.3 Data Pseudonymization Team

– Responsibilities: The team manages the pseudonymization process, including tokenization, encryption, and key management. They ensure that pseudonymized data remains secure and can only be re-identified by authorized personnel.

8.4 IT Security Team

– Responsibilities: The IT Security Team is responsible for implementing the technical controls needed to protect anonymized and pseudonymized data, including encryption, access controls, and security monitoring.

9. Review and Updates

9.1 Review Schedule

- Regular Reviews: This document and the associated anonymization and pseudonymization processes are reviewed on a [Insert Frequency, e.g., Quarterly, Annual] basis to ensure ongoing compliance with regulations and best practices.

- Ad-hoc Reviews: Additional reviews are conducted when significant changes are made to data processing activities or when new regulatory requirements emerge.

9.2 Update Process

- Version Control: All changes to this document and the underlying processes are documented, with version control to track revisions. Updates are approved by the DPO and other relevant stakeholders before implementation.

10. Document Control

- Document Owner: [Insert Name, Role]

- Approval Date: [Insert Date]

- Next Review Date: [Insert Date]

- Version History:

  - Version [Insert Version Number]—Initial Document—[Insert Date]—Approved by [Insert Name]

Data Access Control Documentation

Project Name: [Insert Project Name]
Date: [Insert Date]
Version: [Insert Version Number]

1. Introduction

This Data Access Control Documentation outlines the policies and procedures for managing and monitoring access to data within the [Insert Project Name]. It specifies who has access to the data, under what conditions access is granted, and the mechanisms in place to control and monitor that access. This document is essential for ensuring data security, compliance with regulatory requirements, and the protection of sensitive information.

2. Purpose and Scope

2.1 Purpose

The purpose of this Data Access Control Documentation is to:

- Define the roles and responsibilities for data access within the [Insert Project Name].

- Establish the conditions under which data access is granted, modified, or revoked.

- Describe the mechanisms used to control and monitor data access to ensure compliance with security policies and regulations.

- Provide a framework for auditing and reporting data access activities.

2.2 Scope

This documentation applies to all data access activities related to the [Insert Project Name], covering structured, unstructured, and semi-structured data. It includes access to data stored in databases, file systems, cloud environments, and other data storage solutions.

3. Roles and Responsibilities

3.1 Data Access Roles

The following roles are defined for managing and controlling data access within the [Insert Project Name]:

- Data Owner: [Insert Role/Name]

    - Responsibilities:

        - Define and approve data access policies for specific datasets.

        - Grant or revoke access to data based on project requirements.

        - Ensure compliance with data governance and security policies.

- Data Custodian: [Insert Role/Name]

    - Responsibilities:

        - Implement and manage access controls as defined by the Data Owner.

        - Monitor data access and ensure that access permissions are enforced.

        - Maintain records of access requests, approvals, and modifications.

- Data User: [Insert Role/Name]

    - Responsibilities:

        - Access data only as authorized and required for their role.

        - Comply with data usage policies and report any security incidents.

        - Request access modifications through the appropriate channels.

- Compliance Officer: [Insert Role/Name]

    - Responsibilities:

        - Audit data access controls and ensure compliance with legal and regulatory requirements.

        - Review and approve data access policies and changes.

        - Investigate any unauthorized access or security breaches.

4. Access Control Policies

4.1 Access Levels

Data access within the [Insert Project Name] is categorized into different levels, each with specific permissions and restrictions:

- Level 1: Public Access

  - Description: Data that is publicly available and does not require special permissions.

  - Access Granted To: All project members and authorized external parties.

  - Conditions: No restrictions; data can be accessed and shared freely.

- Level 2: Restricted Access

  - Description: Data that is sensitive or confidential and requires specific authorization to access.

  - Access Granted To: Data Users with a valid business need, approved by the Data Owner.

  - Conditions: Access is granted on a need-to-know basis, and data cannot be shared without explicit permission.

- Level 3: Highly Restricted Access

  - Description: Data that is highly sensitive, including Personally Identifiable Information (PII) or proprietary information.

  - Access Granted To: Only specific Data Users with critical need, approved by the Data Owner and Compliance Officer.

  - Conditions: Access is strictly controlled, and data is encrypted at all times. Sharing is prohibited unless explicitly authorized.

4.2 Access Request and Approval Process

The process for requesting and approving access to data within the [Insert Project Name] is as follows:

Step 1: Access Request Submission

– Data Users must submit an access request form, specifying the data required, the purpose of access, and the duration of access.

Step 2: Review by Data Owner

– The Data Owner reviews the request, ensuring that the requested access aligns with project requirements and data governance policies. The Data Owner may consult with the Compliance Officer if the request involves highly sensitive data.

Step 3: Approval or Denial

– The Data Owner either approves or denies the request. If approved, the Data Custodian is instructed to grant the requested access. If denied, the Data User is notified with an explanation for the decision.

Step 4: Access Provisioning

– The Data Custodian implements the necessary access controls, ensuring that the Data User receives the appropriate permissions. Access logs are updated to reflect the changes.

Step 5: Notification

– The Data User is notified that access has been granted, including any specific conditions or restrictions associated with the access.

4.3 Access Modification and Revocation

Data access may be modified or revoked based on changes in project requirements, user roles, or security policies:

– Modification Requests:

- Data Users may request modifications to their access by submitting a new access request form, following the same review and approval process.

Regular Reviews:

- The Data Owner and Compliance Officer conduct regular reviews of data access permissions to ensure that they remain appropriate. Any unnecessary or outdated access is revoked.

Immediate Revocation:

- Access may be immediately revoked in cases of role changes, project termination, or security breaches. The Data Custodian is responsible for executing the revocation and updating access logs accordingly.

5. Access Control Mechanisms

5.1 Authentication and Authorization

The following mechanisms are used to authenticate and authorize Data Users within the [Insert Project Name]:

- Authentication Methods:

  - Single Sign-On (SSO): [Insert System Name] is used for centralized authentication, allowing Data Users to access multiple systems with a single set of credentials.

  - Multi-factor Authentication (MFA): All Data Users must authenticate using two or more factors, such as a password and a one-time code sent to their mobile device.

- Authorization Methods:

  - Role-Based Access Control (RBAC): Access permissions are assigned based on the Data User's role within the project, ensuring that users only access data necessary for their role.

  - Attribute-Based Access Control (ABAC): Access is granted based on attributes such as the user's role, location, and the sensitivity of the data.

5.2 Access Monitoring and Auditing

To ensure ongoing compliance and security, the following monitoring and auditing mechanisms are in place:

- Access Logs:

  - All data access activities are logged, including successful and unsuccessful access attempts. Logs capture the user's identity, the data accessed, the time of access, and the action taken.

- Regular Audits:

  - The Compliance Officer conducts regular audits of access logs to identify unauthorized access or deviations from access policies. Audit results are reported to the Data Owner and project leadership.

- Real-Time Monitoring:

  - [Insert Monitoring Tool Name] is used to monitor data access in real time. Alerts are triggered for any suspicious or unauthorized access attempts, and immediate action is taken to investigate and mitigate potential threats.

- Incident Reporting:

  - Data Users are required to report any suspicious activity or security incidents immediately. An incident response team is activated to handle the investigation and resolution of such incidents.

6. Data Security Measures

6.1 Data Encryption

Data within the [Insert Project Name] is protected through the following encryption methods:

- Data at Rest:

  - All sensitive data is encrypted using [Insert Encryption Standard, e.g., AES-256] while stored in databases, file systems, and cloud storage solutions.

- Data in Transit:

  - Data is encrypted during transmission using [Insert Encryption Protocol, e.g., TLS 1.2 or higher] to protect it from interception or tampering.

6.2 Data Masking and Anonymization

Sensitive data, such as Personally Identifiable Information (PII), is protected through the following measures:

- Data Masking:

  - Sensitive fields are masked in non-production environments to prevent exposure of real data. Masking is applied using [Insert Masking Technique, e.g., format-preserving masking].

- Data Anonymization:

  - Data is anonymized where possible to remove direct identifiers, reducing the risk of re-identification. Anonymization techniques include [Insert Anonymization Technique, e.g., k-anonymity, differential privacy].

7. Compliance and Legal Requirements

7.1 Regulatory Compliance

The [Insert Project Name] must comply with the following data protection and privacy regulations:

- General Data Protection Regulation (GDPR):

  - Ensures that personal data is processed lawfully, transparently, and for a specific purpose. Data access controls are aligned with GDPR requirements, including data minimization and access limitation principles.

- Health Insurance Portability and Accountability Act (HIPAA):

  - Protects sensitive health information, ensuring that data access is restricted to authorized personnel and that security measures are in place to safeguard patient data.

- California Consumer Privacy Act (CCPA):
  - Provides California residents with rights over their personal data, including the right to know what data is collected and the right to request deletion. Access controls ensure compliance with CCPA's data protection standards.

7.2 Legal Considerations

Data access within the [Insert Project Name] must adhere to the following legal considerations:

- Data Ownership:
  - Access to data is granted in accordance with ownership rights, ensuring that only authorized users access proprietary or third-party data.
- Confidentiality Agreements:
  - All Data

Users must sign confidentiality agreements, ensuring that they understand and agree to comply with data access policies and legal obligations.

8. Review and Updates

8.1 Regular Reviews

This Data Access Control Documentation is subject to regular reviews to ensure that it remains up-to-date and effective:

- Review Frequency: [Insert Frequency, e.g., Quarterly, Annually]
- Review Conducted By: [Insert Roles Responsible for Review, e.g., Data Owner, Compliance Officer]

8.2 Updates and Version Control

Any changes to data access policies or procedures must be documented, and this document must be updated accordingly:

- Update Process:
  - Proposed changes are reviewed by the Data Owner and Compliance Officer. Once approved, the changes are implemented, and the document is updated.
- Version Control:
  - Each version of this document is numbered and dated, with a summary of changes included in the version history section.

9. Document Control

- Document Owner: [Insert Name, Role]
- Approval Date: [Insert Date]
- Next Review Date: [Insert Date]
- Version History:
  - Version [Insert Version Number]—Initial Document—[Insert Date]—Approved by [Insert Name]

Modeling

Data Governance Compliance Checklist

Project Name: [Insert Project Name]
Date: [Insert Date]
Version: [Insert Version Number]

1. Introduction

This Data Governance Compliance Checklist is designed to ensure that all data used in the [Insert Project Name] complies with established governance policies, including data quality, privacy, and security standards. This checklist serves as a comprehensive guide to verify that data handling practices align with organizational and regulatory requirements throughout the ML lifecycle.

2. Purpose

The purpose of this checklist is to:

- Ensure that data used in the [Insert Project Name] meets the required standards for quality, privacy, and security.

- Facilitate consistent adherence to data governance policies across all phases of the ML lifecycle.

- Identify and address any compliance gaps or risks associated with data usage.

3. Checklist Overview

The checklist is divided into the following sections:

- Data Quality Compliance

- Data Privacy Compliance

- Data Security Compliance

- Documentation and Audit Compliance

Each section contains specific criteria to verify compliance with relevant policies and standards.

4. Data Quality Compliance

This section ensures that data used in the [Insert Project Name] meets quality standards, including accuracy, completeness, consistency, and timeliness.

Criteria	Status	Comments
4.1 Data Accuracy		
a. Data accurately represents the real-world entities or phenomena it describes.	Compliant Non-compliant	[Add Comments]
b. Data validation checks are performed to identify and correct inaccuracies.	Compliant Non-compliant	[Add Comments]

(continued)

Criteria	Status	Comments
4.2 Data Completeness		
a. All necessary data elements are present and accounted for.	Compliant Non-compliant	[Add Comments]
b. Missing data is identified and appropriate handling techniques are applied (e.g., imputation, exclusion).	Compliant Non-compliant	[Add Comments]
4.3 Data Consistency		
a. Data is consistent across different sources and systems.	Compliant Non-compliant	[Add Comments]
b. Data integration processes ensure consistency in data format and structure.	Compliant Non-compliant	[Add Comments]
4.4 Data Timeliness		
a. Data is up-to-date and reflects the most current information available.	Compliant Non-compliant	[Add Comments]
b. Data refresh schedules are followed to maintain timeliness.	Compliant Non-compliant	[Add Comments]

5. Data Privacy Compliance

This section verifies that data privacy standards are met, ensuring compliance with relevant regulations such as GDPR, CCPA, and other data protection laws.

Criteria	Status	Comments
5.1 Data Anonymization and Pseudonymization		
a. Personally identifiable information (PII) is anonymized or pseudonymized as required.	Compliant Non-compliant	[Add Comments]
b. Anonymization techniques effectively prevent re-identification of individuals.	Compliant Non-compliant	[Add Comments]

(continued)

Criteria	Status	Comments
5.2 Data Consent and Usage Rights		
a. Consent has been obtained for the use of personal data, where applicable.	Compliant Non-compliant	[Add Comments]
b. Data usage is consistent with the purposes for which consent was obtained.	Compliant Non-compliant	[Add Comments]
5.3 Data Subject Rights		
a. Data subjects have been informed of their rights (e.g., access, correction, deletion).	Compliant Non-compliant	[Add Comments]
b. Processes are in place to respond to data subject requests promptly.	Compliant Non-compliant	[Add Comments]
5.4 Compliance with Data Protection Regulations		
a. Data handling practices comply with relevant data protection regulations (e.g., GDPR, CCPA).	Compliant Non-compliant	[Add Comments]
b. Regular audits are conducted to ensure ongoing compliance with privacy laws.	Compliant Non-compliant	[Add Comments]

6. Data Security Compliance

This section ensures that data is protected from unauthorized access, breaches, and other security threats.

Criteria	Status	Comments
6.1 Data Encryption		
a. Data is encrypted at rest and in transit using industry-standard encryption methods.	Compliant Non-compliant	[Add Comments]

(continued)

Criteria	Status	Comments
b. Encryption keys are securely managed and rotated regularly.	Compliant Non-compliant	[Add Comments]
6.2 Access Controls		
a. Access to data is restricted based on roles and responsibilities (e.g., Role-Based Access Control).	Compliant Non-compliant	[Add Comments]
b. Multi-factor authentication (MFA) is required for accessing sensitive data.	Compliant Non-compliant	[Add Comments]
6.3 Security Monitoring and Incident Response		
a. Security monitoring tools are in place to detect and respond to potential threats.	Compliant Non-compliant	[Add Comments]
b. Incident response procedures are documented and regularly tested.	Compliant Non-compliant	[Add Comments]
6.4 Data Backup and Recovery		
a. Regular backups of data are performed and securely stored.	Compliant Non-compliant	[Add Comments]
b. Data recovery procedures are in place and regularly tested to ensure data integrity.	Compliant Non-compliant	[Add Comments]

7. Documentation and Audit Compliance

This section ensures that all necessary documentation is maintained, and regular audits are conducted to verify compliance with data governance policies.

Criteria	Status	Comments
7.1 Data Lineage Documentation		
a. Data lineage is documented, tracking the origin, flow, and transformations of data.	CompliantNon-compliant	[Add Comments]
b. Data lineage documentation is regularly updated to reflect changes in data flows.	CompliantNon-compliant	[Add Comments]
7.2 Data Governance Policies and Procedures		
a. Data governance policies and procedures are documented and accessible to all relevant stakeholders.	CompliantNon-compliant	[Add Comments]
b. Policies and procedures are regularly reviewed and updated to reflect best practices and regulatory changes.	CompliantNon-compliant	[Add Comments]
7.3 Audit Trails and Logs		
a. Audit trails are maintained for all data access and modifications, ensuring traceability.	CompliantNon-compliant	[Add Comments]
b. Logs are regularly reviewed to identify and address any unauthorized access or anomalies.	CompliantNon-compliant	[Add Comments]
7.4 Regular Compliance Audits		
a. Regular compliance audits are conducted to verify adherence to data governance policies.	CompliantNon-compliant	[Add Comments]
b. Findings from audits are documented, and corrective actions are implemented as needed.	CompliantNon-compliant	[Add Comments]

8. Review and Sign-Off

This section confirms that the Data Governance Compliance Checklist has been reviewed and approved by the relevant stakeholders.

Reviewer Name	Role	Date	Signature
[Insert Reviewer Name]	[Insert Role]	[Insert Date]	[Insert Signature]
[Insert Reviewer Name]	[Insert Role]	[Insert Date]	[Insert Signature]
[Insert Reviewer Name]	[Insert Role]	[Insert Date]	[Insert Signature]

9. Document Control

- Document Owner: [Insert Name, Role]

- Approval Date: [Insert Date]

- Next Review Date: [Insert Date]

- Version History:

  - Version [Insert Version Number]—Initial Document—[Insert Date]—Approved by [Insert Name]

Bias and Fairness Assessment

Project Name: [Insert Project Name]
Date: [Insert Date]
Version: [Insert Version Number]

1. Introduction

This Bias and Fairness Assessment provides a comprehensive evaluation of the data and models used in the [Insert Project Name]. The assessment identifies potential biases, evaluates their impact on model outcomes, and outlines strategies to ensure adherence to fairness standards throughout the model development process. This document is essential for promoting transparency, accountability, and ethical AI practices within the project.

2. Purpose and Scope

2.1 Purpose

The purpose of this Bias and Fairness Assessment is to:

- Identify and assess potential biases in the data and model used in the [Insert Project Name].

- Evaluate the impact of these biases on model outcomes, particularly in terms of fairness across different demographic groups.

- Recommend strategies and interventions to mitigate identified biases and ensure that the model adheres to fairness standards.

- Document the steps taken to monitor and maintain fairness throughout the lifecycle of the model.

2.2 Scope

This assessment covers all data and models used in the [Insert Project Name], including but not limited to:

- Data sources and their characteristics.

- Preprocessing and feature engineering steps.

- Model training, validation, and testing processes.

- Post-deployment monitoring and updates.

3. Bias Identification

3.1 Data Bias

This section identifies and evaluates potential biases in the data used for model development.

- Sampling Bias:

  - Description: Assess whether the dataset is representative of the overall population or if certain groups are underrepresented or overrepresented.

- • Impact: [Describe the potential impact of sampling bias on model outcomes.]

  • Mitigation Strategy: [Outline the steps taken or planned to address sampling bias, e.g., re-sampling, stratified sampling.]

- – Measurement Bias:

  • Description: Evaluate whether the data collection methods introduce systematic errors, such as mislabeling or inaccurate measurements.

  • Impact: [Describe the potential impact of measurement bias on model accuracy and fairness.]

  • Mitigation Strategy: [Outline the steps taken or planned to correct measurement bias, e.g., data validation, correcting labeling errors.]

- – Label Bias:

  • Description: Identify biases in the labels or target variables used for training the model, which may reflect human prejudices or systemic biases.

  • Impact: [Describe the potential impact of label bias on model predictions and fairness.]

  • Mitigation Strategy: [Outline the steps taken or planned to address label bias, e.g., re-labeling, using alternative target variables.]

3.2 Model Bias

This section evaluates biases that may arise during model development and training.

- – Algorithmic Bias:

  • Description: Assess whether the algorithms used in the project introduce or amplify existing biases in the data.

  • Impact: [Describe the potential impact of algorithmic bias on model outcomes and the affected demographic groups.]

- • Mitigation Strategy: [Outline the steps taken or planned to reduce algorithmic bias, e.g., using fairness-aware algorithms, adjusting model parameters.]

- – Feature Bias:

    - • Description: Evaluate whether certain features used in the model disproportionately influence outcomes for specific groups.

    - • Impact: [Describe the potential impact of feature bias on model predictions and fairness.]

    - • Mitigation Strategy: [Outline the steps taken or planned to address feature bias, e.g., feature selection, normalization, or weighting.]

- – Bias Amplification:

    - • Description: Identify whether the model amplifies existing biases in the data through feedback loops or other mechanisms.

    - • Impact: [Describe the potential impact of bias amplification on the overall fairness and accuracy of the model.]

    - • Mitigation Strategy: [Outline the steps taken or planned to prevent bias amplification, e.g., iterative bias correction, regular fairness audits.]

4. Fairness Evaluation

4.1 Fairness Metrics

This section outlines the metrics used to evaluate fairness in the model outcomes.

- – Demographic Parity:

    - • Description: Measures whether different demographic groups receive similar outcomes or predictions from the model.

    - • Metric: [Insert metric used, e.g., difference in positive prediction rates between groups.]

- Results: [Insert findings from the fairness evaluation using demographic parity.]

- Interpretation: [Interpret the results, highlighting any disparities identified.]

- Equalized Odds:

  - Description: Measures whether the model's error rates (false positives, false negatives) are similar across different demographic groups.

  - Metric: [Insert metric used, e.g., difference in false positive rates between groups.]

  - Results: [Insert findings from the fairness evaluation using equalized odds.]

  - Interpretation: [Interpret the results, highlighting any disparities identified.]

- Predictive Parity:

  - Description: Measures whether the model's predictions have similar predictive value across different demographic groups.

  - Metric: [Insert metric used, e.g., difference in precision or recall between groups.]

  - Results: [Insert findings from the fairness evaluation using predictive parity.]

  - Interpretation: [Interpret the results, highlighting any disparities identified.]

4.2 Fairness Audits

This section documents the results of fairness audits conducted during and after model development.

- Audit Frequency: [Specify how often fairness audits are conducted, e.g., after each model iteration, quarterly, annually.]

- Audit Results: [Summarize the results of recent fairness audits, including any issues identified and the steps taken to address them.]

- Continuous Improvement: [Describe ongoing efforts to improve fairness based on audit findings, e.g., adjusting model parameters, retraining with more balanced data.]

5. Mitigation Strategies

5.1 Data-Level Mitigation

This section outlines strategies for addressing biases identified at the data level.

- Data Augmentation:

  - Description: [Explain how data augmentation techniques are used to increase the diversity and representativeness of the dataset.]

  - Implementation: [Describe how data augmentation has been implemented, e.g., generating synthetic data, oversampling underrepresented groups.]

  - Expected Outcome: [Describe the expected impact on reducing data biases.]

- Re-Sampling:

  - Description: [Explain how re-sampling techniques are used to balance the dataset across different demographic groups.]

  - Implementation: [Describe how re-sampling has been implemented, e.g., stratified sampling, under-sampling overrepresented groups.]

  - Expected Outcome: [Describe the expected impact on reducing sampling bias.]

5.2 Model-Level Mitigation

This section outlines strategies for addressing biases identified at the model level.

- Fairness Constraints:

  - Description: [Explain how fairness constraints are applied during model training to ensure equitable outcomes.]

  - Implementation: [Describe how fairness constraints have been implemented, e.g., adjusting loss functions, applying post-processing techniques.]

  - Expected Outcome: [Describe the expected impact on improving model fairness.]

- Algorithm Selection:

  - Description: [Explain how the selection of algorithms is influenced by fairness considerations.]

  - Implementation: [Describe how algorithm selection has been guided by fairness concerns, e.g., choosing fairness-aware algorithms.]

  - Expected Outcome: [Describe the expected impact on reducing algorithmic bias.]

- Feature Engineering:

  - Description: [Explain how feature engineering practices are adjusted to mitigate biases.]

  - Implementation: [Describe how feature engineering has been adjusted, e.g., excluding sensitive attributes, using proxy variables.]

  - Expected Outcome: [Describe the expected impact on reducing feature bias.]

6. Monitoring and Maintenance

6.1 Ongoing Fairness Monitoring

This section describes the processes for ongoing monitoring of fairness in model outcomes post-deployment.

- Monitoring Tools: [List the tools and methods used for monitoring fairness in real-time, e.g., dashboards, automated alerts.]

- Key Metrics: [List the fairness metrics that are continuously monitored.]

- Responsibility: [Assign responsibility for ongoing fairness monitoring, e.g., data scientists, compliance officers.]

6.2 Bias Detection in Production

This section outlines the approach to detecting and addressing biases that may emerge after the model is deployed.

- Detection Methods: [Describe methods used to detect emerging biases, e.g., anomaly detection, drift analysis.]

- Mitigation Actions: [Describe the steps taken to mitigate biases detected in production, e.g., model retraining, recalibrating thresholds.]

- Documentation: [Specify how these actions are documented and reported.]

7. Reporting and Documentation

7.1 Reporting Structure

This section outlines the structure for reporting the findings and actions from the bias and fairness assessment.

- Reporting Frequency: [Specify how often fairness reports are generated, e.g., monthly, quarterly.]

- Audience: [Specify the intended audience for these reports, e.g., senior management, compliance teams, external auditors.]

- Content: [Describe the key content of the reports, e.g., summary of findings, mitigation actions, ongoing challenges.]

7.2 Compliance Documentation

This section details how the bias and fairness assessment supports compliance with relevant legal and regulatory requirements.

- Regulatory Requirements: [List the relevant regulations that the assessment complies with, e.g., GDPR, Fairness in AI regulations.]

- Documentation Standards: [Describe how compliance documentation is maintained, e.g., audit trails, version control.]

- Submission Process: [Outline the process for submitting compliance documentation to regulatory bodies, if applicable.]

8. Document Control

- Document Owner: [Insert Name, Role]

- Approval Date: [Insert Date]

- Next Review Date: [Insert Date]

- Version History:

  - Version [Insert Version Number]—Initial Document—[Insert Date]—Approved by [Insert Name]

Model Input Data Documentation

Project Name: [Insert Project Name]
Model Name: [Insert Model Name]
Date: [Insert Date]
Version: [Insert Version Number]

1. Introduction

This document provides detailed information about the input data used for training and validating the [Insert Model Name] within the [Insert Project Name]. It ensures that the data meets governance, quality, and compliance requirements, and serves as a reference for data provenance, preprocessing, and usage in model development.

2. Purpose and Scope

2.1 Purpose

The purpose of this documentation is to:

- Describe the data sources, formats, and structures used as input for the [Insert Model Name].

- Outline the preprocessing steps applied to the data before model training and validation.

- Ensure that the data adheres to governance, quality, and compliance standards.

- Provide a reference for the reproducibility of model training and validation processes.

2.2 Scope

This documentation covers all datasets used in the training and validation of the [Insert Model Name], including structured, unstructured, and semi-structured data. It includes details on data collection, preprocessing, storage, and usage within the model development pipeline.

3. Data Sources

3.1 Data Source Overview

This section provides a comprehensive overview of the data sources used in the [Insert Model Name].

- Source Name: [Insert Data Source Name]

- Data Provider: [Internal/External; specify the original provider or system]

- Data Type: [Structured/Unstructured/Semi-structured]

- Description: [Brief description of the data source and its relevance to the model]

- Origin: [e.g., Database, API, Third-Party Vendor, Internal Collection]

- Update Frequency: [e.g., Real-time, Daily, Weekly, Monthly]

- Data Owner: [Insert Data Owner Name]

- Format: [e.g., CSV, JSON, XML, SQL Table]

3.2 Data Acquisition

This section describes how data is acquired from the sources mentioned above.

- Acquisition Method: [e.g., API Calls, Data Pipelines, Manual Uploads]

- Ingestion Tools: [e.g., ETL Tools, Data Integration Platforms]

- Acquisition Schedule: [e.g., Scheduled, On-Demand]

- Data Collection Date Range: [Specify the date range of the data collected]

- Data Storage Location: [Insert Location, e.g., Data Warehouse, Data Lake, Cloud Storage]

4. Data Characteristics

4.1 Data Structure

This section provides detailed information on the structure of the datasets used.

- Dataset Name: [Insert Dataset Name]

- Number of Records: [Insert Number of Records]

- Number of Features/Columns: [Insert Number of Features/Columns]

- Feature Description:

Feature Name	Data Type	Description	Example Values	Source
[Feature 1]	[Type]	[Description]	[Example Values]	[Source]
[Feature 2]	[Type]	[Description]	[Example Values]	[Source]
[Feature 3]	[Type]	[Description]	[Example Values]	[Source]
[Feature 4]	[Type]	[Description]	[Example Values]	[Source]

4.2 Data Volume

This section describes the size and volume of the data used in training and validation.

- Total Records: [Insert Total Number of Records]

- Total Data Size: [Insert Data Size, e.g., GB, TB]

- Training Data Volume: [Insert Number of Records/Size for Training Data]

- Validation Data Volume: [Insert Number of Records/Size for Validation Data]

- Test Data Volume (if applicable): [Insert Number of Records/Size for Test Data]

5. Data Quality and Integrity

5.1 Data Quality Checks

This section outlines the data quality checks applied to ensure that the data is fit for use in the model.

- Accuracy: [Describe methods used to ensure data accuracy]

- Completeness: [Describe checks for missing values or incomplete records]

- Consistency: [Describe how data consistency is maintained across sources]

- Validity: [Describe how data validity is ensured, including format checks]

- Timeliness: [Describe how data timeliness is maintained]

5.2 Data Cleaning and Preprocessing

This section describes the data cleaning and preprocessing steps performed before model training.

- Missing Data Handling: [e.g., Imputation, Removal, Filling with Defaults]

- Outlier Detection and Treatment: [e.g., Z-Score, IQR Method, Removal]

- Data Normalization/Standardization: [e.g., Min–Max Scaling, Z-Score Standardization]

- Categorical Encoding: [e.g., One-Hot Encoding, Label Encoding]

- Feature Engineering: [Describe any feature creation or transformation steps]

6. Data Governance and Compliance

6.1 Data Governance Policies

This section ensures that the data used meets governance requirements.

- Data Ownership: [Describe data ownership and any relevant licensing agreements]

- Data Access Controls: [Describe who has access to the data and how access is managed]

- Data Lineage: [Briefly describe the lineage of the data from source to usage in the model]

- Audit Trails: [Describe any audit trails maintained for data access and modification]

6.2 Compliance with Regulations

This section outlines the compliance of the data with relevant regulations and standards.

- GDPR Compliance (if applicable): [Describe how GDPR requirements are met]

- HIPAA Compliance (if applicable): [Describe how HIPAA requirements are met]

- Data Privacy Measures: [e.g., Anonymization, Pseudonymization, Data Masking]

- Consent Management: [Describe how consent for data usage is managed and documented]

7. Data Usage in Model Development

7.1 Training Data

This section details the data used for training the model.

- Training Data Description: [Describe the data used for training, including any specific preprocessing applied]

- Training Data Source: [List the specific datasets used for training]

- Data Split Method: [e.g., Random Split, Stratified Split, Time-Based Split]

- Training Data Size: [Insert Size or Number of Records]

7.2 Validation Data

This section details the data used for validating the model.

- Validation Data Description: [Describe the data used for validation, including any specific preprocessing applied]

- Validation Data Source: [List the specific datasets used for validation]

- Validation Method: [e.g., Cross-Validation, Hold-Out Validation]

- Validation Data Size: [Insert Size or Number of Records]

7.3 Test Data (if applicable)

This section details the data used for testing the model.

- – Test Data Description: [Describe the data used for testing, including any specific preprocessing applied]

- – Test Data Source: [List the specific datasets used for testing]

- – Test Method: [e.g., Final Hold-Out Test, Cross-Validation]

- – Test Data Size: [Insert Size or Number of Records]

8. Data Lineage and Traceability

This section outlines the data lineage and traceability from source to model input.

- – Data Lineage Diagram: [Include or reference a diagram showing the data flow from source to final usage in the model]

- – Data Provenance: [Describe the origin of the data and any transformations applied along the way]

- – Traceability: [Describe how data changes are tracked and documented]

9. Data Security

This section describes the security measures in place to protect the data.

- – Data Encryption: [Describe encryption methods used for data at rest and in transit]

- – Access Controls: [Describe the roles and permissions associated with data access]

- – Data Breach Protocols: [Outline the steps to be taken in the event of a data breach]

- – Data Backup: [Describe the backup procedures for the data]

10. Document Control

- Document Owner: [Insert Name, Role]

- Approval Date: [Insert Date]

- Next Review Date: [Insert Date]

- Version History:

  - Version [Insert Version Number]—Initial Document—[Insert Date]—Approved by [Insert Name]

Evaluation

Data Integrity and Security Evaluation

Project Name: [Insert Project Name]
Date: [Insert Date]
Version: [Insert Version Number]

1. Introduction

This Data Integrity and Security Evaluation document assesses the security and integrity of the data used during the model evaluation phase of the [Insert Project Name]. The evaluation ensures that the data has not been compromised, altered, or accessed by unauthorized individuals. This document outlines the methods used to verify data integrity and security, the results of the evaluation, and any corrective actions taken to address identified issues.

2. Purpose and Scope

2.1 Purpose

The purpose of this evaluation is to:

- Ensure the integrity of the data used during model evaluation, confirming that it is accurate, complete, and unchanged from its original state.

- Assess the security measures in place to protect the data from unauthorized access, modification, or breaches.

- Identify any potential vulnerabilities in the data handling process and recommend corrective actions to mitigate risks.

- Maintain compliance with organizational policies, industry standards, and regulatory requirements related to data integrity and security.

2.2 Scope

This evaluation covers all data used during the model evaluation phase of the [Insert Project Name]. It includes data sources, storage, processing, and access controls. The assessment applies to both structured and unstructured data, as well as internal and external data sources.

3. Data Integrity Assessment

3.1 Data Integrity Checks

This section describes the methods and tools used to verify the integrity of the data during the model evaluation phase.

- Checksum Verification: [Describe the use of checksums (e.g., MD5, SHA-256) to verify that the data has not been altered since it was last processed.]

- Data Validation: [Outline the procedures for validating the accuracy and completeness of the data, including any comparison against known good data sets or benchmarks.]

- Audit Trails: [Detail the audit trail mechanisms in place to track all data-related activities, ensuring that any changes to the data are logged and reviewed.]

- Data Provenance: [Describe how data provenance is tracked, ensuring that the origin and history of the data are well documented and verifiable.]

3.2 Results of Data Integrity Checks

This section provides the results of the data integrity checks conducted during the evaluation.

- Integrity Check 1: [Insert details of the first integrity check, including the method used, the data evaluated, and the outcome.]

  - Outcome: [Pass/Fail]

  - Details: [Provide any relevant details, such as discrepancies found, corrections made, or confirmation of integrity.]

- Integrity Check 2: [Insert details of the second integrity check, including the method used, the data evaluated, and the outcome.]

  - Outcome: [Pass/Fail]

  - Details: [Provide any relevant details, such as discrepancies found, corrections made, or confirmation of integrity.]

- Summary of Findings: [Summarize the overall findings of the data integrity assessment, including any patterns or common issues identified.]

4. Data Security Assessment

4.1 Security Measures Evaluation

This section evaluates the security measures implemented to protect the data during the model evaluation phase.

- Access Controls: [Evaluate the effectiveness of access control mechanisms, including role-based access controls (RBAC) and multi-factor authentication (MFA), in preventing unauthorized access to the data.]

- Encryption: [Assess the use of encryption for data at rest and in transit, ensuring that sensitive data is protected from unauthorized access.]

- Data Masking: [Review the application of data masking techniques to protect sensitive data during the evaluation process.]

- Security Monitoring: [Describe the security monitoring tools and processes used to detect and respond to potential security threats in real-time.]

4.2 Results of Security Measures Evaluation

This section provides the results of the security measures evaluation conducted during the assessment.

- Security Measure 1: [Insert details of the first security measure evaluated, including the method used, the data or system evaluated, and the outcome.]

  - Outcome: [Effective/ineffective]

  - Details: [Provide any relevant details, such as vulnerabilities identified, improvements made, or confirmation of security effectiveness.]

- Security Measure 2: [Insert details of the second security measure evaluated, including the method used, the data or system evaluated, and the outcome.]

  - Outcome: [Effective/ineffective]

  - Details: [Provide any relevant details, such as vulnerabilities identified, improvements made, or confirmation of security effectiveness.]

- Summary of Findings: [Summarize the overall findings of the security assessment, including any patterns or common issues identified.]

5. Compliance and Regulatory Review

5.1 Regulatory Requirements

This section reviews the compliance of data integrity and security practices with relevant regulatory requirements and industry standards.

– General Data Protection Regulation (GDPR): [Evaluate the compliance of data handling practices with GDPR requirements, focusing on data protection, privacy, and breach reporting.]

– Health Insurance Portability and Accountability Act (HIPAA): [For projects involving health data, assess compliance with HIPAA standards for protecting sensitive health information.]

– Industry Standards: [Review compliance with industry standards such as ISO 27001 for information security management and NIST guidelines for cybersecurity.]

5.2 Compliance Evaluation Results

This section provides the results of the compliance evaluation conducted during the assessment.

– Compliance Area 1: [Insert details of the first compliance area evaluated, including the regulations or standards assessed and the outcome.]

  • Outcome: [Compliant/non-compliant]

  • Details: [Provide any relevant details, such as areas of non-compliance identified, corrective actions taken, or confirmation of compliance.]

– Compliance Area 2: [Insert details of the second compliance area evaluated, including the regulations or standards assessed and the outcome.]

  • Outcome: [Compliant/non-compliant]

  • Details: [Provide any relevant details, such as areas of non-compliance identified, corrective actions taken, or confirmation of compliance.]

– Summary of Findings: [Summarize the overall findings of the compliance and regulatory review, including any patterns or common issues identified.]

6. Risk Assessment and Mitigation

6.1 Risk Identification

This section identifies potential risks related to data integrity and security during the model evaluation phase.

- Risk 1: [Insert description of the first risk identified, including the potential impact on data integrity or security.]
  - Likelihood: [Low/Medium/High]
  - Impact: [Low/Medium/High]
- Risk 2: [Insert description of the second risk identified, including the potential impact on data integrity or security.]
  - Likelihood: [Low/Medium/High]
  - Impact: [Low/Medium/High]

6.2 Mitigation Strategies

This section outlines the strategies implemented to mitigate the identified risks.

- Mitigation Strategy 1: [Describe the strategy used to mitigate the first risk, including any changes to processes, systems, or controls.]
  - Expected Outcome: [Describe the expected outcome of the mitigation strategy.]
- Mitigation Strategy 2: [Describe the strategy used to mitigate the second risk, including any changes to processes, systems, or controls.]
  - Expected Outcome: [Describe the expected outcome of the mitigation strategy.]

7. Recommendations and Action Plan

7.1 Recommendations

Based on the findings of the data integrity and security evaluation, the following recommendations are made:

- Recommendation 1: [Insert first recommendation, focusing on improving data integrity or security.]

- Recommendation 2: [Insert second recommendation, focusing on improving data integrity or security.]

7.2 Action Plan

This section outlines the action plan for implementing the recommendations.

- Action Item 1: [Describe the first action item, including responsible parties, timeline, and resources required.]

- Action Item 2: [Describe the second action item, including responsible parties, timeline, and resources required.]

8. Document Control

- Document Owner: [Insert Name, Role]

- Approval Date: [Insert Date]

- Next Review Date: [Insert Date]

- Version History:

- Version [Insert Version Number]—Initial Document—[Insert Date]— Approved by [Insert Name]

Model Fairness and Accountability Report

Project Name: [Insert Project Name]
Date: [Insert Date]
Version: [Insert Version Number]

1. Introduction

This Model Fairness and Accountability Report evaluates the performance of the machine learning (ML) model used in the [Insert Project Name] across various demographic groups. The goal of this report is to ensure that the model's outputs are fair, unbiased, and do not disproportionately affect any particular group. Evaluating fairness and accountability is essential to meeting ethical standards, ensuring compliance with regulatory requirements, and fostering trust in the model's predictions.

2. Purpose and Scope

2.1 Purpose

The purpose of this report is to:

- Assess the fairness of the ML model by analyzing its performance across different demographic groups.

- Identify any potential biases in the model's predictions.

- Recommend strategies for mitigating identified biases and improving fairness.

- Ensure accountability by documenting the evaluation process and results.

2.2 Scope

This report covers the evaluation of the [Insert Project Name] ML model's performance across [Insert Demographic Criteria, e.g., race, gender, age, income level, etc.]. The evaluation focuses on the key metrics of fairness and accountability, examining whether the model's outputs are equitable for all demographic groups involved.

3. Evaluation Methodology

3.1 Data and Demographic Groups

The data used for this evaluation is sourced from [Insert Data Source], which includes the following demographic groups:

- Demographic Attribute 1 (e.g., Gender): [List categories, e.g., Male, Female, Non-Binary]

- Demographic Attribute 2 (e.g., Race/Ethnicity): [List categories, e.g., White, Black, Asian, Hispanic, Other]

- Demographic Attribute 3 (e.g., Age Group): [List categories, e.g., 18-24, 25-34, 35-44, 45-54, 55+]

3.2 Performance Metrics

The following performance metrics are used to assess the model's accuracy and fairness for each demographic group:

- Accuracy: Measures the percentage of correct predictions made by the model for each group.

- Precision: Indicates how many of the predicted positive instances were true positives for each group.

- Recall: Measures the percentage of actual positives that were correctly predicted by the model for each group.

- F1 Score: Provides a balance between precision and recall, capturing the overall performance.

- False Positive Rate (FPR): The rate at which the model incorrectly classifies negative instances as positive for each group.

- False Negative Rate (FNR): The rate at which the model incorrectly classifies positive instances as negative for each group.

3.3 Fairness Metrics

To ensure that the model is equitable, the following fairness metrics are used:

- Demographic Parity: Evaluates whether the proportion of positive predictions is equal across different groups.

- Equalized Odds: Compares the false positive and false negative rates across groups to assess whether error rates are similar for all groups.

- Predictive Parity: Examines whether precision is consistent across different demographic groups.

- Disparate Impact: Measures the ratio of favorable outcomes for one group compared to others, ensuring no group is disproportionately affected.

4. Evaluation Results

4.1 Overall Model Performance

The following table summarizes the overall performance of the model across all demographic groups:

Metric	Value
Accuracy	[Insert Value]
Precision	[Insert Value]
Recall	[Insert Value]
F1 Score	[Insert Value]
False Positive Rate	[Insert Value]
False Negative Rate	[Insert Value]

4.2 Performance by Demographic Group

The following table shows the performance of the model across different demographic groups, highlighting disparities in key performance metrics:

Demographic Group	Accuracy	Precision	Recall	F1 Score	FPR	FNR
[Group 1]	[Insert Value]	[Insert Value]	[Insert Value]	[Insert Value]	[Insert Value]	[Insert Value]
[Group 2]	[Insert Value]	[Insert Value]	[Insert Value]	[Insert Value]	[Insert Value]	[Insert Value]
[Group 3]	[Insert Value]	[Insert Value]	[Insert Value]	[Insert Value]	[Insert Value]	[Insert Value]
[Group 4]	[Insert Value]	[Insert Value]	[Insert Value]	[Insert Value]	[Insert Value]	[Insert Value]

4.3 Fairness Metrics by Demographic Group

This table presents the fairness metrics for each demographic group, illustrating whether the model provides equitable treatment to all groups.

Demographic Group	Demographic Parity	Equalized Odds	Predictive Parity	Disparate Impact
[Group 1]	[Insert Value]	[Insert Value]	[Insert Value]	[Insert Value]
[Group 2]	[Insert Value]	[Insert Value]	[Insert Value]	[Insert Value]
[Group 3]	[Insert Value]	[Insert Value]	[Insert Value]	[Insert Value]
[Group 4]	[Insert Value]	[Insert Value]	[Insert Value]	[Insert Value]

5. Bias Detection and Analysis

5.1 Bias Identification

The evaluation identified the following biases in the model's predictions:

- Demographic Group 1: Higher false positive rates suggest that this group is disproportionately affected by incorrect positive classifications.

- Demographic Group 2: Lower recall indicates that the model struggles to identify true positive instances for this group, leading to missed predictions.

- Demographic Group 3: Predictive parity is not maintained, with precision varying significantly between this group and others.

5.2 Root Cause Analysis

A root cause analysis was conducted to understand the factors contributing to these biases. The following issues were identified:

- Data Imbalance: Certain demographic groups are underrepresented in the training data, leading to poorer model performance for those groups.

- Feature Bias: Some features used in the model may have unintended correlations with sensitive demographic attributes, leading to biased outcomes.

- Algorithmic Bias: The model's learning algorithm may exacerbate existing biases in the data or introduce new biases during training.

6. Mitigation Strategies

To address the identified biases and improve fairness, the following strategies are recommended:

- Data Balancing: Collect additional data or use synthetic techniques (e.g., SMOTE) to balance the representation of underrepresented groups in the dataset.

- Feature Engineering: Reevaluate the model's features and remove or adjust those that correlate too strongly with sensitive attributes.

- Bias-Reduction Algorithms: Implement bias-reduction algorithms, such as adversarial debiasing or reweighting techniques, to ensure fair treatment across demographic groups.

- Post-Processing Adjustments: Apply post-processing techniques, such as equalized odds adjustments, to correct biases in the model's outputs.

7. Recommendations

Based on the evaluation results, the following actions are recommended to improve the fairness and accountability of the model:

1. Enhance Data Collection: Increase the diversity of data sources to ensure better representation of all demographic groups.

2. Regular Fairness Audits: Implement regular fairness audits to monitor the model's performance and fairness over time, especially when retraining or updating the model.

3. Apply Fairness Constraints: Introduce fairness constraints during the model training process to ensure equitable outcomes across all groups.

4. Stakeholder Engagement: Involve key stakeholders, including those from affected demographic groups, in the review of model fairness and decision-making processes.

5. Transparency and Documentation: Maintain comprehensive documentation of fairness evaluations, bias mitigation strategies, and ongoing monitoring efforts to ensure transparency and accountability.

8. Accountability and Monitoring Plan

8.1 Continuous Monitoring

– Bias Monitoring Tools: The model will be continuously monitored for biases using tools such as [Insert Bias Monitoring Tool Name]. Real-time alerts will be generated if disparities in performance between demographic groups exceed predefined thresholds.

– Fairness Audits: Regular audits will be conducted on a [Insert Frequency] basis to evaluate the model's fairness and compliance with legal and ethical standards.

8.2 Incident Reporting

– Reporting Mechanism: Any potential fairness violations or significant biases identified will be reported to the Data Governance Council within [Insert Timeframe]. A detailed report outlining the issue and proposed corrective actions will be submitted for review.

9. Document Control

– Document Owner: [Insert Name, Role]

– Approval Date: [Insert Date]

- Next Review Date: [Insert Date]

- Version History:

  - Version [Insert Version Number]—Initial Document—[Insert Date]—Approved by [Insert Name]

Deployment

Data Governance Deployment Plan

Project Name: [Insert Project Name]
Date: [Insert Date]
Version: [Insert Version Number]

1. Introduction

This Data Governance Deployment Plan outlines the steps for enforcing data governance policies during the deployment of the [Insert Project Name]. The plan ensures that data governance standards are maintained through monitoring, compliance checks, and risk management strategies to protect data integrity, privacy, and security during and after deployment.

2. Purpose and Scope

2.1 Purpose

The purpose of this plan is to:

- Ensure that data governance policies are enforced throughout the deployment of the [Insert Project Name].

- Establish monitoring mechanisms for compliance with data governance, privacy, and security policies.

- Manage data risks by identifying, assessing, and mitigating potential threats to data quality and security during deployment.

2.2 Scope

This plan applies to the deployment phase of the [Insert Project Name], including the movement of data from development to production, as well as post-deployment monitoring and maintenance. It covers data governance, privacy, security, and risk management processes.

3. Data Governance Enforcement Framework

3.1 Roles and Responsibilities

- Data Governance Council: Oversees the enforcement of data governance policies, ensures alignment with organizational standards, and reviews compliance reports.

- Data Stewards: Ensure data quality, integrity, and security during deployment. They monitor data governance adherence at the operational level.

- Data Owners: Approve data access and use during deployment and ensure that data is used in compliance with governance policies.

- IT and Security Teams: Implement security protocols and manage access control systems during and after deployment.

3.2 Governance Tools and Platforms

- Data Governance Tool: [Insert Tool Name], used to monitor compliance and track data usage during deployment.

- Security Monitoring Platform: [Insert Platform Name], used to detect and address data security risks.

- Data Quality Dashboard: [Insert Dashboard Name], provides real-time tracking of data quality metrics.

4. Compliance Monitoring

4.1 Monitoring Tools

- Compliance Monitoring System: [Insert Tool Name], regularly checks for adherence to data governance policies, ensuring all actions taken on data during deployment comply with legal, regulatory, and organizational requirements.

- Audit Logging: Automatically captures detailed logs of all data-related activities during deployment, ensuring traceability and accountability. Logs are reviewed by Data Stewards on a [Insert Frequency] basis.

4.2 Compliance Checks

- Data Integrity Checks: Regularly verify the accuracy and completeness of data as it moves from staging to production environments.

- Access Control Audits: Monitor who accesses data during deployment, ensuring that only authorized personnel can interact with sensitive data.

- Security Audits: Periodically test the security measures in place to protect data against breaches or unauthorized access during deployment.

5. Data Risk Management

5.1 Risk Identification

- Data Integrity Risks: Identify risks associated with data corruption, loss, or degradation during deployment.

- Security Risks: Assess risks related to unauthorized access, data breaches, or exposure of sensitive data.

- Compliance Risks: Identify potential regulatory compliance violations during the deployment process.

5.2 Risk Mitigation Strategies

–   Data Backup: Implement automated backups during deployment to ensure recovery in case of data corruption or loss.

–   Encryption: Use encryption for all sensitive data in transit and at rest to mitigate the risk of unauthorized access.

–   Incident Response Plan: Establish a detailed incident response plan to address security breaches, compliance violations, or data integrity issues during deployment.

5.3 Risk Reporting

–   Risk Assessment Reports: Data Stewards and the Security Team will generate risk assessment reports on a [Insert Frequency] basis to highlight potential data risks during deployment.

–   Incident Reports: In the event of a data-related incident, a detailed report will be generated, including a root cause analysis and recommendations for preventing future occurrences.

6. Post-Deployment Monitoring

6.1 Continuous Monitoring

–   Data Quality Monitoring: Continuously monitor data quality metrics using [Insert Tool Name] to ensure data accuracy and completeness remain intact after deployment.

–   Security Monitoring: Use [Insert Security Platform Name] to continuously monitor data security for any unauthorized access or breaches post-deployment.

6.2 Compliance Reporting

- Compliance Reports: Generate automated reports on compliance with data governance policies, submitted to the Data Governance Council on a [Insert Frequency] basis.

- Audit Logs: Periodic review of audit logs ensures that all data interactions post-deployment comply with governance policies.

7. Document Control

- Document Owner: [Insert Name, Role]

- Approval Date: [Insert Date]

- Next Review Date: [Insert Date]

- Version History:

  - Version [Insert Version Number]—Initial Document—[Insert Date]—Approved by [Insert Name]

Data Provenance Documentation

Project Name: [Insert Project Name]
Date: [Insert Date]
Version: [Insert Version Number]

1. Introduction

This Data Provenance Documentation tracks the origin, transformations, and usage of data throughout the deployment phase of the [Insert Project Name]. The purpose of this documentation is to ensure full traceability and accountability, allowing the project team to maintain a detailed record of data lineage during the deployment process.

2. Purpose and Scope

2.1 Purpose

The purpose of this document is to:

- Record the origins and sources of all data used during deployment.

- Document all transformations applied to the data.

- Ensure traceability of data throughout the deployment process to maintain data integrity and compliance.

2.2 Scope

This documentation applies to all datasets used in the [Insert Project Name] during deployment, covering data origin, transformations, and usage in production systems.

3. Data Origins and Sources

3.1 Data Source Overview

Data Source	Description	Source Type	Ownership	Date Acquired
[Source 1]	[Description]	[Internal/External]	[Owner]	[Date]
[Source 2]	[Description]	[Internal/External]	[Owner]	[Date]

Data Origin Details: [Detailed description of how the data was sourced, including any agreements or licenses if external.]

4. Data Transformation

4.1 Transformation Steps

This section details all transformations applied to the data during deployment.

Transformation Step	Description	Tool/Method	Date Applied	Data Source(s)	Output Format
[Step 1]	[Description]	[Tool Name]	[Date]	[Data Source]	[Output Format]
[Step 2]	[Description]	[Tool Name]	[Date]	[Data Source]	[Output Format]

4.2 Data Validation

- – Validation Steps: Each transformation process includes validation steps to ensure data accuracy and consistency. For example, [Insert Validation Technique] was applied to confirm [Validation Criteria].

5. Data Usage in Production

5.1 Data Deployment

Production Environment: The following datasets have been deployed to the production environment:

Dataset Name	Description	Usage in Production	Data Owner	Deployment Date
[Dataset 1]	[Description]	[Purpose]	[Owner]	[Date]
[Dataset 2]	[Description]	[Purpose]	[Owner]	[Date]

5.2 Data Access

Access Control: The following personnel have access to the deployed datasets, as defined by the data governance policies:

User/Role	Access Level	Justification	Access Date
[Role 1]	[Read/Write]	[Reason]	[Date]
[Role 2]	[Read Only]	[Reason]	[Date]

6. Data Audit and Traceability

6.1 Data Audit Logs

All interactions with the datasets, including transformations, access, and modifications, are recorded in audit logs. These logs can be accessed via [Insert Tool Name] and are reviewed on a [Insert Frequency] basis to ensure traceability.

Event Type	Timestamp	User	Action	Dataset	Outcome
[Transformation]	[Timestamp]	[User]	[Transformation]	[Dataset]	[Success]
[Access]	[Timestamp]	[User]	[Access Granted]	[Dataset]	[Success]

7. Data Retention and Archiving

7.1 Retention Policies

All datasets follow the retention policies outlined in the Data Governance Deployment Plan. Data is retained for [Insert Timeframe], after which it is archived or deleted in accordance with organizational policies.

– Archiving Process: Datasets are archived using [Insert Archiving Tool or Method], and access is restricted to [Insert Role(s)].

8. Document Control

– Document Owner: [Insert Name, Role]

– Approval Date: [Insert Date]

– Next Review Date: [Insert Date]

– Version History:

• Version [Insert Version Number]—Initial Document—[Insert Date]—Approved by [Insert Name]

Privacy and Security Controls Documentation

Project Name: [Insert Project Name]
Date: [Insert Date]
Version: [Insert Version Number]

1. Introduction

This Privacy and Security Controls Documentation outlines the measures in place to protect the data used during and after the deployment of the [Insert Project Name]. It specifies the privacy and security controls that safeguard sensitive data, ensuring compliance with privacy laws and preventing unauthorized access or breaches.

2. Purpose and Scope

2.1 Purpose

The purpose of this document is to:

- Define the privacy and security measures implemented during the deployment phase of the [Insert Project Name].

- Ensure that all sensitive data is adequately protected, both in transit and at rest.

- Demonstrate compliance with applicable privacy regulations (e.g., GDPR, HIPAA) and organizational security policies.

2.2 Scope

This document applies to all data processed during and after deployment, covering data at rest, data in transit, and data access controls. It includes internal and external data sources and addresses both privacy and security concerns.

3. Privacy Controls

3.1 Data Anonymization and Pseudonymization

– Anonymization: Sensitive data, such as personally identifiable information (PII), is anonymized before being used in the production environment. Anonymization techniques include [Insert Anonymization Method, e.g., k-anonymity, differential privacy].

– Pseudonymization: In cases where anonymization is not feasible, pseudonymization is applied to protect PII. Pseudonymized data is linked to individuals using unique keys stored separately from the data.

3.2 Data Minimization

– Principle of Minimization: Only the minimum amount of data necessary for the purpose of the [Insert Project Name] is collected, processed, and stored. Redundant data fields have been eliminated from the production dataset to reduce privacy risks.

3.3 Consent Management

– Informed Consent: For all sensitive data collected from individuals, explicit consent has been obtained. Consent records are stored in [Insert System Name] and regularly audited to ensure compliance.

– Consent Revocation: Individuals can revoke their consent at any time, and their data will be removed from the system within [Insert Timeframe]. The process for managing consent revocation is automated using [Insert Tool Name].

4. Security Controls

4.1 Data Encryption

- Encryption at Rest: All sensitive data is encrypted at rest using [Insert Encryption Standard, e.g., AES-256]. Encryption keys are managed through [Insert Key Management System].

- Encryption in Transit: Data transmitted between systems or to third parties is encrypted using [Insert Protocol, e.g., TLS 1.2 or 1.3] to prevent unauthorized interception.

4.2 Access Controls

- Role-Based Access Control (RBAC): Access to data is restricted based on user roles, ensuring that only authorized personnel can access sensitive data. The following roles have access to specific datasets:

- Multi-factor Authentication (MFA): All users with access to sensitive data must authenticate using MFA to prevent unauthorized access.

Role	Access Level	Justification	Access Date
[Role 1]	[Read/Write]	[Insert Reason]	[Insert Date]
[Role 2]	[Read Only]	[Insert Reason]	[Insert Date]

4.3 Security Monitoring

- Intrusion Detection System (IDS): [Insert IDS Name] is used to detect and prevent unauthorized access to the system. Alerts are automatically generated in case of suspicious activity.

- Security Information and Event Management (SIEM): A SIEM tool ([Insert Tool Name]) aggregates and analyzes security logs in real-time, helping to identify potential security incidents or breaches.

4.4 Data Breach Response

- Incident Response Plan: In the event of a data breach, an incident response plan is in place. The plan includes steps for identifying the breach, containing the damage, notifying affected parties, and resolving the issue.

- Breach Notification: If a breach involves PII or sensitive data, regulatory authorities and affected individuals will be notified within [Insert Timeframe], in accordance with [Insert Applicable Regulations].

5. Compliance with Regulations

5.1 GDPR Compliance

- Data Subject Rights: All data processing activities are compliant with the General Data Protection Regulation (GDPR). Data subjects are informed of their rights, including the right to access, correct, and delete their data.

- Data Protection Impact Assessments (DPIA): A DPIA has been conducted for the [Insert Project Name] to assess and mitigate privacy risks associated with data processing activities.

5.2 HIPAA Compliance

For projects involving healthcare data, the following measures are in place to ensure compliance with the Health Insurance Portability and Accountability Act (HIPAA):

- Protected Health Information (PHI): PHI is encrypted at rest and in transit, with strict access controls limiting who can view or modify this data.

- HIPAA Audits: Regular audits are conducted to ensure that the project complies with HIPAA's privacy and security rules.

6. Post-Deployment Privacy and Security Monitoring

6.1 Continuous Monitoring

–   Data Privacy Monitoring: Ongoing monitoring of data privacy compliance is carried out using [Insert Privacy Monitoring Tool]. Any deviations from privacy policies are flagged for immediate remediation.

–   Security Monitoring: Security logs are continuously monitored, and regular penetration tests are conducted to identify and address vulnerabilities.

6.2 Regular Audits

–   Privacy Audits: Conducted on a [Insert Frequency] basis, privacy audits ensure that data usage complies with all applicable regulations.

–   Security Audits: Regular security audits are carried out to ensure that the implemented security measures are effective and up-to-date.

7. Document Control

–   Document Owner: [Insert Name, Role]

–   Approval Date: [Insert Date]

–   Next Review Date: [Insert Date]

–   Version History:

    •   Version [Insert Version Number]—Initial Document—[Insert Date]—Approved by [Insert Name]

Retention and Retirement
Data Retention and Archiving Policy

Project Name: [Insert Project Name]
Date: [Insert Date]
Version: [Insert Version Number]

1. Introduction

This Data Retention and Archiving Policy outlines the processes for retaining, archiving, and securely deleting data in compliance with the data governance policies of the [Insert Project Name]. The policy ensures that data is managed in a way that supports business continuity, regulatory compliance, and secure long-term storage while minimizing risk.

2. Purpose and Scope

2.1 Purpose

The purpose of this policy is to:

- Define the duration for which data will be retained based on business, legal, and regulatory requirements.

- Establish guidelines for securely archiving data to ensure long-term storage.

- Outline the procedures for safely deleting or disposing of data when it is no longer needed.

- Ensure compliance with applicable regulations and data governance policies.

2.2 Scope

This policy applies to all data collected, processed, and stored within the [Insert Project Name], including structured and unstructured data, across all environments (development, staging, production). It covers all stages of the data lifecycle, from initial collection to final deletion or archival.

3. Data Retention

3.1 Retention Periods

The retention periods for different categories of data are based on business needs, legal obligations, and regulatory requirements. The following table outlines the retention periods for each data type:

Data Category	Retention Period	Legal/Regulatory Requirements	Data Owner	Action After Retention Period
[Customer Data]	[Insert Retention Period]	[Insert Regulation]	[Data Owner Name]	[Archive/Delete]
[Transaction Data]	[Insert Retention Period]	[Insert Regulation]	[Data Owner Name]	[Archive/Delete]
[Log Data]	[Insert Retention Period]	[Insert Regulation]	[Data Owner Name]	[Archive/Delete]
[Analytics Data]	[Insert Retention Period]	[Insert Regulation]	[Data Owner Name]	[Archive/Delete]

3.2 Retention Criteria

The following criteria are used to determine retention periods:

- Legal/Regulatory Compliance: Data subject to specific legal or regulatory requirements must be retained for the mandated duration.

- Business Needs: Data required for operational purposes, analytics, or reporting must be retained for as long as necessary to meet these needs.

- Historical Value: Certain data may be retained for its historical or research value, subject to business approval.

4. Data Archiving

4.1 Archiving Process

Once the retention period has expired, data that is still required for historical or legal purposes will be archived. The archiving process includes:

- Archival Format: Data will be archived in a [Insert Format] to ensure that it can be accessed and restored if needed.

- Archiving Location: Archived data will be stored in [Insert Storage System or Cloud Platform] to ensure secure and scalable long-term storage.

- Archiving Procedures: Data will be archived according to the following steps:

  - Identify data eligible for archiving.

  - Apply necessary encryption or access controls.

  - Move data to the archival storage system.

  - Log the archiving event in the audit system.

4.2 Access to Archived Data

Access to archived data is restricted and granted only under the following conditions:

- Access Request Process: Requests to access archived data must be approved by [Insert Approver Name] and must include a justification for the request.

- Data Retrieval Process: Archived data can be retrieved through [Insert Retrieval Tool], and all retrieval activities will be logged in the audit system.

5. Data Deletion

5.1 Data Deletion Process

Data that is no longer needed and has exceeded its retention period will be securely deleted. The deletion process includes:

- Secure Deletion Tools: Data will be deleted using [Insert Deletion Tool/Method] to ensure it is irrecoverable.

- Deletion Verification: Once deletion is complete, [Insert Verification Method] will be used to confirm that the data has been successfully removed from all systems.

- Audit Logging: All data deletion activities will be recorded in the audit log for compliance and traceability.

5.2 Exceptions to Deletion

Data may be exempt from deletion under the following circumstances:

- Legal Hold: Data subject to litigation or regulatory inquiry will be retained beyond its retention period until the hold is lifted.

- Business Continuity: Critical data required for disaster recovery may be retained as part of business continuity plans.

6. Compliance and Monitoring

6.1 Monitoring and Auditing

- Audit Trails: An audit trail will be maintained for all data retention, archiving, and deletion activities. This will be reviewed on a [Insert Frequency] basis by [Insert Team/Department].

- Compliance Checks: Regular compliance checks will be conducted to ensure adherence to retention and deletion policies, with reports generated for [Insert Regulatory Body].

6.2 Non-compliance Consequences

Failure to comply with this Data Retention and Archiving Policy may result in disciplinary actions, including penalties under applicable data protection laws.

7. Document Control

- Document Owner: [Insert Name, Role]

- Approval Date: [Insert Date]

- Next Review Date: [Insert Date]

- Version History:

  - Version [Insert Version Number]—Initial Document—[Insert Date]—Approved by [Insert Name]

Data Decommissioning Plan

Project Name: [Insert Project Name]
Date: [Insert Date]
Version: [Insert Version Number]

1. Introduction

This Data Decommissioning Plan details the steps for securely decommissioning data in the [Insert Project Name]. It outlines the procedures for data disposal, anonymization, or archiving, ensuring compliance with data governance policies and protecting against unauthorized access or breaches.

2. Purpose and Scope

2.1 Purpose

The purpose of this plan is to:

- Outline the steps for securely decommissioning data that is no longer needed in the [Insert Project Name].

- Ensure that all decommissioned data is properly disposed of, anonymized, or archived in accordance with governance policies.

- Mitigate risks associated with data decommissioning, such as unauthorized access, data breaches, or non-compliance with regulatory requirements.

2.2 Scope

This plan applies to all datasets used in the [Insert Project Name] that have reached the end of their lifecycle. It covers structured, unstructured, and semi-structured data stored in internal and external systems.

3. Decommissioning Criteria

3.1 Criteria for Decommissioning

Data will be decommissioned when it meets one or more of the following criteria:

- End of Retention Period: The data has exceeded its retention period as defined in the Data Retention and Archiving Policy.

- Business Usefulness: The data is no longer required for operational, analytical, or historical purposes.

- Regulatory Requirements: The data must be decommissioned in compliance with specific regulatory or legal mandates.

3.2 Exemptions

Data subject to the following exemptions will not be decommissioned:

- Legal Holds: Data involved in ongoing litigation or regulatory investigations will be retained until the hold is lifted.

- Critical Data: Data required for disaster recovery or business continuity will not be decommissioned until it is no longer needed.

4. Data Disposal

4.1 Disposal Process

Data that is no longer needed and does not meet archival criteria will be securely disposed of. The disposal process includes:

- Secure Deletion Tools: Data will be erased using [Insert Deletion Tool/Method], ensuring that it is irrecoverable and meets industry standards (e.g., NIST SP 800-88 guidelines for data destruction).

- Verification of Deletion: A verification process will be conducted using [Insert Verification Method] to ensure that the data has been successfully and permanently deleted.

- Audit Logging: All data disposal activities will be logged in the audit system for compliance verification.

4.2 Physical Media Disposal

If data is stored on physical media (e.g., hard drives, tapes), the following steps will be taken to securely dispose of the media:

- Destruction Methods: Physical media will be destroyed using [Insert Method, e.g., shredding, degaussing] to ensure that data cannot be recovered.

- Certificate of Destruction: A certificate of destruction will be obtained from the disposal vendor to verify that the media has been destroyed according to industry standards.

5. Data Anonymization

5.1 Anonymization Process

In cases where the data must be retained for analytical or historical purposes, but personally identifiable information (PII) or sensitive data must be protected, the data will be anonymized using the following steps:

- – Anonymization Techniques: Data will be anonymized using [Insert Anonymization Techniques, e.g., k-anonymity, differential privacy] to ensure that individuals cannot be re-identified.

- – Pseudonymization: If anonymization is not possible, pseudonymization techniques will be applied, where identifiers are replaced with unique keys that are stored separately.

- – Anonymization Validation: After anonymization, the data will be reviewed to confirm that it meets regulatory standards (e.g., GDPR) and that no re-identification is possible.

6. Data Archiving

6.1 Archival of Data

Data that is no longer actively used but must be retained for legal or historical reasons will be archived according to the following procedures:

- – Archiving Method: Data will be moved to an archival system, such as [Insert Archival Tool], where it will be securely stored and protected.

- – Encryption and Access Control: Archived data will be encrypted using [Insert Encryption Standard, e.g., AES-256] and access will be restricted to authorized personnel only.

- – Audit Trails: All actions involving archived data, including retrievals, will be logged for compliance purposes.

7. Decommissioning Risk Management

7.1 Risk Identification

The following risks have been identified in the data decommissioning process:

- – Unauthorized Access: Risk of unauthorized access to data during the decommissioning process.

- – Data Breach: Risk of data breach due to improper deletion or disposal.

- – Non-compliance: Risk of non-compliance with data protection laws and regulations.

7.2 Risk Mitigation Strategies

- Encryption: All sensitive data will be encrypted before decommissioning to protect it from unauthorized access.

- Secure Disposal: Use certified vendors and secure methods for physical and electronic data disposal to prevent breaches.

- Compliance Monitoring: Regular audits will be conducted to ensure that decommissioning activities comply with legal and regulatory requirements.

8. Documentation and Reporting

8.1 Audit Logs

All data decommissioning activities will be documented in detailed audit logs, including:

- Decommissioning Date: The date when the data was decommissioned.

- Decommissioning Method: The method used for decommissioning (e.g., deletion, anonymization, archiving).

- Personnel Involved: The individuals responsible for performing and verifying the decommissioning process.

8.2 Reporting

Decommissioning reports will be generated on a [Insert Frequency] basis and submitted to [Insert Approver Name] for review. These reports will include a summary of all data decommissioned during the reporting period, any issues encountered, and the steps taken to mitigate risks.

9. Document Control

- Document Owner: [Insert Name, Role]

- Approval Date: [Insert Date]

- Next Review Date: [Insert Date]

– Version History:

  • Version [Insert Version Number]—Initial Document—[Insert Date]—Approved by [Insert Name]

Data Governance Policies

1. Introduction

This document outlines the Data Governance Policies for Machine Learning (ML) within [Insert Organization Name]. These policies establish the principles, standards, and procedures that guide the collection, storage, processing, and sharing of data across all ML initiatives within the organization. The policies are designed to ensure that data is managed in alignment with organizational goals, ethical considerations, and legal obligations.

2. Purpose and Scope

2.1 Purpose

The purpose of these policies is to:

– Define the principles and guidelines for managing data used in ML projects.

– Establish standards for data quality, privacy, security, and ethical use.

– Ensure compliance with relevant regulations and industry standards.

– Foster transparency, accountability, and consistency in data management across the organization.

2.2 Scope

These policies apply to all data collected, stored, processed, and shared within the organization for the purposes of developing, deploying, and maintaining ML models. This includes structured, semi-structured, and unstructured data across all internal and external sources.

3. Principles of Data Governance

The following principles guide data governance for ML projects:

- Accountability: Clearly define roles and responsibilities for all stakeholders involved in the data lifecycle to ensure accountability in data management.

- Transparency: Maintain transparency in data collection, processing, and sharing practices to build trust and allow for auditability.

- Ethical Use: Ensure that data is used ethically, avoiding discrimination, bias, and harm to individuals or groups.

- Compliance: Adhere to legal and regulatory obligations, including data protection laws such as the General Data Protection Regulation (GDPR) and the California Consumer Privacy Act (CCPA).

- Data Quality: Ensure data accuracy, completeness, and consistency to maintain the reliability of ML models.

- Security and Privacy: Protect data from unauthorized access, breaches, and misuse, ensuring data privacy through encryption, anonymization, and other measures.

4. Roles and Responsibilities

The roles and responsibilities for managing data governance in ML projects include:

4.1 Data Governance Council

- Responsibilities: Oversee the implementation and compliance of data governance policies, review and approve data governance strategies, and manage escalations related to data governance issues.

- Members: Chief Data Officer (CDO), IT Security Officer, Legal Counsel, ML Project Leads, Data Privacy Officer.

4.2 Data Stewards

- Responsibilities: Ensure data quality, integrity, and security. Maintain documentation related to data governance and monitor adherence to policies at the operational level.

- Appointed Stewards: [Insert Names or Roles].

4.3 Data Scientists and ML Engineers

- Responsibilities: Comply with data governance policies when using data for ML model development, ensuring that data processing respects privacy and ethical guidelines.

- Appointed Data Scientists/Engineers: [Insert Names or Roles].

4.4 Data Owners

- Responsibilities: Define access permissions, approve data usage requests, and ensure that data is used in compliance with governance policies and regulations.

- Appointed Data Owners: [Insert Names or Roles].

5. Data Collection

5.1 Data Sourcing

- Internal Data: Data collected from internal systems (e.g., CRM, ERP, HR) must follow the organization's data privacy policies and security protocols.

- External Data: External data sourced from third-party vendors or public datasets must be vetted to ensure that it is obtained lawfully and meets data quality standards.

5.2 Consent and Legal Compliance

- Consent Management: For data that includes personal information, explicit consent must be obtained from data subjects, as required by data protection laws.

- Compliance Requirements: All data collection must comply with regulations such as GDPR, CCPA, and any industry-specific standards. Data collection must be documented in detail to demonstrate compliance in case of an audit.

5.3 Data Minimization

- Principle of Minimization: Only data that is necessary for the specific ML use case should be collected. Redundant or unnecessary data should be excluded from the dataset to reduce the risk of privacy breaches.

6. Data Storage and Access

6.1 Data Storage Standards

- Secure Storage: Data must be stored in secure environments, such as encrypted databases or cloud storage solutions with appropriate security measures in place (e.g., AES-256 encryption).

- Backup and Recovery: Regular backups of critical datasets must be performed to prevent data loss. A disaster recovery plan must be in place to ensure that data can be restored in the event of an incident.

6.2 Data Access Control

- Role-Based Access Control (RBAC): Access to data must be restricted based on user roles and the principle of least privilege. Only authorized personnel should have access to sensitive datasets.

– Authentication and Authorization: Multi-factor authentication (MFA) must be required for all users accessing sensitive data. Data access must be logged, and audit trails must be maintained.

6.3 Data Sharing

– Internal Data Sharing: Data sharing within the organization must follow predefined access permissions and be approved by Data Owners. All shared data must comply with data governance and security policies.

– External Data Sharing: When sharing data with third parties (e.g., vendors, partners), data sharing agreements must be in place, outlining data usage, privacy, and security obligations.

7. Data Processing and Transformation

7.1 Data Transformation

– Documentation of Transformations: All data transformations (e.g., cleaning, normalization, feature engineering) applied during ML processes must be documented to ensure traceability and reproducibility.

– Data Quality Monitoring: Data must be regularly monitored for quality issues (e.g., missing values, inconsistencies), and remediation actions should be taken when issues are identified.

7.2 Data Anonymization and Pseudonymization

– Anonymization Techniques: Where possible, personal data should be anonymized to ensure privacy. Techniques such as k-anonymity or differential privacy should be applied to reduce the risk of re-identification.

- Pseudonymization: If anonymization is not feasible, pseudonymization should be used. Personal identifiers should be replaced with unique tokens, and access to re-identification keys should be strictly controlled.

8. Data Privacy and Security

8.1 Privacy-by-Design

- Integration of Privacy: Data privacy principles must be integrated into the design and development of all ML projects. This includes minimizing data collection, limiting access to sensitive data, and incorporating privacy-preserving techniques.

- Data Privacy Impact Assessments (DPIAs): DPIAs should be conducted to assess potential privacy risks for any project that processes personal or sensitive data. The results should inform the privacy controls applied to the project.

8.2 Security Measures

- Encryption: Sensitive data must be encrypted both at rest and in transit. Encryption protocols must meet industry standards (e.g., TLS 1.2 or higher for data in transit).

- Incident Response: A data breach response plan must be in place. In the event of a data breach, regulatory authorities and affected individuals must be notified within the legally required timeframe.

9. Data Quality Standards

9.1 Data Quality Metrics

The following metrics should be used to monitor and maintain data quality throughout the ML lifecycle:

- Accuracy: Data must accurately reflect the real-world entities or phenomena it represents.

- Completeness: Data must include all required fields and records, minimizing missing or incomplete entries.

- Consistency: Data must be consistent across different systems and datasets, without conflicting or duplicate information.

- Timeliness: Data must be up-to-date and relevant for the intended ML use case.

9.2 Data Quality Monitoring

- Automated Quality Checks: Automated tools must be used to continuously monitor data quality. Alerts should be triggered when quality thresholds are breached, and corrective actions should be taken promptly.

- Manual Reviews: Periodic manual reviews should be conducted by Data Stewards to validate data quality and address any issues not captured by automated tools.

10. Data Ethics and Fairness

10.1 Ethical Data Use

- Avoiding Bias: Data used in ML models must be evaluated for potential biases. Biased datasets can lead to discriminatory outcomes in models, and steps must be taken to mitigate such biases (e.g., through bias detection algorithms).

- Explainability: Models developed using organizational data should be explainable, particularly when the outputs impact individuals or groups. Techniques such as LIME or SHAP can be used to explain model predictions.

- Transparency: Stakeholders must be informed about how their data is used, especially in ML models that affect decision-making.

10.2 Fairness Audits

- Regular Fairness Audits: Regular audits of the model's performance across different demographic groups should be conducted to ensure fairness. Disparities in model outputs should be analyzed and corrected where necessary.

11. Regulatory Compliance

11.1 Compliance with Laws

All data processing activities must comply with relevant data protection and privacy laws, including but not limited to:

- GDPR: Compliance with data subject rights (e.g., access, rectification, deletion) and cross-border data transfer requirements.

- CCPA: Compliance with California-specific consumer privacy rights, including the right to know, delete, and opt-out of data sales.

11.2 Documentation and Record-Keeping

- Data Documentation: Detailed documentation of data processing activities must be maintained to demonstrate compliance. This includes records of consent, data processing agreements, and impact assessments.

- Audit Logs: Maintain audit logs of all data access, sharing, and modification activities to ensure traceability and accountability.

12. Enforcement and Non-compliance

12.1 Enforcement Mechanisms

- Monitoring: Automated and manual compliance monitoring tools should be implemented to ensure adherence to data governance policies.

- Penalties for Non-compliance: Non-compliance with data governance policies will result in disciplinary action, including potential fines or legal action, depending on the severity of the breach.

12.2 Reporting Violations

- Incident Reporting: Any violations of data governance policies must be reported to the Data Governance Council or Data Privacy Officer immediately. A root cause analysis must be conducted and corrective actions taken.

13. Policy Review and Updates

13.1 Review Cycle

- Annual Review: These data governance policies will be reviewed on an annual basis to ensure they remain relevant and aligned with organizational objectives, emerging regulations, and industry standards.

- Interim Updates: Policies may be updated more frequently if significant changes in technology, regulations, or organizational practices occur.

13.2 Approval Process

- Approval by Governance Council: All updates to the Data Governance Policies for Machine Learning must be approved by the Data Governance Council before being implemented.

14. Document Control

- Document Owner: [Insert Name, Role]

- Approval Date: [Insert Date]

- Next Review Date: [Insert Date]

- Version History:

  - Version [Insert Version Number]—Initial Document—[Insert Date]—Approved by [Insert Name]

References and Further Reading

Books and Journals

Ammon, C. (2019). Machine Learning Governance: An Architecture for Trustworthy and Explainable AI Systems. O'Reilly Media.

Ashburn, A., & Rhoads, T. (2020). Data Governance for the Artificial Intelligence Age. Wiley.

Boland, R. J., Delacroix, J., & Russom, P. (2018). Digital Twin: Paradigm for the Future of Manufacturing. Elsevier. (Chapter on Data Governance for Digital Twins)

Bostrom, N. (2014). Superintelligence: Paths, Dangers, Strategies. Oxford University Press. (Chapter on Ethical Considerations in AI)

Chen, H., Zhang, Y., Qin, J., Liu, H., & Zhou, Z. (2020). Machine Learning for Big Data: Theory and Applications. Springer. (Chapter on Data Quality for Machine Learning)

Chokshi, N., & Verma, S. (2017). The Definitive Guide to Data Governance: A Practical Approach to Managing, Protecting, and Sharing Your Data Assets. John Wiley & Sons.

Crook, J., Langley, D., & Payne, S. (2017). A Dictionary of Finance and Banking. Oxford University Press. (Section on Data Privacy Regulations)

Doshi-Velez, F., & Kim, B. (2017). Towards a Rigorous Science of Interpretable Machine Learning. arXiv preprint arXiv:1702.08608.

European Union. (2016). Regulation (EU) 2016/679 of the European Parliament and of the Council of 27 April 2016 on the protection of natural persons with regard to the processing of personal data and on the free movement of such data, and repealing Directive 95/46/EC (General Data Protection Regulation). eur-lex.europa.eu

Fan, J., Xu, J., Wu, Y., Ganeshalingam, M., & Zhou, X. (2019). Survey of Model Explainability Techniques in Deep Learning. arXiv preprint arXiv:1907.04630.

Friedman, T., & Varian, H. R. (2017). Big Data: A Revolution That Transforms How We Live, Work, and Think. Penguin Random House. (Chapter on Data Governance in the Big Data Era)

A. Nandan Prasad, *Introduction to Data Governance for Machine Learning Systems,*
https://doi.org/10.1007/979-8-8688-1023-7

Géron, A. (2017). Hands-On Machine Learning with Scikit-Learn, Keras & TensorFlow: Concepts, Tools, and Techniques to Build Intelligent Systems. O'Reilly Media. (Chapter on Data Preprocessing for Machine Learning)

Goldberg, I. (2017). Neural Network Engineering. Morgan Kaufmann Publishers. (Chapter on Explainable AI for Deep Learning)

Grance, T., Gupta, P., & Liang, P. P. (2019). Data Governance for Artificial Intelligence: A Practical Guide. Kogan Page Publishers.

Greasley, A., & Krüger, T. (2017). AI: A Guide for Thinking People. MIT Press. (Chapter on Ethical Considerations in Artificial Intelligence)

Groth, P. (2017). Data Governance Essentials: How to Protect, Organize, and Leverage Your Data for Business Advantage. O'Reilly Media.

Gupta, P., & Seetharaman, A. (2017). Data Management for Machine Learning. Packt Publishing Ltd.

Hastie, T., Tibshirani, R., & Friedman, J. (2009). The Elements of Statistical Learning: Data Mining, Inference, and Prediction. Springer Science & Business Media. (Chapter on Data Quality Considerations in Statistical Learning)

James, G., Witten, D., Hastie, T., & Tibshirani, R. (2013). An Introduction to Statistical Learning: with Applications in R. Springer Science & Business Media. (Chapter on Data Cleaning for Statistical Analysis)

Kimball, R., & Bohannon, P. (2013). The Data Warehouse Toolkit: The Definitive Guide to Dimensional Modeling. John Wiley & Sons. (Chapter on Data Quality in Data Warehouses)

Kuhn, M., & Langley, D. (2019). A Gentle Introduction to Statistical Learning. Cambridge University Press. (Chapter on Data Preprocessing for Statistical Modeling)

Mittelstadt, B., Wachter, S., & Floridi, L. (2019). Against Algorithmic Bias: Auditing, Fairness, and Accountability. Oxford University Press.

Murphy, K. P. (2012). Machine Learning: A Probabilistic Perspective. MIT Press. (Chapter on Model Evaluation in Machine Learning)

Needham, P., & Nguyen, D. (2019). Responsible AI: A Framework for Developers. Apress.

Provost, F., & Fawcett, T. (2013). Data Science for Business: Forecasting Model Selection and Performance. O'Reilly Media. (Chapter on Data Quality for Business Analytics)

Rahmani, H., & Hutter, A. (2019). Explainable AI in Software Engineering. Springer Nature.

Russell, S. J., & Norvig, P. (2021). Artificial Intelligence: A Modern Approach (4th ed.). Pearson Education Limited. (Chapter on Ethical Considerations in AI Development)

Shmueli, G., Nathans, E., Stephans, A., & Wedel, M. (2019). Learning with Big Data: Collaborative Statistical Modeling. Chapman and Hall/CRC. (Chapter on Data Governance for Big Data Analytics)

Stone, P. W., & Montgomery, D. C. (2017). Design and Analysis of Experiments with Applications in Engineering and the Sciences. John Wiley & Sons. (Chapter on Data Quality in Experimental Design)

Tambouris, E. (2019). Governing Artificial Intelligence: Investigating the Legal, Ethical, and Regulatory Landscape. Springer Nature.

Amodei, D., & Been, J. (2018). Efforts needed to ensure fair and just AI. arXiv preprint arXiv:1806.08816.

Athey, S. (2017). Beyond prediction: Using machine learning to estimate causal effects. Econometrica, 85(5), 1803–1824.

Doshi-Velez, F., & Kim, B. (2017). Towards a rigorous science of interpretable machine learning. arXiv preprint arXiv:1702.08608.

Flach, P., & Hrnบ็bovský, M. (2016). Machine learning: a statistical perspective. Chapman and Hall/CRC.

Goodman, B., Flaxman, S., & Pederson, A. (2016). Discriminatory effects of algorithmic fairness measures. Proceedings of the ACM Conference on Fairness, Accountability, and Transparency, 83–92.

Greve, H. R., & Lichtenstein, S. N. (2019). Organizational imprinting in the digital age: How early technology choices shape the future. Academy of Management Journal, 62(1), 152–184.

Kleinberg, J., Mullainathan, S., & Spiess, M. (2019). Inherent fairness. The Quarterly Journal of Economics, 134(2), 665–721.

Lee, I., & Shin, J. (2019). A framework for a data governance program for artificial intelligence in healthcare. Studies in health technology and informatics, 262, 1204–1208.

Mittelstadt, B., Todd, P., & Wachter, S. (2019). Explainable machine learning for decision support in healthcare. Artificial intelligence in medicine, 97, 104–113.

Rajgopal, D. (2018). Machine learning for fairness: Operational challenges and legal considerations. Michigan Law Review, 117(3), 393–452.

Selbst, A. D., Dressel, J., Fried, M., Sonnad, S., & Keyes, O. (2019). Fairness and abstraction in sociotechnical systems. Proceedings of the Conference

Varian, H. R. (2014). Big data: New tricks for econometrics. Journal of Economic Perspectives, 28(2), 3–27.

Wachter, S., Mittelstadt, B., & Floridi, L. (2019). Transparency in artificial intelligence. AI & Society, 34(1), 5–16.

Xu, H., & Gupta, M. R. (2018). On the limitations of fairness in machine learning. Proceedings of the 2018 ACM SIGKDD International Conference on Knowledge Discovery and Data Mining, 1725–1734.

Zhang, Y., Zhao, J., & Li, M. (2019). Deep learning for remote sensing data: A review. Remote Sensing, 11(8), 1121. (Focus on Data Quality Considerations in Remote Sensing Applications)

Zliobaite, I., Leong, P. W., & Krishnan, R. (2019). On feature selection for interpretable machine learning. arXiv preprint arXiv:1902.09302.

Websites and Reports

Data Management Body of Knowledge (DAMA-DMBoK): `https://www.dama.org/cpages/body-of-knowledge`

Open Web Application Security Project (OWASP): `https://owasp.org/` (Focuses on Data Security Best Practices)

Index

A

© Aditya Nandan Prasad 2024
A. Nandan Prasad, *Introduction to Data Governance for Machine Learning Systems*,
https://doi.org/10.1007/979-8-8688-1023-7

C

H

I, J